METHODS IN MOLECULAR BIOLOGY™

Series Editor
John M. Walker
School of Life Sciences
University of Hertfordshire
Hatfield, Hertfordshire, AL10 9AB, UK

For other titles published in this series, go to
www.springer.com/series/7651

Human Embryonic Stem Cell Protocols

Second Edition

Edited by

Kursad Turksen

Ottawa Hospital Research Institute, Ottawa, Ontario, Canada

Editor
Kursad Turksen
Ottawa Hospital Research Institute
 (OHRI)
Sprott Centre for Stem Cell Research
501 Smyth Road, Ottawa ON K1H 8L6
Canada
kturksen@ohri.ca

ISSN 1064-3745 e-ISSN 1940-6029
ISBN 978-1-60761-368-8 e-ISBN 978-1-60761-369-5
DOI 10.1007/978-1-60761-369-5

Library of Congress Control Number: 2009935009

© Humana Press, a part of Springer Science+Business Media, LLC 2006, 2010
All rights reserved. This work may not be translated or copied in whole or in part without the written permission of the publisher (Humana Press, c/o Springer Science+Business Media, LLC, 233 Spring Street, New York, NY 10013, USA), except for brief excerpts in connection with reviews or scholarly analysis. Use in connection with any form of information storage and retrieval, electronic adaptation, computer software, or by similar or dissimilar methodology now known or hereafter developed is forbidden.
The use in this publication of trade names, trademarks, service marks, and similar terms, even if they are not identified as such, is not to be taken as an expression of opinion as to whether or not they are subject to proprietary rights.

Printed on acid-free paper

springer.com

Preface

Embryonic stem (ES) cells, and more specifically human ES cells, remain promising tools for understanding questions of lineage commitment and for exploring potentialities for regenerative medicine applications.

In this second edition, I have tried to bring together a number of new protocols that will extend the topics and reflect new developments since publication of the first edition.

I would like to take this opportunity to thank all the contributors for very graciously providing their protocols for this volume. Without them and their willingness to share protocol details, this new volume would not have materialized.

I would also like to thank Dr. John Walker, the Editor-in-Chief of the Methods in Molecular Biology series, for his continued support.

Patrick Marton, the Editor of the Methods in Molecular Biology series at Springer, also deserves thanks for always being available to answer my questions, for patiently listening to my suggestions, and for supporting this volume during its maturation stages. Thanks also to David Casey for his invaluable help during the production stages of this volume.

Kursad Turksen

Contents

Preface .... *v*
Contributors .... *ix*

1. Human Embryo Culture and Assessment for the Derivation of Embryonic Stem Cells (ESC) 1
   A. Henry Sathananthan and Tiki Osianlis

2. Derivation of Human Embryonic Stem Cell Lines from Vitrified Human Embryos 21
   Teija T. Peura, Julia Schaft, and Tomas Stojanov

3. Growth of Human Embryonic Stem Cells Using Derivates of Human Fibroblasts 55
   Carmen Escobedo-Lucea and Miodrag Stojkovic

4. In Vitro Neural Differentiation of Human Embryonic Stem Cells Using a Low-Density Mouse Embryonic Fibroblast Feeder Protocol 71
   John A. Ozolek, Esther P. Jane, James E. Esplen, Patti Petrosko, Amy K. Wehn, Teresa M. Erb, Sara E. Mucko, Lyn C. Cote, and Paul J. Sammak

5. Optimization of Physiological Xenofree Molecularly Defined Media and Matrices to Maintain Human Embryonic Stem Cell Pluripotency 97
   Isabelle Peiffer, Romain Barbet, Antoinette Hatzfeld, Ma-Lin Li, and Jacques A. Hatzfeld

6. Serum-Free and Feeder-Free Culture Expansion of Human Embryonic Stem Cells 109
   Katherine E. Wagner and Mohan C. Vemuri

7. Single Cell Enzymatic Dissociation of Human Embryonic Stem Cells: A Straightforward, Robust, and Standardized Culture Method 121
   Catharina Ellerström, Johan Hyllner, and Raimund Strehl

8. Monitoring Stemness in Long-Term hESC Cultures by Real-Time PCR 135
   Amparo Galán and Carlos Simón

9. Culture and Preparation of Human Embryonic Stem Cells for Proteomics-Based Applications 151
   Charles C. King

10. A Two- and Three-Dimensional Approach for Visualizing Human Embryonic Stem Cell Differentiation 179
    Christian B. Brøchner, Peter S. Vestentoft, Niels Lynnerup, Claus Yding Andersen, and Kjeld Møllgård

11. Immunoflourescence and mRNA Analysis of Human Embryonic Stem Cells (hESCs) Grown Under Feeder-Free Conditions 195
    Aashir Awan, Roberto S. Oliveri, Pernille L. Jensen, Søren T. Christensen, and Claus Yding Andersen

12. Study of Gap Junctions in Human Embryonic Stem Cells 211
    Raymond C.B. Wong and Alice Pébay

13. hESC Engineering by Integrase-Mediated Chromosomal Targeting 229
    *Ying Liu, Uma Lakshmipathy, Ali Ozgenc, Bhaskar Thyagarajan,
    Pauline Lieu, Andrew Fontes, Haipeng Xue, Kelly Scheyhing,
    Chad MacArthur, and Jonathan D. Chesnut*

14. Human Embryonic Stem Cell Differentiation on Periodontal Ligament
    Fibroblasts 269
    Y. Murat Elçin, Bülend İnanç, and A. Eser Elçin

15. Generation of Neural Crest Cells and Peripheral Sensory Neurons from Human
    Embryonic Stem Cells 283
    Ronald S. Goldstein, Oz Pomp, Irina Brokhman, and Lina Ziegler

16. Embryonic Stem Cells as a Model for Studying Melanocyte Development ... 301
    Susan E. Zabierowski and Meenhard Herlyn

17. In Vitro Derivation of Chondrogenic Cells from Human Embryonic
    Stem Cells 317
    Wei Seong Toh, Eng Hin Lee, Mark Richards, and Tong Cao

18. Vascular Differentiation of Human Embryonic Stem Cells in Bioactive
    Hydrogel-Based Scaffolds 333
    Sharon Gerecht, Lino S. Ferreira, and Robert Langer

19. Differentiation of Neural Precursors and Dopaminergic Neurons from Human
    Embryonic Stem Cells 355
    Xiao-Qing Zhang and Su-Chun Zhang

20. Transplantation of Human Embryonic Stem Cells and Derivatives to the Chick
    Embryo 367
    Ronald S. Goldstein

21. Genetic Manipulation of Human Embryonic Stem Cells 387
    Silvina Epsztejn-Litman and Rachel Eiges

22. Genetic Manipulation of Human Embryonic Stem Cells in Serum
    and Feeder-Free Media 413
    Stefan R. Braam, Chris Denning, and Christine L. Mummery

23. Human-Induced Pluripotent Stem Cells: Derivation, Propagation, and Freezing
    in Serum- and Feeder Layer-Free Culture Conditions 425
    *Hossein Baharvand, Mehdi Totonchi, Adeleh Taei, Ali Seifinejad,
    Nasser Aghdami, and Ghasem Hosseini Salekdeh*

Index 445

Contributors

NASSER AGHDAMI • *Department of Stem Cells, Cell Science Research Center, Royan Institute, ACECR, Tehran, Iran*

CLAUS YDING ANDERSEN • *Laboratory for Reproductive Biology, Copenhagen University Hospital, Rigshospitalet, Copenhagen Ø, Denmark*

AASHIR AWAN • *Laboratory for Reproductive Biology, University Hospital of Copenhagen, Copenhagen, Denmark; Department of Biology, University of Copenhagen, Copenhagen, Denmark*

HOSSEIN BAHARVAND • *Department of Stem Cells, Cell Science Research Center, Royan Institute, ACECR, Tehran, Iran; Department of Developmental Biology, University of Science and Culture, ACECR, Tehran, Iran*

ROMAIN BARBET • *Centre National de la Recherche Scientifique, Institut André Lwoff, Villejuif, France*

STEFAN R. BRAAM • *Department of Anatomy & Embryology, Leiden University Medical Centre, Leiden, The Netherlands*

CHRISTIAN B. BRØCHNER • *Department of Cellular and Molecular Medicine, University of Copenhagen, The Panum Institute, Copenhagen N, Denmark*

IRINA BROKHMAN • *Gonda Research Center, Faculty of Life Sciences, Bar-Ilan University, Ramat-Gan, Israel*

TONG CAO • *Stem Cell Laboratory, Faculty of Dentistry, National University of Singapore, Singapore*

JONATHAN D. CHESNUT • *Invitrogen Corporation, Carlsbad, CA, USA*

SØREN T. CHRISTENSEN • *Department of Biology, University of Copenhagen, Copenhagen, Denmark*

LYN C. COTE • *Magee-Womens Research Institute, Pittsburgh, PA, USA*

CHRIS DENNING • *Department of Anatomy & Embryology, Leiden University Medical Centre, Leiden, The Netherlands*

RACHEL EIGES • *Stem Cell Research Laboratory, Medical Genetics Unit, Shaare Zedek Medical Center, Jerusalem, Israel*

A. ESER ELÇIN • *Tissue Engineering, Biomaterials and Nanobiotechnology Laboratory, Ankara University, Ankara, Turkey*

Y. MURAT ELÇIN • *Tissue Engineering, Biomaterials and Nanobiotechnology Laboratory, Ankara University, Ankara, Turkey*

CATHARINA ELLERSTRÖM • *Cellartis AB, Göteborg, Sweden*

SILVINA EPSZTEJN-LITMAN • *Stem Cell Research Laboratory, Medical Genetics Unit, Shaare Zedek Medical Center, Jerusalem, Israel*

TERESA M. ERB • *Department of Obstetrics, Gynecology, and Reproductive Sciences, University of Pittsburgh School of Medicine, Pittsburgh, PA, USA*

CARMEN ESCOBEDO-LUCEA • *Centro de Investigación Príncipe Felipe (CIPF), Valencia, Spain*

JAMES E. ESPLEN • *Department of Pathology, University of Pittsburgh School of Medicine, Pittsburgh, PA, USA*

LINO S. FERREIRA • *Center of Neurosciences and Cell Biology, Department of Zoology, University of Coimbra, Coimbra, Portugal; Biocant – Center of Innovation in Biotechnology, Cantanhede, Portugal*

ANDREW FONTES • *Invitrogen Corporation, Carlsbad, CA, USA*

AMPARO GALÁN • *Stem Cell Bank, Prince Felipe Research Center (CIPF), Valencia, Spain*

SHARON GERECHT • *Department of Chemical and Biomolecular Engineering, Johns Hopkins University, Baltimore, MD, USA*

RONALD S. GOLDSTEIN • *Gonda Research Center, Faculty of Life Sciences, Bar-Ilan University, Ramat-Gan, Israel*

ANTOINETTE HATZFELD • *Centre National de la Recherche Scientifique, Institut André Lwoff, Villejuif, France*

JACQUES A. HATZFELD • *Centre National de la Recherche Scientifique, Institut André Lwoff, Villejuif, France*

MEENHARD HERLYN • *The Wistar Institute, Philadelphia, PA, USA*

JOHAN HYLLNER • *Cellartis AB, Göteborg, Sweden*

BÜLEND INANÇ • *Tissue Engineering, Biomaterials and Nanobiotechnology Laboratory, Ankara University, Ankara, Turkey*

ESTHER P. JANE • *Department of Neurosurgery, University of Pittsburgh School of Medicine, Pittsburgh, PA, USA*

PERNILLE L. JENSEN • *Laboratory for Reproductive Biology, University Hospital of Copenhagen, Copenhagen, Denmark; Department of Biology, University of Copenhagen, Copenhagen, Denmark*

CHARLES C. KING • *Department of Pediatrics, University of California, San Diego, La Jolla, CA, USA*

UMA LAKSHMIPATHY • *Invitrogen Corporation, Carlsbad, CA, USA*

ROBERT LANGER • *Department of Chemical Engineering, Massachusetts Institute of Technology, Cambridge, MA, USA*

ENG HIN LEE • *Department of Orthopaedic Surgery, Yong Loo Lin School of Medicine, NUS Tissue Engineering Program, National University of Singapore, Singapore*

MA-LIN LI • *Key Laboratory of Yunnan of Pharmacology for Nature Products, Kunming Medical University, Kunming, China*

PAULINE LIEU • *Invitrogen Corporation, Carlsbad, CA, USA*

YING LIU • *Invitrogen Corporation, Carlsbad, CA, USA*

NIELS • *Lynnerup Laboratory of Biological Anthropology, Department of Forensic Pathology, University of Copenhagen, The Panum Institute, Copenhagen N, Denmark*

CHAD MACARTHUR • *Invitrogen Corporation, Carlsbad, CA, USA*

KJELD MØLLGÅRD • *Department of Cellular and Molecular Medicine, University of Copenhagen, The Panum Institute, Copenhagen N, Denmark*

SARA E. MUCKO • *Magee-Womens Research Institute, Pittsburgh, PA, USA*

CHRISTINE L. MUMMERY • *Department of Anatomy & Embryology, Leiden University Medical Centre, Leiden, The Netherlands*

ROBERTO S. OLIVERI • *Laboratory for Reproductive Biology, University Hospital of Copenhagen, Copenhagen, Denmark*

Tiki Osianlis • *Senior Embryologist, Monash IVF, Clayton, Victoria, Australia*
Ali Ozgenc • *Invitrogen Corporation, Carlsbad, CA, USA*
John A. Ozolek • *Department of Pathology, University of Pittsburgh School of Medicine, Pittsburgh, PA, USA*
Alice Pébay • *Centre for Neuroscience and Department of Pharmacology, The University of Melbourne, Parkville, Victoria, Australia*
Isabelle Peiffer • *Centre National de la Recherche Scientifique, Institut André Lwoff, Villejuif, France; CNRS, Institut de Génétique Humaine, Montpellier, France*
Patti • *Petrosko – Magee-Womens Research Institute, Pittsburgh, PA, USA*
Teija T. Peura • *Australian Stem Cell Centre QLD, Brisbane, Australia*
Oz Pomp • *Gonda Research Center, Faculty of Life Sciences, Bar-Ilan University, Ramat-Gan, Israel*
Mark Richards • *Applied Research Group, School of Chemical & Life Sciences, Nanyang Polytechnic, Singapore*
Ghasem Hosseini • *Salekdeh Department of Stem Cells, Cell Science Research Center, Royan Institute, ACECR, Tehran, Iran*
Paul J. Sammak • *Department of Obstetrics, Gynecology, and Reproductive Sciences, University of Pittsburgh School of Medicine, Pittsburgh, PA, USA; Magee-Womens Research Institute, Pittsburgh, PA, USA*
A. Henry Sathananthan • *Monash Immunology and Stem Cell Laboratories, Faculty of Medicine and Nursing, Monash University, Clayton, Victoria, Australia*
Julia Schaft • *Sydney IVF, Sydney, Australia*
Ali Seifinejad • *Department of Stem Cells, Cell Science Research Center, Royan Institute, ACECR, Tehran, Iran*
Kelly Sheyhing • *Invitrogen Corporation, Carlsbad, CA, USA*
Carlos Simón • *Stem Cell Bank, Prince Felipe Research Center (CIPF), Valencia, Spain; IVI Foundation, Instituto Universitario IVI, University of Valencia, Valencia, Spain*
Tomas Stojanov • *Sydney IVF, Sydney, Australia*
Miodrag Stojkovic • *Centro de Investigación Príncipe Felipe (CIPF), Valencia, Spain*
Raimund Strehl • *Cellartis AB, Göteborg, Sweden*
Adeleh Taei • *Department of Stem Cells, Cell Science Research Center, Royan Institute, ACECR, Tehran, Iran*
Bhaskar Thyagarajan • *Invitrogen Corporation, Carlsbad, CA, USA*
Wei Seong Toh • *Stem Cell Laboratory, Faculty of Dentistry, National University of Singapore, Singapore*
Mehdi Totonchi • *Department of Stem Cells, Cell Science Research Center, Royan Institute, ACECR, Tehran, Iran*
Mohan C. Vemuri • *Primary and Stem Cell Systems, Invitrogen Corporation, Grand Island, NY, USA*
Peter S. Vestentoft • *Department of Cellular and Molecular Medicine, University of Copenhagen, The Panum Institute, Copenhagen N, Denmark*
Katherine E. Wagner • *Primary and Stem Cell Systems, Invitrogen Corporation, Grand Island, NY, USA*
Amy K. Wehn • *Magee-Womens Research Institute, Pittsburgh, PA, USA*

RAYMOND C.B. WONG • *Department of Biological Chemistry, University of California Irvine, Irvine, CA, USA*

HAIPENG XUE • *Invitrogen Corporation, Carlsbad, CA, USA*

SUSAN E. ZABIEROWSKI • *The Wistar Institute, Philadelphia, PA, USA*

SU-CHUN ZHANG • *Departments of Anatomy and Neurology, School of Medicine, The Stem Cell Research Program, Waisman Center, and the WiCell Institute, University of Wisconsin-Madison, Madison, WI, USA*

XIAO-QING ZHANG • *Departments of Anatomy and Neurology, School of Medicine, The Stem Cell Research Program, Waisman Center, and the WiCell Institute, University of Wisconsin-Madison, Madison, WI, USA*

LINA ZIEGLER • *Gonda Research Center, Faculty of Life Sciences, Bar-Ilan University, Ramat-Gan, Israel*

Chapter 1

Human Embryo Culture and Assessment for the Derivation of Embryonic Stem Cells (ESC)

A. Henry Sathananthan and Tiki Osianlis

Abstract

The culture and critical assessment of early human embryos during the first week of human development are reviewed for the derivation of ESC. Both normal and abnormal features are assessed by phase contrast microscopy of whole embryos and in serial sections of fixed material by light and electron microscopy (TEM). Normal embryos follow a time table of development and have equal blastomeres with minimal fragmentation and nuclear defects. Abnormal embryos show more fragmentation and nuclear aberrations such as micronucleation and multinucleation, reflected by aneuploidy, polyploidy, and mosaicism. The selection of normal embryos and the hardiest of embryos that survive to blastocysts is recommended for the derivation and culture of ESC.

Key words: Human, preimplantation embryos, blastocysts, culture, assessment, ESC, TEM.

1. Introduction

Theoretically it is possible to derive human ESC from any embryonic cleavage stage (2-cell to 8-cell) prior to blastulation – **Figs. 1.1** and **1.2** (1, 2). This is because the blastomeres have not yet differentiated to other cell lineages and are believed to be "totipotent" – each cell capable of giving rise to a complete individual, which is now debatable (3). Recently, the morula (~32 cells) has been used to produce ESC (4). However, morulae have internalized cells within those at the surface, destined to become ICM and trophoblast in blastocysts. Hence most of these early embryonic cells seem to have the potential to generate ESC including cells of the extra-embryonic cell lineages.

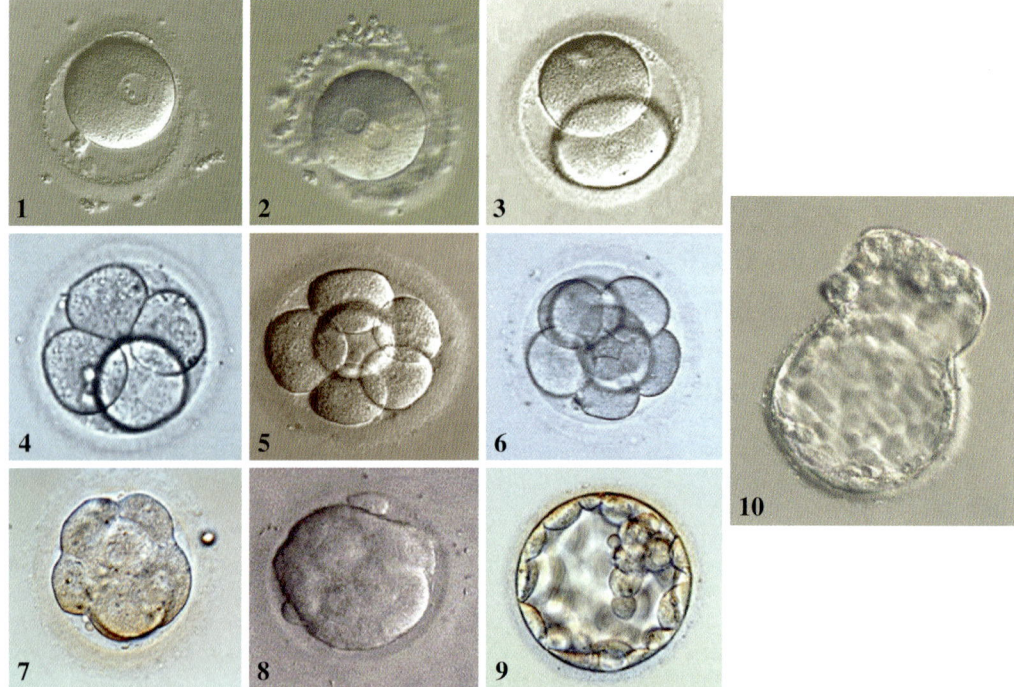

Fig. 1.1. Normal whole embryos – 1-cell to blastocyst (Phase contrast). 1. Activated oocyte; 2. Fertilized ovum (2PN); 3. 2-cell; 4. 4-cell; 5. 6-cell; 6. 8-cell; 7. Compaction; 8. Morula; 9. Blastocyst; 10. Hatching Blastocyst (9).

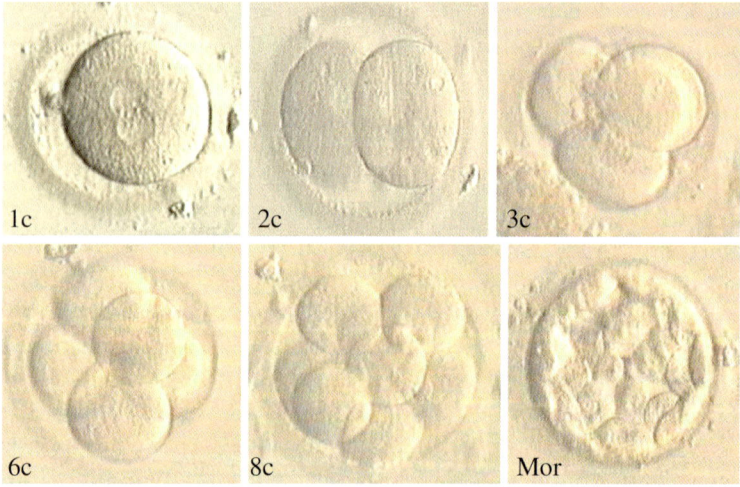

Fig. 1.2. Normal human embryos – 1-cell to morula (phase-contrast). The cleavage embryos have equal blastomeres and minimal cytoplasmic fragmentation, except the 3-cell embryo ×400 (20).

The preferred stage used to derive ESC is the blastocyst on day 5, where the ICM (embryoblast) has already differentiated from a peripheral trophoblast, which becomes the placenta (**Fig. 1.2**). The ICMs of blastocysts have been successfully used in the derivation of ESC lines, which is now a routine procedure (5–7). These cells are pluripotent and can differentiate into all 3 germ layers: endoderm, ectoderm, and mesoderm, from which all tissues of the human body are formed (8). A time table of embryo development in week 1 is shown in **Table 1.1**.

Table 1.1
Normal embryonic growth from day 2 to 7

Day	Embryo	Appearance/hours
D1	Fertilized ovum	2 PN (12 h) and Syngamy (18–24 h)
D2	Cleaving embryo	2–6 cells: rounded blastomeres
D3	Cleaving embryo Compacting embryo	8–10 cells or rounded blastomeres Blastomeres show evidence of adhesion
D4	Compacted morula Early cavitating	Blastomeres show increased adhesion Beginning of blastocoele formation
D5	Early blastocyst Mid-blastocyst Expanding blastocyst	Blastocoele formed ICM, trophoblast, and blastocyst clearly seen Trophoblast expanding, Zona thinning out. Embryo growing, blastocoele much increased
D6/7	Late blastocyst Hatching blastocyst Hatched blastocyst	Expanded ~150–200 cells; diameter ~215 mm Trophoblast hatching out of zona Trophoblast and ICM hatched out of empty zona

An embryo that develops to this time table is likely to be more viable than the one which shows delayed growth. Modified from (9)

This chapter deals with the culture and assessment of viable preimplantation embryos from pronuclear ova to blastocysts (**Figs. 1.1** and **1.2**), since it is important to select the most normal and vigorous embryos for ESC culture. The assessment is morphological and noninvasive using an inverted phase microscope with advanced Hoffman optics and images recorded on video with a digital camera. If you have already an established IVF laboratory, culturing blastocysts for ESC is not too difficult.

2. Materials

2.1. Culturing Embryos
(see Notes 1–9)

1. 4-well nunclon dishes (Becton Dickinson, Lane Cove, NSW, Australia)
2. 1-, 5-, and 10-ml falcon sterilogical pipettes (Becton Dickinson)

3. Art pipette tips (In vitro Technologies, Noble Park, Vic, Australia)
4. Glass Pasteur Pipettes (Pacific Lab Products, Blackburn, Vic, Australia) (*see* **Note 5**)
5. SIVF Cleaveage and Blastocyst media and Paraffin oil (COOK Medical, Eight Mile Planes, Qld, Australia). All culture media are stored at 2–8°C (*see* **Note 2**)
6. MINC incubators used for embryo culture (COOK Medical, Eight Mile Planes, Qld, Australia) (*see* **Note 4**)
7. Sanyo Trigas incubators (Quantum Scientific, Australia) used for culture media equilibration
8. Nitrogen, CO_2, and Special Mix (6% CO_2, 5% O_2, 89% N_2) cylinders (BOC, Preston, Vic, Australia)

3. Methods

Some of the embryos cultured and assessed noninvasively in the laboratory were generated at Monash IVF, Melbourne, a premier assisted reproductive technology (ART) center for the treatment of infertility, with over 20,000 successful pregnancies (take-home babies): website www.monashivf.com (*see* **Section 3.1**). Others were developed in well-established IVF hospitals in Singapore and Bangalore. Both in vitro fertilization (IVF) and intracytoplasmic sperm injection (ICSI) procedures were used to generate these embryos in our laboratories here and overseas (9–13).

3.1. Culturing Human Embryos In Vitro Involves

3.1.1. Fertilization Assessment

Oocytes are viewed for fertilization 14–18 h post sperm insemination. Oocytes exhibiting two pronuclei are deemed normally fertilized zygotes (monospermic) and are suitable for culture (**Figs. 1.3–1.6**). More than two pronuclei, usually 3, means that abnormal fertilization has taken place and the embryos are not suitable for culture (e.g., dispermy or digyny). These embryos might grow quite successfully but have an abnormal (triploid) complement of chromosomes. Oocytes with no pronuclei are indicative of failure to fertilize and those with one pronucleus are either activated or show abnormal fertilization – asynchrony of pronuclear formation or silent fertilization, when sperm heads do not expand to form male pronuclei. These oocytes are not suitable for culture (*see* **Note 6**).

3.1.2. Cleavage Stage Culture

On the day of insemination (day 0), cleavage stage culture dishes are prepared in a clean laminar flow (**Figs. 1.7–1.15**):
1. Using a displacement pipette 2×10 µl of cleavage media is dispensed into each well of a labeled 4-well nunclon dish.

Human Embryo Culture and Assessment 5

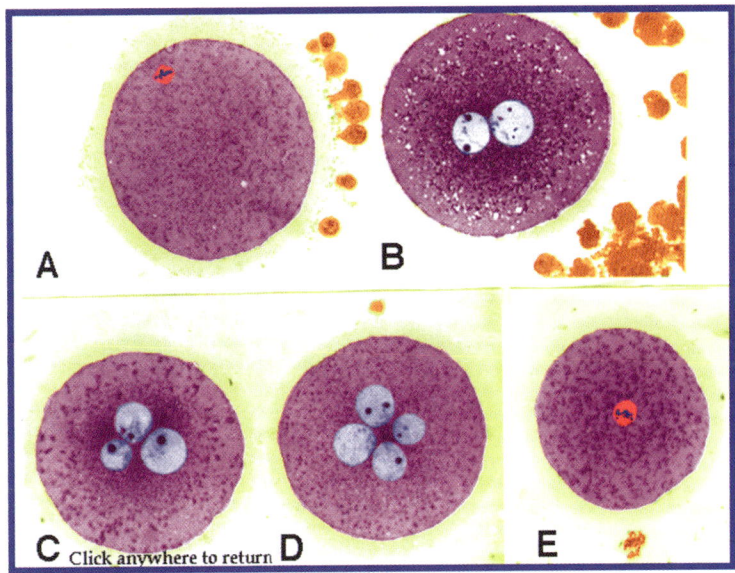

Fig. 1.3. Fertilization: oocyte to pronuclear ova and syngamy – sections (LM). **a**. Mature oocyte (Metaphase II spindle). **b**. Bipronuclear ovum. **c**. Tripronuclear ovum (Dispermy). **d**. Tetrapronuclear ovum. **e**. Syngamy – first mitotic spindle – ×400 (22).

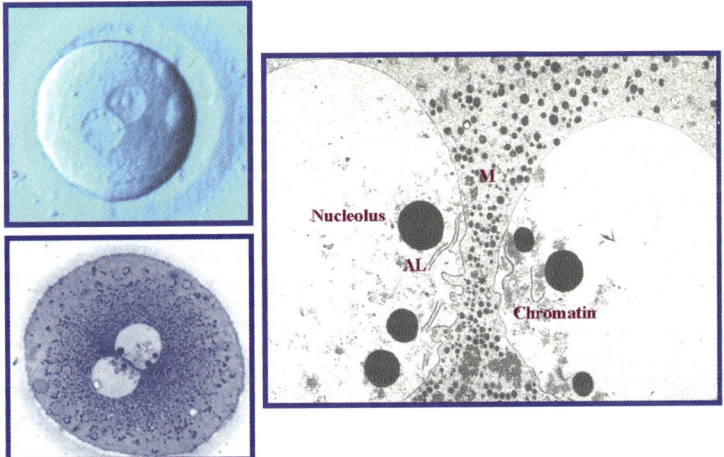

Fig. 1.4. Normal bipronuclear ova (Phase, LM, and TEM). These bipronuclear, after monospermic fertilization, seem normal. Note alignment of nucleoli adjacent to apposing pronuclear membranes. What is more significant is the alignment of chromatin, associated with nucleoli, which would condense to form the male and female chromosomes at syngamy ×400, ×35,700 (22).

2. Paraffin oil is over-laid, approximately 0.5 ml with either a 1-ml displacement pipette or using a 1- or 5-ml serological pipette and an electric pipette aid.

3. Up to 8 zygotes can be accommodated per dish so enough dishes should be made for the number of zygotes being cultured for individuals.

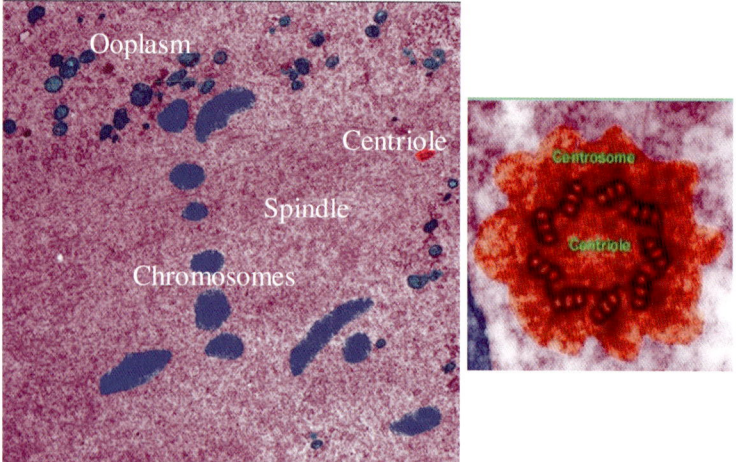

Fig. 1.5. Syngamy: the first mitotic spindle at metaphase TEM. Syngamy is the culmination of fertilization when maternal and paternal chromosomes come together on a bipolar spindle. Note displaced chromosomes outside spindle. The centriole is a descendant of the sperm centrosome in a 8-cell embryo (22).

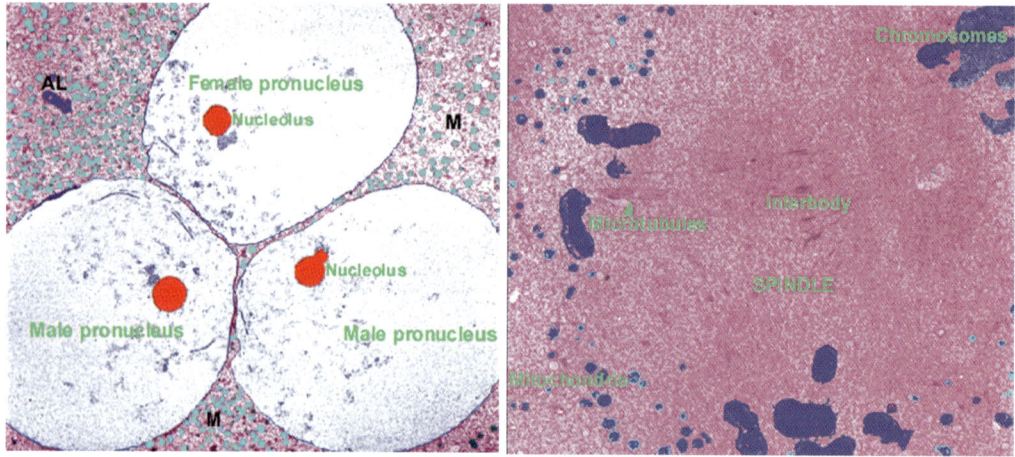

Fig. 1.6. Dispermic tripronuclear (3PN) ova (TEM). Classical images of 3PN ova at the pronuclear stage and syngamy. Note chromatin (blue specks) and nucleoli located toward adjacent membranes of the pronuclear envelopes. The spindle is tripolar enabling the ovum to divide into 3 cells, instead of 2 cells. ×5000, ×8000 (22).

4. Once dishes are complete, they are placed into a CO_2 incubator (*see* **Note 3**). Dishes are equilibrated for a minimum of 4 h in a 6% CO_2 environment (may vary depending on culture media being used) and 5% O_2 in nitrogen and 37°C. Having a 5% O_2 environment is not vital for dish equilibration. However, it is important for culturing embryos.

5. After fertilization assessment (day 1), individual zygotes are moved from the fertilization dish to the cleavage culture dish (day 1–3 culture dish) using a glass flame pulled Pasteur

Human Embryo Culture and Assessment

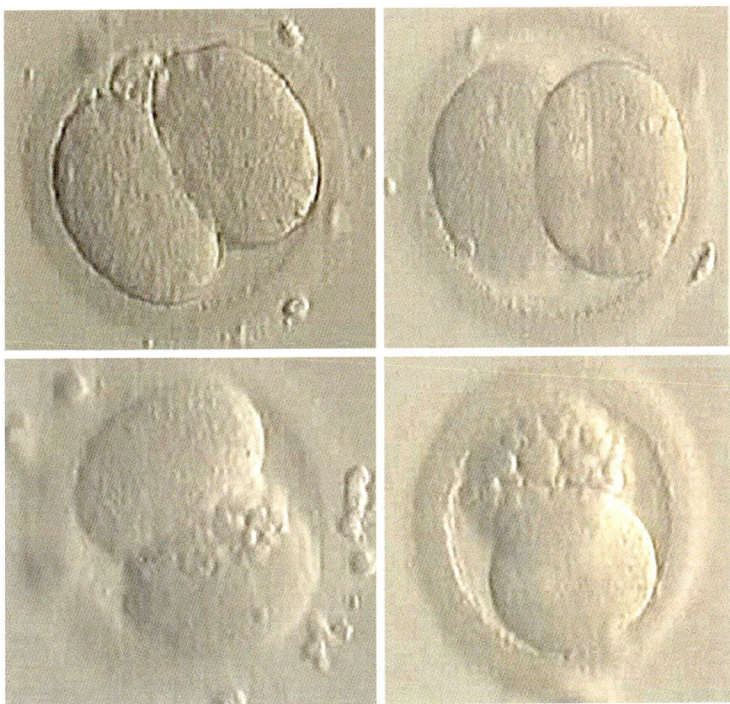

Fig. 1.7. Two-cell embryos – normal and fragmented (phase contrast). Both normal and abnormal embryos are evident. Fragments appear over the cleavage furrow or a whole cell can fragment totally (Apoptosis?) ×400 (20).

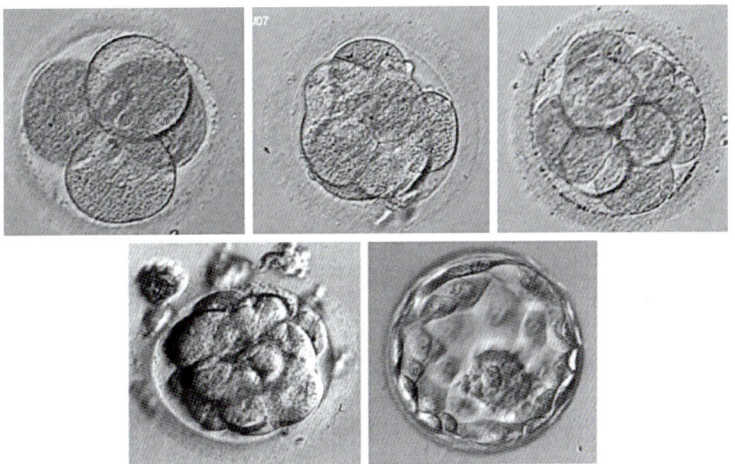

Fig. 1.8. Normal cleaving embryos – preimplantation (phase contrast). Four-cell, 8-cell, 9-cell, morula, and early blastocyst. Note almost equal blastomeres and no fragmentation. The blastocyst shows an ICM and trophoblast. Embryos cultured by Monash IVF. ×400.

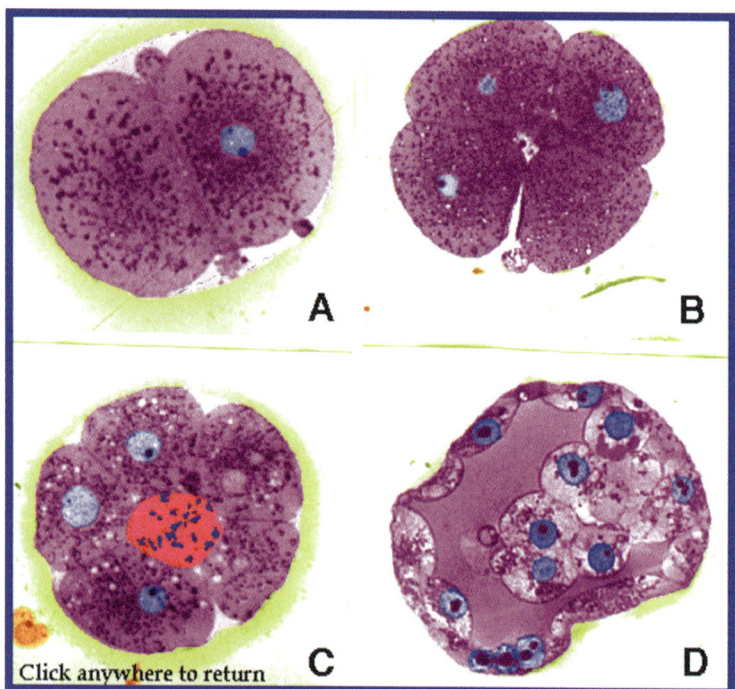

Fig. 1.9. Normal bipronuclear (2PN) embryos: 2-cell to blastocyst – sections (LM). These are embryos developed after monospermic fertilization. **a**. 2-cell. **b**. 4-cell. **c**. 6-cell; dividing cell (*red*). **d**. Early blastocyst with inner cell mass, trophoblast, and blastocoele. ×400 (22).

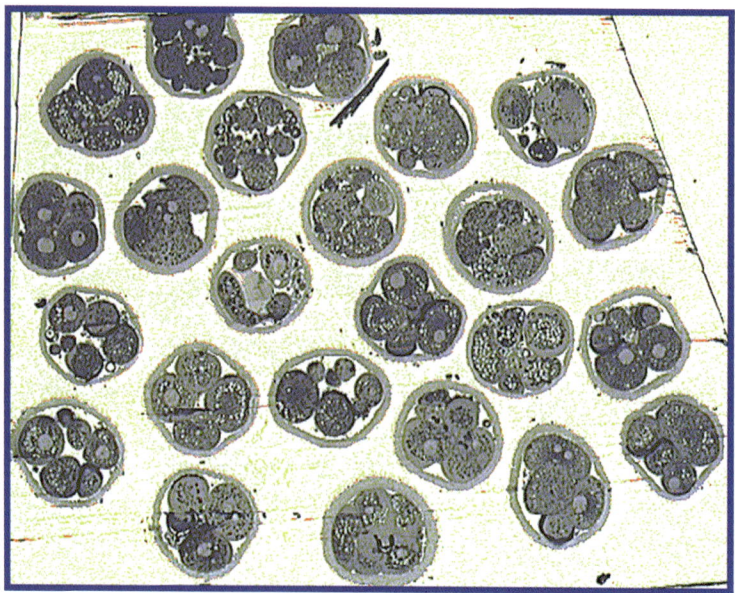

Fig. 1.10. Three- to 6-cell dispermic embryos – sections (LM). Both normal and abnormal embryos are shown. Vacuolated blastomeres are degenerating. Note variation in cell size, few fragments, and multinucleated blastomeres ×100 (19).

Human Embryo Culture and Assessment

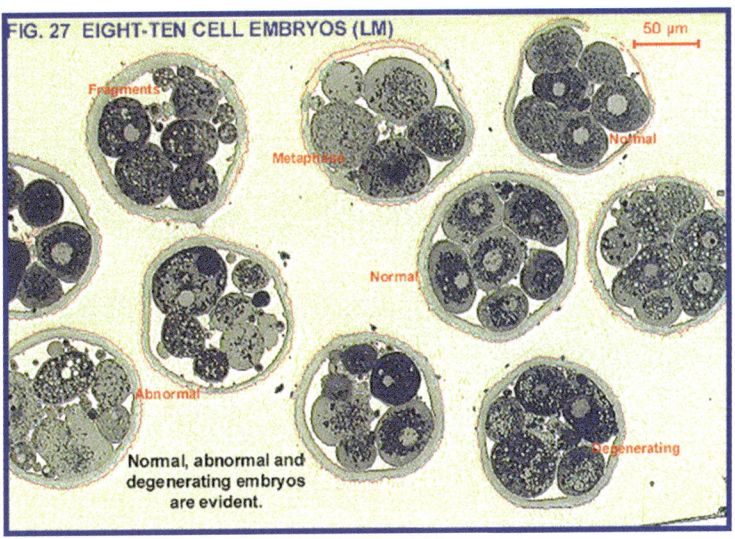

Fig. 1.11. Eight- to 10-cell dispermic embryos – sections (LM). Both normal and abnormal embryos are evident. Blastomeres with clear vacoules are degenerating. Note unequal-sized blastomeres in abnormal embryos and few fragments ×200 (19).

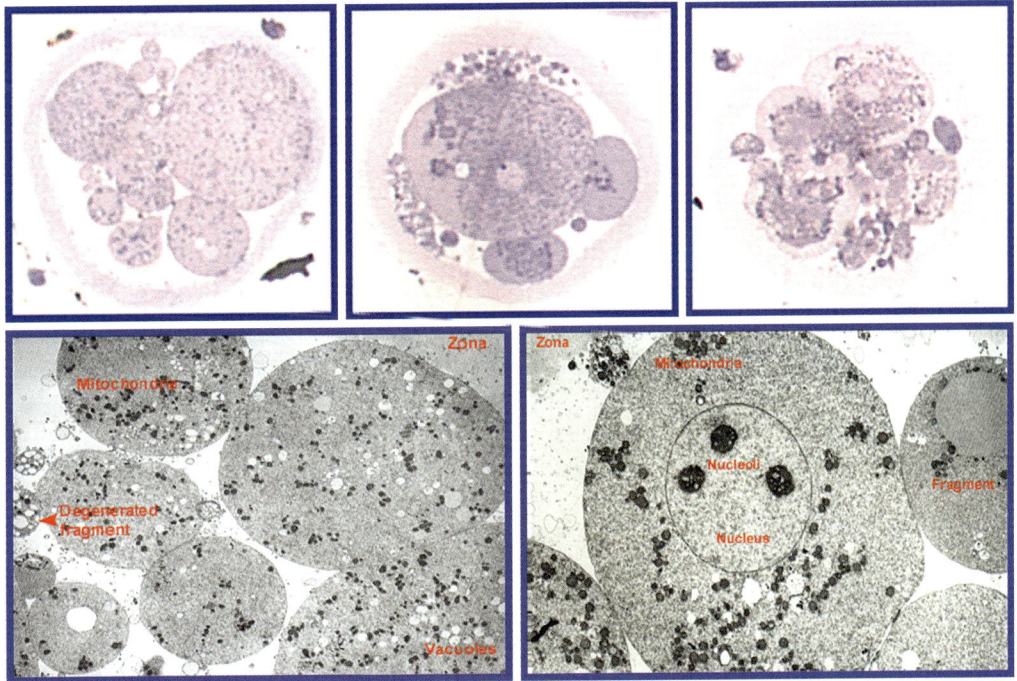

Fig. 1.12. Fragmented dispermic embryos (LM and TEM). Fragmentation is a common occurrence in early human embryos. These fragments are devoid of nuclear material – compare with a normal blastomere (*right*). Four- to 8-cell and 5-cell (*left*) and a 10-cell embryo (*right*). ×400, ×3500, ×6000 (11, 19).

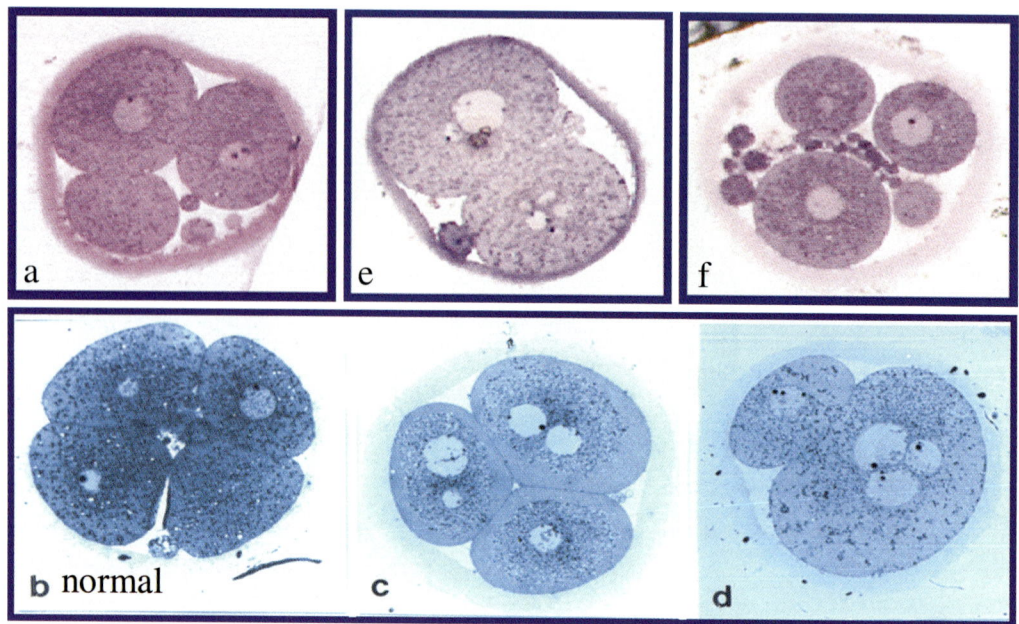

Fig. 1.13. Normal and abnormal 2- to 4-cell embryos – sections (LM). The 4-cell is normal. All other embryos show multiple nuclei and micronuclei and minimal fragmentation (11, 22).

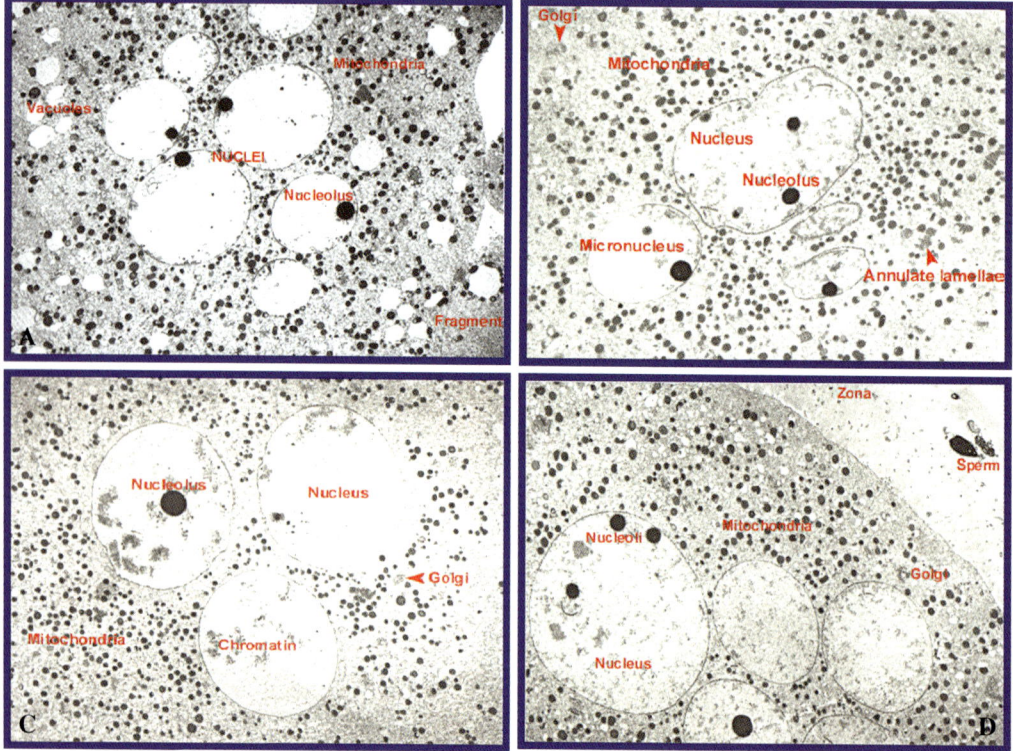

Fig. 1.14. Abnormal multinucleated dispermic embryos (TEM). **a**. 1-cell (fragmented) **b**. 2-cell (micronucleated), and **c**,**d**. 3-cell embryos (multinucleated) ×6000, ×4000 (22).

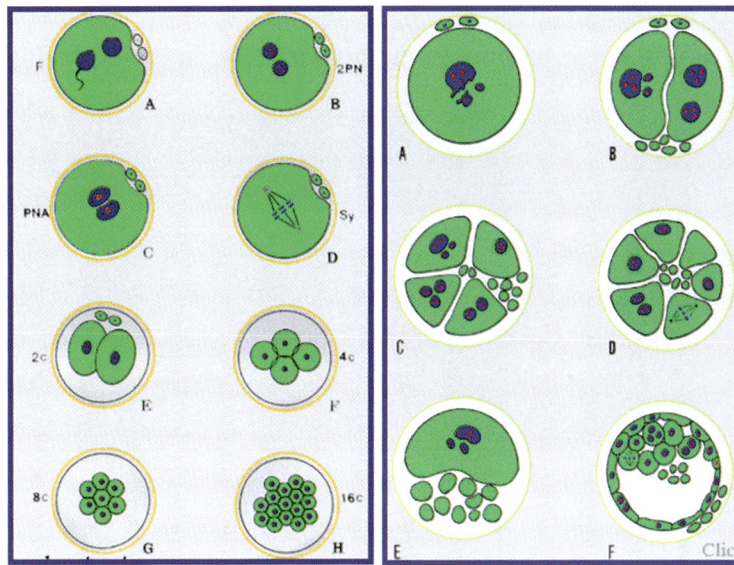

Fig. 1.15. Normal and abnormal embryos – diagrams. Both normal and abnormal blastomeres are seen in normal and abnormal embryos. Cytoplasmic fragmentation, multinucleation, micronucleation are the main abnormalities. Fragments may be internalized or external in the PVS, few to many (11).

pipette (*see* **Notes 5 and 11**). One zygote per droplet is transferred to the day 1–3 culture dish. Only normally fertilized zygotes are transferred. Zygotes are incubated in a 6% CO_2, 5% O_2, 89% nitrogen environment and 37°C temperature (*see* **Note 3**).

6. Using an inverted microscope or stereomicroscope, 24-h post sperm insemination zygotes are viewed and classified as 2PN if 2 pronuclei are still visible, syngamy if no pronuclei are evident, and early cleavage if embryos are at the 2-cell stage at the 24-h check.

7. Day 2 embryos are examined for developmental status and degree of fragmentation. Embryos are expected to be between 2- and 4-cell stages at 42–44 h post sperm insemination.

8. Day 3 embryos are once again viewed under inverted microscope for development and degree of fragmentation (*see* **Note 7**). Embryos are expected to be between 6- and 8-cell stages of development between 64 and 66 h post insemination (*see* **Tables 1.1 and 1.2**).

3.1.3. Extended Culture to Blastocysts

1. On day 2, the blastocyst stage culture dishes (day 3–6 culture dishes) are prepared (*see* **Notes 1 and 2**) (**Figs. 1.15–1.18**).

2. A displacement pipette is used to dispense 2 × 10 μl of blastocyst media into each well of a labeled 4-well nunclon dish.

Table 1.2
Embryo grading for embryo transfer in the laboratory

Grade 1: Blastomeres of equal size and no fragmentation
Grade 2: Blastomeres of equal size and minor fragmentation (<10%)
Grade 3: Blastomeres of unequal size and variable fragmentation
Grade 4: Blastomeres of equal or unequal size and significant fragmentation (>10%)
Grade 5: Few blastomeres of any size and severe fragmentation (>50%)
(Grades 1 and 2 have a greater potential of establishing a clinical pregnancy)
Modified from (14).

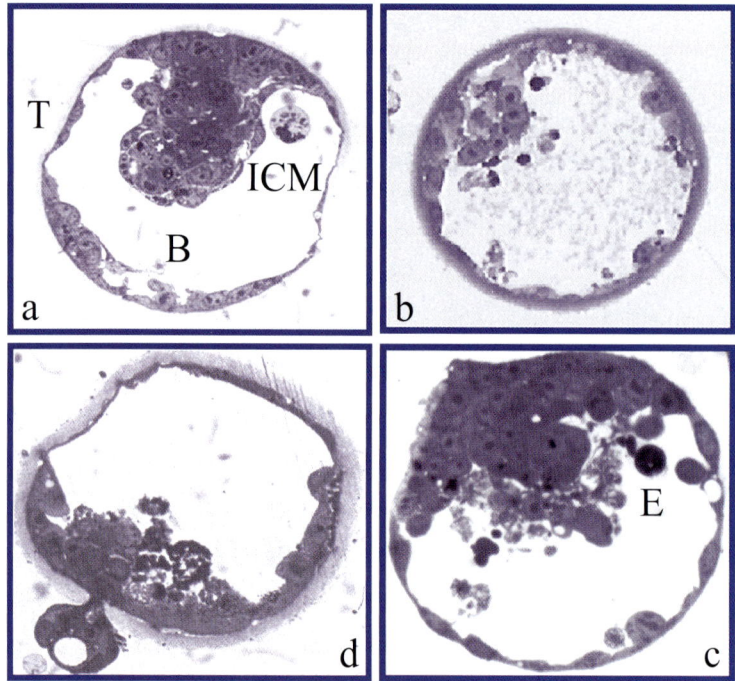

Fig. 1.16. Normal and abnormal blastocysts after ICSI (LM). **a.** Normal blastocyst with trophoblast (T), inner cell mass (ICM), and blastocoele (B). **b.** Disorganized ICM. **c.** Disorganized endoderm (E). **d.** Failed hatching – degenerating. ×400 (15).

3. Paraffin oil is overlayed, approximately 0.5 ml using a 1-ml displacement pipette or a seriological pipette and electric pipette aid.
4. A wash dish is prepared by placing 0.5 ml of blastocyst media into all 4 wells of a 4-well nunclon dish and overlaying with ∼ 0.5 ml oil.

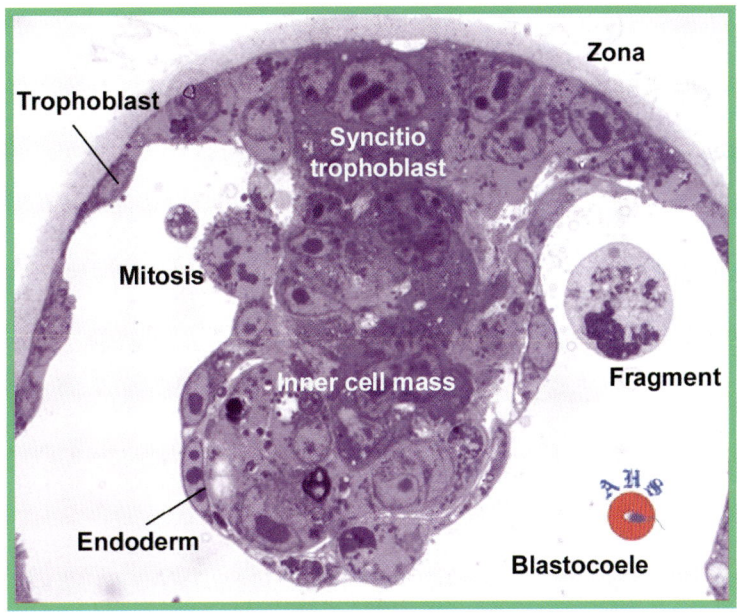

Fig. 1.17. Human expanded blastocyst – ICM on day 6 (LM). Embryonic stem cells are derived from the inner cell mass of blastocysts. Note the different cells that are associated with the ICM ×1000 (17).

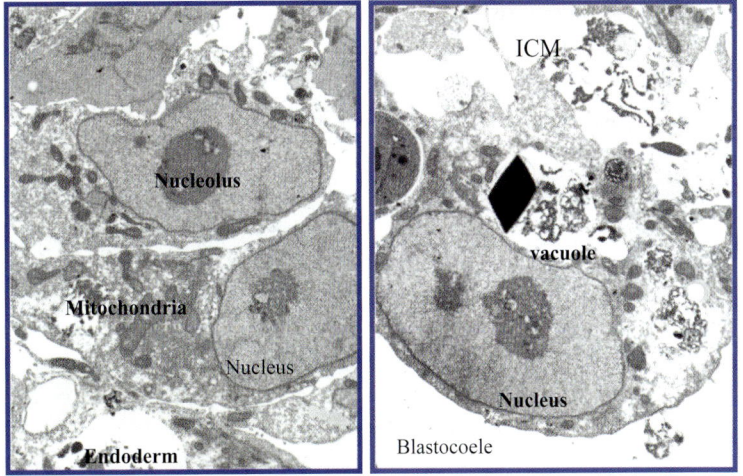

Fig. 1.18. Normal blastocyst – two ICM cells and an endoderm cell (TEM). The ICM cells are differentiated cells with mitochondria, RER, Golgi, etc. Endoderm cells are clearly phagocytic with vacuoles. Compare with ESC (15).

5. Dishes are equilibrated for a minimum of 4 h in a 5–6% CO_2 environment (depending on culture media being used) and 5% O_2 in nitrogen and 37°C (*see* **Note 3**).

6. After the embryos have been assessed on day 3, they are washed in the blastocyst wash dish by transferring each embryo in all four wells of the wash dish, followed by

transferring the embryo to the blastocyst culture dish. This is performed using a flame-pulled glass Pasteur pipette (*see* **Note 5**).

7. Day 4 embryos are examined for development and should be at the morula stage.

8. Day 5 embryos should be at the blastocyst stage. A cavity (blastocoele) should be evident and a defined inner cell mass (ICM) (*see* **Table 1.1**).

9. Embryos that have not developed to the blastocyst stage on day 5 and progress to blastocysts on day 6 are still developmentally viable although they are a day behind schedule. However, embryos that have not progressed to the blastocyst stage of development by day 6 do have compromised developmental competency. If the blastocysts do not hatch on days 6/7, they are not viable.

Contamination is a problem during culture of embryos and should be prevented (*see* **Notes 9–11**).

3.2. Embryo Assessment (Morphology)

This is based on multifocal images obtained by phase contrast microscopy of live embryos (PM) and light microscopy (LM) and transmission electron microscopy (TEM) of serial sections, some already published (**Figs. 1.3–1.20**). For more images of gametes, fertilization, and embryos see www.sathembryoart.com, Sathananthan et al. (10–13), Veeck (14).

3.2.1. Fertilization

First it is important to verify normal, monospermic fertilization at the pronuclear stage when 2 pronuclei, male and female, are demonstrable in the central cytoplasm. The position and alignment of nucleoli within pronuclei are now assessed to predict normal development (10). More significantly, these nucleoli are associated with chromatin that will condense to form the maternal and paternal chromosomes at syngamy. This chromatin cannot be seen with the laboratory microscope. Dispermy or digyny (after ICSI) will result in tripronuclear, triploid embryos, which should be discarded.

1. A normal bipronuclear *(2PN)* ovum shows (**Figs. 1.3** and **1.4**):
    a) Two pronuclei, male and female, associated within the ooplasm
    b) Two polar bodies – PB1 with chromosomes and PB2 with a nucleus
    c) Nucleoli aligned close to apposing pronuclear envelopes
    d) No cortical granules (CG) or few beneath oolemma after IVF
    e) Crowding of organelles, mostly mitochondria, around PN
    (Delayed CG exocytosis has been observed after ICSI at the 2-cell stage)

2. An embryo at syngamy before cleavage (**Fig. 1.5**) will have:
   a) Two polar bodies and a centralized bipolar spindle with centrioles (first mitosis)
   b) Paternal and maternal chromosomes aligned on the equator of the spindle
   c) Two centrioles at each pole of the spindle aligned perpendicular to one another
   d) No asters at spindle poles, but with a sperm tail attached to one pole

   (All except the 2 polar bodies cannot be seen with the phase microscope)

3. An abnormal dispermic pronuclear ovum will show (**Fig. 1.6**):
   a) Three pronuclei (2 male and 1 female)
   b) Two polar bodies (PB1 and PB2) in the perivitelline space (PVS)
   c) A triploid chromosome complement 69XXY or 69XXY
   d) Two male centrosomes and 2 sperm asters after sperm incorporation
   e) A bipolar or tripolar spindle at syngamy

   (Digynous ova have 2 female and 1 male pronucleus and only PB1 in the PVS)

3.2.2. Cleavage Stage Embryo Assessment

The most important morphological parameters to assess cleavage embryo normality in the laboratory are blastomere appearance, fragmentation, and multinucleation. Those with equal blastomeres with minimal cytoplasmic fragmentation and few multinucleated cells have a better prospect of implantation. The fate of each embryo should be monitored right up to blastocyst hatching (**Fig. 1.1**). Extensively fragmented embryos, commonly seen in arrested or slow-developing embryos, should be discarded. Fragmentation, regarded now as an apoptotic phenomenon, goes hand in hand with multinucleation and micronucleation expressed by chromosomal or genetic abnormalities (10–12). Aneuploidy, haploidy, polyploidy, and mosaicism are the chief causes of early embryonic loss which approaches 60% in ART. Of course, some of these abnormalities could be detected in ESC cell cultures by karyotyping, which is now a routine procedure. Embryos are usually graded in the laboratory for embryo transfer (*see* **Table 1.2**).

1. Normal embryos (**Figs. 1.7–1.9**) have:
   a) Rounded equal blastomeres, except when cells are dividing
   b) Blastomeres with well-defined outlines – cell membranes
   c) Cells with centralized, single nuclei
   d) No fragments or minimal fragmentation (>10%)

   (Embryos should develop according to the time frame without delay – *see* **Table 1.1**)

2. Abnormal embryos (10–16), (**Figs. 1.7**) show:
   a) Extensive cytoplasmic fragmentation of blastomeres (30–50%)
   b) Spontaneous fragmentation of whole blastomeres (apoptosis?)
   c) Unequal or fused blastomeres of varying sizes
   d) Multinucleation of blastomeres – polyploidy, mosaicism
   e) Micronuclei in blastomeres beside normal nuclei – aneuploidy
3. Arrested or degenerating embryos (**Figs. 1.10–1.15**) show:
   a) Dark granular blastomeres with central aggregation of organelles
   b) Extensive vacuolation of blastomeres – increases cell density
   c) Cells with eccentrically located nuclei
   d) Multinucleated cells, many fragments, fused or unequal cells
   e) Lack of compaction of cells in later embryos and morulae

3.2.3. Blastocyst Assessment

Blastocysts cultured in vitro need to be assessed for their viability and suitability for generating ESC. Blastocysts are classified according to their age, growth, activity, and morphology (15, 16):

1. Early blastocysts (day 5) show cavitation and formation of the blastocoele
2. Mid blastocysts (day 5/6) show growth with increase in cell number and a thick zona
3. Expanding blastocysts (day 6) have a large blastocoele, distinct ICM, stretched trophoblast, and a thin zona
4. Hatching blastocysts (day 6/7) show emergence of the embryo and a breached zona
5. Hatched blastocysts (day 7) are expanded or contracted with no zona

As blastocysts grow, their total cell numbers increase to about 150–250, determined by DAPI staining (16). Optimally, a normal blastocyst on day 5 should be used for isolation of ICM cells. Expanded late and hatched blastocysts have a clump of multinucleated syncytiotrophoblast at the embryo pole, which invades the ICM and may interfere with its isolation. The best stages for isolation are day 5 or 6.

Here are some characteristics of blastocysts.

6. Normal blastocysts (**Figs. 1.15–1.18**) have a:
   a) Distinct trophoblast, ICM, and blastocoele
   b) Well-defined, compacted ICM with many cells
   c) Trophoblast forming a continuous epithelium
   d) Large fluid-filled blastocoele or cavity
   e) Few early cleavage stage fragments in the blastocoele or PVS

7. Abnormal blastocysts (**Fig. 1.16**):
   a) Have no ICM or have a small or dispersed ICM
   b) Fail to expand and hatch on day 6/7 – are moribund or dead
   c) Arrest in development and eventually degenerate
   d) Have many early-stage fragments in PVS – interferes with hatching
   e) Show multinucleated cells in ICM, trophoblast, and endoderm
8. Hatched postblastocysts (**Fig. 1.19**) cultured to day 9 (17):
   a) Show the proliferation of ESC from the ICM and a peripheral trophoblast
   b) The ESC have large nuclei and scanty cytoplasm, compared to ICM cells
   c) Have a multinucleated syncytiotrophoblast (implants in the endometrium)
   d) Have no zona pellucida or shell
   e) Show an amniotic-like cavity at one pole
      (The exact origins of ESC is possibly by dedifferentiation of ICM cells.)
9. Undifferentiated ESC derived from ICM (**Fig. 1.20**) in saucer-shaped colonies present (18):

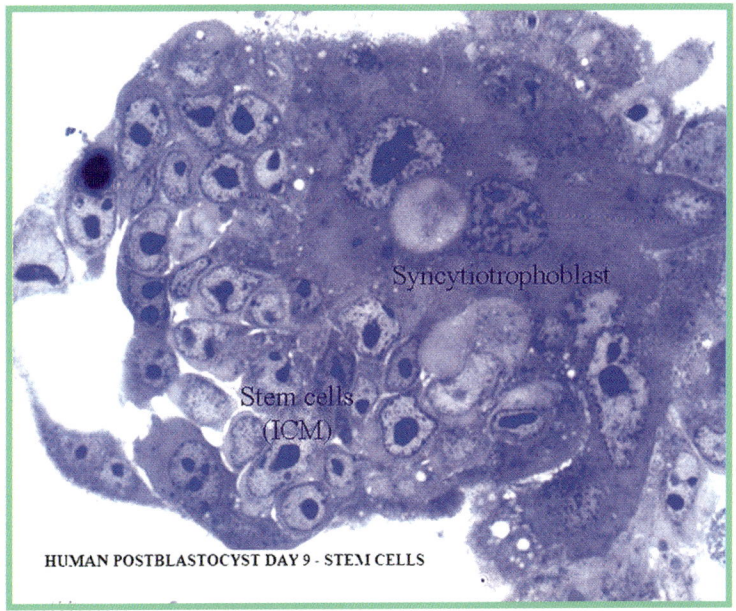

Fig. 1.19. Human postblastocyst in culture on day 9 (LM). The stem cells have proliferated within the ICM and show a more simplified structure than ICM. Note amniotic-like cavity (left) ×1000 (17).

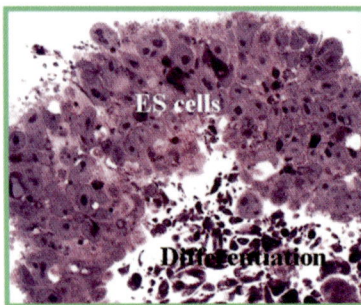 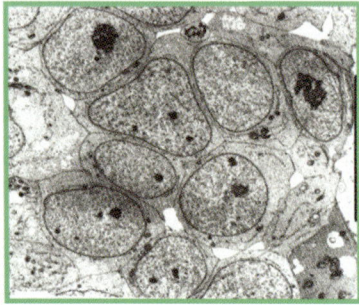

Fig. 1.20. Undifferentiated ESC in a colony – passages 35 and 14 (LM and TEM). The ESCs have large nuclei with dark nucleoli and scanty cytoplasm. Differentiating protein-synthesizing cells are found within colony ×1000, ×7000 (17, 18).

a) A simplified structure compared to ICM cells (dedifferentiation)

b) Scanty cytoplasm with fewer cellular organelles

c) Large nuclei with reticulated nucleoli

d) Few mitochondria and typical somatic centrosomes

e) Few cell junctions (desmosomes) and mitotic cells

For more images and embryo assessment in practical human ART, *see* references (19–25).

4. Notes

1. *Single or group culture of embryos:* There are two methods of culturing embryos, a single embryo in a single droplet or several embryos in a single droplet. There are advantages to both systems. Single embryo culture allows you to monitor individual embryo progress and identify developmental competencies within the embryo depending on their growth and when they reach those milestones. Group culture allows you to group like embryos, and if there are positive autocrine/paracrine effects then these will be amplified.

2. *Culture media:* There are numerous human embryo culture media that culture embryos satisfactorily. It is important to choose a complex sequential culture media, a media that has the addition of amino acids and protein and takes into consideration the different requirements of the embryos at different stages.

3. *Environment:* It is extremely important that the embryos are cultured in a stable environment. The incubators do need to be monitored and the readings on the incubators do need to be verified with thermometers or probes in the chambers that house the embryos and if using large CO_2 incubators for culture,

the CO_2 environment needs to be monitored. It is recommended to do daily checks of equipment along with 24-h long checks to make sure there are no great variations during off peak times that would not be seen. It is also extremely beneficial to view or manipulate embryos in a chamber that is set at 37°C and 5–6% CO_2, an IVF chamber similar to a humidity crib. This ensures the embryo's environment is as stable as possible during examination and changing over from one dish to the other.

4. *Incubators:* Everyone has their incubator of choice. When choosing an incubator for your laboratory think about how you will use the incubator, how many times you will need to go in and out of it, and what you will be using it for – just for embryo culture, will you be equilibrating, etc.

5. *Pipettes for embryo movement:* There are numerous pipettes that can be used to move embryos from one dish to the other. Glass Pasteur pipettes that have been heated and pulled into size over a burner are just one pipette. Displacement pipettes can also be used along with pipettes that are specifically designed to perform this function that come with a barrel and disposable tips of varying diameters.

6. *If all embryos are not growing normally:* Change media. It is extremely important to make sure that the embryos are exposed to culture media appropriate to their developmental stage. Also make sure that there are no toxins in the air, i.e., is the air free of cleaning agents, fixatives, or other contaminants.

7. *If there is excessive fragmentation of embryos:* Check that the temperatures the embryos are exposed to are correct.

8. *Not all embryos are progressing to the blastocyst stage:* This is expected. Approximately 40–50% of zygotes will progress to the blastocyst stage.

9. *Contamination:* If clean culture media dispensing or manipulation techniques are not adhered to, then unwanted growth may occur in the culture media and in the droplets the embryos are cultured in.

10. If contamination is present, use unopened culture media bottles to make up new change over dishes and wash dishes. Once the media has equilibrated, wash the embryos thoroughly. To avoid this, make sure not to use pipette tips that have touched nonsterile environments or to continually dip into the culture media bottle – aliquot media from the bottle into a tube or dish and dispense 10-μl droplets from aliquot.

11. When transferring embryos from one dish to the other, if your pipette touches a nonsterile environment or is suspected to have touched, throw the pipette and continue with a clean pipette. If in doubt, throw it out!

References

1. Fong, C-Y., Richards, M. and Bongso, A. (2006) Unsuccessful derivation of human embryonic stem cell lines from pairs of human blastomeres. *Reprod. BioMed. Online* **13**, 295–300.
2. Klimanskaya, I., Chung, Y., Becker, S. et al. (2006) Human embryonic stem cell lines derived from single blastomeres. *Nature* doi,1038/nature 05142, 1–4.
3. Sathananthan, A.H. (2007) The derivation of embryonic stem cells for therapy: new technologies. *Reprod. BioMed. Online* **14**, 635–39.
4. Strelchenko, N. and Verlinsky, Y. (2006) Embryonic stem cells from Morula. *Meth. Enzymol.* **418**, 93–108.
5. Bongso. A., Fong C-Y., Ng, S.C. et al. (1994) Isolation and culture of inner cell mass cells from human blastocysts. *Hum. Reprod.* **9**, 2110–17.
6. Hoffman, L.M. and Carpenter, M.K. (2005) Characterization and culture of human embryonic stem cells. *Nat. Biotechnol.* **23**, 699–708.
7. Trounson, A. (2006) The production and directed differentiation of human embryonic stem cells. *Endocr. Rev.* **27**, 208–19.
8. Larsen, W.J. (1998) *Essentials of Human Embryology*. Churchill Livingstone, New York, pp. 394.
9. Gunasheela, S. (2005) *The A-Z Encyclopedia on Male and Female Infertility*. Jaypee, New Delhi, pp. 146.
10. Sathananthan, A.H. and Gunasheela, S. (2007) Human oocyte and embryo assessment for ART, In: Elder, K., Cohen, J. (eds) *Human Preimplantation Embryo Selection*, 1st ed, Informa, Healthcare, pp. 1–14.
11. Sathananthan, A.H., Ng, S.C., Bongso, A. et al. (1993) *Visual Atlas of Early Human Development for Assisted Reproductive Technology*. National University & Serono, Singapore, pp. 209.
12. Munne, S. (2006) Chromosome abnormalities and their relationship to morphology and development of human embryos. *Reprod. Biomed. Online* **12**, 234–53.
13. Sathananthan, A.H. (ed) (1996) *Visual Atlas of Human Sperm Structure and Function for Assisted Reproduction*. National University & Serono, Singapore, pp 279.
14. Veeck, L. (1999) *An Atlas of Human Gametes and Conceptuses*. Parthenon, London, pp. 215.
15. Sathananthan, A.H., Gunasheela, S. and Menezes, J. (2003) Critical evaluation of human blastocysts for assisted reproduction techniques and embryonic stem cell biotechnology. *Reprod. Biomed. Online* 7, 219–27.
16. Bongso, A. (1999) *Handbook on Blastocyst Culture*. Sydney Indusprint, Singapore, pp. 93.
17. Sathananthan, A.H. (ed) (2007) Embryonic Stem Cells (DVD): A multi-author production of digital images from Monash Immunology & Stem Cell Laboratories. Presented at the Monash booth at the 5th ISSCR Annual Meeting, Cairns, Australia.
18. Sathananthan, H., Pera, M. and Trounson, A. (2002) The fine structure of human embryonic stem cells. *Reprod. Biomed. Online* **4**, 56–61.
19. Sathananthan, A.H., Tarin, J.J., Gianaroli, L. et al. (1999) The dispermic human embryo. *Hum. Reprod. Update* **5**(5) CD-ROM.
20. Sathananthan, A.H., Menezes, J. and Gunasheela, S. (2003) Preimplantation human development – video of live embryos on DVD.
21. Rao, K.A., Brinsden, P.R., and Sathananthan, A.H. (eds.) (2004) *The Infertility Manual* (2nd ed), Jaypee Brothers, India, pp. 560.
22. Sathananthan, A.H. (2002–2008) Early human development. (Gonads, gametes, fertilization, embryos, organelles, ultrastructure, www.sathembryoart.com
23. Gasser, R. (ed.) www.virtualhumanembryo.lsuhsc.edu (Stages 1–4) Carnegie collection (Online).
24. Elder, K. and Cohen, J. (Eds) (2007) *Human Preimplantation Embryo Selection*. Informa Healthcare, London, pp. 370.
25. For trouble shooting in human ART contact embryomail@dlist.anri.barc.usda.gov

Chapter 2

Derivation of Human Embryonic Stem Cell Lines from Vitrified Human Embryos

Teija T. Peura, Julia Schaft, and Tomas Stojanov

Abstract

Human embryonic stem cell lines are usually derived from human embryos that have become excess to clinical needs in assisted reproduction programs, whether because the couple in question has completed their family or because the embryo was found to be clinically unsuitable for transfer due to severe genetic condition (in case of pre-implantation genetic diagnosis, PGD). Culturing embryos to a blastocyst stage (5–6 days after IVF) before embryo transfer or cryopreservation instead of earlier commonly used 8-cell stage (3 days after IVF) calls for new methods for embryo cryopreservation and allows higher efficiencies for the actual stem cell derivation. Despite the vast advances in other fields of embryonic stem cell research, methods for derivation of new lines have not changed much over the years, mainly due to scarcity of embryos limiting experimentation. We describe here methods required to derive new embryonic stem cell lines starting from the initial cryopreservation of an embryo and finishing with a new cell line. We cover embryo cryopreservation and warming using a highly efficient vitrification method, the production of feeder cells and feeder plates, as well as embryo handling, plating and critical early passages, including earliest possible cryopreservation of putative stem cells using vitrification.

Key words: Human, embryonic stem cells, derivation, culture.

1. Introduction

Human embryonic stem cells (hESC) were first derived in 1988 by Thomson et al. (1) and have since been under intense investigation throughout the scientific community. Attempts to explore their enormous potential in different areas of biomedical research have led to deepening understanding of the basic biology of these cells. Clinical applications of these cells are expected to range from being used as tools for in vitro investigation of cellular processes and drug discovery, to being a source of cells for tissue generation and cell replacement therapies. Their unique characteristics include the ability to grow in vitro indefinitely, while retaining their capacity to differentiate into specialised somatic cell types. The source of hESC lines is most commonly excess human embryos from assisted reproduction programs, although novel ways of producing hESCs without harming the embryo have been recently suggested (2). The actual methods of hESC derivation have not changed greatly since the first reports of primate embryonic stem cell generation by Thomson et al. (3), still relying on cultivation of cells on top of a feeder cell layer and mechanical passaging. Although the main emphasis over the last few years has been to establish defined conditions for culture of hESCs in order to reduce the effects of unknown biological factors in basic research and to facilitate their downstream clinical use, the scarcity of embryos has meant much more cautious approach towards any changes in the actual derivation methods.

This chapter covers the steps from the moment a human embryo is assigned for storage, to be later possibly used for stem cell derivation, up until a new embryonic stem cell line has been generated and secured by cryopreservation. This involves embryo cryopreservation using a highly efficient fibre plug vitrification method, the production of feeder cells, either as initiated from the primary tissues or from the commercially sourced already-established cell lines, and finally, the embryo plating, initial embryonic outgrowth passages and early cryopreservation to obtain and secure a stem cell line. The emphasis is in the methods directly related to handling of embryos, embryonic outgrowths and putative ES lines. The methods for human IVF are not described, as they fall outside the aims of this chapter. And because once the putative stem cell line has been obtained, it can be treated like any other hESC line following the protocols of choice for its expansion and characterisation, these methods have also been omitted.

2. Materials

2.1. Embryo Vitrification and Warming

1. Stereomicroscope (e.g. Olympus, SZX7).
2. Heated stage for the stereomicroscope (LEC instruments, Scoresby, VIC, Australia, Cat No. LEC944) and warm stage controller (LEC Instruments, Cat No. LEC916).
3. Sterile disposable 4-well plates (NUNC, Cat No. 144444).

4. Vitrification kit (Cryologic, Mulgrave, VIC, Australia, Cat No. MKIT) consisting of CVM vitrification block with handle and lid, 200 CVM fibre plugs and sleeves (sterile), CVM cryobath with lid, CVCup, foam mat, foam handrest, pipette (Finnpipette 0.3–3 μL), pipette tips (Finitips, sterile).

5. Vitrification Solutions kit (COOK IVF, Eight Mile Plains, QLD, Australia, Cat No. K-SIBV-5000) consisting of a Cryobase, Vitrification Solutions 1 and 2 and DMSO.

6. Vitrification warming kit (COOK IVF, Cat No. K-SIWV-5000) consisting of a Cryobase and Warming Solutions 1, 2 and 3.

7. Liquid nitrogen (ACE Cryogenics, Sydney, Australia).

8. Canes (Genetics Australia, Bacchus Marsh, VIC, Australia, Cat No. AMPCANES 13 mm).

9. Goblets (Genetics Australia, Cat No. APA003).

10. Laboratory timer, suitable for counting seconds (Sigma Chemical Company, St. Louis, MO, USA, Cat No. Z672076-1EA).

11. Liquid nitrogen Dewar for embryo storage (e.g. Taylor–Wharton, 34HC).

12. Tube rack.

13. Blastocyst culture media (Cook IVF, Cat No. K-SIBM-20), store at +4°C for a few weeks.

14. Embryo culture grade oil (COOK IVF, K-SICO-200), store at +4°C for a few weeks.

15. K-MINC-1000 mini-incubator (Cook IVF, Cat No. K-MINC-1000-US).

2.2. Feeder Cell Preparation

1. EtOH (Crown Scientific, Cat No. EA042/2.5), diluted to 70% v/v with water.

2. Human tissue sample, aseptically and freshly collected, preferably at least few grams (or approx. 0.5 × 0.5 cm in size).

3. Sterile forceps small and large (Promedica, R29.009), washed and autoclaved in a sterile packaging before use.

4. Scissors (e.g. dissecting scissors, straight, 115 mm, Cat No. T104, ProSciTech, Thuringowa, QLD, Australia), washed and autoclaved in a sterile packaging before use.

5. Sterile scalpel blades (e.g. scalpel blades, carbon steel, profile 10, Cat No. LSB10, ProSciTech).

6. Scalpel blade handles (scalpel handle No. 3, 125 mm, Cat No. T133, ProSciTech).

7. Magnesium- and calcium-free PBS(−) (Invitrogen Australia, Mount Waverley, VIC, Australia, Gibco, Cat No. 14190-144), store at room temperature.

8. 35-mm culture dishes (BD Falcon, Cat No. 351008, BD Australia, North Ryde, NSW, Australia).

9. T12.5 or T25 cell culture flasks (BD Falcon, Cat Nos. 353018 and 353014, resp.).

10. Glass Pasteur pipettes (Sigma, Cat No. P 1736) sterilised by autoclaving or dry oven sterilisation in sterile packaging before use.

11. Feeder medium consisting of Dulbecco's Modified Eagle's Medium (DMEM high glucose, no L-glutamine, no sodium pyruvate, Invitrogen Gibco 11960-044) supplemented with 10% foetal calf serum (Invitrogen Gibco 16141-079), 1× MEM non-essential amino acids (Invitrogen Gibco 11140-050), 2 mM glutamine (Invitrogen Gibco 25030-149), 1 mM sodium pyruvate (Invitrogen Gibco 11360-070) and 50U/50 µg penicillin/streptomycin (Invitrogen Gibco 15140-122). Once prepared, the media is sterile filtered and stored at +4°C preferably for 2–3 weeks, maximum for 1 month.

12. Disposable sterile filters, 0.22 µm (Millipore, Cat No. SVGP01015, Millipore Australia, North Ryde, NSW, Australia).

13. Trypsin 0.25% (Invitrogen Gibco, Cat No. 25200-056), thawed, aliquoted to suitable aliquots, e.g. 1–5 mL in sterile disposable tubes and re-frozen. Stored at −20 for over a year. Avoid thawing and re-freezing!

14. Disposable sterile tubes for aliquots (BD Falcon 5 mL tubes, Cat No. 352002).

15. 1, 5 and 10 mL disposable sterile pipettes (BD Falcon, 357521, 357543 and 357551, resp.).

16. 15- and 50-mL sterile disposable centrifuge tubes (BD Falcon, 352095 and 352070, resp.).

17. Liquid waste container or aspirator system (e.g. Vacsax® system, Medeleq, Nerang, QLD, Australia).

18. Haemocytometer, Neubauer Improved (ProSciTech, SVZ4NIOU).

19. 96-well plate (e.g. BD Falcon, Cat No. 353916), for Trypan Blue staining of cells.

20. Trypan Blue (Sigma, Cat No. T 8154), store at room temperature.

21. 1.2-mL cryovials (Nalgene, Cat No. NAL50000012, Nalgene Nunc International, Rochester, NY, USA).

22. DMSO (Sigma, Cat No. D 2650). For feeder cells, a bottle of 100 mL can be used, stored at room temperature.

23. Permanent marker surviving freezer and liquid nitrogen conditions.

24. Alternatively, a cryo label printer (Brady, TLS2200) and cryolabels (Soanar Plus, Cat No. PTL-19-427).

25. Liquid nitrogen Dewar (e.g. Taylor-Wharton, 5LD (for storing in canes) or LS3000A (for storing in boxes), Taylor-Wharton Australia, Albury, NSW, Australia).

26. Cryoboxes (e.g. 5026-0909A 2 ML tube boxes, Taylor-Wharton) or canes (Genetics Australia, Cat No. AMPCANES 13 mm) for storing cryovials.

27. Mitomycin C (Sigma, Cat No. M 4287), 2-mg vial is reconstituted in 2 mL of PBS (with or without Ca and Mg) to make a 100X stock solution, which is then aliquoted to suitable aliquots in sterile disposable tubes and frozen. The final working concentration is 10 μg/mL. Store stock solution at −20°C for over a year and use working solutions fresh. Avoid thawing and re-freezing!

28. Gelatin (Sigma, Cat No. G 1890), prepared into 0.1% solution in MilliQ water, sterile filtered and stored at +4°C for 2–3 weeks. Dissolving gelatin can be aided by warming the water and mixing it regularly, or even by autoclaving it.

29. Culture vessels for feeder plating, e.g. 1-well organ culture dishes (BD Falcon Cat No. 353037), 4- or 6-well plates (NUNC Cat No. 176740 and 140675, resp.), T12.5, T25, T75 or T175 flasks (BD Falcon, Cat No. 353018, 353014, 353024 and 353028, resp.)

2.3. Embryo Plating and Outgrowth Passaging

1. Stereomicroscope with an extension allowing for the extension of eyepieces, making it suitable for working inside a Class II Biological Safety Cabinet (e.g. Olympus, SZX7).

2. Heated stage for the stereomicroscope (LEC instruments, Cat No. LEC944) and warm stage controller (LEC Instruments, Cat No. LEC916).

3. EtOH (Crown Scientific, Cat No. EA042/2.5), diluted to 70% v/v with water.

4. Feeder cells, either in-house prepared or commercially sourced, e.g. CCL-110 human fibroblast line (also called Detroit 551) from ATCC, Manassas, VA, USA.

5. Feeder plates (*see* **Section 3.2.5**), plated preferably 1–7 days prior to use. The feeder plates can be utilised up to 2 weeks after plating, if absolutely necessary.

6. Blastocyst culture media (Cook IVF, Cat No. K-SIBM-20).

7. Embryo-handling media (Cook IVF, K-SIFB-100).

8. Derivation media (same as hESC culture media), consisting of knock-out DMEM (Invitrogen Gibco 10829-018) supplemented with 2 mM glutamine (Invitrogen Gibco

25030-081), 50 U/ml penicillin and 50 mg/ml streptomycin (Invitrogen Gibco 15140-122), 1× MEM nonessential amino acids (Invitrogen Gibco 11140-050), 0.1 mM β-mercaptoethanol (Invitrogen Gibco 21985-023) and 20% Knock-out Serum Replacement (KSR, Invitrogen Gibco 10828-028). Once prepared, the media is sterile filtered and stored at +4°C preferably for 2–3 weeks, maximum for 6 weeks. Working aliquots of the media are to be equilibrated in incubator for several hours or overnight before using for culture. On the day of use, add 4 ng/ml bFGF (see below).

9. bFGF (Chemicon, Cat No. GF-003) 25 μg/mL stock solution is made by dissolving the contents of the 50-μg vial in 2 ml of Ca^{2+} and Mg^{2+} free PBS and aliquoted in 20-μl aliquots in sterile tubes and frozen. 1.6 μl of this stock solution is added into each 10 ml of complete medium on the day of using the media.

10. Pronase (Sigma, Cat No. P 8811), made into 5× stock solution of 20 mg/mL in DMEM-Hepes (no protein), sterile filtered, aliquoted to suitable aliquots of 50 μl in sterile disposable tubes and frozen. Stored at −20°C for up to a year (or longer). Avoid thawing and re-freezing! Working Solution of 4 mg/mL can be thawed and stored in the refrigerator for about a week.

11. Acidic Tyrode's Solution (Sigma, Cat No. T 1788), thawed, aliquoted to suitable aliquots, e.g. 1–5 mL in sterile disposable tubes and re-frozen. Stored at −20°C for over a year. Avoid thawing and re-freezing!

12. Embryo culture grade oil (Cook IVF, K-SICO-200), stored at +4°C for a few weeks.

13. 35-mm culture dish (Becton Dickinson, Cat No. 351008).

14. Embryo transfer pipettes, e.g. Glass Pasteur pipettes (Sigma, Cat No. P 1736), sterilised by autoclaving or dry oven sterilisation, flame-pulled and fire-polished just before using.

15. Indirect mouth hose (Cook IVF, K-MALT-50.0-DPIP) to be used with flame-pulled Pasteur pipettes. This type of mouth hose does not allow for an open airway contact between the solution and the mouth, as the aspirating and pulling pressure is mediated by a thin membrane inside the hose.

16. P20 pipette 2.0–20 μl (Eppendorf, Cat No. 3111 000.130).

17. 20-μl tips (Fronine, Cat No. 0030 075 005).

18. Alternatively, disposable sterile embryo transfer pipettes (0.29–0.31 mm, Swemed Int., Cat No. H-290-310) can also be used.

19. Ultra Sharp splitting blades (Bioniche Animal Health, Cat No. ESE020). These can be used several times until becoming blunt and have to be washed by brief soaking in detergent (e.g. 1% 7X cleaning solution (MB Biomedicals, Cat No. 76670)), flushing with sterile water and finally sterilising with 70% EtOH and allowing to dry. Just prior to use, the blade can be flushed again with 70% EtOH, allowed to dry and finally flushed in the embryo-handling (or other appropriate) media. Alternatively, they can be packaged and sterilised by irradiation. However, due to their plastic handles, these blades cannot be autoclaved or dry heat sterilized.

2.4. Vitrification and Warming of Putative Early hES Cells

1. Stereomicroscope with an extension allowing for the extension of eyepieces, making it suitable for working inside a Class II Biological Safety Cabinet (e.g. Olympus, SZX7).

2. Heated stage for the stereomicroscope (LEC instruments, Cat No. LEC944) and warm stage controller (LEC Instruments, Cat No. LEC916).

3. Cryo label printer (Brady, TLS2200) and cryolabels (Soanar Plus, Cat No. PTL-19-427) for labelling of outer straws.

4. 0.5-ml outer straws (Minitube Australia, Smythes Creek, VIC, Australia, Cat No. STR0001).

5. Open pulled straws (OPS) (Minitube Australia, Cat No. STR0501).

6. Tube rack.

7. Sterile disposable 4-well plates (Nalgene Nunc, Cat No. 144444).

8. Sucrose Solution: 2 M sucrose solution is prepared by dissolving 6.846 g of sucrose (Sigma-Fluka, Cat No. 84097) in 5 mL of DMEM-HEPES (see below). Allow sucrose to dissolve in the incubator or warm block (this will take approx. 2 h, but can be speeded up by occasional shaking). When dissolved, make the solution up to a final volume of 10 mL and filter sterilise. Store at +4°C for 2 weeks to a month.

9. Sterile filters, 0.22 µm (Millipore, Cat No. SVGP01015).

10. DMEM-HEPES without phenol red (Invitrogen Gibco, Cat No. 1063-029).

11. Foetal calf serum (Invitrogen Gibco, Cat No. 10439-024).

12. DMSO (Sigma, Cat No. D-2650). Preferably use small 5-mL ampoules, as the DMSO for vitrification should ideally be fresh, less than 4 weeks since the ampoule had been opened. Close the opened ampoule tightly with Parafilm and store at room temperature.

13. Ethylene glycol (Sigma, Cat No. E 9129).

14. Straw putty (BactoLab).
15. Goblets clear plastic 13 mm (Genetics Australia, Cat No. APA003).
16. Canes (Genetics Australia, Cat No. AMPCANES 13 mm).
17. Cane tags (Genetics Australia, Cat No. canetabsp).
18. Liquid nitrogen.
19. Liquid nitrogen container (Taylor-Wharton, 5LD).
20. Sterile forceps, small and large (Promedica, R29.009).
21. P20 pipette 0.1–20 µl (Eppendorf).
22. 20-µl tips (Fronine, Cat No. 0030 075 005).
23. Glass Pasteur pipettes (Sigma, Cat No. P 1736).
24. Ultra sharp splitting blade (Bioniche Animal Health, Belleville, ON, Canada, Cat No. ESE020).
25. Mitomycin-treated feeder cell plate with bFGF-supplemented hESC media.

3. Methods

3.1. Embryo Vitrification and Warming

Vitrification is an alternative method to the more traditional controlled rate freezing, which relies on slow, controlled reduction of temperature of embryos loaded into standard 0.25- or 0.5-mL straws in the solution containing relatively low levels of cryoprotectants in a solution (e.g. 10% of glycerol). But as cells are made up mostly of water, they are prone to damage caused by ice crystal formation in this type of freezing. Vitrification avoids ice crystal formation by using high concentration of cryoprotectants and ultra rapid cooling, causing the expulsion of water from the cells and rendering cells to a "glass-like" consistency with no crystal formation. However, high concentration of cryoprotectants increases the toxicity to the embryo and therefore the timing of exposure of embryo to cryoprotectants must be minimal, making this method as demanding as it is successful when properly executed. The methods here utilise a commercial CVM^{TM} vitrification kit from Cryologic (Mulgrave, VIC, Australia) and a commercial kit of vitrification solutions from Cook IVF (Eight Mile Plains, QLD, Australia). Some parts of the kits can be replaced with equivalent products from elsewhere, but especially unique fibre plugs and straws as well as the vitrification block, which have been specifically designed for this purpose, cannot be replaced if wanting to use this specific method.

All protocols involving embryos should be performed in a biological safety cabinet under sterile conditions using aseptic techniques and preferably disposable consumables. All cell culture media should be sterile filtered after preparation and stored in appropriate conditions at +4°C. Embryo incubation should be performed in a controlled humidified atmosphere of 6% CO_2, 5% O_2 and 89% N_2 at 37°C, preferably in a mini-incubator with separately controlled culture chambers, and embryos should be removed from these conditions only for a minimum length of time required for a given procedure.

3.1.1. Embryo Vitrification

1. Start preparing a vitrification plate by first pipetting 500 μl of Cryobase into wells #1 and #2 of a 4-well plate.

2. Then pipette 460 μl Vitrification Solution 1 into well #3, add 40 μl DMSO into the same well and mix thoroughly by pipetting up and down several times (*see* **Note 1**).

3. Finally pipette 420 μl Vitrification Solution 2 into well #4, add 80 μl DMSO into the same well and mix thoroughly.

4. Let the plate warm at 37°C warm stage for a minimum of 10 min.

5. In the meantime, label individual fibre plugs, still attached to their respective straws, for each embryo to be frozen. Also label the canes where the fibre plugs will be stored, two canes per batch of fibre plugs and a goblet. Check that the cane intended to be on top has a small hole in it so that it will not float when placed into liquid nitrogen (*see* **Note 2**).

6. Fill the cryobath with liquid nitrogen to approximately half the height of the vitrification block (*see* **Note 3**).

7. Fill a liquid nitrogen container with liquid nitrogen and place the labelled canes into the bucket to pre-cool.

8. Slowly lower the vitrification block with lid in place into the bath. Take care, as the liquid nitrogen will boil rapidly. Allow the block to cool to liquid nitrogen temperature, which takes approximately 5–6 min. As the block approaches liquid nitrogen temperature, the liquid around the block boils more vigorously for a few moments and then becomes relatively calm.

9. In the meantime, remove fibre plugs from their straws, placing them on a tube rack at room temperature inside the biological safety hood.

10. After the vitrification block has cooled, remove its lid and place the straws in the slots in the block to pre-cool for at least 3 min. The straws must be pre-cooled to avoid de-vitrification of the beads when they are placed into the straws.

11. Replace the lid of the block over straws to reduce condensation build-up on the surface of the block (see **Note 4**).
12. Pull several glass Pasteur pipettes of differing diameters for mechanical collapsing of embryos, diameters depending on the stage of embryo, whether early blastocyst, blastocyst, expanded blastocyst or hatched blastocyst (see **Note 5**).
13. For transferring embryos, set a pipette to 2 μl and use a clean tip for each embryo.
14. A maximum of 3–4 embryos to be vitrified can be transferred into well #1 at one time.
15. With the allocated fibre plug ready, move the specific embryo to be vitrified to well #2 and set the timer for 3 min.
16. Transfer the embryo to well #3 (Vitrification Solution 1) and start the timer. Carefully pipette the embryo up and down the suitable size pipette and observe as it starts to collapse. Change the pipette to a smaller size as the embryo collapses further. Take care not to shear any hatching cells from hatching blastocysts. The embryo may not fully collapse, but it should collapse at least partially (see **Note 6**).
17. When the timer has 1 min 10 s remaining (i.e. after 1 min 50 s in Vitrification Solution 1), remove the lid from the vitrification block.
18. With 1 min remaining (i.e. after 2 min in Vitrification Solution 1), transfer the embryo into well #4 (Vitrification Solution 2). Do this by first flushing the tip of the pipette in well #4 and then pipetting the embryo using that tip from well #3 into well #4. Expel the embryo and pipette it few times up and down in order to thoroughly but quickly mix it with the vitrification solution.
19. Aspirate approximately 1.5 μl of this solution, then the embryo and finally any remaining volume (of 2 μL) of Vitrification Solution 2 into the pipette tip, avoiding bubbles.
20. Take the fibre plug in the other hand and bring it close to the pipette tip over the top of well #4, holding the fibre plug so that the hook is horizontal.
21. Expel the droplet containing the embryo onto the end of the pipette tip while holding the pipette as vertical as possible, ensuring no bubbles occur.
22. Wipe the droplet onto the hook of the fibre plug.
23. Check the pipette tip under microscope to ensure the embryo has left the tip. If the embryo has not left the pipette tip or is stuck to it, re-aspirate it into the tip, remove the droplet without embryo from the fibre plug and repeat the previous steps.

24. Carefully and quickly transfer the fibre plug to the vitrification block and place the hook and the droplet onto its surface. The droplet will vitrify into a glassy bead instantly (*see* **Note 7**). The overall aim is to expose the embryo to the Vitrification Solution 2 for no shorter than 30 s and no longer than 45 s.

25. Leave the hook and the droplet on the surface of the vitrification block until it has become a glass bead, minimum of 5 s.

26. Moving the fibre plug by the handle, quickly but carefully insert the fibre into a pre-chilled straw and gently press the plug down.

27. Raise the straw by the fibre plug, just high enough to allow the top of the straw to be held without touching the block.

28. Between the thumb and forefinger of one hand, squeeze the top of the straw while firmly pushing the fibre plug into place with the other hand to seal the straw. Holding the straw between the thumb and forefinger warms the top of the straw, which gives it flexibility and allows the fibre plug to make a secure seal.

29. Move the straw to the pre-labelled cane in liquid nitrogen container for storage.

30. Repeat the previous steps for remaining embryos.

31. When all embryos have been vitrified, they can be transferred to long-term storage. This should be done as quickly as possible without subjecting the straws to temperature fluctuations (*see* **Note 8**).

32. Decant the remaining liquid nitrogen back to a storage tank and place the block on its side with the straw holding holes facing downwards to allow the liquid nitrogen to run from the holes in the block.

33. After the block has warmed sufficiently, clean it with water and dry using a lint-free cloth. Cover the block with aluminium foil, label with the date and add a sticker with heat label. Sterilise the block by dry heat at either 140°C for 6 h or 160° for 2 h and let cool completely before using it again.

3.1.2. Embryo Warming

1. One day before the planned warming, prepare blastocyst culture plate(s) by pipetting 0.7 mL of blastocyst culture medium to well #1 of each 4-well plate and cover with 0.3 mL of equilibrated mineral oil. To each of wells #2, #3 and #4, add 10 µL of blastocyst culture media and cover with 0.7 mL of equilibrated mineral oil. Equilibrate the plate overnight in the incubator.

2. On the day of warming, prepare the warming plates; one plate for all embryos of one patient or several plates if embryos from more than one patient are to be warmed (*see* **Note 9**).

3. Prepare the warming plate by pipetting 500 μl Warming Solutions 1 and 2 into wells #1 and #2 of a 4-well plate, then pipette 500 μl of Warming Solution 3 into well #3 and 500 μl of Cryobase into well #4.

4. Let the plate warm at 37°C warm stage for at least 10 min.

5. Fill a small container with liquid nitrogen. Identify the embryo(s) (i.e. fibre plug(s)) that you wish to warm and place them in the liquid nitrogen container.

6. Focus the microscope view on the bottom of well #1 of the warming plate.

7. Place the liquid nitrogen container as close to the microscope as possible to reduce time and distance between the container and the warming plate. Set the timer for 5 min.

8. Locate and correctly identify the fibre plug you wish to warm. Lift the straw by the fibre plug handle until the top of the straw is just above the level of liquid nitrogen. Between the thumb and forefinger of one hand, squeeze the top of the straw while gently twisting the fibre plug with the other hand. Holding the straw between thumb and forefinger warms the top of the straw allowing it to regain flexibility and break the seal with the fibre plug (the vitrified embryo is not warmed by this action).

9. When the fibre plug loosens, immediately extract the hook and the bead from the straw and place it into well #1, stirring it in a small circle until the bead dissolves. With a pipette set on 5 μl, quickly move the embryo to well #2 (wash well) and start the timer. During this time you should see the embryo shrinking to a fully collapsed stage.

10. After 5 min, transfer the embryo from well #2 to well #3. Do this by first flushing the tip of the pipette in well #3 and then pipetting the embryo using that tip from well #2 into well #3. Expel the embryo and pipette it few times up and down in order to thoroughly but quickly wash it in the new solution. Re-set the timer for another 5 min.

11. After 5 min, transfer the embryo from well #3 to well #4 as above. Re-set the timer for another 5 min. Embryo may start to show signs of re-expansion.

12. After 5 min, transfer the warmed embryo into the prepared blastocyst culture plate and allow to recover at least for a few hours. If desired, embryo can be left in culture overnight before plating. However, this has to be determined on a case-by-case basis, depending on the appearance, the developmental stage and age of the embryo.

13. If another embryo from the same patient is to be thawed, the same warming plate can be re-used for up to four embryos.

3.2. Feeder Cell Preparation

Although embryonic stem cells can be nowadays routinely cultured in semi-defined extracellular matrices, the unique demands of a new line derivation – supporting the proliferation of only the few initial pluripotent cells within the embryo to a fully established homogenous stem cell line – reduce the temptation of trying new conditions for derivation attempts. No doubt defined matrices will take over from the feeder cell-supported cultures soon enough, but in the meantime the protocols for feeder preparation may still become useful. Depending on specific circumstances, the feeders can be initiated from the primary tissue samples or from the commercially sourced already established primary cell lines, and both approaches have been described here. Whether using commercial or in-house prepared feeders, it is advisable to prepare a "master stock" of feeders at early passages (e.g. between passages 3 and 5, or as early as possible) and freeze them down in suitable batches to serve as a starting point for later expansion of "working stocks". Please note that these protocols refer to handling of human fibroblasts, which however differ only slightly from the handling of mouse fibroblasts, protocols for which are widely available.

All protocols involving cell culture work should be performed in a biological safety cabinet under sterile conditions using aseptic techniques and preferably disposable consumables. All cell culture media should be sterile filtered after preparation and stored in appropriate conditions at +4°C. Cell culture incubation should be performed in controlled humidified atmosphere of 5% CO_2 in air at 37°C.

3.2.1. Establishing Primary Feeder Cell Line

1. Obtain the sample of the desired tissue in a way that ensures its sterility and allows its processing as soon as possible after collection/biopsy (*see* **Note 10**).

2. Using sterile forceps, transfer the piece(s) of freshly cut tissue to a sterile 35-mm Petri dish containing 3–4 mL of sterile PBS(−) (*see* **Note 11**) to submerge and wash the tissue pieces(s).

3. Agitate the tissue gently trying to remove as much as possible any possibly adhering blood, mucus, etc.

4. Transfer the tissue to a new PBS-filled Petri dish for another wash and continue washing this way until satisfied that the piece of tissue is as clean as possible (usually 2–4 washes are enough), then finally transfer the tissue piece(s) on a new empty (dry) Petri dish (the tissue should still be moist with PBS).

5. Using a sterile scalpel blade, finely cut the tissue into pieces of at least 1 mm^2 or as small as possible. Ideally the pieces should be of reasonably equal size, allowing the attachment to occur at same rate.

6. Transfer at least 6–8, but preferably many more small pieces (*see* **Note 12**) of tissue to the bottom of an empty (dry) horizontally lying T25 cell culture flask with a flame-pulled Pasteur pipette or another type of sterile pipette. Replace the cap on the flask and let it stand upright for 15–20 min (*see* **Note 13**). Prepare at least a few flasks this way to ensure enough material for the cultures and to safeguard against accidental contamination or other loss of a single flask.

7. After 15–20 min, pipette 5 ml of warm feeder medium to the bottom of each flask while flask still remains in an upright position. DO NOT dispense medium over the tissue pieces stuck to flask.

8. Slowly and gently tilt the flask until the first pieces touch the media – if the pieces remain attached, continue tilting until the flask lies in a horizontal position, covering all the tissue pieces with the medium. Then transfer the flask into the incubator very, very carefully, avoiding any media movement that may lodge the pieces. The flasks can also be transferred to an incubator immediately after adding the media, so that they can be just left in the position where they were tilted without need for further movement.

9. However, if the first pieces detach when media touches them, return the flask to the upright position and let the remaining pieces dry another 5 min before trying to tilt the flask again. If necessary, continue trying this way until the pieces remain attached. Ideally all pieces would be stuck to their position so firmly during the drying period that they would remain in same positions even after being covered with media. Having just a few pieces detaching and floating is not detrimental to the subsequent culture, but if the source material is scarce, these pieces can be removed from the flask and transferred to another empty flask for a new attempt to attach them.

10. Flasks should not be disturbed for a minimum of 48 h, but you can observe them without moving them to see if the pieces remain attached.

11. After 48 h check for attachment of the pieces and the establishment of fibroblast growth under microscope. This can be seen as a visible outgrowth of cells from the original attached piece. Depending on the type of original tissue, other cell types can be observed as well (e.g. epithelial).

12. If after 48 h the pieces of tissue have attached and cells are growing out, change the culture medium using 5–7 mL of fresh media for a T25 flask in order to remove any possible free floating pieces, red blood cells or debris.

13. If the pieces have not attached before or after 48 h, you may try to re-attach them by repeating the procedure of drying them as described above. However, this will have less success the "older" the pieces get, but in case of a valuable, scarce material, it might still be worthwhile to try.

14. If culturing for longer than 7–8 days, change the media at least once more or approximately every 6 days after the initial media change 2 days after plating.

15. When the cells originating from the original attached pieces are approaching approximately 70% confluency, they are ready to be either passaged and/or frozen. Reaching this may take between 5 and 10 days or even longer, depending on the amount of the original tissue material plated.

3.2.2. Passaging Feeder Cells

1. Cells are ready to be passaged when they have reached maximum of 70–80% confluence and are still in the exponential growing phase. View the cells and determine the split ratio. For expanding fibroblast cells, this is usually between 1:2 and 1:5 (i.e. one flask of cells split to two to five new flasks).

2. Remove the required number of 0.25% trypsin aliquots from the freezer and warm in a bench warmer or in the incubator.

3. Warm also the required amount of PBS(−) and warm (and if desired, gas-equilibrate) the required amount of feeder media.

4. Transfer first all required reagents and consumables to a biological safety hood, then the feeder flask(s) to be passaged. Aim to handle no more than a few flasks at the time (*see* **Note 14**).

5. Aspirate medium from flask(s) to a waste container and add enough warm PBS(−) to cover the cells completely (*see* **Table 2.1**).

6. Agitate the flask gently, then aspirate PBS(−) and add enough warm trypsin to cover the cells completely (*see* **Table 2.1**).

Table 2.1
Volumes of PBS(−), trypsin and feeder media needed for passaging human fibroblasts in the most common culture flask sizes

Flask type	Vol of PBS(−) for washing (mL)	Volume of trypsin (mL)	Volume of feeder media added to flask (mL)	Total volume for flask (mL)
T25	5	1	5	7
T75	10	1.5	7	20
T175	15	3	12	50

7. Incubate cells with trypsin for 2–3 min on a warm surface or in the incubator. Then tap the side of flask several times to dislodge the cells and check under microscope to ensure that cells have detached properly.

8. Once cells have detached, add feeder medium to the flask(s) (*see* **Table 2.1**) and mix with the cells and collect the cell suspension into a 50-mL Falcon tube.

9. Add approximately ½ of the previous volume of feeder media to collect the remaining cells and add this too to the 50 mL tube. Depending on how many flasks are being handled, the cells from several flasks can be combined to one or more tubes.

10. Centrifuge the tube(s) for 4 min at 300g.

11. Remove supernatant being careful not to disrupt the cell pellet.

12. Resuspend cells in a pre-calculated volume of feeder media. Usually this is calculated on the basis that 1–2 mL of final cell suspension is needed for each new flask. Transfer this amount of cell suspension to each new flask.

13. Add enough feeder media to new flasks to make up the final volume of the culture (*see* **Table 2.1**). Alternatively, if splitting cells only to a few flasks, you can resuspend the cells in the final volume of the feeder media allocated to all flasks and transfer this cell suspension into new flasks in one step.

14. Transfer newly plated flasks into incubator and leave until cells are ready for another passage or freezing.

15. Usually there is no need to change the culture media during this period (*see* **Note 15**). However, if culturing for longer than 7–8 days, the media can be changed once.

3.2.3. Freezing Feeder Cells

1. Cells are ready to be frozen when they have reached 70–80% confluence and are still in an exponential growing phase. Before freezing determine how many cells are to be frozen per tube and in how many batches, as depending on the amount of flasks to be frozen, the freezing may have to be done in several batches (*see* **Note 16**).

2. In preparation for freezing, label the estimated required amount of cryovials with cell line name and passage number, the freezing date and the number of cells intended for each tube and place in a tube rack (*see* **Note 17**). Usually cells are frozen at 5–10×10^6 cells/mL in 0.5–1 mL volume, but depending on specific needs, these figures can be modified. Pre-cool the tubes by storing the rack at −20°C until needed.

3. Estimate the amount of feeder media needed for final dilution and place at room temperature.

4. Trypsinise the cells as per passaging protocol (**Section 3.2.2**).

5. After centrifugation, resuspend cells in a suitable volume of feeder medium. This volume can be determined according to the estimated numbers of cells/number of flasks that cells are collected from so that the final cell suspension would not be too dilute or too concentrated to facilitate an accurate counting. For example, for one T75 flask containing 70–80% confluent cells, the volume could be between 1 and 3 mL.

6. Determine the *total* number of cells in the cell suspension by haemocytometer.

7. Once the total cell number in the tube is known, calculate the exact volume needed to dilute cells to a desired concentration. Taking into account the original volume of feeder media in which the cells are currently suspended and that 10% of the final volume is to consist of DMSO, add the remaining volume of room temperature feeder media into cell suspension and mix gently. (For example, if you have 25×10^6 cells in a 4 mL of cell suspension, and you want to freeze the cells as 0.5×10^6 cells/0.5 mL aliquot, you aim for a final cell concentration of 1×10^6 cells/mL. For this the final volume of cell suspension to be frozen has to be 25 mL. As the cells are already in 4 mL volume, and 2.5 mL (10%) of the final volume has to be reserved for DMSO, you need to add 25 mL − 4 mL − 2.5 mL = 18.5 mL of the feeder media.)

8. Place the tube of newly resuspended cells on ice while retrieving the labelled cryovials from the freezer. Loosen the lids of the cryovials so that they are ready for filling with cell suspension.

9. Add calculated volume of DMSO to the tube containing the cell suspension so that the final freezing media composition is 90% feeder medium and 10% DMSO.

10. Mix thoroughly with the pipette, keeping the tube on ice from this moment onward.

11. Pipette appropriate volume of final cell suspension to each cryovial, mixing the remaining cells in the tube frequently, e.g. after every 3–5 aliquots, to make sure that the overall aliquoted cell number stays the same throughout the process.

12. Work quickly keeping the cell mixture on ice and the cryovials in the pre-cooled rack on ice at all times, and handling them as little as possible so the tubes will not get warmed by handling.

13. When all of the cell suspension has been used, quickly transfer the vials into a −80°C freezer overnight or maximum for few days. Although the cells can be stored in these conditions for

several weeks or even months, their viability is eventually going to be affected by the storage temperature and generally it is recommended that the storage will take place in liquid nitrogen at −196°C.

14. Finally transfer the vials into liquid N_2 in suitable containers, e.g. canes or boxes. Do this as quickly as possible so that the vials do not have a chance to warm up as that would be harmful for the cells.

3.2.4. Thawing Feeder Cells

1. Determine the number of vials to be thawed and their location.

2. In preparation for thawing, fill a suitable container with 37–40°C water or set up a water bath to this temperature range.

3. Remove the cell cryovial(s) from LN_2 tank and transfer to the laboratory in a suitable container filled with LN_2, dry ice or nothing, depending on the distance to the lab and the time it takes to get there. A vial can be transported in room temperature for a minute if it will be processed immediately upon arrival. Depending on the number of vials to be thawed, thawing may have to be done in several batches (*see* **Note 18**).

4. Pipette a minimum of 7 mL of warm equilibrated feeder medium per every 0.5 mL of frozen cells to be thawed into a 50-ml Falcon tube.

5. Place the cryovial(s) into warm water and allow to thaw until only a sliver of ice can be seen in the vial. This will usually take a few minutes.

6. When thawed, spray the cryovial(s) with 70% ethanol and wipe with a tissue while transferring to the biological safety hood to remove any residual water from the water bath from the outside of the tubes.

7. Pipette the contents of the vial slowly and drop-wise into the tube containing feeder medium.

8. Pipette 1 mL of the feeder-media cell suspension back to the cryovial to recover any remaining cells and transfer them back to the 50-mL tube. Depending on the number of vials thawed, the cells from several vials can be collected to one or more 50-mL tubes.

9. Centrifuge the tubes(s) for 4 min at $300g$.

10. Remove supernatant being careful not to disrupt the cell pellet.

11. Resuspend cells in a pre-calculated volume of feeder media. Usually this is calculated on the basis that 1–2 mL of final cell suspension is needed for each flask to be plated. Transfer this amount of cell suspension to each new flask.

12. Add enough feeder media to new flasks to make up the final volume of the culture (*see* **Table 2.1**). Alternatively, if splitting cells only to a few flasks, you can resuspend the cells in the final volume of the feeder media allocated to all flasks and transfer this cell suspension into new flasks in one step.

13. Transfer newly plated flasks into incubator and leave until cells are ready for another passage or mitomycin treatment.

14. If required, change the media in the cell culture flask(s) 1 day after thawing to remove dead cells. Often there is no need for this, assuming cell viability is within acceptable limits of 80–100%. However, if culturing for longer than 7–8 days, the media can be changed once (*see* **Note 15**).

3.2.5. Mitomycin C Treatment of Feeder Cells and Feeder Cell Plating

1. Cells are ready to be mitomycin C treated when they have reached 70–80% confluence and are still in an exponential growing phase.

2. Remove mitomycin C stock solution aliquot(s) from the freezer and thaw and warm in a bench warmer or in the incubator.

3. Add mitomycin C stock solution to the flask containing the feeder cells to make 1:100 dilution (e.g. add 100 μL mitomycin C to a flask containing 10 mL of feeder medium).

4. Note the time of mitomycin C addition and return the flask(s) to incubator. Leave for 2–3 h.

5. Add enough 0.1% gelatin solution to each vessel to coat the surface (*see* **Table 2.2**) and leave at least for 10 min (can be left for longer, even up to several hours).

6. Remove 0.1% gelatin-solution from vessels and allow to dry in a biological safety cabinet.

7. If using organ culture dishes or 4-well plates, add PBS(−) to the moat of the dishes and to the centre of the 4-well plate (*see* **Note 19**).

8. Approximately half an hour before the end of the mitomycin C incubation time, remove required number of 0.25% trypsin aliquots from the freezer and thaw and warm in a bench warmer or in the incubator.

9. At the completion of the incubation, remove mitomycin C containing feeder medium from the feeder cell flask(s) and dispose to waste appropriately (*see* **Note 20**).

10. Wash the cells at least 3 times with PBS(−), disposing the wash solution to waste using same consideration as for other mitomycin C-containing solutions.

11. From this point onwards, trypsinise the cells as described for feeder cell passaging in **Section 3.2.2**, using volumes as described in **Table 2.2**.

Table 2.2
Volumes of 0.1% gelatin, trypsin and feeder media needed for passaging human fibroblasts in the most common culture flask sizes

Vessel	Growth area (cm^2)	Volume of 0.1% Gelatin (mL)	Volume of Trypsin (mL)	Concentration of cells	Cells per well/vessel ($\times 10^6$)	Volume of media (mL)
4-well plate	1.9	0.3	n/a	High Low	0.13 0.45	0.8
Organ culture dish	2.9	0.3	n/a	High Low	0.2 0.07	1
6-well plate	10	1–2	n/a	Low	0.23	3
T25 flask	25	1–2	1	Low	0.6	7
T75 flask	75	2–4	1.5	Low	1.8	20
T175 flask	175	8–10	5	Low	5.4	50

12. After the total cell number of the cell suspension has been counted, calculate the volume of cell suspension required for each vessel to be seeded with the correct number of cells, as per **Table 2.2** (*see* **Note 21**). However, these cell numbers can be modified to suit particular feeder line characteristics.

13. Pipette the required amount of cell suspension to each vessel, top them up with feeder media as indicated in **Table 2.2** for the final culture volume and transfer the vessels to incubator.

14. A sample of the newly seeded plates/flasks should be checked the next day to ensure successful plating procedure. The feeder vessels will be ready the next day to use for hESC culture and up to and including 7 days after plating.

3.3. Embryonic Stem Cell Derivation

The last chapter starts by describing the critical techniques of embryo handling, including zona pellucida removal, embryo bisection and plating. Instead of the use of immunocytochemistry for isolating the inner cell mass from the rest of the embryo, we describe here a mechanical approach and an approach where such isolation is not utilised at all. The methods for first critical passages of early embryonic outgrowths are followed by vitrification and warming of early putative stem cells to ensure the survival of a cell line in case of adversity. Nevertheless, exactly same methods can be used for cryopreservation of any established stem cell lines. The methods described here have been used for successful derivation of over 30 human embryonic stem cell lines, including GMP-level clinical grade lines (4) and PGD-derived lines (5).

All protocols involving embryo and cell culture work should be performed in a biological safety cabinet under sterile conditions using aseptic techniques and preferably disposable consumables. All cell culture media should be sterile filtered after preparation and stored in appropriate conditions at +4°C. Cell culture incubation should be performed in controlled humidified atmosphere of 5% CO_2 in air or in 6% CO_2, 5% O_2 and 89% N_2 at 37°C.

3.3.1. Embryo Plating

1. In preparation for plating, make sure you have suitable feeder plates ready, plated ideally at least 1 day but no more than 2–3 days before. The most practical plates from handling point of view (allowing easy access for manipulation) are the 1-well organ culture dishes.

2. Assess the embryo microscopically and confirm that it is suitable for plating (*see* **Note 22**). Depending on the quality and developmental stage of the embryo, determine if it needs to be released from zona pellucida and/or bisected (*see* **Note 23**).

3. Replace the feeder media in the feeder plate with warm gas-equilibrated hES media (*see* **Note 24**). Leave the dish in incubator until needed.

4. If the embryo has already hatched and no zona pellucida removal is needed, transfer the embryo directly from the culture plate to a Petri dish containing 3 mL embryo-handling media (*see* **Note 25**).

5. If the embryo is still enclosed inside zona pellucida, prepare few drops (20–30 μL) of 4 mg/mL Pronase on a 3.5-cm Petri dish and cover with oil, allow to warm on a warm stage. Prepare also four 3.5-cm Petri dishes with 3 mL of embryo-handling media each, allow to warm.

6. Transfer the embryo from the culture plate to a Pronase drop in a minimum amount of media using a suitable transfer pipette.

7. Leave the embryo in Pronase for a few minutes while observing occasionally. Once the zona pellucida appears clearly thinned, undulated and/or expanded, transfer the embryo to the first Petri dish. The zona has not dissolved completely at this stage, but will do so within the next few minutes.

8. Wash the embryo through three dishes allowing embryo to spend at least 30 s in each dish, each time in a minimum volume of media in order to dilute the Pronase effectively (*see* **Note 26**).

9. After the last wash, transfer the embryo to the last Petri dish for bisection.

10. Alternatively, if wanting to avoid the use of Pronase, zona pellucida can be removed by using acidic Tyrode's solution or by mechanical cutting with a blade.

11. For removal of zona by Tyrode's acidic solution, transfer the embryo from the culture plate to the drop of acidic Tyrode's solution, prepared on a 3.5-cm Petri dish under oil as the Pronase drops. Observe the embryo constantly, as the effect of acidic solution is very fast, taking usually only 5–10 s. Once the zona pellucida appears clearly thinned, transfer the embryo immediately to the first Petri dish and wash through the dishes as previously (*see* **Note 27**).

12. After the last wash, transfer the embryo to the last Petri dish for bisection.

13. For removal of zona mechanically with a splitting blade, lower the blade (which has been sterilised with 70% EtOH, allowed to dry and rinsed in a dish containing embryo-handling media) to the dish and using the tip of the blade, draw several parallel grooves to the bottom of the dish very close to each other (approx. 10–20 µM apart). Do not cut too deeply, as it will create too deep ridges and furrows and encourages the formation of loose plastic spirals. Do this by observing under the stereomicroscope.

14. Using the tip of the blade, but not actually touching the embryo, rotate the embryo above the created grooves using the force of media flow the blade movement is creating. Try to get the embryo to a suitable position, where it lies on top of the grooves and a comfortably executed bisection line would not cut through the inner cell mass, but leave it either on top or below of the cutting line (*see* **Fig. 2.1**).

15. Press the tip of the blade slightly below the embryo, very gently, and press the rest of the blade down directly on top of the middle of the embryo. The zona pellucida will flatten, but will not be cut cleanly due to its elasticity. Once the zona has flattened to at least half of its original height, move the blade sideways carefully, intention being to roll the embryo only very slightly to its side, putting the blade into a position where it is pressing down diagonally down the zona (*see* **Fig. 2.1**). Keep the tip of the blade gently pressed down at the bottom of the dish at all times to give it support and to secure its position.

16. Now press the blade down, slicing a diagonal piece of zona away. The grooves at the bottom of the dish keep the embryo in place, preventing it from slipping away. The best position to do the cut is away from the inner cell mass, and preferably in the area where the perivitelline space between the embryo and the zona is the widest.

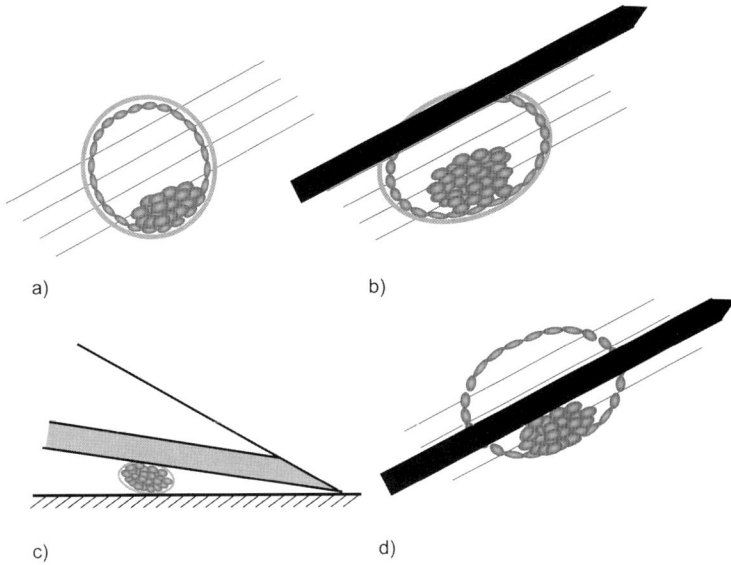

Fig. 2.1. Graphic presentation of the embryo and bisection blade positions for mechanical removal of zona pellucida and embryo bisection. (**a**) The blastocyst placed on top of the grooves drawn on the bottom of the petri dish; (**b**) The blastocyst where the blade is squashing it and where it has been slightly rotated so that the blade can slice the zona diagonally; (**c**) The position of the tip of the blade as viewed from the side for zona slicing and embryo bisection; (**d**) The position of the blade for embryo bisection.

17. Once the zona has been sliced open, the embryo can be agitated by a pipette or gently pressed by the blade to extract the embryo out of the remains of the zona. Sometimes it is useful to slice the zona a bit more to widen to opening. Removing the zona by bisection is a very delicate operation and, unless carefully executed, may damage the embryo. Hence this technique is recommended only to those with superior embryo-handling skills and steady hands. If the integrity of the blastocoele cavity is compromised, an expanded blastocyst may collapse, making subsequent embryo bisection (if intended) difficult due to inability to see the ICM clearly. However, as the total collapsing usually takes at least a minute or so, by acting quickly even an embryo with compromised blastocoele cavity can still be bisected.

18. Once the embryo has been released from the zona pellucida, it can be either plated whole or bisected to separate ICM containing polar trophectoderm from only the trophectoderm cell containing mural trophectoderm.

19. If embryo is bisected, it is done using the blade as described above for the mechanical removal of zona. The main difference is that the position of embryo over the grooves is even

more critical, as the intention is to protect the ICM and make sure that it is left completely intact. As the embryo without zona is not as elastic as the embryo surrounded by zona, the bisection is considerably easier and may even be done on a flat surface at the bottom of the dish and not on top of the grooves, as the embryo is less likely to slip under the blade.

20. The embryo can be bisected right through the middle of the embryo, or asymmetrically, trying to cut as close as possible to the ICM (*see* **Fig. 2.1**). After bisection the embryo halves may be slightly attached to the bottom of the dish, especially if bisection was performed on top the grooves. If so, let the dish stand for a minute or so and very gently agitate with a flow of medium to release the halves. If needing to resort to the blade or a pipette, be careful not to scrape any plastic fragments or spirals from the surface of the dish as they only make it harder to release the halves. If the embryo half is really stuck, use a glass pipette whose end has been fire-polished and very gently tease the embryo half away from the plate. If any remnants of the zona pellucida can be seen remaining around the embryo halves, remove them by the blade and/or gently pipetting.

21. Ideally, after the bisection, only the part containing ICM is transferred to the feeder plate; however, if in any doubt about which part is which, both parts can be moved and plated on a same plate some distance from each other.

22. Transfer the whole, non-bisected embryo or one or two halves of a bisected embryo to a feeder plate and transfer to the incubator (*see* **Note 28**).

23. Change the media in the plates every day or every second day (*see* **Note 29**).

3.3.2. Passaging Early Embryonic Outgrowths

1. After the embryo plating, observe the outgrowths every day or every second day to assess the optimal time for passaging. The criteria for a right time to passage depend partly on whether the embryo was plated whole or after bisection, and how much trophectodermal cells were plated together with the ICM.

2. If the embryo was plated whole or with lots of TE cells, try to observe especially the epiblast-like outgrowth ("bud") among the TE-outgrowth cells (*see* **Fig. 2.2a**). The two types of cells differ by appearance, ICM-derived cells showing smaller cells more tightly packed together, whereas TE-derived cells tend to grow flatter, larger and more spread out, giving out a "reticular" appearance.

3. The optimal time for passaging depends on several factors, the most critical being the abundance and location of TE-derived cells surrounding the ICM-derived "bud". Ideally the "bud" is

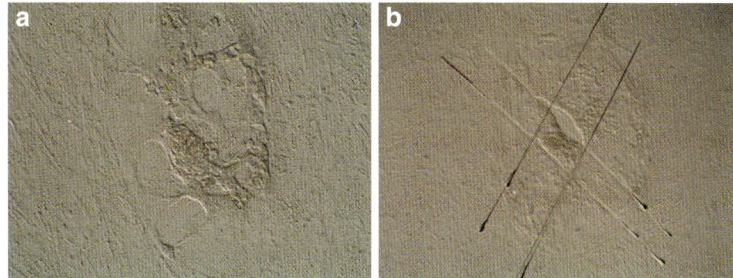

Fig. 2.2. Early embryonic outgrowths. (**a**) Clearly visible epiblast-type outgrowth ("bud") in the middle of trophectodermal cell-derived outgrowth; (**b**) Clearly visible "bud" and the cutting lines indicating where the first passage will be done.

easily distinguished, compact and consisting of tens or hundreds of cells. As it is approaching such a stage, if it is not completely surrounded by TE-derived cells, it can be left to grow up to 10 days or longer. However, if it is tightly surrounded by TE-derived cells, it should be removed from that environment and passaged to a new dish, sometimes even as early as 4 days after plating. Usually the first passage is however done between 5 and 8 days after plating. The age of feeders plays a role in the decision as well; ideally the outgrowth should not be left on the feeders older than 10–14 days maximum, as by that time they start to peel away and presumably lose their ability to support pluripotent cell growth.

4. The passaging is done simply by cutting the "bud" free of the surrounding cells, whether feeder cells or TE-derived cells, using either a cutting blade or a sharp-edged glass pipette (*see* **Fig. 2.2b**). The use of blade ensures the maximum control of operations.

5. The "bud" is then gently teased away from the plate using a pipette tip or a flame-polished glass pipette and transferred to a fresh feeder plate, whose feeder media has been replaced with the derivation media sometime prior to passaging. It is recommended that the rest of the outgrowth will be passaged as well, and in case of a large outgrowth area, it can be cut into few pieces, and all pieces distributed evenly in the new dish, away from the original ICM-derived bud. It is also advisable to keep the original plating dish in culture for a few days longer and continue observing it, just in case if any loose ICM-derived cells would still be growing in it.

6. If it is not possible to see any ICM-derived clear "bud" in the outgrowth even after culture of 8–10 days, it is advisable to passage the outgrowth anyway, especially if lot of TE-derived outgrowth is present. In this case, the whole outgrowth area is cut to few pieces and the pieces distributed evenly into the new dish.

7. The media change routine is continued as before, and the development of the passage 1 outgrowth (or outgrowths, in case if the original outgrowth was cut and transferred as several pieces) observed every day or every second day.

8. The second passage is performed essentially as the first one, with the exception that usually there are much less or none of the TE-derived cells present in the plate. At second passage, the outgrowth(s) can be cut to few pieces or left intact as one piece, depending on their appearance and size. It is difficult to give any absolute rules for these early passages, as any actions are highly dependent on the appearance of the outgrowths. Some examples of the outgrowths at early passages can be seen in **Fig. 2.2**.

9. In rare cases the emergence of the embryonic stem cells can be observed as early as passage 1; however, more commonly this occurs in passages 2–3 – depending of course on how frequently the outgrowths have been passaged. The emergency of a stem cell line from the time of plating can happen anytime between 14 and 21 days. An example of a very early stage of clear ES line growing out of the outgrowth can be seen in **Fig. 2.3**.

10. Once the early ES line is being observed, it can be started to treat similarly as any other newly passaged ES lines, passaging and culturing it as any other hESC line. However, as at this stage the existence of the whole cell line is based on only one or few colonies, it is advisable to vitrify even the smallest fragment of the colonies to ensure the survival of the line in case of loss of culture for any reason (vitrification described in **Section 3.4**). As vitrification

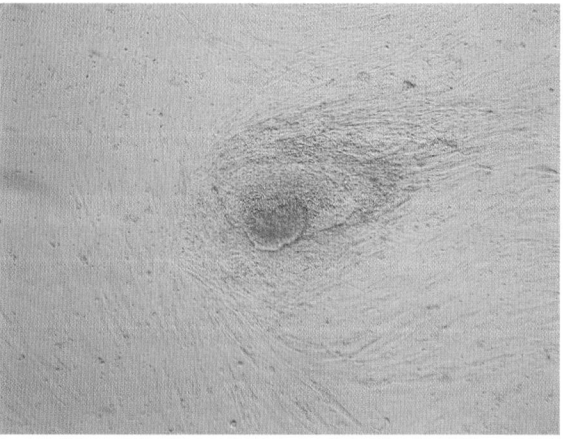

Fig. 2.3. Very early manifestation of a putative embryonic stem cell line merging from the embryonic outgrowth.

method ensures the survival rate of over 90% of the vitrified fragments, this is in optimal way to preserve the line at the earliest possible stage.

3.4. Vitrification and Warming of Early Putative hES Cells

Unlike feeder cells and mouse embryonic stem cells, human ES cells should not be slow frozen in single cell suspension, but vitrified as small cell clumps or fragments. This cell vitrification method follows the same principle as embryo vitrification and greatly enhances the survival rate of cryostored and warmed hESC lines. The protocol involves the stepwise introduction of the cells into the vitrification solution with a high concentration of cryoprotectants (DMSO and ethylene glycol) together with sucrose to increase the osmolality of the solution. Cell fragments immersed in this vitrification solution will then be loaded into open pulled straws and immediately plunged into liquid nitrogen for a long-term storage. The warming protocol involves a stepwise dilution of the toxic cryoprotectants and re-introduction of the cells into culturing medium.

3.4.1. Vitrification of Putative hES Cells

1. In preparation for vitrification, prepare correct number of cryolabels for all outer straws to be used for vitrification, indicating cell line, passage number, date, initials of the operator and any other additional information deemed necessary. Label outer straw with prepared cryolabels.

2. Also, write the cell line and passage number information on the top part of an OPS straw with a permanent marker and draw a continuous line around the very top of each straw (this allows for viewing the straw when encased in the outer straw). For ease of handling, stack all the bundled straws upright in a tube rack.

3. Prepare one vitrification plate per cell line, if only under eight or so straws are to be vitrified. If more than eight straws are to be vitrified, prepare more plates accordingly. Prepare the plates on 4-well plates as follows:
   Well #1 (Contains "Bench medium", i.e. DMEM-Hepes + 20% FCS)
       800 µL DMEM-HEPES without phenol red
       200 µL FCS
   Well #2 (As well #1)
       800 µL DMEM-HEPES without phenol red
       200 µL FCS
   Well #3 (contains Vitrification Solution 1, i.e. 10% DMSO and 10% EG in Bench medium)
       600 µL DMEM-HEPES without phenol red
       200 µL FCS
       100 µL DMSO
       100 µL ethylene glycol

Well #4 (contains Vitrification Solution 2, i.e. 0.5 M sucrose, 20% DMSO and 20% EG in Bench medium)
   150 µL DMEM-HEPES without phenol red
   250 µL 2 M sucrose solution
   200 µL FCS
   200 µL DMSO
   200 µL ethylene glycol

4. Let the plate warm on a 37°C warm stage or in the incubator for at least 20 min.

5. During this time, using a flame-pulled glass pipette and an ultra sharp splitting blade, cut colonies of hES cells to be vitrified into small fragments approximately 1×1 mm.

6. When the vitrification plate has thoroughly warmed, transfer hESC fragments to wells #1 and/or #2 and allow them to equilibrate in the bench medium for at least 1 min (timing of this is not critical, fragments can be left there for several minutes).

7. Position the liquid nitrogen container, forceps, straws and putty in the biological safety hood so that all are within easy reach. Place the goblet(s) in the cane(s), label the cane tag(s) appropriately and immerse the cane(s) in the liquid nitrogen.

8. Pipette a 20-µL aliquot of Vitrification Solution 2 from well #4 to the upturned lid of the plate to make a drop.

9. Set the pipette to 5 µL and using a new tip, transfer 1–6 hESC fragments to well #3. Do this by first flushing the tip of the pipette in well #3 and then pipetting the fragments using that tip from well #1 or well #2 into well #3. Expel the fragments and pipette them few times up and down in order to thoroughly but quickly mix them with the vitrification solution. Leave the fragments in well #3 for 25 s, starting from the time when fragments first entered the well.

10. Transfer the fragments from well #3 to the 20-µL drop in a similar manner, in a minimum volume, and mix thoroughly.

11. Pipette the fragments in approximately 1–2 µL volume to a clean area adjacent to the 20-µL drop, thus making a smaller drop containing all the fragments.

12. Stick the narrow end of an OPS straw into the smaller drop to allow the capillary action of the straw to aspirate the fragments inside the straw.

13. Plunge the narrow end of the straw into liquid nitrogen, but keep hold on the straw. The time the fragments spend in the two drops and the straw (i.e. in the Vitrification Solution 2) prior to plunging must not exceed 25 s in total!

14. Using forceps and working in the nitrogen vapour, transfer the OPS straw inside the labelled outer straw. Seal the end of the outer straw with the putty, then place the straw in the assigned goblet and cane. Try to avoid getting liquid nitrogen inside the outer straw.

15. Continue until all straws are used and all hESC fragments have been vitrified.

16. Place the goblet in liquid nitrogen Dewar for long-term storage. Until the hESC fragments are thawed, they should not be subjected to temperature fluctuations as these can cause them to de-vitrify and experience chilling injuries.

3.4.2. Warming of Putative hES Cells

1. In preparation of hESC plating after the warming, replace the feeder media in a suitable feeder plate with warm gas-equilibrated hES media. Leave the dish in incubator until needed.

2. Prepare one warming plate per cell line, if only under eight straws are to be warmed. If more than eight straws are to be warmed, prepare more plates accordingly. Prepare the plates on 4-well plates as follows:

   Well #1 (contains Warming Solution 1, i.e. 0.2 M sucrose in Bench medium)
       100 µL sucrose solution
       700 µL DMEM-HEPES without phenol red
       200 µL FCS

   Well #2 (contains Warming Solution 2, i.e. 0.1 M sucrose in Bench medium)
       50 µL sucrose solution
       750 µL DMEM-HEPES without phenol red
       200 µL FCS

   Well #3 (contains Bench medium, i.e. DMEM-HEPES and 20% FCS)
       800 µL DMEM-HEPES without phenol red
       200 µL FCS

   Well #4 (As well # 3)
       800 µL DMEM-HEPES without phenol red
       200 µL FCS

3. Let the plate warm on a 37°C warm stage or in the incubator for at least 20 min.

4. Transfer the straws to be warmed from the liquid nitrogen Dewar into a goblet held in the liquid nitrogen container and place the container in the biological safety hood.

5. Focus the microscope view on the bottom of well #1 of the warming plate.

6. Remove one straw from the goblet with forceps and while working in the liquid nitrogen vapour, cut off the end of the straw containing the putty with scissors, being careful not to cut the OPS straw inside (which can be located by looking for the black ring on the top of the OPS straw).

7. Remove the OPS straw from the outer straw with fine-nosed forceps, still working in the liquid nitrogen vapour.

8. Remove the OPS straw from the outer straw completely and let it warm in air for 1–3 s before submerging the narrow end of the straw into well #1 (*see* **Note 30**).

9. Once submerged, immediately cover the top of the straw with a finger in order to prevent solutions to be drawn deeper into the straw by capillary action. Contrarily, blocking the straw allows the solution containing hESC fragments to drift out of the straw and into the solution. Observe this under the microscope to make sure all the fragments are coming out.

10. After 1 min, transfer the fragments to well #2.

11. After 5 min, transfer the fragments to well #3.

12. After 5 min, transfer the fragments to well #4.

13. After 5 min, transfer the fragments into prepared feeder cell plate, label the plate with appropriate information and carefully transfer to incubator for culture.

14. Change the media in the plates every day or every second day, as described in **Section 3.3.1**.

4. Notes

1. Wear gloves at all times when handling DMSO.

2. Straws can be stored in a goblet, covered with another goblet attached to the same cane, thus preventing straws floating out of their storage cane. Goblet (and cane) sizes are determined by the amount of straws to be stored in one cane, but ideally straws from only one or few patients will be stored in the same cane.

3. Protective eyewear and cryogloves have to be worn at all times when handling liquid nitrogen.

4. If the conditions are very humid, the use of air-conditioning can reduce the amount of condensation build-up over the duration of the vitrification. Use a sterile mixing cannula to clear the top of any condensation build-up.

5. Different embryonic stages represent roughly the following diameters: early blastocyst and blastocyst 120 μM, expanding and hatched blastocyst between 120 and 300 μM. The diameters of the flame-pulled Pasteur pipettes should be marginally smaller. However, the suitable pipette is best found empirically rather than measuring the exact sizes.

6. Embryo collapsing is important in order to allow rapid entering of the Vitrification Solution 1 into the embryonic blastomeres from all sides, inside and outside. Intact blastocyst cavity might slow this process, making equilibration of the embryo with the vitrification solution in a short time allotted to it not complete.

7. The fibre plug should be moved through the liquid nitrogen vapour above the surface as quickly as possible to prevent slow cooling and possible chilling injury to the embryos, the aim being to reach the drop of temperature from +37°C to −196°C as promptly as possible.

8. Once vitrified the specimens should not be subjected to temperature fluctuations as they can cause the specimens to devitrify and experience chilling injury.

9. Two embryos may be thawed simultaneously; however, they should be staggered 5 min apart.

10. The human tissue in question can be a piece of foreskin, a piece of skin (with subcutaneous layer attached), a piece of aborted foetus or placental tissue or almost any kind of piece of surgical human tissue.

11. Whether PBS contains Ca and Mg ions is not critical at this stage; however, for convenience only PBS(−) is used throughout these procedures, as it is needed for cell washing later when passaging the cells.

12. The more pieces of tissue can be plated down at this stage, the less population doublings the cells have to go through in order to obtain numbers sufficient for master and working banks. As the primary cell lines will eventually start senescencing after 50 or more passages, and generally it is considered that early passages of primary cell lines are more suited for providing feeder support (although credible evidence to support this is generally lacking), it is better to use the cells at earlier rather than later population doublings.

13. Allowing pieces of tissue to partially dry promotes their attachment to the bottom of the flask where fibroblasts have better chances of growing, as opposed to if the pieces are left to just float in the medium.

14. The number of flasks an operator can comfortably handle at one time depends on the skill and speed of the operator and the size of the culture flasks – larger flasks generally taking more time to process. Ideally the cells should not be left for too long to dry in between the washes or other solution changes (no more than 1–2 min) or left to incubate too long in trypsin solution (no more than 6–8 min), so the decision about how many flasks to handle should be based on assessment of whether these times would be exceeded. Usually a reasonably skilled operator would not try to handle more than 2–4 T75 flasks at a time.

15. If there are considerable numbers of dead cells present in the cultures, the media can be changed to remove them. However, generally a simple passaging leaves majority of cells (90% or more) viable, thus negating the need for media change. Only if the cells initiate their culture from very low concentration, requiring extended culture period (beyond 7 days), it is advisable to change the media every 5–6 days to ensure the continuing availability of nutrients.

16. As with passaging, the number of flasks an operator can comfortably handle at one time depends on the skill and speed of the operator and the size of the culture flasks. The added complication in cell freezing is the final stage of incubating cells in cryoprotectant solution and in cold, which can be harmful to cells if prolonged. There are no hard and fast rules about what is an acceptable time period the cells can be exposed to these adverse conditions, and the operator has to judge how important the optimal viability of the cells after freeze-thaw is, and how it compares against the practicality of freezing cells in one vs. several batches. However, a usual rule of thumb for a single operator is not to freeze more than four T75 flasks as one batch. Nevertheless, if more than one operator is involved in the process, more flasks can be handled as each stage of the process can be shared and staggered to make the overall procedure faster.

17. Labelling can be done by writing directly into the tube with a permanent marker surviving freezer and liquid nitrogen conditions or by using a dedicated labeller and tape suitable for liquid nitrogen storage.

18. As the DMSO-containing cryoprotectant solution is harmful to cells, the cells should be removed from it as soon as possible. This dictates the number of vials that can be handled at any one time, so that cells in any vial would not be left in the cryopotectant solution for longer than 1–2 min after thawing. Hence, usually the number of vials comfortably thawed at one time is less than 10 vials per person. Nevertheless, if more than

one operator is involved in the process, more vials can be handled as each stage of the process can be shared and staggered to make the overall procedure faster.

19. This is to compensate for the evaporation from these plates with limited volume of media. Six-well plates have larger volume of media per well and do not necessarily need this humidification compensation any more.

20. Mitomycin C is a carcinogenic agent and should be handled and disposed of with utmost care, making sure that all consumables (flasks, pipettes, tubes) as well as media used to wash mitomycin C-exposed cells are also discarded appropriately through a professional hazardous waste disposal system.

21. Although there are no hard and fast rules about the optimal feeder density for embryonic stem cell derivation and culture, and the densities used are highly dependent on the individual characteristics of a particular feeder cell line, we have routinely used higher feeder cell density in the initial derivation of a new line, and after the cell line has been established, moved it to a lower density culture for subsequent expansion and maintenance.

22. In practice, almost any embryo is worth plating, just to see if it would have a few live cells that possibly could initiate the stem cell line. However, if the embryo is clearly thoroughly dead, as indicated by the loss of cellular membrane integrity of practically all blastomeres (cell lysis) accompanied with the complete collapsing of a blastocoele and dark appearance of all cells, it may not be worth plating. Please note however that the criteria for embryo viability in this case are quite different to the criteria applied to embryos to assess their viability for embryo transfer purposes: for foetal development the presence of viable trophectodermal cells and viability of most ICM cells is required, whereas for stem cell derivation the presence of few viable ICM cells may be enough.

23. For bisection, the ICM has to be clearly visible, and hence usually only blastocysts with a clear blastocoele cavity and identifiable ICM can be bisected. If the embryo has collapsed, it can be either allowed to continue its culture for an additional hour or so to re-expand, or it can be plated whole.

24. Gas equilibration can be achieved by leaving an appropriate media aliquot in a container with a lid slightly ajar in the incubator overnight before plating. Using equilibrated media eliminates the pH fluctuation in the culture, as compared to using a media that is only warmed but not gas equilibrated. Although no definitive evidence exists whether this has critical effect on derivation success, as it is well known that embryos themselves require this kind of steady

conditions, it is plausible to assume that the sensitive early stages of stem cell derivation would likewise benefit. Moreover, there is no reason to believe that this would have adversary effect either, so overall it might be better to err on the side of caution.

25. Try to avoid transferring oil droplets from the oil covering the culture drops to the petri dish, as they have a tendency to float on the media surface and interfere with the view. This can be done by transferring the embryo through a wash drop made at the time of transfer, for example, on the lid of the receiving petri dish utilising the media from the petri dish, or alternatively, by inserting the pipette tip (and the embryo) to the very edge of the petri dish and removing the tip immediately, so that any adhering oil droplets stay on the edge.

26. Washing all traces of Pronase from embryos is critical, as even minute traces of it, if transferred to the feeder plate, may have effect on feeder cells, which can consequently start peeling off.

27. Although using acidic Tyrode' solution is quicker, it is also more prone to mistakes and less forgiving if something goes wrong, possibly leaving the whole embryo utterly lysed and destroyed. Hence this method should be used with a special caution.

28. Gas atmosphere of 5–6% CO_2, 5% O_2 and 89–90% N_2 are recommended, being the same as for optimal embryo culture.

29. Media change can be done either every day or every second day without it having considerable effect on success. It is advisable however to add fresh bFGF on the days when media is not changed, as its half-life in solutions is notoriously short.

30. Handle the straw carefully and especially avoid the liquid nitrogen shooting out of the wider end of the straw once it is lifted off the liquid nitrogen. It can damage the eyes, so wearing safety goggles is highly recommended for this step.

References

1. Thomson, J. A., Itskovitz-Eldor, J., Shapiro, S. S., Waknitz, M. A., Swiergiel, J. J., Marshall, V. S. and Jones, J. M. (1998) Embryonic stem cell lines derived from human blastocysts. *Science* **282**, 1145–7.
2. Klimanskaya, I., Chung, Y., Becker, S., Lu, S. J. and Lanza, R. (2006) Human embryonic stem cell lines derived from single blastomeres. *Nature* **444**, 481–5.
3. Thomson, J. A., Kalishman, J., Golos, T. G., Durning, M., Harris, C. P., Becker, R. A. and Hearn, J. P. (1995) Isolation of a primate embryonic stem cell line. *Proc Natl Acad Sci USA* **92**, 7844–8.
4. Crook, J. M., Peura, T. T., Kravets, L.,-Bosman, A. G., Buzzard, J. J., Horne, R., Hentze, H., Dunn, N. R., Zweigerdt, R., Chua, F., Upshall, A. and Colman, A. (2007) The generation of six clinical-grade human embryonic stem cell lines. *Cell Stem Cell* **1**, 490–4.
5. Peura, T., Bosman, A., Chami, O., Jansen, R. P., Texlova, K. and Stojanov, T. (2008) Karyotypically normal and abnormal human embryonic stem cell lines derived from PGD-analyzed embryos. *Cloning Stem Cells* **10**, 203–16.

Chapter 3

Growth of Human Embryonic Stem Cells Using Derivates of Human Fibroblasts

Carmen Escobedo-Lucea and Miodrag Stojkovic

Abstract

The majority of human embryonic stem cell (hESC) lines have been derived and grown using mouse or human feeder cells, or using Matrigel®, an animal derivative rich in extracellular matrix (ECM) proteins. However, reliance on feeder layers and animal products limits the manipulation and clinical application of hESC. Alternatively, human fibroblasts produce an ECM which could be employed to coated plates and be easily sterilized. We have shown that hESC grown on this matrix and in the presence of medium conditioned by fibroblast cells maintain markers of pluripotency, including expression of cell surface proteins (SSEA3, SSEA4, TRA-1-60, TRA-1-81), alkaline phosphatase activity, and specific intracellular markers (*NANOG*, *OCT*, *REX1*). Moreover, hESC cultured on this novel human-derived ECM display a normal karyotype. This growth system reduces exposure of hESC to feeder layers and animal ingredients, thereby limiting the risk of pathogenic contamination and additionally facilitating manipulation of hESCs.

Key words: Human embryonic stem cells, in vitro growth, extracellular matrix.

1. Introduction

Human embryonic stem cells (hESCs) are cells derived from early human embryos and can be maintained in vitro as immortal pluripotent cells, but remain responsive to many differentiation-inducing signaling pathways. The majority of hESC lines have been derived and grown using mouse or human feeder cells, or using Matrigel®, a murine derivative rich in extracellular matrix (ECM) proteins (1, 2). Use of feeder layers and animal products limits research and clinical applications of hESC. Therefore, it is necessary to avoid the utilization of animal products during derivation, growth, and differentiation of hESC. An attractive alternative to the complications posed by the use of feeder cells is the development of an animal-free ECM or a functional combination of its components to support the culture of hESC. The ECM is a complex entity composed of structural molecules (collagen and elastin), glycoproteins (fibronectin, laminin), and proteoglycans (hyaluronic acid, dermatan sulfate, chondroitin sulfate, heparin, heparan sulfate, and keratan sulfate). Additionally, the ECM harbors growth factors or cytokines in order to protect against degradation (3), as well as matrix degrading enzymes and their inhibitors. The diversity of ECM composition, organization, and distribution is regulated by various mechanisms including differential splicing and post-translational modifications of matrix proteins and the effects of calcium (4), growth factors, or cytokines on the production and secretion of ECM components by cells (5).

Here we describe a very simple protocol for the long-term growth of undifferentiated hESC using human-derived ECM (hdECM) and conditioned medium, which have both been derived from human foreskin feeder cells.

2. Materials

2.1. Culture of Human Foreskin Fibroblast for Derivation of Conditioned Media and hdECM

1. Human foreskin fibroblast (ATCC, Cat No. CRL-2429, Passages 11–18).
2. Iscove's medium (Sigma, St. Louis, MO, USA), supplemented with 10% human serum (HS), 1% glutamax (GIBCO, Invitrogen, Carlsbad, CA, USA).
3. Gelatin (1.5%; Sigma). In a sterile bottle, add embryo-tested water (Sigma) to gelatin. Warm the mixture at 37°C in water bath using shaker. Store at 4°C or make aliquots and keep frozen at –20°C. These aliquots can be stored for up to 6 months. Before proceeding with the culture, thaw the aliquots and dilute them to 0.01% with sterile Dulbecco's PBS (DPBS, Invitrogen). Filter the solution through 0.22 μm filter (Nalgene, Hereford, UK). Coat the culture surfaces by pipetting 1 ml/well into 6-well plates (BD, San Jose, CA, USA) or 12 ml into a 75 cm^2 flask (Iwaki, Ibaraki, Japan). Allow the gelatin to settle at 37°C for 30 min. Plates may be used immediately or stored at 4°C to prevent evaporation.
4. Mitomycin C (Fluka, Buchs, Switzerland) is dissolved in Iscove's medium at 1 mg/ml, stored at 4°C, and then added to the cultures as required.
5. Triple Select (GIBCO, Invitrogen) is used instead of trypsin to detach cells from tissue culture plates.

6. DPBS without Ca^{2+} or Mg^{2+}, pH 7.4 (GIBCO, Invitrogen).
7. Trypan Blue (Sigma) to count and evaluate cell viability (*see* **Note 1**).
8. Neubauer hemocytometer (Brand, Wertheim, Germany).

2.2. Derivation of Conditioned Media

As previously described (6), TESR1 medium represents the basic medium for derivation of medium conditioned by human foreskin fibroblasts.

2.3. Preparation of Human-Derived Extracellular Matrix

1. RIPA buffer (Pierce Biotechnology, Rockford, IL): 25 mM Tris–HCl pH 7.6 (Sigma), 150 mM NaCl, 1% NP-40, 1% sodium deoxycholate, 0.1% SDS (all from Sigma).
2. Halt Protease Inhibitor Cocktail (Pierce Biotechnology).
3. Cell scrapers (Falcon, BD Biosciences, Madrid).
4. DPBS.
5. TERS1 conditioned media: after collection, media could be frozen at $-80°C$ up to 6 months.

2.4. Culture and Maintenance of Undifferentiated hESC Lines

1. H9 and H1 hESC lines (WiCell, Madison, WI, USA).
2. TERS1 conditioned media.
3. TGFβ1 (Invitrogen): reconstituted with sterile 4 mM HCl (Sigma) containing 1 mg/ml of human serum albumin (Sigma) to a final concentration in stock solution of 40 ng/ml. Store in 50 µl aliquots at $-20°C$.
4. Human recombinant bFGF (Invitrogen) is dissolved in 1 ml DMEM Knockout Medium (Invitrogen) and stored in 100 µl aliquots at $-20°C$.

2.5. Analysis of Undifferentiated hESC Markers

2.5.1. Staining of Pluripotency Cell Surface Markers by Immunocytochemistry

Antibodies that detect specific cell-surface hESC markers are commercially available from Chemicon. Antibodies for detection of ECM component fibronectin are from Sigma. All secondary antibodies are available from Invitrogen. See **Tables 3.1** and **3.2** for recommended dilutions and providers.

1. DPBS without Ca^{2+}, Mg^{2+} (GIBCO).
2. 4% paraformaldehyde (*see* recipe at **Section 2.5.2, point 2**).
3. 0.05% sodium azide (Sigma) in DPBS.
4. Triton-X-100 (Sigma): prepare a dilution of 1% Triton-X-100 in DPBS to permeabilize the hESC.
5. Blocking solution: 4% serum in DPBS. Serum for the blocking solution should be of the same origin/animal as the secondary antibody.
6. Prolong gold anti-fade reagent with DAPI (Invitrogen). Try to avoid the bubbles in the sample.

Table 3.1
List of primary antibodies

Primary antibody	Isotype	Working dilution	Catalogue number	Provider
SSEA1	IgM	1:100	MAB 4301	Chemicon
SSEA4	IgG3	1:100	MAB 4304	Chemicon
Tra 1-81	IgM	1:100	MAB 4381	Chemicon
Tra 1-60	IgM	1:100	MAB 4360	Chemicon
Fibronectin	IgG1	1:100	F0916	Sigma

Table 3.2
List of secondary antibodies

Secondary antibody	Working dilution	Catalogue number	Provider
Alexa Fluor goat against IgG	1:500	A11029	Invitrogen
Alexa Fluor against IgM	1:500	A21042	Invitrogen
Alexa Fluor against IgG1	1:500	A21124	Invitrogen

2.5.2. Alkaline Phosphatase (AP) Detection Kit **(Chemicon Millipore, Billerica, MA)**

1. This kit provides two components for AP determination: Fast Red Violet solution (0.8 g/L stock) and napthol AS-BI phosphate solution (4 mg/mL) in AMPD buffer (2 mol/L), pH 9.5.

2. Paraformaldehyde 4% in PBS: Prepared fresh with distilled water. To prepare 100 ml, heat 50 ml distilled water to 60°C on hot plate in fume hood (do not exceed 65°C) and add 4 g of paraformaldehyde powder. Stir the solution until it becomes clear (few drops of NaOH can be added). After that, filter the solution through a 0.22-μm filter and add 50 ml of sterile PBS at pH 7.4. Do not use until the solution reaches room temperature and adjust the pH if necessary.

3. TBST 1X Rinse Buffer: Prepared fresh with 20 mM Tris–HCl, pH 7.4, 0.15 M NaCl, 0.05% Tween-20 (all Sigma).

2.5.3. Intracellular Markers by RT-PCR Analysis

1. High Pure RNA Isolation Kit (Roche, Barcelona, Spain) used for total RNA extraction. Additional reagents required: absolute ethanol (Merck, Darmstadt, Germany), standard tabletop microcentrifuge capable of $13,000 \times g$ centrifugal force, sterile microcentrifuge tubes of 1.5 ml (Eppendorf, Hamburg, Germany). (*See* **Table 3.3** for primers and reaction conditions.)

Table 3.3
List of used primers and conditions

Gene	Character	Primers name	Primers sequence	Tm (°C)	Fragment (pbs)
OCT 4	Pluripotency	OCT4.F OCT4.R	CCTGTCTCCGTCACCACTCT CAAAAACCCTGGCACAAACT	60.31 60.01	128
TERT	Pluripotency	TERT.F TERT.R	GACCTCCATCAGAGCCAGTC CGCAAGACCCCAAAGAGTT	59.80 60.24	84
FGF4	Pluripotency	FGF4.F FGF4.R	GGCGTGGTGAGCATCTTC CGTAGGCGTTGTAGTTGTTGG	60.37 60.59	139
FOXD3	Pluripotency	FOXD3.F FOXD3.R	CAGAGCCCGCAGAAGAAG CGAAGCAGTCGTTGAGTGAG	59.81 59.77	133
NANOG	Pluripotency	NANOG.F NANOG.R	GGTGGCAGAAAAACAACTGG CATCCCTGGTGGTAGGAAGA	60.53 59.92	100
AFP	Endoderm	AFP.F AFP.R	TGCGTTTCTCGTTGCTTACA GCTGCCATTTTTCTGGTGAT	60.58 60.08	81
beta-TUBULIN3	Ectoderm	TUB3.F TUB3.R	AGTATCCCGACCGCATCAT CATCCGTGTTCTCCACCAG	60.32 60.10	120
b2-Microglobulin	Housekeeping	B2M.F B2M.R	CTCGCGCTACTCTCTCTTTCTGG GCTTACATGTCTCGATCCCACTAAA	59.00 60.10	335

PCR reaction is performed at 60°C and for 35 cycles, except for *b2-microglobulin* (55°C and 35 cycles).

2.5.4. Telomerase Activity

1. TRAPEZE Telomerase Detection Kit (Chemicon, Billerica, MA, USA)
2. PBS without Ca^{2+}, Mg^{2+} (GIBCO)
3. 10–20% nondenaturing polyacrylamide precasted gels
4. 10X TBE (BioRad, Hercules, CA, USA)
5. SYBR green (Molecular Probes)
6. Loading buffer 10X (BioRad)
7. Nondenaturing 10–20% polyacrylamide gel (15% precasted gels from BioRad)

2.6. Analysis of Karyotype Stability

1. Colchicine (Serva, Heidelberg, Germany): dilute in standard culture medium without serum.
2. KCl (0.56%, Merck).
3. Fixative = methanol:glacial acetic acid (3:1), both from Merck.

2.7. Preparation of Samples for Scanning Electron Microscopy (SEM)

1. PBS (GIBCO).
2. Glutaraldehyde (2.5%): dilute glutaraldehyde from a stock of 25% (Electron Microscopy Science, Hattfield, PA) (v/v) in PBS.
3. Petri dish (Falcon, BD).
4. Disposable Pasteur pipettes (Falcon).
5. Osmium (1%; Sigma): Prepared diluting osmium in 70% ethanol.
6. Preparation of alcohol gradient: Use the appropriate volume of absolute ethanol (Merck), mixed with distilled water to prepare 30, 50, and 70% solutions.
7. Autosamdry is employed to dehydrate the samples in a controlled atmosphere.

3. Methods

Extracellular matrix compounds play important roles in cell-adhesion, attachment, cell interactions, and proliferation (7, 8). Previous studies have demonstrated that the components of ECM support undifferentiated growth of hESC (2). Extracellular matrix is generally organized into a three-dimensional fibrous structure and therefore, it is crucial to extract the ECM samples from human fibroblasts rapidly and at 4°C. This prevents conformational changes and degradation of the proteins.

To validate functionality, it is very important to evaluate the ability of extracted ECM to maintain hESC in the undifferentiated state. This can be assessed by long-term growth of hESC on the extracted ECM and the routine analysis of pluripotency using the following assays: RT-PCR, immunocytochemistry, determination of telomerase and AP activities, karyotype stability, and analysis of the ability to spontaneously differentiate (9).

3.1. Preparation of Fibroblast Cells and Conditioned Media

1. Human foreskin fibroblasts are grown in Iscove's Medium and the cells are split using Triple Select every 5–7 days.
2. When confluent, the cells were inactivated by mitomycin C at 37°C and 5% CO_2 for 3.0 h.
3. Then, they are washed with DPBS three times (5 min), digested, and counted (see **Note 1**).

4. Fibroblasts are seeded at a density of 6×10^6 cells in T75 flask coated with 0.1% gelatine and cultured at 37°C and 5% CO_2 for 24 h before adding TERS1 medium.

5. TERS1 conditioned medium is collected every day until day 7, stored at −80°C for 6 months, and then thawed for further use (*see* **Notes 2** and **3**).

3.2. Preparation of Extracellular Matrix

1. Human foreskin fibroblasts are grown in an appropriate medium and the cells are split using Triple Select every 5–7 days. When the cells reach 100% confluency, they are inactivated by treatment with mitomycin C for 3.0 h, and then washed three times with DPBS.

2. Cells are detached, counted, and seeded (2×10^5 cells/per well) in a 6-well plate, coated with 0.1% gelatin, and cultured at 37°C and 5% CO_2 for 24 h.

3. Inactivated cell cultures are maintained during a week, changing media every second day.

4. Cells are washed twice with DPBS without Ca^{2+} and Mg^{2+} and subsequently lysed with RIPA buffer (1.5 ml per 10^6 cells).

5. After addition of RIPA buffer, cells are incubated 20 min on ice using an orbital shaker, after which time the RIPA buffer is removed by aspirating.

6. Lysed cells are eliminated from the plates by rinsing six times (5 min each) with DPBS. Plates containing the remaining hdECM are stored at 4°C or dried. An example of the SEM of foreskin fibroblast cells and ECM samples are shown in **Fig. 3.1A** and **B**, respectively.

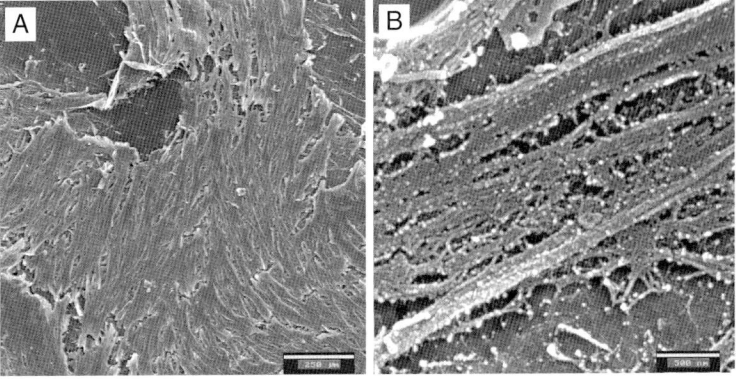

Fig. 3.1. Scanning electron microscopy of foreskin fibroblast cells (**A**) and ECM (**B**) derived from human fibroblasts. Foreskin fibroblasts were cultured before lysis by RIPA buffer. (**B**) Note the presence of ECM layer, which covers entire growth surface. *Scale bars* 250 μm (**A**) and 500 nm (**B**).

3.3. Growth of Undifferentiated hESC Using Feeder-Free Conditions and hdECM

1. Once ECM-coated plates are ready for use, conditioned medium which has been collected previously should be added.
2. Add 2 ml of conditioned media which contains 50 and 100 ng/ml of fresh TGFβ and bFGF, respectively. Place the 6-well plates in the incubator at 37°C with 5% CO_2 for at least 3 h before seeding the new hESC colonies.
3. Wash the hESC colonies maintained on human feeder twice with pre-warmed DMEM medium.
4. Dissect hESC colonies mechanically and collect them in a 15-ml polypropylene tube (*see* **Notes 4** and **5**).
5. Centrifuge the cells for 3 min at $200 \times g$ and remove the supernatant.
6. Gently resuspend the pellet with conditioned medium supplemented with fresh TGF and bFGF to the 15-ml tube.
7. Add 1 ml of media containing 10–20 colonies to each well of ECM-coated plates.
8. Incubate the plates at 37°C with 5% CO_2 and change the media every second day.
9. Undifferentiated hESC cells should be transferred to new plates every 5–7 days and maintained at 37°C with 5% CO_2.

3.4. Detection of Cell Surface Pluripotency Markers by Immunocytochemistry

1. Remove culture medium.
2. Wash the cells once with DPBS without Ca^{2+} and $Mg2^+$. It is important to add the DPBS very gently and not directly to the cells.
3. Fix hESC in 4% paraformaldehyde for 15 min at room temperature by adding 1.5 ml/well to the 6-well plates.
4. Wash 4 times with DPBS.
5. To permeabilize the hESC, add 1.5 ml per well of 1% Triton-X-100, followed by an incubation of 10 min at room temperature.
6. Wash twice (5 min each) with DPBS.
7. Apply 4% of blocking solution for 45 min at room temperature. Serum for the blocking solution should be of the same species as the secondary antibody.
8. Remove blocking solution but do not wash the cells.
9. Dilute primary antibodies blocking solution (Table I).
10. Add 1 ml of the corresponding primary antibody dilution to the cultures at least 1 h at RT or overnight at 4°C.
11. Wash 3 times with DPBS. Cells can be left overnight before adding the secondary antibody.
12. Dilute secondary antibodies in DPBS and add 1 ml per well.

13. Incubate for 60 min at room temperature and in darkness as exposure to light may cause bleaching of fluorescent labels.
14. Wash the cells 4 times with DPBS.
15. If the cells are attached to a coverslip, mount it on a slide using prolong gold antifade reagent with DAPI. Remove any bubbles that may have formed during mounting.
16. Let the slides dry for 15 min in conditions where they are protected from light.
17. Keep the samples at room temperature. After 90 min, the samples can be observed by fluorescence microscope. Examples of the morphology and cell surface markers of undifferentiated hESC grown on plastic dishes coated by ECM derived from human fibroblasts and in the presence of conditioned medium are presented in **Fig. 3.2A–E**.

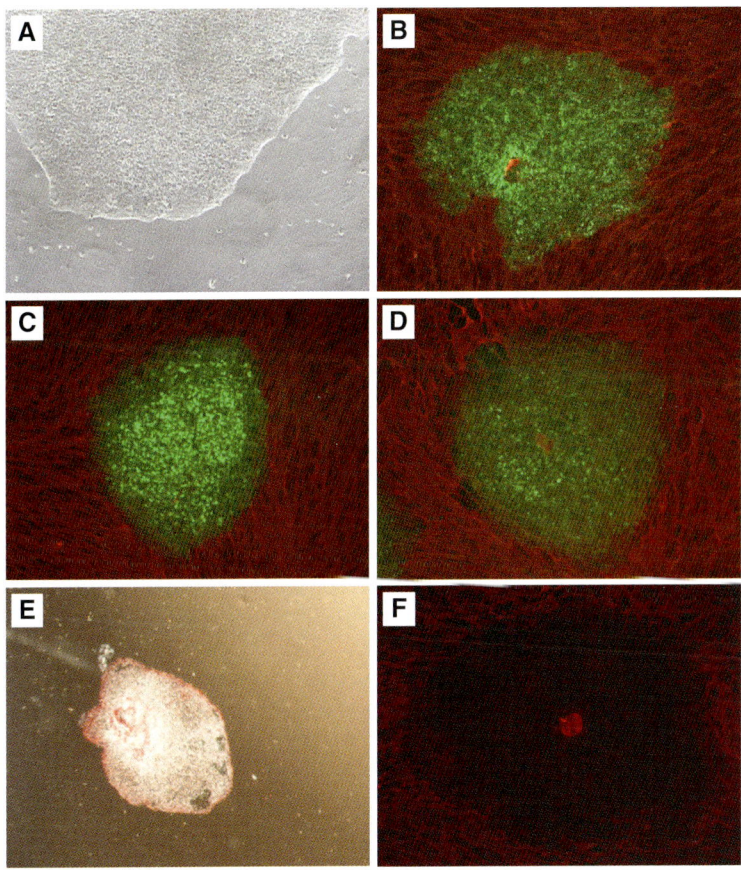

Fig. 3.2. (a) Morphology and cell surface markers of undifferentiated hESC grown for 15 passages on plastic dishes coated by ECM derived from human fibroblasts and in the presence of conditioned medium. Note round and compacted hESC colony (**A**). The presence of SSEA4 (**B**, *green*), TRA-1-60 (**C**, *green*), TRA-1-81 (**D**, *green*), and AP (**E**, *red*) markers of pluripotency but not of SSEA1 (**F**) was observed. Note the presence of specific fibronectin staining (**B–D, F**). Images were obtained using Zeiss Axiovert 200 M microscope (magnifications ×100).

3.5. Preparation of Samples for SEM

1. Wash samples twice with PBS during 2–3 min.
2. Fix for 30 min at 37°C with 2.5% glutaraldehyde. Add glutaraldehyde carefully until the samples are totally covered.
3. Incubate for 30 min and store it at 37°C, keeping the samples inside a box or Petri dish to avoid release of carcinogenic vapor.
4. Aspirate glutaraldehyde with a disposable Pasteur pipette.
5. Wash 3–4 times during 3 min with PBS.
6. Cover the samples with PBS and maintain them at 4°C.
7. Fixation and contrasting: add 1 ml of 1% osmium to cover the culture for 1 h.
8. Wash three times (5 min each) with distilled water at 4°C (do not exceed 15 min).
9. Dehydration procedure: all steps are performed at 4°C.
    9.1. Add 30% alcohol for 5 min.
    9.2. Wash with 50% alcohol (10 min) at 4°C. Do not exceed time since protein can be damaged.
    9.3. Wash twice with 70% alcohol (10 min each).
10. Keep the samples in 70% alcohol at 4°C until the moment of critical point of dehydration has been achieved (*see* **Note 6**).
11. Observe the sample under scanning electron microscope.

3.6. Preparation of Samples for RT-PCR Analysis

Analysis of mRNA expression by reverse transcription (RT)-PCR is carried out using standard protocols. Primer sequences, PCR conditions, and DNA fragments are available (*see* **Table 3.3**).

Transcripts encoding the following proteins OCT4, TERT, NANOG, FGF4, FOXD3, AFP, and beta-Tubulin III are assessed for both undifferentiation and specific lineage markers. b2-Microglobulin is used as housekeeping.

3.6.1. Extraction and Quantification of RNA

Total RNA extraction can be performed using 20 hESC colonies maintained on ECM. As a positive control, undifferentiated hESC maintained on feeder cells should be used. We use provider recommended instructions (Roche) with some modifications.

1. Cut the cells from the culture with a needle and collect them in a centrifuge tube.
2. Centrifuge for 3 min at $250 \times g$. Remove carefully the supernatant and discard it.
3. Resuspend the pellet in 100 µl of PBS.
4. Add 200 µl of lysis buffer and vortex for 15 s.

5. Transfer the entire sample to a filter tube and insert it in its corresponding collection tube.
6. Insert tubes in a tabletop centrifuge and centrifuge for 15 s at 8000×g.
7. Remove the filter tube from the collection tube, discard the flowthrough liquid, and combine again with the collection tube.
8. In another sterile centrifuge tube, mix 90 μl DNAse incubation buffer with 10 μl per sample. After mixing the solution, add 100 μl per sample to the upper reservoir of the filter tube and incubate for 15 min at room temperature.
9. Add 500 μl wash buffer to the upper reservoir of the filter tube and centrifuge for 15 s at 8000×g. Discard the flowthrough and combine filter tube with the used collection tube.
10. Add 500 μl of the second wash buffer to the upper reservoir of the filter tube and centrifuge for 15 s at ×8000g. Discard the flowthrough and add 200 μl of the second wash buffer again and centrifuge for 2 min at maximum speed (×13000g) to remove wash buffer.
11. Discard the collection tube and insert the filter tube into a clean, sterile 1.5-ml microcentrifuge tube.
12. Elute the RNA adding 100 μl of elution buffer to the upper reservoir of the tube. Centrifuge for 1 min at ×8000g. The microcentrifuge tube contains the eluted RNA.
13. Quantify the extracted RNA and evaluate its quality using Nanodrop or RNA integrity gel. Either use the eluted RNA directly in RT-PCR or store the eluted RNA at −80°C for later analysis.

3.6.2. cDNA Synthesis

We use 50 μl from each sample to obtain cDNA (following high-capacity cDNA archive kit recommendations).

3.7. Telomerase Activity Assay

Telomerase activity is assayed by telomeric repeat amplification protocol using the Trapeze Kit (Chemicon) and according to the manufacturer's protocol with some modifications.

Sample Preparation:
1. For stem cell analysis, collect 30–100 colonies. Pellet cells at ×400g for 5 min at 4°C. It is highly recommended to use positive and negative controls provided with the kit and a negative control (a differentiated cell line and/or heat-inactivated immortal/stem cells).

2. Wash cell pellet with sterile Ca^{2+} and Mg^{2+} PBS free for 5 min at $\times 400g$ at 4°C.
3. Resuspend cell pellet with 5–20 µl CHAPS lysis buffer for stem cell colonies.
4. Place on ice for 30 min.
5. Pellet cells at high speed (12.000 rpm) for 20 min at 4°C.
6. Transfer supernatant to a new tube. At this point supernatants are kept at –80°C and can be stored up to one year to be used for telomerase detection.

PCR reaction:

1. Use 2–4 µl of sample per PCR reaction.
2. Negative controls by heat inactivation must be subjected to 85°C for 10 min. Then use the same volume for PCR reaction 2–4 µl.
3. Prepare a master mix containing all components but tem plates, all components are provided by the kit except for Taq polymerase:

10X TRAP reaction buffer	5 µl
50X dNTP mix	1 µl
TS primer	1 µl
TRAP primer mix	1 µl
Taq polymerase (5 U/µl)	0.4 µl
dH$_2$O	29.5 µl
Template	2 µl

4. PCR program
   1 cycle: 30°C   30 min
   30–33 cycles: 94°C 30 s
                    59°C 30 s

3.7.1. Separation by Electrophoresis

1. For sample electrophoresis, use 10–15 µl PCR reaction and 2–5 µl loading buffer 10X (BioRad).
2. Load in a non-denaturing 10–20% polyacrylamide gel (we use 15% precasted gels from BioRad).
3. Run the gel in TBE 0.5X until both color bands are out of the gel.
4. Prepare SYBR green solution (1/10,000) in TBE 1X.
5. Stain the gel for 15–20 min in the dark.
6. Visualize bands in a transilluminator (same wavelength as ethidium bromide). Examples of specific intracellular hESC markers are presented in **Fig. 3.3A, B**.

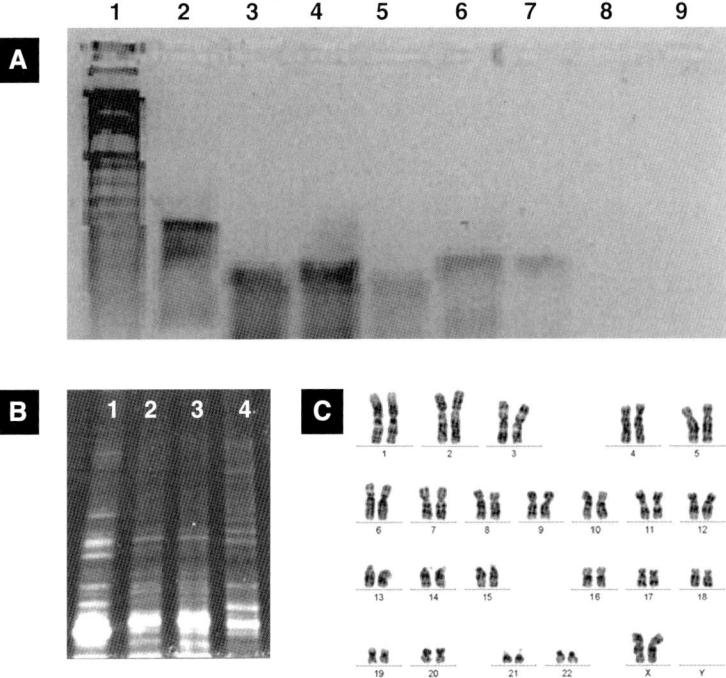

Fig. 3.3. (**A**) Analysis of specific molecular hESC markers. hESC were grown on ECM and in the presence of conditioned medium for more than 20 passages. Lines 1. Ladder with 1 Kb; 2. Corresponds to the housekeeping gene *b-2 microglobulin*. Cells express all the typical hESC markers 3. *NANOG*; 4. *OCT4*; 5. *TERT*; 6. *FOXD3*; 7. *FGF4*. After 20 passages over ECM, cells are negative for differentiation markers *Tuj1* and *AFP* (lines 8 and 9, respectively). (**B**) Colonies retained telomerase activity after 20 passages on hdECM (line 1). Lines 2 and 3 are negative control and foreskin feeder cells. Line 4 is H9 p52 colonies maintained over feeder. (**C**) hESC colonies grown on human ECM and in the presence of conditioned medium show normal (46, XX) karyotype after 20 passages.

3.8. Cytogenetic Analysis

Cytogenetic analysis is conducted before and after long-term culture of hESC grown on ECM.

1. Dividing cells are arrested at metaphase by the addition of 0.1 ml of 50 µg/ml of colchicine solution to a culture flask with 5 ml of total volume.
2. Incubate for 2 h.
3. For cell harvesting, transfer the cells of each sample to a conical centrifuge tube and centrifuge on a clinical benchtop centrifuge ($\times 400g$) for 10 min.
4. Remove supernatant and add gently 2–3 ml of warm (37°C) 0.56% KCl. Gently resuspend cells by pipetting.
5. Return tubes to incubator for 15 min. Tubes may now stand upright.
6. Centrifuge at $\times 500g$.

7. Remove supernatant without disrupting pellet. Gently add 3–4 ml of fixative solution: methanol:glacial acetic acid (3:1 v/v). Move tube gently but rapidly to prevent cell clumping. Incubate for at least 30 min. The procedure may be interrupted at this point for 1–2 h. Refrigerate if tubes are to be held for more than 30 min (see **Note 7**).

8. Centrifuge the cells at $\times 400g$ and resuspend them in fresh fixative solution. Repeat this step twice. Cells may be left refrigerated for several days once this step is completed. However, the cells must be washed at least twice before leaving overnight.

9. Giemsa/trypsin/leishman (GTL) banding techniques are employed for karyotypic analysis.

10. An average of 50 banded metaphase spreads are karyotyped and analyzed for chromosomal rearrangements. GTL-banded metaphase spreads are captured and arranged using Cytovision software from Applied Imaging (Santa Carla, CA, USA). **Figure 3.3C** represents normal karyotype of hESC grown on human ECM and in the presence of conditioned medium.

4. Notes

1. Cell counting. Take 20 µl of cell suspension and dilute with 70 µl of culture medium. Add 10 µl of trypan blue solution, mix, and incubate for 1 min before counting viable (round, clear cells) and nonviable (blue) cells using a Neubauer Hemocytometer. 6×10^6 cells per flask and 2×10^5 cells per well are harvested per 75 cm^2 and 6-well plates, respectively. More than 75% of them are viable.

2. Before usage of conditioned TESR1 medium, add 100 ng/ml of bFGF and 50 ng/ml of TGF-β1. Conditioned medium could be kept at +4°C during 1 week or stored at −80°C for 6 months.

3. No differences between frozen and fresh conditioned media are observed.

4. Human embryonic stem cells were cultured and maintained as previously described (10). Cells were passaged mechanically and replated on ECM-coated plates.

5. The optimal size of hESC before splitting procedure is when the colonies totally cover the magnification field ($\times 10$).

6. Critical point of dehydration to obtain dehydrated and stable sample: use controlled conditions, CO_2, 31°C, 73 apm in an Autosamdri machine.
7. Do not remove buffy coat on top of pellet.

Acknowledgments

The authors would like to thank Dr. Deborah Burks and Tugce Pehlivan for critical reading of the manuscript and Dario Melguizo, Sonia Prado, Dr. Angel Ayuso-Sacido, Dr. X. Chen Xiong, and Petra Stojkovic for technical support. This work was supported by funds for research in the field of Regenerative Medicine from the Regional Government Health Department (Generalitat Valenciana), the Instituto Carlos III belonging to the Spanish Ministry of Health and Consumer Affairs.

References

1. Klimanskaya, I., Chung, Y., Meisner, L., Johnson, J., West, M.D., and Lanza, R. (2005) Human embryonic stem cells derived without feeder cells. *Lancet* **365**, 1636–1641.
2. Xu, C., Inokuma, M.S., Denham, J., et al. (2001) Feeder-free growth of undifferentiated human embryonic stem cells. *Nature Biotechnol* **19**, 971–974.
3. Kagami, S., Kondo, S., Loster, K., et al. (1998) Collagen type I modulates the platelet-derived growth factor (PDGF) regulation of the growth and expression of beta1 integrins by rat mesangial cells. *Biochem Biophys Res Commun* **252**, 728–732.
4. Sjaastad, M.D., and Nelson, W.J. (1997) Integrin-mediated calcium signaling and regulation of cell adhesion by intracellular calcium. *Bioessays* **19**, 47–55.
5. Schuppan, D., Cramer, T., Bauer, M., Strefeld, T., Hahn, E.G., and Herbst, H. (1998) Hepatocytes as a source of collagen type XVIII endostatin. *Lancet* **352**, 879–880.
6. Ludwig, T.E., Levenstein, M.E., Jones, J.M., et al. (2006) Derivation of human embryonic stem cells in defined conditions. *Nature Biotechnol* **24**, 185–187.
7. Werb, Z., and Chin, J.R. (1998) Extracellular matrix remodeling during morphogenesis. *Ann NY Acad Sci* **857**, 110–118.
8. Cukierman, E., Pankov, R., Stevens, D.R., and Yamada, K.M. (2001) Taking cell-matrix adhesions to the third dimension. *Science* **294**, 1708–1712.
9. Thomson, J.A., Itskovitz-Eldor, J., Shapiro, S.S. et al. (1998) Embryonic stem cell lines derived from human blastocysts. *Science* **282**, 1145–1147.
10. Stojkovic, P., Lako, M., Stewart, R., et al. (2005) An autogeneic feeder cell system that efficiently supports growth of undifferentiated human embryonic stem cells. *Stem Cells* **23**, 306–314.

Chapter 4

In Vitro Neural Differentiation of Human Embryonic Stem Cells Using a Low-Density Mouse Embryonic Fibroblast Feeder Protocol

John A. Ozolek, Esther P. Jane, James E. Esplen, Patti Petrosko, Amy K. Wehn, Teresa M. Erb, Sara E. Mucko, Lyn C. Cote, and Paul J. Sammak

Abstract

Human embryonic stem cells (hESCs) have the capacity to self-renew and to differentiate into all components of the embryonic germ layers (ectoderm, mesoderm, endoderm) and subsequently all cell types that comprise human tissues. HESCs can potentially provide an extraordinary source of cells for tissue engineering and great insight into early embryonic development. Much attention has been given to the possibility that hESCs and their derivatives may someday play major roles in the study of the development, disease therapeutics, and repair of injuries to the central and peripheral nervous systems. This tantalizing promise will be realized only when we understand fundamental biological questions about stem cell growth and development into distinct tissue types. In vitro, differentiation of hESCs into neurons proceeds as a multistep process that in many ways recapitulates development of embryonic neurons. We have found in vitro conditions that promote differentiation of stem cells into neuronal precursor or neuronal progenitor cells. Specifically, we have investigated the ability of two federally approved hESC lines, HSF-6 and H7, to form embryonic and mature neuronal cells in culture. Undifferentiated hESCs stain positively for markers of undifferentiated/pluripotent hESCs including surface glycoproteins, SSEA-3 and 4, and transcription factors Oct-3/4 and Nanog. Using reduced numbers of mouse embryonic fibroblasts as feeder substrates, these markers of pluripotency are lost quickly and replaced by primarily neuroglial phenotypes with only a few cells representing other embryonic germ layer types remaining. Within the first 2 weeks of co-culture with reduced MEFs, the undifferentiated hESCs show progression from neuroectodermal to neural stem cell to maturing and migrating neurons to mature neurons in a stepwise fashion that is dependent on both the type of hESCs and the density of MEFs. In this chapter, we provide the methods for culturing pluripotent hESCs and MEFs, differentiating hESCs using reduced density MEFs, and phenotypic analyses of this culture system.

Key words: Embryonic stem cells, neural mouse feeder.

1. Introduction

Neuroglial differentiation of embryonic stem cells (ES cells) is of great importance in the study of central and peripheral nervous system tissue regeneration, repair, and development. ES cells that have been differentiated to neuroglial phenotypes are being used for repair of nervous system injury secondary to stroke (1, 2), spinal cord injury (3–6), and even in the enteric nervous system to repopulate missing enteric neurons and glial stroma in experimental models of Hirschsprung disease (7). Very specific and directed differentiation to produce populations of dopaminergic neurons has been achieved in the hope of transplantation and eventual cure of Parkinson's disease (8–12). Neural stem cells are being utilized to study other degenerative diseases of the central nervous system including Alzheimer's disease and other metabolic degenerative conditions (13). Many protocols have been investigated and developed for neuroglial differentiation of human embryonic stem cells (hESC). These include the use of inducing agents such as noggin, retinoic acid, sonic hedgehog, and various neurobasal medium (14–18). Nat et al. have demonstrated consistent and programmed expression of neuroglial markers in six different hESC lines using neurobasal medium under adherent and nonadherent conditions (17). The study of early nervous system development will be greatly enhanced by studying the behaviors of ES cells in culture, as they differentiate toward a neuroglial phenotype, and in vivo after transplantation in early embryos. This avenue of investigation may benefit from a cell culture system that methodically allows for slow and progressive differentiation of ES cells toward primarily neuroglial phenotypes without extensive genetic or exogenous manipulation of the cells.

Traditionally, hESCs can be maintained as pluripotent stem cells (cells that have the capability to differentiate into all three embryonic germ layer cell types and the capacity for self-renewal) by growing them on a feeder substrate, usually mouse embryonic fibroblasts (MEFs) and routinely passaging the cells to prevent unwanted spontaneous differentiation. The main factors affecting the behavior of hESCs in vitro (excluding direct manipulations of the cell by siRNA, transfection, or other genetic manipulation) include all of the basic elements of cell culture including composition of the medium, incubator conditions, method of passaging, substrate, and initial cell type. Of these, the medium composition and substrate (including feeder-free systems) have been the focus of attention in attempts to generate directed differentiation toward these neural/neuroglial phenotypes.

Generally, hESCs when grown on an MEF feeder substrate require a "high" density of feeder cells per well to maintain pluripotency. It is interesting that a great body of literature examining the specific effects of changing feeder density on the basic cellular behavior of the various available hESC lines does not exist. In one study that specifically addresses this issue, Heng et al. describe the detrimental effects of growing hESCs on very high feeder densities (30,000 cells/cm^2 and above) with the speculation that the feeder cells use most of the nutrients and oxygen in the culture depriving the hESCs (19). Empiric evidence suggests that feeder density does affect differentiation with both higher and lower quantities of MEF feeders contributing to enhance spontaneous differentiation (personal communication, Mark Clements). The observation that approximately 200,000 feeders per well of a 6-well culture plate (21,000 cells/cm^2) may be appropriate for maintaining ideal pluripotent colony morphology is dependent on the hESC line with H1 hESCs requiring higher feeder densities and H9 hESCs lower densities for ideal colony morphology. Generally, we and others in our laboratory have used between 175,000 (18,400 cells/cm^2) and 225,000 (23,700 cells/cm^2) MEF feeder cells per well of a 6-well cell culture dish plate. Between these densities, hESCs maintain a high proportion of cells that are morphologically and phenotypically pluripotent. The cells within the central portions of the pluripotent colonies are then manually or enzymatically passaged every 7 days, replated, and thus a population of pluripotent cells is maintained. If left undisturbed, however, differentiation begins to occur primarily at the edge of colonies after several days and the cells will continue to grow. We have shown previously that in comparison to using a mitotically inactivated tumor cell line (U87 glioblastoma) as a feeder substrate, low-density (50,000–150,000 cells/well; 5,200–15,600 cells/cm^2) mitotically inactivated MEFs seem to promote greater differentiation of HSF-6 hESCs toward a neural phenotype (20). Perhaps of greater significance, this range of low-density MEF feeders [approximately 100,000 cells/well (10,400 cells/cm^2)] seemed to promote the least proliferative and apoptotic activities in the differentiating HSF-6 hESCs as measured by BrDU uptake and TUNEL assays, respectively. In comparison, proliferative and apoptotic activities were increased among differentiating hESCs on MEF feeder densities at the lower and upper ends of the low-density spectrum (5,200 and 15,600 cells/cm^2, respectively) (20).

In this chapter, we describe a method of "naturally" directing hESCs into primarily neuroglial phenotypes with between 50 and 80% efficiency by using a low-density MEF feeder approach. We will introduce data from previous experiments with the HSF-6 hESC line as well as data using the H7 hESC line. With the approach described herein, cells begin to express neuroepithelial/neural stem cell markers (nestin, Pax-6) in the first week with continued expression into the second week with little expression of other germ layer markers. Markers of maturing neural cells such as β-tubulin, NCAM, and DCX are present in the first week, but expression is accelerated between the first and second weeks and NeuN, a mature neural phenotypic marker, is seen sparsely but more so in the second and third weeks in culture. With this unencumbered approach, 50–80% of cells express markers of developing and maturing/mature neural lineages. For both of these hESC lines, an MEF density of 10.4 K cells/cm^2 seems to promote an orderly and rapid progression to neural phenotypes. Very reduced densities (5.2 K cells/cm^2) promote increased proliferative and programmed cell death and perhaps delayed and aberrant emergence of neural phenotypes, particularly in the H7 line. Embryoid bodies appear already differentiated toward neuroglial phenotypes even after a short time in culture and this program is maintained on the low-density protocol. For specific cell isolation purposes, growing undifferentiated hESCs on low-density feeders may provide faster and easier access to differentiating cells than EBs mostly due to the bulk and thickness of the EBs. An important caveat is that the MEFs must be early in their lifespan and used very soon after treating with mitomycin C for best results.

2. Materials

2.1. Cell Culture and Cryopreservation: Mouse Embryonic Fibroblast and Human Embryonic Stem Cells

1. Mouse embryos: CF1 mouse strain available from commercial vendors such as The Jackson Laboratory, Bar Harbor, Maine.

2. Phosphate buffered saline (PBS): PBS suitable for cell culture is prepared in deionized distilled water from premade packets (Sigma-Aldrich Corp., St. Louis, MO, USA) and filter sterilized (0.2 μm filters) or alternatively Dulbecco's phosphate buffered saline can be used for cell culture (Invitrogen, Corp., Carlsbad, CA, USA).

3. Mouse embryonic fibroblast (MEF) medium (DMEM-MEF): Dulbecco's Modified Eagle's Medium (DMEM) supplemented with 10% fetal bovine serum, 2 mM L-glutamine, nonessential amino acids, 100U/ml Penicillin, 100 μg/ml Streptomycin (all from Gibco/Invitrogen Corp., Carlsbad, CA, USA).

4. Trypsin/EDTA: TrypLE Express (Gibco/Invitrogen, Invitrogen Corp., Carlsbad, CA, USA).

5. Amphotericin-B, Mitomycin-C, DMSO (Sigma-Aldrich Corp., St. Louis, MO, USA).

6. Ultrapure water with 0.1% gelatin (Chemicon, Billerica, MA, USA).

7. Glass (9 in.) Pasteur pipettes (ISC Bioexpress, Kaysville, UT, USA).
8. Hemocytometer (Fisher Scientific, Pittsburgh, PA, USA).
9. Prepare mitomycin-C stock solution by dissolving 2 mg of mitomycin-C in 10 ml of PBS to create a 200 μg/ml stock. Filter sterilize through 0.2 μm filter. Prepare mitomycin-C medium (10 μg/ml) by adding 500 μl of mitomycin-C stock to each 10 ml of MEF medium.
10. HESCs designated HSF-6 and H7 provided by the National Stem Cell Bank, WiCell Research Institute (Madison, WI, USA).
11. HES maintenance medium: Knockout DMEM supplemented with 20% Knockout serum replacement, 2 mM L-glutamine, nonessential amino acids, 100 U/ml Penicillin, 100 μg/ml Streptomycin, 4 ng/ml bFGF (all from Invitrogen), and 2.5 μg/ml Amphotericin-B (Sigma).
12. Coverslips for differentiation and subsequent immunocytochemistry: Nunc™Thermanox® coverslips; Cole-Parmer Instrument Company, Vernon Hills, IL.
13. Culture flasks (Corning Inc., Corning, NY, USA).

2.2. Immunocytochemistry

1. 1% Paraformaldehyde: A 4% stock can be prepared by dissolving 4 g of paraformaldehyde powder (Sigma) in PBS while applying heat and stirring. The paraformaldehyde will go into solution at around 70°C and with the addition of a few drops of 10 M sodium hydroxide until the pH is 8.0. The solution is passed through a filter flask to remove residual particulates. The solution can then be brought to pH 7.4 with addition of HCl. After diluting 1:4 to make a 1% solution, aliquots can be made and stored at –20°C for approximately 1 month.
2. CAS-block (Invitrogen).
3. 0.1% (v/v) Triton X 100 (Sigma) and 0.05% (v/v) Tween-20 (Sigma) are prepared in PBS.
4. Hoechst 33258 (Sigma).
5. TOTO®-3 (Molecular Probes, Invitrogen) solution: A 1-ml stock is prepared by adding 1 μl TOTO®-3 into 100 μl RNase (Sigma) and 900 μl PBS.
6. Vectashield with DAPI (Vector Laboratories, Burlingame, CA, USA).
7. Gelvatol is prepared by adding 5 g polyvinyl alcohol to 100 mL PBS every hour for 4 h with constant stirring. Cover and stir overnight at 4°C until dissolved. Add 3 g polyvinyl alcohol and stir until dissolved. Add 1 crystal sodium azide (a broad spectrum biocide). Add 50 mL glycerol and stir overnight (at

least 16 h). Centrifuge to remove undissolved particles (16,000×*g* in a microcentrifuge in 1-ml microcentrifuge tubes for 45 min). Aliquots can be stored at 4°C. All chemicals can be acquired from Sigma.

8. Primary antibodies: Rabbit anti-β-Tubulin Class III (Sigma) 1:200, Guinea Pig anti-Doublecortin (Abcam Inc., Cambridge, MA, USA) 1:1000, Rabbit anti-Glial Fibrillary Acidic Protein (GFAP) (Millipore) 1:500, Mouse anti-NCAM (neural cell adhesion molecule) (AbCam) 1:200, Rabbit anti-Nestin (Millipore) 1:200, Mouse anti-Neuronal Nuclei (NeuN) (Millipore) 1:200, Mouse anti-Pax-6 (Santa Cruz Biotechnology, Inc., Santa Cruz, CA, USA) 1:200, Mouse anti-Oct-3/4 (Santa Cruz) 1:100, Goat anti-Nanog (R&D Systems, Minneapolis, MN) 1:50.

9. Secondary antibodies are all Alexa Fluor® (Molecular Probes, Invitrogen) conjugates with species and fluorophores as follows: goat anti-mouse IgG 546; goat anti-rabbit IgG 488; goat anti-guinea pig IgG 488; donkey anti-goat IgG 488.

2.3. Imaging and Image Analysis

1. Phase contrast: All phase contrast images were acquired using a Nikon Eclipse TS100-F body (with 100% visual/photo trinocular head, 45 degree inclination, 50–75 mm interpupillary adjustment; quintuple nosepiece, built-in transformer; plain stage; coarse and fine focus) with 10X widefield eyepiece, CFI PLAN FLUOR 4× phase objective (N.A. 0.13; 17.1 mm working distance), CFI PLAN FLUOR DLL 10× objective (N.A. 0.3; 16 mm working distance), CFI PLAN FLUOR DLL 20× objective (spring loaded; N.A. 0.5; 2.1 mm working distance), TI-CELWD extra long working distance condenser (NA 0.3 with built-in aperture diaphragm), T1-SNCP noncenterable phase slider with phase ring phl (4X) and Ph1 (ADL 10, 20, 40xf) and Nikon DS-Fi1 digital camera with U2 controller (Nikon, Inc., Melville, NY, USA).

2. Confocal Immunofluorescence: The majority of the fluorescence images presented here were acquired on a Leica DMIRE2 laser scanning inverted confocal microscope (Leica Microsystems GmbH, Wetzlar, Germany) equipped with Ar/ArKr (456/488/514 nm), Gre/Ne (HeNe 543/594 nm), HeNe (633 nm), and 405 nm blue diode lasers. This microscope is equipped with 10X, 20X, 40X (dry and oil), and 63X oil objectives. The Leica confocal software package version 2.61 (Build 1537) was used for driving the microscope, image acquisition, and some image processing.

3. Spinning Disk Confocal: Some images represented in **Figs. 4.2** and **4.9** were acquired using a Nikon TE2000-E inverted microscope (Nikon, Inc.) with Yokogawa spinning

disk confocal head, 3 line krypton-argon laser (blue, green, red excitation) and Mercury Arc lamp (violet excitation) equipped with a Photometrics CoolSNAP HQ (Photometrics®, A Division of Roper Scientific, Inc., Tucson, AZ) CCD camera. Nikon Elements is the software used for image acquisition and processing. A 40X, 1.3NA oil objective was used for most image acquisition.

4. Image Analysis Software: Adobe Photoshop CS2 (Adobe System, Inc., San Jose, CA, USA).

3. Methods

3.1. Preparation and Inactivation of Mouse Embryonic Fibroblasts

1. Isolate mouse embryonic fibroblasts (**Fig. 4.1**) from 12.5 day mouse embryos.

2. Cut open each embryonic sac with a scissors and release the embryos into the dish.

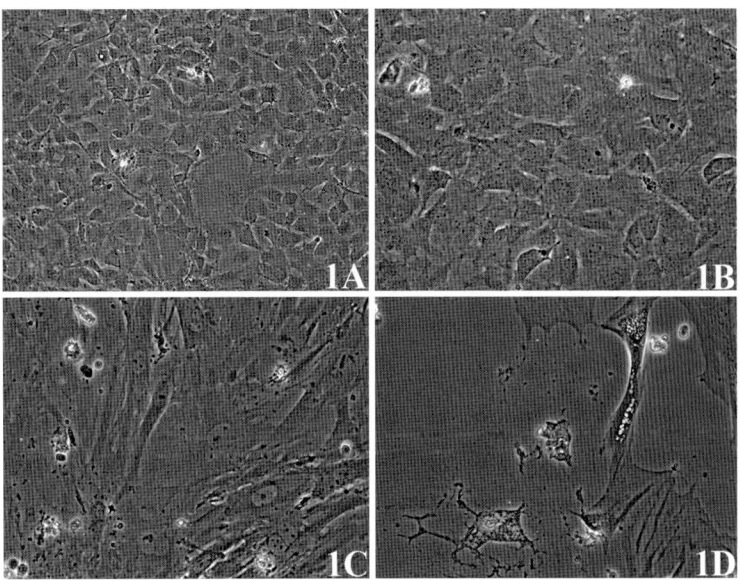

Fig. 4.1. Phase contrast of MEFs. (**A**, **B**) Confluent and healthy appearing MEFs at passage 5. Many cells have a spindle configuration with long processes. Note the large, round to oval, plump nucleus of uniform size and consistency. Cyotplasmic granularity is minimal and the cytoplasm has an overall homogeneous appearance. These feeders were derived from the CF1 mouse embryo using the protocol described in the text (10X, 20X). (**C**) Mitomycin-treated MEFs after passage 3. The cells are more defined by their elongated and spindle shape but retain the cytoplasmic and nuclear characteristics of their untreated counterparts (20X). (**D**) Mitomycin-treated MEFs after 27 days. These MEFs have extreme variability of shape and size, nuclear irregularity, and cytoplasmic vacuoles. These feeders are unsuitable for propagation or differentiation of hESC (20X).

3. Using two forceps, remove the placenta and membranes from the embryo. Once they have been removed, dissect out the head and visceral tissue (i.e. anything that is dark in color). Place the embryos in a clean petri dish and wash three times with 10 ml PBS.

4. With curved scissors, finely mince the tissue. Add 2 ml TrypLE and continue to mince. Add an additional 5 ml of TrypLE and incubate at 37°C (for about 20 min). Pipette the embryos in TrypLE vigorously, until few chunks remain.

5. Neutralize the TrypLE with about 20 ml culture medium (regular DMEM) and transfer the contents of the dish to a 50-ml conical tube. Mix the contents of the tube well, and evenly add to T175 culture flasks containing 25 ml DMEM-MEF and 2.5 µg/ml amphotericin-B (There should be approximately 3 embryos per T175.).

6. Place these flasks in a 37°C/5%CO_2 incubator overnight. The next day, change the medium to eliminate debris (*see* **Note 1**).

3.2. Inactivation of MEFs and Preparation of Culture Wells for Seeding of hESCs

1. Aspirate existing medium from the MEF flasks (*see* **Note 2**).

2. Add 10 ml (75 cm flask) or 20 ml (150 cm flask) of Mitomycin-C medium to each flask and incubate for 2–3 h. Wrap tube of Mitomycin-C stock in foil and store at 4°C.

3. Aspirate Mitomycin-C medium.

4. Wash flasks 3 times with PBS.

5. Add prewarmed TrypLE to each flask and incubate for 2–5 min. Look under microscope to make sure cells have detached from the plate.

6. Add 10 ml (75-cm flask) or 20 ml (150-cm flask) of DMEM-MEF medium to neutralize TrypLE and transfer to 50-ml tube. Wash flask once with 5–10 ml of medium to get all cells.

7. Count cells using a hemocytometer (cells/ml) (*see* **Note 3**).

8. To plate MEFs, pipette 0.1% gelatin solution onto dish with or without a plastic coverslip on bottom of well, enough to cover the bottom (*see* **Note 4**).

9. After 3 h at 37°C, aspirate excess gelatin and seed MEFs into gelatin-coated well at the appropriate predetermined density as defined above and allow to adhere in MEF medium for 24 h.

10. Prior to seeding stem cells, aspirate MEF medium and wash wells one time with PBS to remove any remaining serum. Wells are now ready for hESC seeding. HESC should be seeded relatively soon after mitomycin treated MEFs are plated (within 48 h) (*see* **Note 5**).

3.3. Growth of Pluripotent Human Embryonic Stem Cells (hESCs)

1. Both hESC types were grown in MEF plates as described above and maintained in hES medium (*see* **Note 6**) (**Figs. 4.2 and 4.3**).
2. When ready (*see* **Note 7**), the medium was refreshed and colonies were manually dissociated into clusters of 10–15 cells using pulled glass Pasteur pipettes while visualized under a 4X objective (phase contrast).
3. Clusters were then gently triturated several times using a 5-ml pipette and seeded into new MEF plates. HES maintenance media was changed every two days.
4. Colonies were typically passaged every 7 days.

3.4. Cryopreservation of hESCs

1. HESCs can be cryopreserved and stored in liquid nitrogen indefinitely.
2. Undifferentiated stem-cell colonies are manually dissociated as described above and centrifuged at $\times 200g$ for 5 min.
3. The supernatant is aspirated and the pellet gently resuspended in hES maintenance medium with the addition of 30% fetal bovine serum and 10% DMSO.
4. Aliquots of 1 ml are made into appropriate cryovials, which are then put into an isopropanol cell-freezing container. This container is placed into a –80°C freezer overnight. The cryovials are removed to liquid nitrogen storage the next day.
5. To reculture frozen cells, the cryovials are quickly thawed in a 37°C waterbath and the cell suspension diluted into 5 ml of hES maintenance medium. This is then centrifuged at $\times 200g$ for 5 min, the supernatant aspirated, and the pellet gently resuspended in 3 ml hES maintenance medium.
6. The cell suspension is distributed equally among 3 wells of a 6-well plate containing MEF cells as described above. An additional 2 ml of hES maintenance medium is added to each well and the plate is incubated under standard conditions (37°C, 5% CO_2).

3.5. Preparation of Embryoid Bodies

1. To form embryoid bodies (EBs), H7s were dissociated manually as above (In **Section 3.3**) and clusters were seeded into tissue culture flasks or plates that have not been surface treated or coated and without a supporting MEF cell monolayer. Cultures of hESC line HSF-6 were mechanically passaged and colonies allowed to form embryoid bodies in suspension for 8 days (for the early experiments using HSF-6 hESCs) (**Fig. 4.4**).
2. EBs were grown in hES medium described above for growing pluripotent hESCs. Spherical embryoid bodies formed within a few days and continued to enlarge over time.

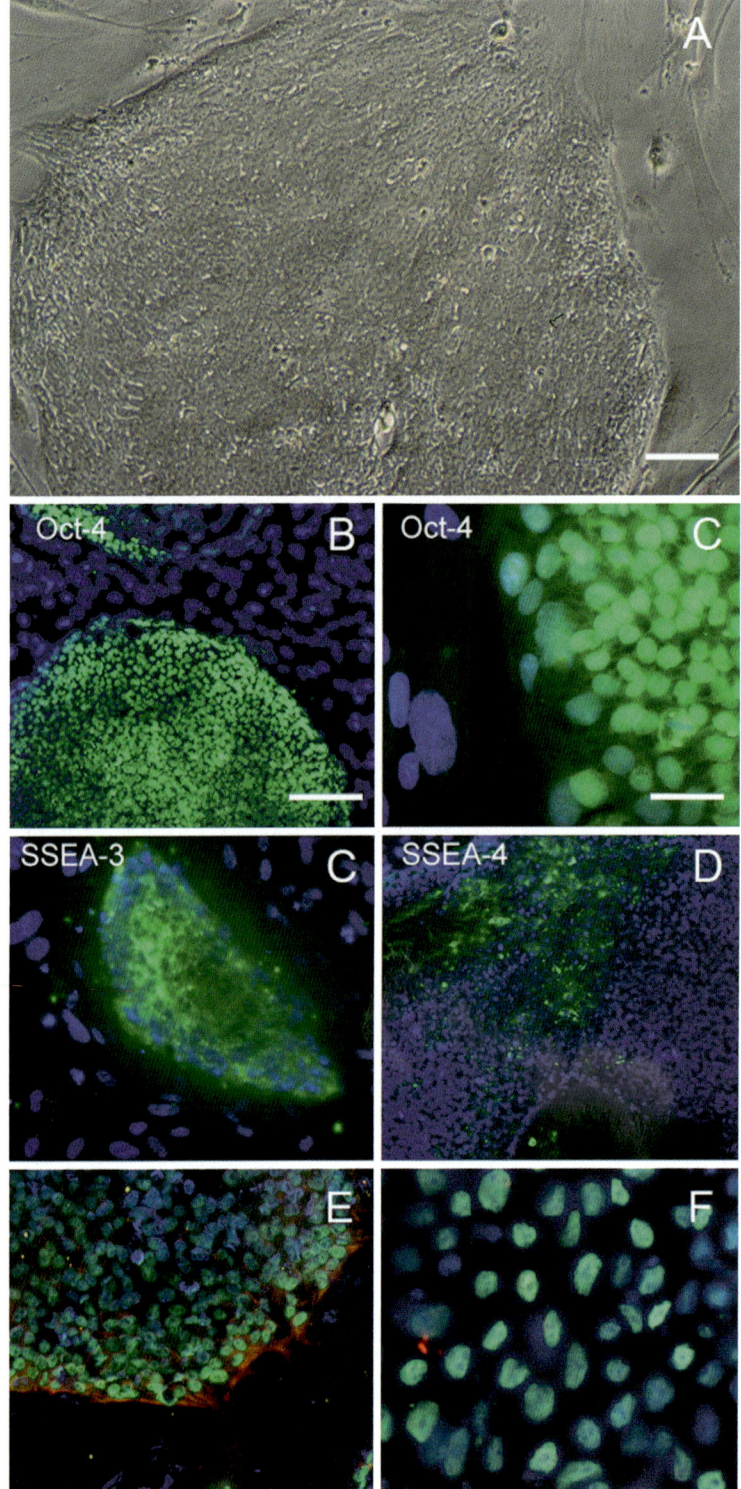

Fig. 4.2. Pluripotent hESCs. HSF-6 and H7 grown on mouse embryonic fibroblasts feeder cells at density of 5.2K cells/cm^2. (**A**) Phase contrast image of HSF-6 pluripotent colony. (**B**, **C**) Oct-3/4 staining (*green*) of undifferentiated HSF-6.

In Vitro Neural Differentiation of hESCs   81

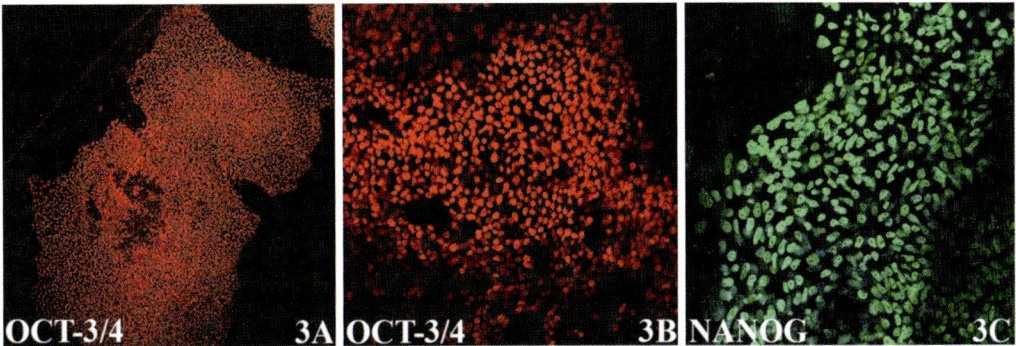

Fig. 4.3. Immunocytochemical staining of pluripotent H7 colony with markers commonly associated with pluripotent ES cells, Oct-3/4 and Nanog. (**A**, **B**) Oct-3/4 (*red*) shows diffuse bright staining within all nuclei of colony (10X and 40X respectively). (**C**) Nanog (*green*) staining is intense and bright in most nuclei of this H7 pluripotent colony (10X).

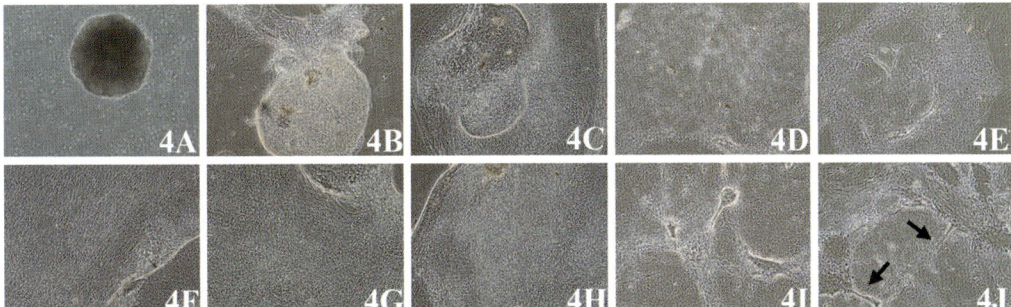

Fig. 4.4. Phase contrast images of H7 hESC. (**A**) H7 EB in suspension at day 8. (**B–E**) Day 15 on feeders; Left to right: H7EB/LDMEF, H7 EB/NDMEF, H7/LDMEF, H7/NDMEF. EBs at day 15 after seeding still maintain thick centers with complex formations within the central portions while differentiated cells spread in a corona from the center. EBs on NDMEF are spreading from the center of the EB compared to EB on LDMEF. (**F–J**) Day 30 on feeders; Left to right: H7EB/Gelatin only, H7EB/LDMEF, H7 EB/NDMEF, H7/LDMEF, H7/NDMEF. At day 30, EBs are plated out in sheets of differentiated cells. H7s on LD and ND MEFs show spreading of differentiated cells at day 15 and form more complex tubular structures at day 30 (*arrows*). All images taken using 4X objective. EB = embryoid body, LDMEF = low-density mouse embryonic fibroblast, NDMEF = normal-density mouse embryonic fibroblast.

3. Medium was changed every 3 days by pelleting embryoid bodies ($\times 200g$ for 3 min) and gently resuspending in fresh medium before returning to the same culture vessel.

4. Alternatively, EBs can be allowed to settle by gravity into the bottom of a conical tube before being resuspended in fresh medium. Centrifugation proved to be faster and allowed for collection of all EBs including smaller ones.

Fig. 4.2 (continued) Feeders are Oct-3/4 negative. Note the nuclear size difference (Hoechst DNA- blue) between cells within the colony and fibroblast-like feeder cells. (**D**) SSEA-3 staining of an entire undifferentiated HSF-6 colony. Note that the cell surface proteoglycan partially covers the colony. (**E**) H7 hESC colony on mouse embryonic fibroblast feeders stained with Oct-3/4 (*green*) and the neurectodermal marker, nestin (*red*). Differentiation is commonly seen in cells at the colony periphery and nestin is found at low levels in cells within the colony. (**F**) On feeder fibroblasts, Oct-3/4 positive colonies (*green*) are negative for the trophectodermal marker CDX-2 (*red*). *Bar* in (**A**) 100 μm; (**B**) 50 μm; (**C**) 25 μm.

3.6. Differentiation of Human Embryonic Stem Cells Using Low-Density Mouse Embryonic Fibroblasts

3.6.1. Differentiation of H7 EBs

1. Clusters from manually dissociated H7 colonies maintained in our lab were taken at passage 106 and seeded into EB culture conditions as described above (**Fig. 4.5**).

2. After 30 days embryoid bodies were picked under a dissecting microscope for uniformity of size (approximately 1 mm in diameter, spherical in shape) and placed into one of three experimental culture conditions in 24-well plates with 13 mm plastic coverslips (Nunc™ Thermanox®; *see* **Section 2**) precoated with 0.1% gelatin and placed at the bottom of each well (*see* **Note 8**).

3. EBs were cultured on gelatin alone, on normal density MEF monolayer (18.5×610^3 cells/cm^2), or on a very reduced density MEF monolayer (5.2×10^3 cells/cm^2). In all three conditions, the EBs quickly attached to the surface and were thereafter maintained as normal adherent cultures and allowed to grow for 5, 15, or 30 days with hES medium changes every 2 days.

4. On days 5, 15, or 30 medium was aspirated from the wells and the cells washed once with PBS before being fixed with 1% paraformaldehyde in PBS for 30 min at room temperature.

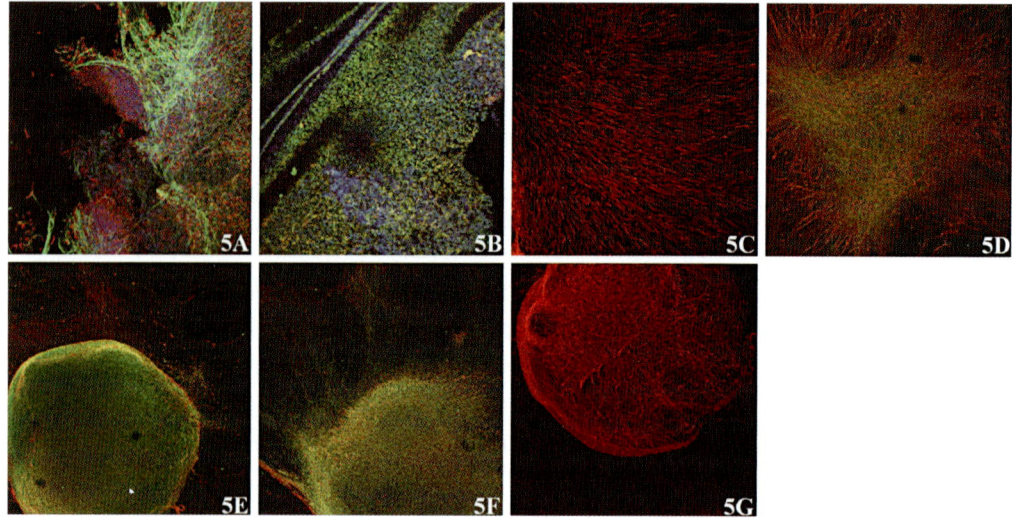

Fig. 4.5. H7EB grown on gelatin, LDMEF, and NDMEF: (**A**) Nestin (*red*)/DCX (*green*) double labeling of H7EB/LDMEF after 5 days (20X). (**B**) NCAM (*red*)/Class III β-tubulin (*green*) double labeling of H7EB/LDMEF (20X). (**C**) H7EB/LDMEF-D15: Diffuse staining with Class III-B-tubulin (*red*) and weaker staining with DCX (*green*) (10X). (**D**) H7EB/LDMEF-D30: Persistence of labeling with Class III β-tubulin (*red*) and DCX (*green*) (10X). (**E, F**) Both H7EB/NDMEF and EB/Gelatin at day 15 show widespread appropriate expression of Class III β-tubulin (*red*) and DCX (*green*) (10X). (**G**) H7EB/Gelatin at day 15 shows diffuse expression of GFAP (*red*) and no Oct-3/4 staining (10X). EB = embryoid body, LDMEF = low-density mouse embryonic fibroblasts, NDMEF = normal-density mouse embryonic fibroblasts, D15 = day 15 after plating on feeders, D30 = day 30 after plating on feeders, DCX = doublecortin, GFAP = glial fibrillary acidic protein.

3.6.2. Differentiation of H7 Human Embryonic Stem Cells

1. Clusters from manually dissociated H7 colonies maintained in our lab were taken at passage 73 and seeded into 0.1% gelatin-coated coverslips containing either 100,000 MEF cells/well (10,400 cells/cm^2) or 150,000 MEF cells/well (15,600 cells/cm^2). Colonies were grown for 5 or 10 days with hES medium changes every two days. On days 5 or 10 medium was aspirated from the wells and the cells washed once with PBS before being fixed with 1% paraformaldehyde in PBS for 30 min at room temperature (**Figs. 4.6 and 4.7**).

2. In addition, clusters from manually dissociated H7 colonies were taken at passage 109 and seeded on coverslips at a normal density MEF monolayer (18.5×10^3 cells/cm^2) or a very reduced density MEF monolayer (5.2×10^3 cells/cm^2). Colonies were grown for 15 or 30 days with hES medium changes every two days. On days 15 or 30 medium was aspirated from the wells and the cells washed once with PBS before being fixed with 1% paraformaldehyde in PBS for 30 min at room temperature.

3.6.3. Differentiation of HSF-6 Human Embryonic Stem Cells

1. For HSF-6 hESC cells, clusters from manually dissociated HSF-6 colonies maintained in our lab were taken at passages 65 and 66 and seeded into 0.1% gelatin-coated coverslips (**Figs. 4.8 and 4.9**).

2. Seeding densities were approximately 50,000 cells/well (5.2×10^3 cell/cm^2), 100,000 cells/well (10.4×10^3 cells/cm^2), and 150,000 cells/well (15.6×10^3 cells/cm^2).

3. Cells were fixed at days 7, 12, 19, and 42 (5.2×10^3 MEF cells/cm^2 density only).

3.7. Immunocytochemistry

1. Cells grown under differentiation conditions as described above were fixed with 1% paraformaldehyde for 30 min at room temperature after which cells were washed three times in PBS then permeablized using 0.1% Triton-X 100 for 10 minutes at room temperature.

2. Nonspecific background was reduced by incubating for 30 min in CAS-Block at room temperature.

3. All primary antibodies were diluted in CAS-Block and incubated overnight at 4°C (*see* **Note 9**).

4. After removing unbound primary antibody with three 5-minute washes of PBS, the cells were incubated for 1 h at room temperature with fluorophore-conjugated secondary antibodies of appropriate species diluted 1:200 in CAS-Block.

5. After three 5-minute washes with 0.05% Tween-20, nuclei were counterstained with 1 μg/ml Hoechst 33258. Prior to the addition of the UV laser on the laser scanning confocal microscope, we used TOTO®-3 iodide (*see* **Note 10**).

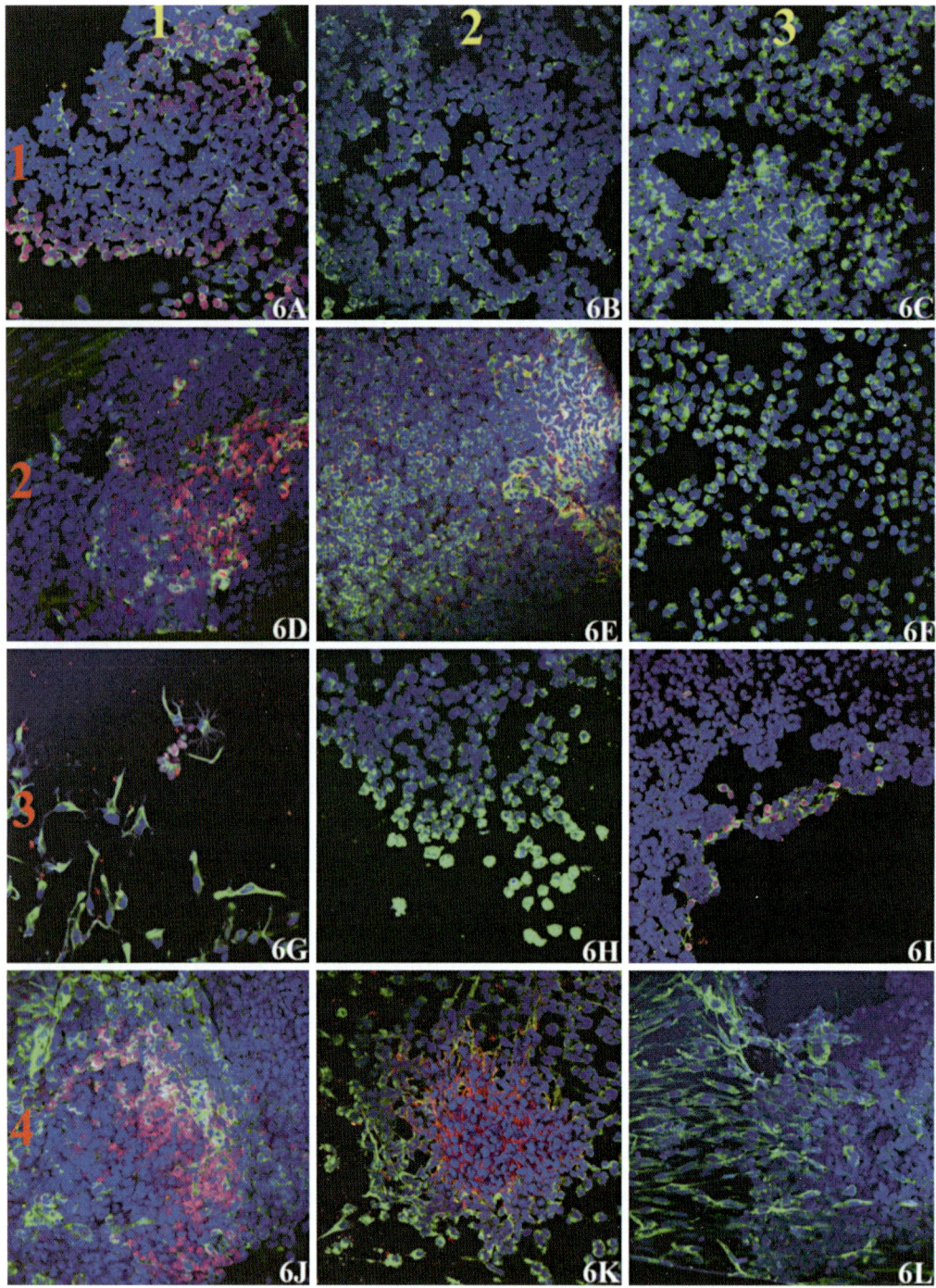

Fig. 4.6. H7s grown on LDMEF at day 5 and 10 in culture. Rows 1–4 (*red numbers*): D5/10.4K, D5/15.6K, D10/10.4K, D10/15.6K respectively; Columns 1–3 (*yellow numbers*): Nestin (*green*)/Oct-3/4 (*red*), Nestin (*green*)/DCX (*red*), Class III β-tubulin (*green*)/NeuN (*red*), respectively. Widespread staining for nestin was apparent at both days and feeder densities. Oct-3/4 was focally present at day 5 and almost gone by day 10 in culture. The staining for Oct-3/4 seen at day 10 at 150K density was mostly in the cytoplasmic compartment. Staining for DCX was consistently present in greater amount when H7s were grown on 150K density at both day 5 and day 10. Class III β-tubulin was present at both time points and feeder densities. All images using 40X objective. Feeder density represented as MEFs/cm^2 (10.4K = 10,400 MEFs/cm^2). LDMEF = low-density mouse embryonic fibroblast, DCX = doublecortin.

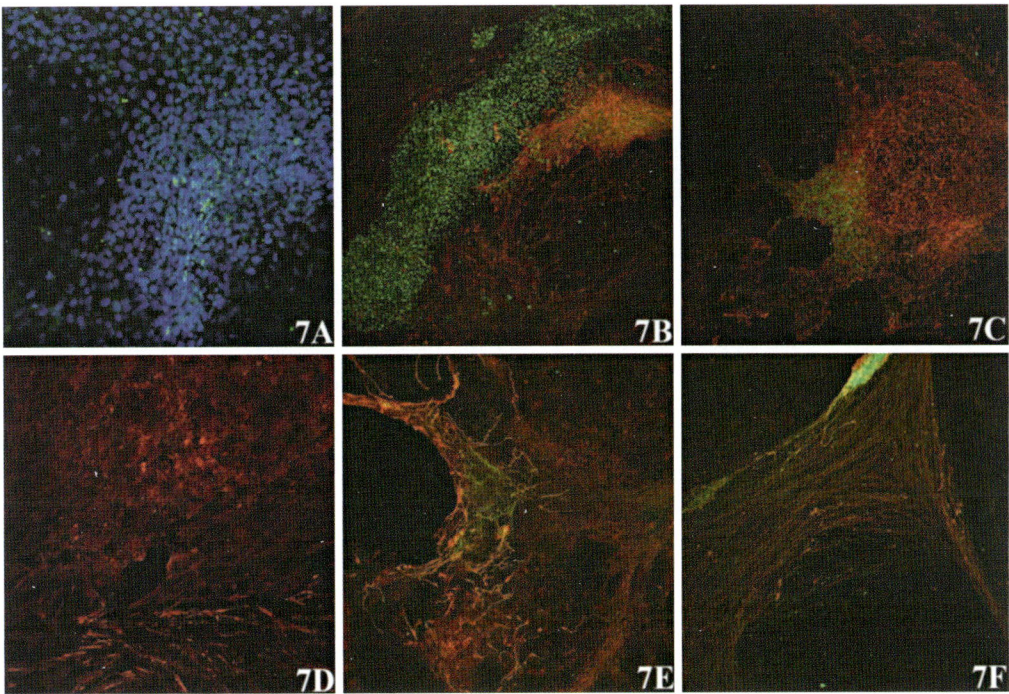

Fig. 4.7. H7s on LD- and NDMEF at days 15 and 30 in culture. (**A, B, C**) Nestin (*red*) and Pax-6 (*green*) expression. Sparse expression of nestin and Pax-6 at day 15 on LDMEF (**A**) (40X). Very diffuse staining for nestin and Pax-6 at day 30 on LDMEF (**B**) with decreasing expression at day 30 on NDMEF (**C**). H7/LDMEF at both days 15 and 30 demonstrated widespread expression of Class III β-tubulin (*red*) with more noticeable expression of DCX (*green*) at day 30 (**D, E**). Day 30 expression of Class III β-tubulin is more structured with cells containing long processes connecting to each other in networks. Similar expression of Class III β-tubulin and DCX is seen for H7/NDMEF at day 30 (**G**). All images except (**A**) using 10X objective. LDMEF = low-density mouse embryonic fibroblasts, NDMEF = normal density mouse embryonic fibroblasts, DCX = doublecortin.

6. Since the plastic coverslips used in this experiment tend to scatter fluorescent signals, a sandwich was made wherein the plastic coverslip was mounted on a glass microscope slide with the cells facing up. A second glass coverslip is then mounted using gelvatol over the cell surface. Double-labeling of cells was performed by diluting primary antibodies of different species together in CAS-Block and incubating simultaneously. Similarly appropriate secondary antibodies are also diluted and incubated together (*see* **Notes 11** and **12**).

3.8. Imaging and Analysis

1. Phase contrast: Phase contrast images were acquired through and processed using Nikon NIS-Elements D version 2.30, SP1 (Build 325) software (Nikon, Inc., Melville, NY, USA).

2. Fluorescence imaging: For laser-scanning confocal microscopy, pinhole aperture was set at 1.00 and kept constant for all imaging. Laser power was not adjusted for any image acquisitions (blue diode, 17%; 488, 41%; 543, 42%).

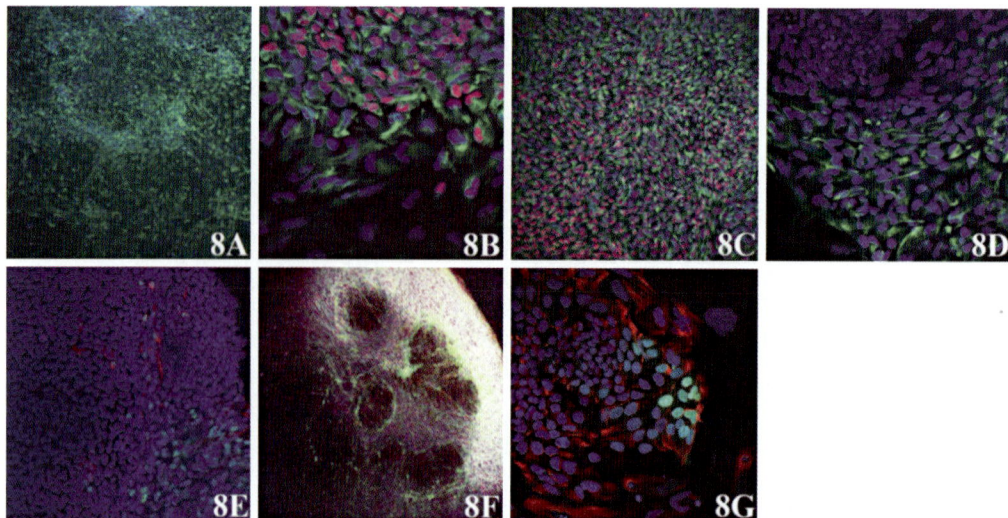

Fig. 4.8. HSF-6 hESC grown on LDMEF. (**A, B**) On 5.2K at day 7 in culture there is widespread nestin (*green*) and little to no Pax-6 (*red*) expression (**A**, 10X) with only focal expression of Pax-6 at the edges of colonies (**B**, 40X). By day 7, nestin expression is widespread particularly at the differentiated edges of colonies. However, when grown on 10.4K MEFs, Pax-6 expression peaks at day 7 (**C**, 20X) and very little is seen at day 12 at 10.4K MEF (**D**, 40X). Sparse staining for mature neuronal markers such as NeuN is seen at any density or time point and very little Class III β-tubulin is present at 10.4K MEF at day 7 (**E**, 20X). Widespread Class III β-tubulin and NCAM expression is seen by day 12 again most prevalent in cells grown on 10.4K MEF (**F**, 10X). Oct-3/4 staining is minimal at day 7 on 10.4K MEF (**G**, 40X). MEFs = mouse embryonic fibroblasts, MEF density given as cells/cm^2 (10.4K = 10,400 cells/cm^2).

Assessment for appropriate staining was done over the entire coverslip using epifluorescence (visual mode), and images were acquired using the laser-scanning feature (scan mode). Each channel was acquired sequentially to limit "crossover or bleedthrough" particularly between DAPI and FITC channels. Gain (PMT adjustment) for each channel (DAPI [UV/blue], FITC [green], TRITC[red], Cy5 [far red]) was adjusted to maximize but not oversaturate areas of appropriate staining. Z-position was adjusted to acquire the maximum number of cells in a given field. Line-averaging at the time of acquisition was used to reduce fluorescing pixels that did not correspond to true cellular fluorescence. A line-average of 4 or 8 was generally employed in most experiments. Most images were acquired in a 1024 × 1024 format, which produced an image size of approximately 3 MB (1 MB for each channel [blue,

Fig. 4.9. (continued) marker, brachyury (*red*, **D**) or extra-embryonic marker, GATA-4 (*red*, **E**). HSF-6 cells were differentiated on 0.1% gelatin for 42 days (**A**) or 36 days (**C**) before fixing and imaging. While both sox-1 and brachyury staining was observed within tight colonies, no individual cells outside of the groups exhibited staining for brachyury or GATA-4. H7 cells differentiated for 1 week also show high levels of nestin (*red*, **F**), typically over 60% of cells in culture. Rare cells were positive for the trophectodermal marker, CDX-2 (*red*, **G**).

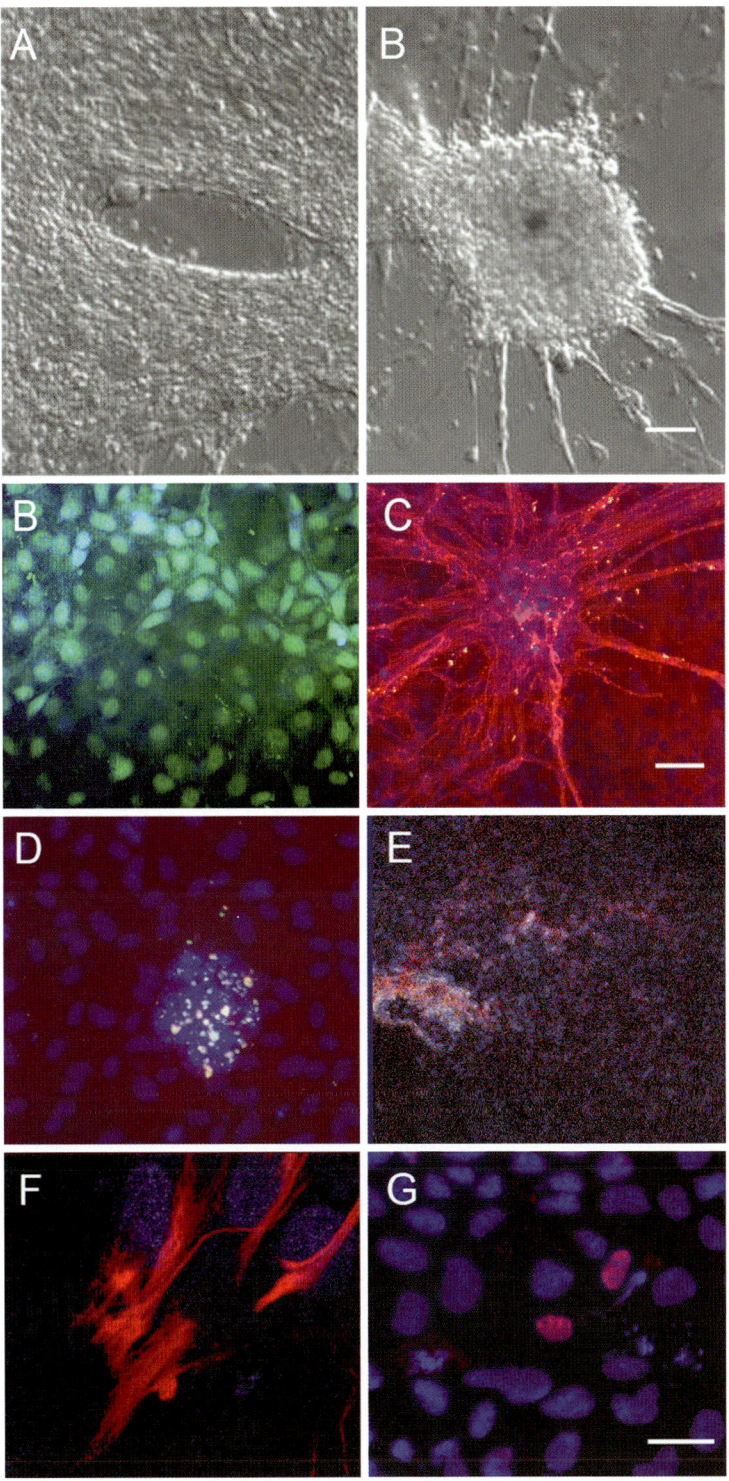

Fig. 4.9. Neurectodermal lineages from adherent HSF-6 hESC and H7 on low-density feeders (5.2 K cells/cm^2). HSF-6 cells cultured on low-density feeders form either epithelioid sheets (**A**) or foci with radial fibers (**B**). Colonies exhibit positive staining with neurectodermal markers sox-1 (*green*), mushashi-1 (*red*, **C**) and sparse staining with mesenchymal

green, red]). For semiquantitative analysis, entire coverslips were scanned at medium power to identify areas with the greatest numbers of cells staining for a particular neuroglial or pluripotency marker. For neural differentiation, images were generally taken using the 10X dry and 40X oil immersion objectives. Lower magnification images were used where fluorescence intensity was bright to demonstrate global staining pattern and numbers of cells staining for a particular marker. In the captions of figures, the objective magnification is listed. For details of the semiquantitative approach used previously, please refer to Ozolek et al. (20). For the analysis presented here, we used a more global approach commonly employed in the pathology literature where intensity is graded on a scale of 0–3 where 0 is no staining, 1+ is mild or low intensity, 2+ is moderate intensity, and 3+ is strong or bright intensity. Similarly, the percentage of cells staining is estimated on a 0–3 scale where 0 is no cells staining, 1+ is less than or equal to 10% of cells staining, 2+ is 10–50% of cells staining, and 3+ is greater than 50% cells staining. In **Table 4.2**, this is reflected as two numbers; the first representing intensity grading and the second representing percentage of cells staining. **Tables 4.1** and **4.3** show only the percentages of staining cells. EB thickness was obtained by using the Z-position function on the Leica confocal. The thickest portion of the EB was determined by DAPI intensity under epifluorescence conditions and the upper and lower limits obtained under scanning conditions by

Table 4.1
Semiquantitative analysis of neural markers at days 5 and 10 in differentiating H7 hESCs on low-density MEF

	D5		D10	
	10.4K MEF	15.6K MEF	10.4K MEF	15.6K MEF
Nestin	3*	3	2–3	2
β-tubulin	3	3	1	1
NeuN	0	0	Rare	0
DCX	0	0	2	2
Oct-3/4	2	1	1	2

Nestin shows early and widespread expression at days 5 and 10 while β-tubulin shows early expression that wanes by day 10. DCX expression is more prevalent at day 10. Oct-3/4 is present at both time points and feeder densities.
*0 = no cells positive for marker, 1 = ≤10% of cells staining, 2 = 10–50% of cells staining, 3 = >50% of cells staining; 10.4K/15.6K = 10,400 or 15,600 MEFs/cm^2; D5/D10 = day 5 or day 10 in culture after seeding on feeders; DCX = doublecortin.

Table 4.2
Semiquantitative analysis of neuroglial markers in H7 grown on low and normal density mouse embryonic fibroblasts

	EB-LDMEF-D15	EB-LDMEF-D30	EB-NDMEF-D15	EB-NDMEF-D30	EB-GEL-D15	EB-GEL-D30	H7-LDMEF-D15	H7-LDMEF-D30	H7-NDMEF-D15	H7-NDMEF-D30
Nestin	0-1/1*	2/3	0/0	2/3	1/1	2/3	0/0	2-3/3	0/0	1-3/3
Pax-6	2/1	1/1	1/1	2/2	2/1	2/1	1/1	2-3/3	2-3/2	2-3/1
GFAP	2/2	2/2	2/3	2-3/3	2/3	1-2/2	3/1	2-3/1	3/1	1-3/2
Oct-3/4	0/0	0/0	0/0	0/0	0/0	0/0	0/0	0/0	0/0	0/0
β-tubulin	2/3	2/3	2/3	2-3/3	2/3	2-3/3	1-3/2	1-3/2	2-3/2	1-3/3
DCX	1/1	2/2	2/3	2-3/2	2/3	2/3	1/1	2/2	1/1	1-3/2
Max EB thickness (μm)	41.7±21.5	32.3±2.2	40.9±6.3	33.0±6.6	47.5±9.0[a]	32.2±4.8[a]	—	—	—	—

For EBs on low or normal density feeders, the most widespread expression of the neuroglial markers examined was in both LDMEF and NDMEF at day 30 with nestin, GFAP, β-tubulin, and DCX expression quite diffuse. This pattern was also seen in EBs grown on gelatin alone. It is interesting that nestin appears to show more intense and widespread expression at day 30 rather than earlier at day 15 and GFAP and β-tubulin expression is high and constant at both day 15 and 30. This pattern was similar in the H7 grown on LDMEF and NDMEF at day 30 with the added high level expression of Pax-6 at day 30. Not surprisingly, EBs at day 15 had greater maximum thickness with the difference for EBs grown on gelatin being statistically significant [a]. No Oct-3/4 staining was present.

* (A/B): A = Staining intensity: 0 = no staining, 1 = mild intensity, 2 = moderate intensity, 3 = strong intensity; B = Number of cells staining: 0 = no cells, 1 = ≤10%, 2 = 10–50%, 3 = >50%; LDMEF = low-density mouse embryonic fibroblasts at 5.2K cells/cm^2; NDMEF = normal-density mouse embryonic fibroblasts at 18.5K cells/cm^2, D15/D30 = day 15 or day 30 in culture after seeding on feeders, EB = embryoid body, GFAP = glial fibrillary acidic protein, DCX = doublecortin.

Table 4.3
Semiquantitative analysis of neuroglial markers at days 7, 12, and 19 in differentiating HSF6 hESCs grown on low-density mouse embryonic fibroblasts

	D7			D12			D19		
	5.2K	10.4K	15.6K	5.2K	10.4K	15.6K	5.2K	10.4K	15.6K
Nestin	3*	3	3	3	3	3	3	2	3
Oct-3/4	0–1+ (rare)	0–1+ (rare)	0–1+	0	0	0	0	0	0
Pax-6	2	2	2	0	1	1	0	0	0
β-tubulin	1	1	1	2	3	3	2	3	2
NCAM	1	1	1	1	2	3	2	3	3
NeuN	0	1	0	1	1	1	1	1	1

Differentiating cells demonstrated widespread nestin expression at all time points and densities. Oct-3/4 was seen rarely at day 7 and not at all at days 12 or 19. Pax-6 was present only at day 7 while β-tubulin and NCAM expression was lower at day 7 but more widespread by days 12 and 19 being seen in more than 50% of cells. NeuN followed a similar pattern to β-tubulin and NCAM but was seen in only a few cells usually in association with Class III β-tubulin cytoplasmic staining.
*0 = no cells positive for marker, 1 = ≤10% of cells staining, 2 = 10–50% of cells staining, 3 = >50% of cells staining; 5.2K/10.4K/15.6K = 5,200/10,400/15,600 cells/cm^2, respectively; D7/D12/D19 = day 7, 12, 19 in culture after seeding on feeders; NCAM = neural cell adhesion molecule.

determining the plane where the nuclei of cells could barely be discerned. Once the limits were determined, the spatial difference on the Z-axis between the upper and lower limits was the Z-thickness (in microns, μm) corresponding to the EB thickness. All images were imported into Adobe Photoshop CS2 for final image composition, resizing, and adjustments (Please see detail below). Most images were adjusted using only the "Auto Levels" function to enhance brightness and contrast. Some of the lower magnification images required additional brightness and contrast adjustments.

When using the spinning disk confocal, some general principles are important for acquiring images of good quality that portray biological significance. Image intensity depends in sequence on (1) illumination intensity, (2) density and quantum efficiency of sample label, (3) detector gain and exposure, and (4) digital display settings. The best image is produced by adjusting parameters 1–4 in sequence. First ensure that the excitation light is as bright as the sample will take, given limitations on photobleaching and photodamage of the sample. This step is essential to reduce photon and statistical noise in the image that can be produced if unnecessary camera gain is used to compensate for a dim sample. Next, optimization of sample staining is key. Sample signal is influenced by density of antigen, antibody binding, choice of fluorophore, and antibleaching agents. Magnification must be matched to the resolution of the detector. We use a 40x oil immersion objective. The highest point brightness is obtained in practice with the highest numerical aperture objective. Among high NA objectives, the lowest magnification objective produces the highest point brightness. Lower magnification objectives also increase the field of view, giving large sample selection per image. Amplification of signal by the camera can be modified, but in practice, image acquisition is standardized with high gain and 2×2 binning, given sample intensity and illumination. If sample brightness permits, 1×1 binning produces higher resolution image that can be cropped and enlarged for final presentation. Twelve-bit cameras produce grey levels from 0 to 4095. Noise from the camera is approximately 100 grey levels. The image quality is related to the ratio of the signal level to noise level. Adjust exposure time to produce a peak signal between 1000 and 2000 grey levels. Highlights should be below 4000 [(If underexposed can increase exposure time (i.e., from 1 s to 8 s), if overexposed can decrease exposure time (i.e., from 1 s to 300 ms)]. Quantitative comparison of image intensity is difficult since illumination, sample preparation, and system gain must be standardized – a nontrivial task. However, comparisons within a given experiment are relatively easy. Adjust the exposure so the brightest portion of the sample is not saturated and maintain constant light, exposure, and display settings.

3. Image Analysis: Data presentation in montages is critical to telling the visual story. Images are archived by date, cell type, treatment, and immunostain. Images are reviewed with the shareware program Irfranview, which produces thumbnail views and quickly scrolls through high-resolution views of many image formats including 16-bit tiff or raw images and directly links to photoediting software such as Adobe Photoshop. After selecting the set of images, a standard frame is selected and each image is cropped to that frame size by copying a fixed marquee size. An extra large canvas is created and each image is pasted in a separate layer and positioned on the canvas. Snap to grid allows positioning with uniform white borders. Zooming in can allow correction of border width. Final canvas size is created by cropping around the set of images. The montage of unedited images is saved and further changes in labeling and contrast are made on a copy of the file. Image resolution for publication is 300 dots per inch (dpi) with an image width from 2 to 8 in. Photoshop default resolution is

set to the computer display, 72 dpi. Under images/image size, uncheck resample image and change resolution from 72 to 300 and make note of the physical image width. The resolution of primary images and the requirement for 300 dpi final resolution will limit the degree of cropping and enlargement that can be produced. Further enlargement is produced by interpolation between real pixels and should be avoided. Contrast is adjusted by using the levels command. If intensity is to be compared among panels, the contrast adjustment must be standardized. If morphology of different intensity areas is to be compared contrast can be adjusted differently. Start on the brightest image and adjust the upper display maximum to the brightest grey level to avoid clipping. Set the display minimum for the dimmest image to avoid clipping the shadows. Dark areas between cells can contain some absolute black, but avoid too much clipping of the black level at edges of cells. The midlevel intensity is adjusted with the grey slider in the levels dialog box. The optimum is often within the bright shoulder of the intensity histogram. Adjusting the middle level produces a nonlinear contrast adjustment, which is acceptable to major journals as long as the adjustment is applied uniformly to each panel. Adjusting subregion contrast differently (dodging and burning) is no longer acceptable image processing. Other adjustments such as median filtering to reduce speckle noise or unmask sharpening to enhance edges in low-contrast images are acceptable only if specifically described in the methods. Magnification bars are prepared from images of a stage micrometer containing calibrated lines at 10 μm intervals. Use the marquee tool to measure the magnification in μm/pixel or μm/total image width. The width of cropped panels can be determined by the ratio of panel width to full image width in pixels. The line tool is used to create magnification bar about 1/5 of the panel width.

4. Notes

1. When flasks are becoming 80–90% confluent and still in the log growth phase, it is time to freeze them. In general, this happens about the second day after preparing the embryos, but may occur sooner or later. MEFs should be thawed at least one week before they are needed for co-culture with hESCs. Medium exchange is necessary to remove any residual DMSO and cellular debris present after cryopreservation.

2. MEF proliferation is halted by mitomycin-C treatment. This is to prevent overgrowth of the MEF when co-cultured with hESC.

3. We use a hemocytometer with an Improved Neubauer ruling pattern. For density measurements, take number of MEFs required per cm^2 and multiply by area of culture well bottom (culture well dimension can be obtained at Corning website in a PDF format (www.corning.com/lifesciences/products_services/product_literature/us_canada_index.asp#Lab_plasticware). Take this total number of MEFs and divide by the density of MEFs obtained by the hemocytometer (cells/ml) to establish volume of mitomycin-treated MEFs needed to seed well.

4. For later immunophenotyping and microscopy, growing the cells on a coverslip is convenient.

5. Since MEFs are primary cells they have a limited lifespan in culture. If they begin to elongate and doubling time increases significantly, they are beyond their useful state. They need to be carefully monitored to avoid over growing the culture, which results in early senescence. We have also found that using fresh MEFs (less than passage 12) promotes better and more uniform growth of hESCs under pluripotent conditions and neural differentiation under low-density conditions.

6. Initial HSF-6 experiments (*see* reference *20*) did not use supplemental bFGF (basic fibroblast growth factor) in the hESC medium. Subsequent experiments with H7 hESC used supplemental 4 ng/ml bFGF. The addition of bFGF provided more stable pluripotency in colonies. No other supplements (i.e., retinoic acid, noggin, sonic hedgehog protein) were added.

7. Colonies are ready as determined by size (generally colonies 1.5–2.5 mm in diameter) and homogeneity (colony composed of small indistinguishable cells with minimal numbers of larger differentiating cells at the edges (Please see phase contrasts images, **Fig. 4.4**)).

8. HESCs do not adhere or grow well on glass coverslips, hence the use of the plastic coverslips. Using these allows for direct transfer to a glass slide after immunocytochemistry for immediate microscopy.

9. Antibodies were carefully chosen to reflect a progression of neural maturation: neuroepithelial – neural stem cell – maturing and migrating neurons – mature neurons. Anti-β-Tubulin Class III is a maturing and mature neural marker; expected staining pattern: cytoplasmic, filamentous; positive control: fetal brain-13 weeks. Anti-Doublecortin is a marker of migrating neuroblasts; expected staining pattern: cytoplasmic, filamentous;

positive control: fetal brain-13 weeks. Anti-Glial Fibrillary Acidic Protein is a mature astroglial marker; expected staining pattern: cytoplasmic, filamentous; positive control: fetal brain-13 weeks. Anti-NCAM (neural cell adhesion molecule) is a maturing and mature neural marker; expected staining pattern: cytoplasmic, filamentous; positive control: fetal brain-13 weeks. Anti-Nestin is a neuroepithelial/neural stem marker; expected staining pattern: cytoplasmic, filamentous; positive control: fetal brain-13 weeks. Anti-Neuronal Nuclei (NeuN) is a mature neural marker; expected staining pattern: nuclear and cytoplasmic; positive control: fetal brain-13 weeks. Anti-Pax-6 is a neuroepithelial marker/ventral neuroectoderm; expected staining pattern: nuclear; positive control: fetal brain-13 weeks. Anti-Oct-3/4 is pluripotency marker; expected staining pattern: nuclear; positive control: pluripotent H7 hESC. Anti-Nanog is a pluripotency marker; expected staining pattern: nuclear; positive control: pluripotent H7 hESC.

10. TOTO-3 iodide is a long-wavelength red spectrum fluorophore (642/660 excitation/emission spectrum) for nuclear detection. If TOTO®-3 is used, Hoechst is still added or alternatively an antifade mounting medium containing DAPI is added (such as Vectashield with DAPI from Vector Laboratories) so that the nuclei are visible under epifluorescence parameters.

11. Store slides at 4°C in the dark in a sealed slide box. In the past, we have added DABCO as an antifade agent to the gelvatol, although, in our experience this has caused increased background scatter of fluorescent light. Other antifade reagents can be purchased through Molecular Probes or other vendors and may not be necessary depending on the type of microscopy to be performed. For general scanning and capture of images, an antifade is probably not necessary. For example, the Alexa Fluor® conjugates available through Molecular Probes have a long shelf life particularly when slides are properly stored in slide boxes with the lids tightly closed at 4°C.

12. All immunocytochemical analyses were performed with the appropriate positive and negative controls listed above. Negative controls consisted of incubating tissue/cells with labeled secondary antibodies only. All stain positive and negative controls in these studies were appropriate. Controls are critical for accurate and consistent analysis using immunocytochemical or immunohistochemical staining particularly when separate experiments are done over extended time periods as primary and secondary antibodies lose specificity and fluorescence and precipitate leading to potentially false positive and negative results.

References

1. Bliss, T., Guzman, R., Daadi, M., Steinberg, G.K. (2007) Cell transplantation therapy for stroke. *Stroke* **38**, 817–26.
2. Kondziolka, D., Wechsler, L. (2008) Stroke repair with cell transplantation: neuronal cells, neuroprogenitor cells, and stem cells. *Neurosurg Focus* **24**, E13.
3. Bambakidis, N.C., Butler, J., Horn, E.M., Wang, X., Preul, M.C., Theodore, N., Spetzler, R.F., Sonntag, V.K. (2008) Stem cell biology and its therapeutic applications in the setting of spinal cord injury. *Neurosurg Focus* **24**, E20.
4. Coutts, M., Keirstead, H.S. (2008) Stem cells for the treatment of spinal cord injury. *Exp Neurol* **209**, 368–77.
5. Fitzgerald, J., Fawcett, J. (2007) Repair in the central nervous system. *J Bone Joint Surg Br* **89**, 1413–20.
6. Takeuchi, H., Natsume, A., Wakabayashi, T., Aoshima, C., Shimato, S., Ito, M., Ishii, J., Maeda, Y., Hara, M., Kim, S.U., Yoshida, J. (2007) Intravenously transplanted human neural stem cells migrate to the injured spinal cord in adult mice in an SDF-1- and HGF-dependent manner. *Neurosci Lett* **426**, 69–74.
7. Gershon, M.D. (2007) Transplanting the enteric nervous system: a step closer to treatment for aganglionosis. *Gut* **56**, 459–61.
8. Bjorklund, A., Dunnett, S.B., Brundin, P., Stoessl, A.J., Freed, C.R., Breeze, R.E., Levivier, M., Peschanski, M., Studer, L., Barker, R. (2003) Neural transplantation for the treatment of Parkinson's disease. *Lancet Neurol* **2**, 437–45.
9. Dass, B., Olanow, C.W., Kordower, J.H. (2006) Gene transfer of trophic factors and stem cell grafting as treatments for Parkinson's disease. *Neurology* **66**, S89–103.
10. Newman, M.B., Bakay, R.A. (2008) Therapeutic potentials of human embryonic stem cells in Parkinson's disease. *Neurotherapeutics* **5**, 237–51.
11. Takahashi, J. (2006) Stem cell therapy for Parkinson's disease. *Ernst Schering Res Found Workshop*, 229–44.
12. Yang, D., Zhang, Z.J., Oldenburg, M., Ayala, M., Zhang, S.C. (2008) Human embryonic stem cell-derived dopaminergic neurons reverse functional deficit in parkinsonian rats. *Stem Cells* **26**, 55–63.
13. Heese, K., Low, J.W., Inoue, N. (2006) Nerve growth factor, neural stem cells and Alzheimer's disease. *Neurosignals* **15**, 1–12.
14. Baharvand, H., Mehrjardi, N.Z., Hatami, M., Kiani, S., Rao, M., Haghighi, M.M. (2007) Neural differentiation from human embryonic stem cells in a defined adherent culture condition. *Int J Dev Biol* **51**, 371–8.
15. Dottori, M., Pera, M.F. (2008) Neural differentiation of human embryonic stem cells. *Methods Mol Biol* **438**, 19–30.
16. Nat, R., Hovatta, O. (2004) In vitro neural differentiation of human embryonic stem cells. *J Cell Mol Med* **8**, 570–1.
17. Nat, R., Nilbratt, M., Narkilahti, S., Winblad, B., Hovatta, O., Nordberg, A. (2007) Neurogenic neuroepithelial and radial glial cells generated from six human embryonic stem cell lines in serum-free suspension and adherent cultures. *Glia* **55**, 385–99.
18. Zhou, J.M., Chu, J.X., Chen, X.J. (2008) An improved protocol that induces human embryonic stem cells to differentiate into neural cells in vitro. *Cell Biol Int* **32**, 80–5.
19. Heng, B.C., Liu, H., Cao, T. (2004) Feeder cell density – a key parameter in human embryonic stem cell culture. *In Vitro Cell Dev Biol Anim* **40**, 255–7.
20. Ozolek, J.A., Jane, E.P., Krowsoski, L., Sammak, P.J. (2007) Human embryonic stem cells (HSF-6) show greater proliferation and apoptoses when grown on glioblastoma cells than mouse embryonic fibroblasts at day 19 in culture: comparison of proliferation, survival, and neural differentiation on two different feeder cell types. *Stem Cells Dev* **16**, 403–12.

Chapter 5

Optimization of Physiological Xenofree Molecularly Defined Media and Matrices to Maintain Human Embryonic Stem Cell Pluripotency

Isabelle Peiffer, Romain Barbet, Antoinette Hatzfeld, Ma-Lin Li, and Jacques A. Hatzfeld

Abstract

We describe in this chapter the development of a xenofree molecularly defined medium, SBX, associated with xenofree matrices, to maintain human embryonic stem cell (hESC) pluripotency as determined by phenotypic, functional and TLDA studies. This simple, inexpensive, and more physiological culture condition has been chosen because (1) it is xenofree and molecularly defined; it is devoid of albumin, which is a carrier of undefined molecules; (2) it maintains pluripotency, but very significantly reduces differentiation gene expression during hESC self-renewal, as compared to the widely used culture conditions tested so far; and (3) it can be further improved by replacing high concentrations of expensive additives by physiological concentrations of new factors. Xenofree molecularly defined media and matrices represent valuable tools for elucidating still unknown functions of numerous embryonic genes using more physiological culture conditions. These genes encode potential new factors controlling hESC self-renewal and pluripotency.

Key words: Defined culture media optimization, physiological xenofree culture media, matrices, pluripotency, transcriptomics, human embryonic stem cells, human mesenchymal stem cells, clinical grade culture media.

1. Introduction

Optimization of physiological, xenofree, molecularly defined culture conditions for hESCs or induced pluripotent stem (iPS) cells is an important step, which will permit further characterization of genes controlling self-renewal and pluripotency. The functions of about 40% of the genes expressed by hESCs are still unknown (1). These genes might encode for important cytokines, receptors, extracellular matrix (ECM) components, or transcription factors. Any long-term strategy to grow hESCs in physiological conditions must take into account the existence of these currently unknown molecules.

Many studies have been performed to optimize hESC culture conditions (*see* review (2)). Xenogenic cell-feeder layers have been replaced by extracellular matrices of MEF (3), by various types of human cells (4–15), or by other extracellular matrices such as xenogenic matrigel (16), ECM from human serum (7, 17), or defined ECM (18–23, 33). Fetal calf serum has been replaced by serum replacement (3, 5, 8–11, 19, 24–28), complemented with high physiological concentrations of cytokines (20, 29–32). Some so-called "defined media" have also been described (21, 22, 26, 33–36). However, none of these media are molecularly defined, as they contain albumin in various forms, which are not recombinant, and may carry undefined impurities including cytokines, hormones, or growth inhibitors (37). Serum, albumin, or physiological concentrations of cytokines can promote physiological expression or down-modulation of receptors (38).

In the strategy we present here to use and develop further physiological molecularly defined media for hESCs, we first prepared and used ECM produced by hMSCs as a reference to replace them progressively by molecularly defined matrices. Serum albumin-containing media are replaced by a xenofree, molecularly defined liquid medium without albumin and transferrin. This medium called SBX contains only defined components, including synthetic lipid and iron carriers, which do not introduce any undefined molecules to the culture.

However, xenofree molecularly defined media alone are not sufficient to define physiologically relevant culture conditions for pluripotent hESCs or iPS. It is indeed important to use Taqman Low Density Array cards with genes representative of (a) hESC self-renewal; (b) commitment and differentiation to the three embryonic and extra-embryonic germ layers; and (c) other important genes controlling human embryonic development, such as the TGF-β superfamily and related genes. This strategy has been described recently (39) to study the physiological role of the TGF-β superfamily to control hESC pluripotency. We recommend this strategy, in addition to the method described in this chapter, for those who would like to study new genes or molecules to improve pluripotent self-renewal together with the lowest possible level of differentiation. We will mention in the notes the necessary interaction between the development of a physiological xenofree molecularly defined medium and TLDA cards with well-selected genes (*see* **Notes 1 and 10**).

2. Materials

2.1. Chemicals and Plastic

1. 200 mM L-glutamine, 100X (Invitrogen). Before use, make sure the white precipitate, visible when L-glutamine is thawed, is completely dissolved. If necessary, warm up the L-glutamine solution for about 5–10 min in a 37°C water bath (*see* **Note 2**). Aliquot and store at −20°C.
2. Phosphate buffered saline, 1X PBS, without $CaCl_2$ and $MgCl_2$ (Invitrogen).
3. Cell lysis buffer (SBL, AxCell, France).
4. Mitomycin C (Sigma). This is toxic and should be handled with gloves. Stock solution: 2 mg/ml in PBS. Mitomycin C, when in solution, is extremely light sensitive. Liquid aliquots should be wrapped in aluminum foil and can be stored at 4°C for up to 4 weeks.
5. Collagenase type IV (Invitrogen) 1 mg/ml (100 µl) + 100 mM $CaCl_2$ (Sigma) (100 µl) + 9.8 ml PBS for 10 ml.
6. Standard medium for routine maintenance of human mesenchymal stem cells (hMSC): MEMα (minimum essential medium) (Invitrogen) supplemented with 10% FBS (fetal bovin serum) (Biowest), L-glutamine, and 1 ng/ml bFGF. Protect the medium from light to prevent peroxide formation.
7. Trypsin-ethylenediaminetetraacetic acid (EDTA): 0.25% trypsin-EDTA (Invitrogen) is diluted 1:5 in PBS to passage hMSCs.
8. Soybean trypsin inhibitor (Worthington) used according to the manufacturer's procedure.
9. SBX-M: xenofree molecularly defined medium for hMSCs (AxCell, France) (*see* **Note 3**).
10. SBX: xenofree molecularly defined medium for hESCs (AxCell, France) (*see* **Note 3**).
11. Human basic fibroblast growth factor (bFGF) (Peprotech).
12. Activin A (Peprotech). SBX-A: SBX supplemented with 10 ng/ml Activin A.
13. TGF-β (R&D). SBX-T: SBX supplemented with 100 pg TGF-β/ml.
14. 50 ml polypropylene conical tubes (Falcon).
15. 14 ml polypropylene round-bottom tubes (Falcon).
16. Multiwell 6-well, 12-well, and 24-well tissue culture-treated plates (Falcon).
17. Tissue culture flasks: 25 cm^2, 75 cm^2, and 150 cm^2 (TPP).

2.2. Cells

1. hESCs: start with an easy-growing hESC line without chromosomal abnormalities (*see* **Note 4**).
2. hMSCs (*see* **Note 5**).

2.3. Transcriptomics

1. RNeasy kit (Qiagen). Store RNAs at −80°C.
2. High Capacity cDNA Archive Kit (Applied Biosystems). Store cDNAs at −20°C.

3. Methods

3.1. hESC or iPS lines

To become familiar with the method, choose an easy-growing hESC or iPS line without chromosomal abnormalities (*see* **Note 4**) and with closely knit growing cells.

3.2. hMSC Matrix Preparation

1. Prepare your hMSC stock culture yourself or buy hMSCs (*see* **Note 5**).
2. Plate the cells in 150 cm^2 flasks at a cell density of 2×10^4 cells/cm^2 and leave the flasks at 37°C in a 5% CO$_2$ incubator until subconfluency (2–3 days).
3. Treat subconfluent hMSC cultures with mitomycin C. Dilute the stock solution in the culture medium to obtain a final concentration of 10 µg/ml. Use reduced light when adding the stock solution. Leave in contact with hMSCs for 2 h 30 min in the CO$_2$ incubator. The mitomycin C-containing supernatant should be discarded as a toxic waste.
4. Wash the culture three times with PBS.
5. Add 3 ml 0.05% trypsin-EDTA and incubate at 37°C until the cells detach. Then stop trypsin action by adding soybean trypsin inhibitor.
6. Collect the cells in a tube and centrifuge at 200g for 5 min.
7. Aspirate the supernatant and resuspend the cells in 10 ml of SBX-M.
8. Perform cell count using a Malassez chamber after staining with 0.5% trypan blue.
9. Plate mitomycin C-inactivated hMSCs at the concentration of 2.5×10^5 cells/cm^2 (*see* **Note 5**) in standard hMSC medium or SBX-M.
10. Equilibrate in SBX medium (*see* **Note 3**) for one night.
11. Wash once with double distilled water and treat with the SBL lysis buffer (SBL, AxCell, France) for 10 min.
12. Wash twice with PBS without damaging the matrix.

13. Store the plates with PBS or SBX at 4°C for a few days or use the plate immediately after one more wash with SBX at room temperature.

3.3. Bulk Scraping Method (BS) of hESC Culture

3.3.1. If There Are Differentiated Cells in the Culture

1. Wash the hESC plates with PBS.
2. Use collagenase solution for less than 10 min. Monitor the collagenase treatment progress under an inverted microscope with high magnification.
3. As soon as differentiated cells start to detach, stop the reaction by washing with PBS, SBX, or a collagenase inhibitor. Eliminate the differentiated cells. Add SBX medium. Then follow the procedure for the undifferentiated cells.

3.3.2. If There Are Very Few Differentiated Cells

Collagenase treatment is avoided and the BS method becomes equivalent to a fast cut-and-paste method without enzymatic treatment (see **Note 6**) (**Fig. 5.1c**).

1. In order to obtain large colony fragments, scrape off the culture by scoring the entire surface with a 10-ml pipette held perpendicularly to the plate (using a to and fro movement in one direction followed by the same movement at 90° to the first).
2. Aspirate gently the suspension of colony fragments with a 10-ml pipette. (Save the plates after scraping and add culture medium. Put these plates in the incubator to obtain "Regenerated Cultures" after 1 week with daily feeding (see **Note 7**).)
3. To seed new plates, pour gently the suspension of colony fragments into a 15- or 50-ml tube.
4. Leave the tube in upright position for 1–3 min on the bench, the shorter, the better. The large colony fragments deposit first.
5. Remove immediately and gently the supernatant containing single cells and small colony fragments (less than 300 cells) as soon the large ones have deposited. Leave the pellet with large pieces.
6. Re-suspend carefully the large fragments in enough medium to seed 2–3 plates out of one. When the method becomes well adapted to your cell line, you will be able to seed 6 plates out of 1 (see **Note 8**).
7. Put the plates in a humidified incubator equilibrated at 37°C and 5% CO_2, with perfectly horizontal trays and no vibrations.
8. Change the SBX medium daily for 1 week before passaging the cells.

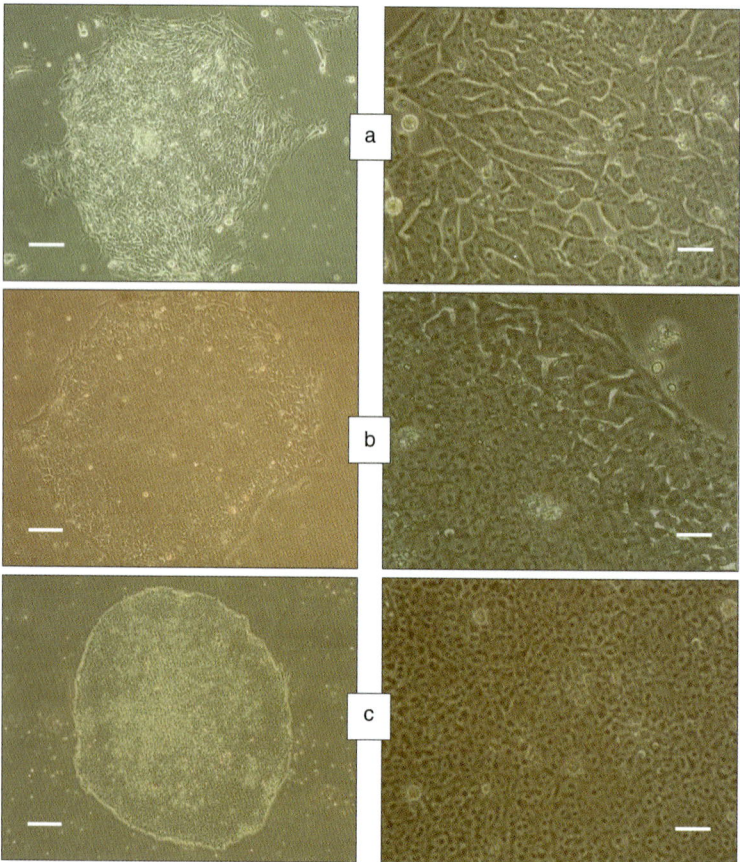

Fig. 5.1. hESC colony morphology at different steps of SBX and xenofree matrix optimization. *Bars*: 250 μm (*left panels*), 30 μm (*right panels*). **a**. With SBX without Activin-A or low TGF-β concentration, colonies are clearly differentiating on a defined matrix with laminin, collagen IV, fibronectin, and vitronectin. **b**. With SBX-A and a optimized molecularly defined matrix (laminin, collagen IV, fibronectin, and vitronectin), the rim of the colony presents differentiated cells, which can be eliminated by a short, mild collagenase treatment. **c**. Typical morphology of undifferentiated pluripotent hESCs self-renewing on hMSC matrix with SBX-A: no enzymatic treatment is required before bulk scraping (BS) passaging. TLDA analysis demonstrates a good correlation between a low commitment/differentiation gene expression and the colony morphology, with small closely knit cells with a prominent nucleolus and a high nuclear/cytoplasmic ratio (39).

3.4. Checking for Pluripotency

3.4.1. Morphology

This is the easiest and fastest control to do (*see* **Note 9**).

Look for critical details: the small (12 μm), closely knit cells with no spaces between them, with very high nucleo-cytoplasmic ratio, very little cytoplasm and a huge, dark, nucleolus (**Fig. 5.1c**). The colonies should have a clear-cut edge and be constituted, as far as possible, by small cells all of the same type. Larger cells with spaces between them and a small nucleo-cytoplasmic ratio represent a hallmark of differentiation (**Fig. 5.1a and b**), which can be confirmed more easily by TLDA than by microarrays (*see* **Note 10**). If the morphology is satisfactory, checking pluripotency markers should give the expected results.

3.4.2. Taqman Low Density Arrays

We recommend Taqman Low Density Arrays (TLDA, Applied Biosystems) for their sensitivity and the possibility to check rapidly a representative selection of gene markers for hESC self-renewal or commitment/differentiation toward all embryonic and extra-embryonic germ layers (39) (*see* **Note 10**).

1. Prepare extracts and purify total RNA with the RNeasy kit from Qiagen, following the manufacturer's instructions, and store RNA at −80°C until use.
2. Prepare cDNA with the cDNA Archive kit, following the manufacturer's instructions and store cDNA samples at −20°C.
3. Establish the list of the genes to be studied or use the TLDA cards and program already published (39) (*see* **Note 1**).

3.4.3. Membrane, Cytoplasmic, and Nuclear Markers

More information will be obtained by checking accepted markers of pluripotency such as SSEA-3, SSEA-4, TRA-1-60, TRA1-80, OCT4, or alkaline phosphatase staining (40) (*see* **Note 11**).

4. Notes

1. Queries or suggestions concerning SBX media, the methodology to optimize these media, collaborations to test new molecules, requests for the TLDA program (*see* **Note 10**), production of clinical grades, etc., can be mailed to the academic address: sbxa2000@yahoo.com.
2. When preparing stock solutions for new molecules, take care to dissolve the product correctly in the appropriate solvent, at the right pH and temperature (check in the Merck Index). Never shake solutions containing ECM molecules – they will stick together.
3. SBX xenofree molecularly defined liquid media.
   A serum-free medium is not necessarily a defined medium, and a medium which contains recombinant bovine albumin is neither xenofree nor defined – it varies from batch to batch. A xenofree molecularly defined medium contains no animal molecules, no serum, and no carriers, such as albumin, containing undefined molecules. All the components are defined.
   All SBX media are devoid of albumin and transferrin. These lipid or iron carriers bring into the culture undefined amounts of hormones, growth factors or their inhibitors. In contrast,

all SBX media contain a synthetic carrier to which defined fatty acids and cholesterol are added. Other defined lipid components may be added on request. Transferrin is replaced by a synthetic iron carrier.

SBX and SBX-M both contain recombinant human insulin and b-FGF, HEPES, selenium, pyruvate, 2-dithiothreitol, and glutamine.

SBX media may differ by the basal liquid medium: DMEM-F12 for hESCs (SBX) or MEMα with nucleosides for hMSCs (SBX-M). They may also differ by the addition of specific cytokines and hormones, NEAA, and specific rare metals.

In this study, as a starting medium for hESCs, we recommend the SBX medium to which addition of 10 ng/ml Activin A (SBX-A) provides a satisfactory self-renewal with low differentiation. However, we suggest improving the composition of SBX by replacing the high, physiological concentration of Activin A, which is expensive, by physiological concentrations (100 pg/ml) of TGF-β, which is cheaper. This might require optimizing other cytokines or ECM molecules. ECM components are important to synergize with cytokines. The xenofree nondefined hMSC-derived matrix provides a good reference matrix to develop new xenofree molecularly defined matrices.

SBX is a medium which has been developed to allow each experimenter to test new molecules. Depending on the molecules to be tested, Axcell will provide the appropriate SBX liquid medium. SBX media can be developed on request for cell therapy, following the international regulations (*see* **Note 1**).

4. Do not use a late-passaged cell line, especially if it has been treated enzymatically for passaging. Such lines easily become transformed. Passage the cells by the Bulk Scraping (BS) procedure described above, which maintains the integrity of large groups of cells, similarly to the cut-and-paste method (39). Practice the BS procedure with the culture condition to which the cell line has adapted, before passaging the cells with SBX and xenofree matrices.

5. You may also try various types of human cell lines to produce new matrices. Mitomycin C treatment can thus be avoided, since the cells will be lysed. Test various cell densities to be lysed in order to prepare an efficient matrix. In our hands, hMSC matrices with SBX-A medium have provided as good a colony morphology and twice less differentiation gene expression (as demonstrated by TLDA) than MEF feeders with SR medium. hMSC matrices represent therefore a good reference xenofree matrix to develop further molecularly defined matrices.

6. This technique represents the best way to self-renew hESCs for years without inducing chromosomal abnormalities.

7. "Regenerated cultures".

    We report here a "trick" to increase the hESC harvest when the BS method is used without collegenase treatment. We observed that if we re-feed daily the culture plates after removal of colony pieces, the empty colony spaces are repopulated within 1 week by a homogenous, thin layer of regenerated hESCs. Cell morphology, phenotypic characteristics (not shown), and differentiation gene expression (as demonstrated by TLDA) of the regenerated cultures are similar to those of the culture from which they regenerate.

8. Beginners have a strong tendency, when passaging, to seed colony fragments that are too small or in insufficient amounts. We strongly suggest that, when adapting the method to your cell line, you plate very large pieces with over 5000 cells. Seeding too many plates out of one reduces growth and increases differentiation.

9. To obtain the maximum of information from an inverted microscope in phase contrast at magnifications 20× and 40×, it is extremely important that the microscope be correctly adjusted (this is rarely the case in cell culture laboratories). Moreover, the microscope setting should be checked every day. This only takes a few seconds, once one is familiar with the operation. This will save a lot of time and enables one to observe important details.

    The correct handling of a good inverted microscope in phase contrast is not only important to check the quality of pluripotency. By observing the membrane morphology, quality of cell adhesion, growth rate, etc., you will be able to deduce which kind of new molecule to add: other lipids, ECM components, growth factors, etc. SBX medium is perfect to improve synergies between new molecules and assess their physiological significance by phenotypic, functional and TLDA studies.

10. In a recent publication (39), we have provided a list of 96 genes, including a calibrator gene. These genes are representative of hESC self-renewal, commitment, and differentiation toward the 3 embryonic germ layers and the trophoblast. They also include genes of the TGF-β, Wnt, Notch FGF, and ECM families and genes involved in gametogenesis and apoptosis. TLDA was used either to check hESC expression of self-renewal and commitment/differentiation genes, or to perform a transcriptomic assay of pluripotency with embryoïd bodies (EBs) (39). We clearly demonstrated that commitment/differentiation gene expression was reduced by 50% during hESC self-renewal on xenofree matrices with SBX

medium as compared to hESC culture on MEF feeders with SR medium. The TLDA we described, together with a computer program to compare experimental samples thanks to calibrator genes and reference samples, represents a powerful tool to study the effect of new molecules in the control of hESC or iPS self-renewal and pluripotency.

11. Remember that none of the pluripotency phenotypic markers are specific when taken independently. They can still be widely expressed in cultures that start to express a significant number of commitment/differentiation genes (*see* **Note 10**).

Acknowledgments

Many thanks to Dr. Mary Osborne-Pellegrin for her help in editing the manuscript.

References

1. Skottman, H., Mikkola, M., Lundin, K., Olsson, C., Stromberg, A. M., Tuuri, T., et al. (2005) Gene expression signatures of seven individual human embryonic stem cell lines. *Stem Cells* **23**, 1343–1356.
2. Mallon, B. S., Park, K. Y., Chen, K. G., Hamilton, R. S., and McKay, R. D. (2006) Toward xeno-free culture of human embryonic stem cells. *Int J Biochem Cell Biol* **38**, 1063–1075.
3. Klimanskaya, I., Chung, Y., Meisner, L., Johnson, J., West, M. D., and Lanza, R. (2005) Human embryonic stem cells derived without feeder cells. *Lancet* **365**, 1636–1641.
4. Amit, M., Margulets, V., Segev, H., Shariki, K., Laevsky, I., Coleman, R., and Itskovitz-Eldor, J. (2003) Human feeder layers for human embryonic stem cells. *Biol Reprod* **68**, 2150–2156.
5. Cheng, L., Hammond, H., Ye, Z., Zhan, X., and Dravid, G. (2003) Human adult marrow cells support prolonged expansion of human embryonic stem cells in culture. *Stem Cells* **21**, 131–142.
6. Choo, A., Padmanabhan, J., Chin, A., Fong, W. J., and Oh, S. K. (2006) Immortalized feeders for the scale-up of human embryonic stem cells in feeder and feeder-free conditions. *J Biotechnol* **122**, 130–141.
7. Ellerstrom, C., Strehl, R., Moya, K., Andersson, K., Bergh, C., Lundin, K., et al. (2006) Derivation of a xeno-free human embryonic stem cell line. *Stem Cells* **24**, 2170–2176.
8. Genbacev, O., Krtolica, A., Zdravkovic, T., Brunette, E., Powell, S., Nath, A., et al. (2005) Serum-free derivation of human embryonic stem cell lines on human placental fibroblast feeders. *Fertil Steril* **83**, 1517–1529.
9. Hovatta, O., Mikkola, M., Gertow, K., Stromberg, A. M., Inzunza, J., Hreinsson, et al. (2003) A culture system using human foreskin fibroblasts as feeder cells allows production of human embryonic stem cells. *Hum Reprod* **18**, 1404–1409.
10. Inzunza, J., Gertow, K., Stromberg, M. A., Matilainen, E., Blennow, E., Skottman, et al. (2005) Derivation of human embryonic stem cell lines in serum replacement medium using postnatal human fibroblasts as feeder cells. *Stem Cells* **23**, 544–549.
11. Koivisto, H., Hyvarinen, M., Stromberg, A. M., Inzunza, J., Matilainen, E., Mikkola, et al. (2004) Cultures of human embryonic stem cells: serum replacement medium or serum-containing media and the effect of basic fibroblast growth factor. *Reprod Biomed Online* **9**, 330–337.

12. Lee, J. B., Lee, J. E., Park, J. H., Kim, S. J., Kim, M. K., Roh, S. I., and Yoon, H. S. (2005) Establishment and maintenance of human embryonic stem cell lines on human feeder cells derived from uterine endometrium under serum-free condition. *Biol Reprod* **72**, 42–49.

13. Richards, M., Fong, C. Y., Chan, W. K., Wong, P. C., and Bongso, A. (2002) Human feeders support prolonged undifferentiated growth of human inner cell masses and embryonic stem cells. *Nat Biotechnol* **20**, 933–936.

14. Stojkovic, P., Lako, M., Stewart, R., Przyborski, S., Armstrong, L., Evans, J., et al. (2005) An autogeneic feeder cell system that efficiently supports growth of undifferentiated human embryonic stem cells. *Stem Cells* **23**, 306–314.

15. Yoo, S. J., Yoon, B. S., Kim, J. M., Song, J. M., Roh, S. I., You, S., and Yoon, H. S. (2005) Efficient culture system for human embryonic stem cells using autologous human embryonic stem cell-derived feeder cells. *Exp Mol Med* **37**, 399–407.

16. Kleinman, H. K., McGarvey, M. L., Liotta, L. A., Robey, P. G., Tryggvason, K., and Martin, G. R. (1982) Isolation and characterization of type IV procollagen, laminin, and heparan sulfate proteoglycan from the EHS sarcoma. *Biochemistry* **21**, 6188–6193.

17. Stojkovic, P., Lako, M., Przyborski, S., Stewart, R., Armstrong, L., Evans, J., Zhang, X., and Stojkovic, M. (2005) Human-serum matrix supports undifferentiated growth of human embryonic stem cells. *Stem Cells* **7**, 895–902.

18. Amit, M., and Itskovitz-Eldor, J. (2006) Maintenance of human embryonic stem cells in animal serum- and feeder layer-free culture conditions. *Methods Mol Biol* **331**, 105–113.

19. Amit, M., Shariki, C., Margulets, V., and Itskovitz-Eldor, J. (2004) Feeder layer- and serum-free culture of human embryonic stem cells. *Biol Reprod* **70**, 837–845.

20. Beattie, G. M., Lopez, A. D., Bucay, N., Hinton, A., Firpo, M. T., King, C. C., and Hayek, A. (2005) Activin A maintains pluripotency of human embryonic stem cells in the absence of feeder layers. *Stem Cells* **23**, 489–495.

21. Liu, Y., Song, Z., Zhao, Y., Qin, H., Cai, J., Zhang, H., et al. (2006) A novel chemical-defined medium with bFGF and N2B27 supplements supports undifferentiated growth in human embryonic stem cells. *Biochem Biophys Res Commun* **346**, 131–139.

22. Ludwig, T. E., Levenstein, M. E., Jones, J. M., Berggren, W. T., Mitchen, E. R., Frane, J. L., et al. (2006) Derivation of human embryonic stem cells in defined conditions. *Nat Biotechnol* **24**, 185–187.

23. Xu, C., Inokuma, M. S., Denham, J., Golds, K., Kundu, P., Gold, J. D., and Carpenter, M. K. (2001) Feeder-free growth of undifferentiated human embryonic stem cells. *Nat Biotechnol* **19**, 971–974.

24. Ding, V., Choo, A. B., and Oh, S. K. (2006) Deciphering the importance of three key media components in human embryonic stem cell cultures. *Biotechnol Lett* **28**, 491–495.

25. Draper, J. S., Moore, H. D., Ruban, L. N., Gokhale, P. J., and Andrews, P. W. (2004) Culture and characterization of human embryonic stem cells. *Stem Cells Dev* **13**, 325–336.

26. Lu, J., Hou, R., Booth, C. J., Yang, S. H., and Snyder, M. (2006) Defined culture conditions of human embryonic stem cells. *Proc Natl Acad Sci U S A* **103**, 5688–5693.

27. Skottman, H., Stromberg, A. M., Matilainen, E., Inzunza, J., Hovatta, O., and Lahesmaa, R. (2006) Unique gene expression signature by human embryonic stem cells cultured under serum-free conditions correlates with their enhanced and prolonged growth in an undifferentiated stage. *Stem Cells* **24**, 151–167.

28. Gui-An, C., and Hong-Mei, P. (2005) Serum-free medium cultivation to improve efficacy in establishment of human embryonic stem cell lines. *Hum Reprod* **21**, 217–222.

29. Levenstein, M. E., Ludwig, T. E., Xu, R. H., Llanas, R. A., Vandenheuvel-Kramer, K. et al. (2005) Basic FGF support of human embryonic stem cell self-renewal. *Stem Cells* **24**, 568–574.

30. Wang, G., Zhang, H., Zhao, Y., Li, J., Cai, J., Wang, P., et al. (2005) Noggin and bFGF cooperate to maintain the pluripotency of human embryonic stem cells in the absence of feeder layers. *Biochem Biophys Res Commun* **330**, 934–942.

31. Xiao, L., Yuan, X., and Sharkis, S. J. (2006) Activin A maintains self-renewal and regulates fibroblast growth factor, Wnt, and bone morphogenic protein pathways in human embryonic stem cells. *Stem Cells* **24**, 1476–1486.

32. Xu, C., Rosler, E., Jiang, J., Lebkowski, J. S., Gold, J. D., O'Sullivan, C., et al. (2005) Basic fibroblast growth factor supports undifferentiated human embryonic stem

33. Li, Y., Powell, S., Brunette, E., Lebkowski, J., and Mandalam, R. (2005) Expansion of human embryonic stem cells in defined serum-free medium devoid of animal-derived products. *Biotechnol Bioeng* **91**, 688–698.

34. Rajala, K., Hakala, H., Panula, S., Aivio, S., Pihlajamaki, H., Suuronen, R., et al. (2007) Testing of nine different xeno-free culture media for human embryonic stem cell cultures. *Hum Reprod* **22**, 1231–1238.

35. Vallier, L., Alexander, M., and Pedersen, R. A. (2005) Activin/Nodal and FGF pathways cooperate to maintain pluripotency of human embryonic stem cells. *J Cell Sci* **118**, 4495–4509.

36. Yao, S., Chen, S., Clark, J., Hao, E., Beattie, G. M., Hayek, A., and Ding, S. (2006) Long-term self-renewal and directed differentiation of human embryonic stem cells in chemically defined conditions. *Proc Natl Acad Sci USA* **103**, 6907–6912.

37. Curry, S. (2002) Beyond expansion: structural studies on the transport roles of human serum albumin. *Vox Sang* **83** Suppl 1, 315–319.

38. Wolfe, R. A., Wu, R., and Sato, G. H. (1980) Epidermal growth factor-induced down-regulation of receptor does not occur in HeLa cells grown in defined medium. *Proc Natl Acad Sci USA* **77**, 2735–2739.

39. Peiffer, I., Barbet, R., Zhou, Y-P., Li M-L., Monier, M.N., Hatzfeld, A. and Hatzfeld J. (2008) Use of xenofree matrices and molecularly-defined media to control human embryonic stem cell pluripotency: effect of low, physiological, TGF-β concentrations. *Stem Cells Dev* **17**, 519–533.

40. Mitalipova, M., and Palmarini, G. (2006) Isolation and characterization of human embryonic stem cells. *Methods Mol Biol* **331**, 55–76.

cell growth without conditioned medium. *Stem Cells* **23**, 315–323.

Chapter 6

Serum-Free and Feeder-Free Culture Expansion of Human Embryonic Stem Cells

Katherine E. Wagner and Mohan C. Vemuri

Abstract

Human embryonic stem cells (hESCs) are pluripotent stem cells derived from the inner cell mass of human blastocysts. hESCs have become a great asset to studying human diseases and genetic functions of healthy organisms. The rate at which hESCs are being used in laboratories is exponentially increasing, and with that, the need for xeno-free hESCs is also increasing. Xeno-free grade hESCs, cells that have not come into contact with any animal-derived components except those of human origin, are critical for eventual drug therapy, cell therapy, and disease treatment in humans. However, advances toward a xeno-free hESC environment are still being developed. Replacement of murine feeder layers with extracellular matrix proteins has advanced the research, and some advances toward a serum-free and feeder-free environment for hESCs are described in this chapter.

Key words: Human embryonic stem cell, serum-free, feeder-free, STEMPRO® hESC SFM.

1. Introduction

Human embryonic stem cells (hESC), first derived and established in 1998 (1), have historically been isolated and maintained on a layer of murine embryonic fibroblasts (MEF), often called a feeder layer. Further passages of these cells, therefore, have been performed on MEFs. MEFs have been used with hESCs as a simple transition from earlier work with murine embryonic stem cells (mESCs) in which MEFs were used as a feeder layer (2). More recently, human feeder layers have been used for hESC derivation (3, 4). With an increased interest in embryonic stem cell therapy, the need for hESCs free of murine feeder cells becomes paramount. Steps toward a feeder-free environment for maintenance of hESCs have stimulated a generation of feeder-free and serum-free medium development (5–7). STEMPRO® hESC SFM is a new, feeder-free, and serum-free medium optimized for hESC maintenance that has been validated with over 15 hESC lines. To enable hESC growth in a feeder-free and xeno-free environment, the current use of Matrigel™ may be replaced with CELLstart™, a defined, xeno-free matrix designed specifically for stem cells. Here, reliable methods for culturing hESCs in STEMPRO® hESC SFM have been described, along with methods for using CELLstart™ and other matrices.

2. Materials

2.1. Medium Preparation for hESC Expansion and Maintenance

1. STEMPRO® hESC SFM kit (Cat. No. A10007-01, Invitrogen Corporation, Grand Island, NY, USA) (*see* **Note 1**).
2. Basic fibroblast growth factor (bFGF) (Cat. No. 13256-059, Invitrogen) dissolved to 10 μg/mL in a 0.1% BSA solution in DPBS, aliquotted, and stored at −20°C.
3. 55 mM 2-Mercaptoethanol (Cat. No. 21985-023, Invitrogen).

2.2. Coating Dishes

1. Geltrex™ hESC Qualified (Cat. No. A10480-01, Invitrogen) coated 60-mm tissue culture-treated cell-culture dishes (Cat. No. 353004, Falcon, Bedford, MA, USA). Geltrex™ hESC Qualified must be diluted 1:30 in DMEM/F12 with GlutaMAX™ (Cat. No. 10565-018, Invitrogen) (*see* **Note 2**).
2. For a xeno-free alternative, CELLstart™ (Cat. No. A10142-01, Invitrogen) coated 60-mm tissue culture-treated cell-culture dishes (Cat. No. 353004, Falcon). CELLstart™ must be diluted 1:50 in DPBS with calcium and magnesium (Cat. No. 14040-133, Invitrogen) (*see* **Note 3**).

2.3. Passaging hESC

1. Collagenase Type IV (Cat. No. 17104-019, Invitrogen) dissolved to a 10 mg/mL solution in DMEM/F12 with GlutaMAX™ (Cat. No. 10565-018, Invitrogen), aliquotted, and stored at −20°C.
2. Cell scraper (Cat. No. 353085, Falcon).
3. As an alternative to enzymatic passaging, STEMPRO® EZPassage™ Tool (Cat. No. 23181-010, Invitrogen).
4. DPBS without calcium and magnesium (Cat. No. 14190-144, Invitrogen).

2.4. Differentiation Removal	1. Dissecting microscope within a laminar flow hood
2. 22-gauge needle and 10 cc syringe (*see* **Note 4**).
2.5. Freezing hESC	1. Dimethyl sulfoxide (DMSO) (Cat. No. 472301, SIGMA, Louis, MO, USA)
2. KnockOut™ Serum Replacement (KSR) (Cat. No. 10828-028, Invitrogen)
3. DMEM/F12 with GlutaMAX™ (Cat. No. 10565-018, Invitrogen)
2.6. Thawing hESC	1. StemPro® hESC SFM complete medium
2. Dishes precoated with Geltrex™ hESC Qualified or CELLstart™
2.7. Embryoid Bodies (EBs)	1. 100-mm Petri dishes (Cat. No. 351029, Falcon)
2. DMEM/F12 with GlutaMAX™ (Cat. No. 10565-018, Invitrogen)
3. KnockOut™ Serum Replacement (KSR) (Cat. No. 10828-028, Invitrogen)
4. Non-Essential Amino Acids (NEAA) (Cat. No. 11140-050, Invitrogen)
5. GlutaMAX™-1 Supplement (Cat. No. 35050-061, Invitrogen)
6. 55-mM 2-Mercaptoethanol (Cat. No. 21985-023, Invitrogen)

3. Methods

hESCs require a daily change of medium in the culture dish, as the growth and expansion of these cells require fresh growth factors (8) and a removal of floating cells or cell debris. Human ESCs should be maintained in a designated incubator where no other cell types are stored. Every day, the cells must be inspected for growth and presence of differentiated cells. If differentiated cells are visible, they can either be removed each day prior to medium replacement, or prior to a passage, depending on the severity of the differentiation (*see* **Section 3.4**).

StemPro® hESC SFM can be prepared as a 1X medium and used for 1 week when stored at 2–8°C. 2-Mercaptoethanol is the only component which must be added fresh each day to the volume needed (refer to **Table 6.1**). At first use, the StemPro® hESC SFM supplement should be thawed, aliquotted, and refrozen to avoid repeated freeze-thaws.

Table 6.1
STEMPRO® hESC SFM volumes needed

Culture container	Surface area (cm²)	Volume of STEMPRO® hESC SFM
6-well plate	9.6	2 mL per well
12-well plate	3.2	1 mL per well
24-well plate	2.0	0.5 mL per well
35-mm dish	3.5	2 mL
60-mm dish	6.0	4 mL
100-mm dish	10.0	10 mL

3.1. Medium Preparation for hESC Expansion and Maintenance

1. Prepare 100 mL of STEMPRO® hESC SFM by adding 90.8 mL of DMEM/F12 with GlutaMAX™ into a 100 mL container.
2. Add 2.0 mL of STEMPRO® hESC supplement.
3. Add 7.2 mL of 25% BSA.
4. Add 80 μL bFGF from Step 2.1.2 and mix thoroughly.
5. Aliquot appropriate volume needed for the day to a centrifuge tube and add 2-mercaptoethanol at a 1:550 dilution, giving a final concentration of 0.1 mM.
6. Incubate centrifuge tube with cap slightly loosened for at least 20 min at 37°C, 5% CO_2 in humidified air to equilibrate the medium. The remainder of the medium should be stored at 2–8°C.

3.2. Coating Dishes

3.2.1. Geltrex™ hESC Qualified

1. Thaw 1 tube of Geltrex™ hESC Qualified at 2–8°C overnight.
2. Remove DMEM/F-12 from 2–8°C storage and add 29 mL of cold DMEM/F-12 to the 1 mL of Geltrex™ hESC Qualified. Mix gently.
3. Cover the whole surface of each culture dish with the Geltrex™ hESC Qualified solution (refer to **Table 6.2**).
4. Incubate in a 37°C, 5% CO_2 incubator for 1 h.
5. Transfer each dish to a laminar flow hood and allow it to equilibrate to room temperate (about 1 h). For unused dishes, seal each dish with parafilm to prevent drying. Store the Geltrex™ hESC Qualified-treated dishes at 2–8°C for up to 1 week.

Table 6.2
Geltrex™ hESC Qualified and CELLstart™ volumes needed

Culture container	Surface area (cm^2)	Volume of diluted substrate
6-well plate	9.6	750 μL per well
12-well plate	3.2	250 μL per well
24-well plate	2.0	160 μL per well
35-mm dish	3.5	1.5 mL
60-mm dish	6.0	3 mL
100-mm dish	10.0	5 mL

6. Pipette out diluted Geltrex™ hESC Qualified from culture container and discard. The culture vessel is now ready for addition of cells. It is not required to rinse off Geltrex™ hESC Qualified from the culture container after removal.

7. Cells can be passaged directly into STEMPRO® hESC SFM onto Geltrex™ hESC Qualified-coated culture containers.

3.2.2. CELLstart™

1. Dilute CELLstart™ 1:50 in DPBS with calcium and magnesium.
2. Cover the whole surface of each culture dish with the CELLstart™ solution (refer to **Table 6.2**).
3. Incubate in a 37°C, 5% CO$_2$ incubator for 2 h.
4. It is recommended to coat the culture container the day of use or the day before. If precoating the day before, the culture container should be wrapped in parafilm to avoid drying, and stored at 2–8°C.
5. Pipette out diluted CELLStart™ from culture container and discard. Culture vessel is now ready for addition of cells. It is not required to rinse off CELLStart™ from the culture container after removal.
6. Cells can be passaged directly into STEMPRO® hESC SFM onto CELLstart™-coated culture containers.

3.3. Passaging hESC

Optional: Dissect out the differentiated portions of human embryonic stem cell culture using a 22-gauge needle (*see* **Section 3.4**).

3.3.1. Collagenase Type IV

1. Equilibrate sufficient complete STEMPRO® hESC SFM (refer to **Table 6.1**) and 6–8 mL wash medium per 60-mm dish (*see* **Note 5**).
2. Rinse the dish of hESCs with DPBS.

3. Add 0.5 mL Collagenase IV solution to a 60-mm dish and incubate for 3 min. Adjust volume of Collagenase IV for various dish sizes, i.e., 35-mm dishes need 0.25 mL Collagenase IV.

4. After incubation, remove Collagenase IV from dish and rinse the dish with 3 mL DPBS.

5. Apply 3–4 mL wash medium to the dish and scrape cells off the dish using a cell scraper.

6. Using a 5-mL serological pipette, gently transfer cell suspension into a 15-mL centrifuge tube.

7. Wash the dish with 3 mL of wash medium and add to the tube.

8. Centrifuge the cells at $200 \times g$ for 2 min at room temperature.

9. After the cells are spun, pour off supernatant. Tap tube with finger to loosen the cell pellet from the bottom of the tube.

10. Using a 5-mL serological pipette, gently resuspend the cells in preequilibrated STEMPRO® hESC SFM.

11. Break up the cell colonies to desired size (usually 50–100 cells per piece) and transfer cell colonies to new CELLstart™ or Geltrex™ hESC Qualified-coated dishes at desired density.

12. Place the dishes into an incubator set at 37°C with 5% CO_2 in air. Move dishes gently to evenly spread out cells.

3.3.2. STEMPRO® EZPassage™ Disposable Cell Passaging Tool

1. Equilibrate sufficient complete STEMPRO® hESC SFM (refer to **Table 6.1**).

2. Change the medium in the dish to fresh, complete STEMPRO® hESC SFM.

3. Pull open packaging and remove EZPassage™ under a laminar flow hood.

4. Hold the culture vessel in one hand and pull (roll) the EZPassage™ Tool across the entire dish in one direction (ex. left to right). Apply enough pressure so the entire roller blade touches the dish and maintains uniform pressure during the rolling action (*see* **Note 6**).

5. Keep pulling (rolling) the EZPassage™ parallel to the first pass until the entire dish has been covered.

6. Rotate the culture vessel 90°, and then repeat steps D and E (*see* **Fig. 6.1**).

7. Using a serological pipette, rinse the dish using medium on the dish so that the cut colonies are suspended in the medium.

8. Transfer the medium containing colonies to a 15-mL tube. A rinse of the dish with STEMPRO® hESC SFM is optional after this step.

Fig. 6.1. An hESC colony that has been cut by the STEMPRO® EZPassage™ Tool (40× magnification).

9. Transfer cell colonies to new CELLstart™ or Geltrex™ hESC Qualified-coated dishes at desired density.
10. Place the dishes into an incubator set at 37°C with 5% CO_2 in air. Move dishes gently to evenly spread out cells.
11. Discard EZPassage™ Tool after use. Do not reuse.

3.4. Differentiation Removal

1. Equilibrate 6–8 mL wash medium per 60-mm dish.
2. Set up hESC dish on a dissecting microscope in a bio-safety cabinet or laminar flow to comfortably observe colonies.
3. Aspirate the medium and gently add 3–4 mL of pre-equilibrated wash medium per 60-mm dish.
4. Cut out and remove any overtly differentiated colonies with a 22-gauge needle (*see* **Fig. 6.2**). Face the bevel of the needle toward you and use the long edge of the needle tip to scrape.

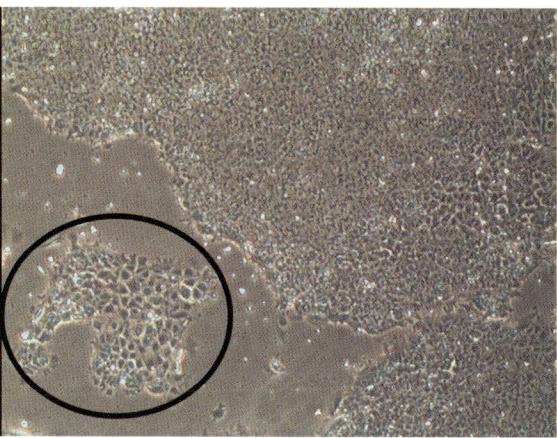

Fig. 6.2. hESC grown in STEMPRO® hESC SFM. The cells enclosed in the *black circle* are differentiated and need to be removed (40× magnification).

Avoid using the tip of the needle to scrape, as that will scratch the plastic dish. It is generally easier to scrape the far side of the dish than the close side. Rotate the dish as needed, leaving your hand and needle in its comfortable position.

5. Aspirate the medium and gently add 3–4 mL of wash medium.
6. The cells are now ready to be fluid changed or passaged.
7. Return to Passaging hESC, Steps 3.3.1 or 3.3.2.

3.5. Freezing hESC

1. Prepare freezing medium as below. The freezing medium consists of two solutions; do not mix solutions A and B together. The total volume needed is 1 mL for each vial to be frozen. If the cells are 90–100% confluent in a 60-mm dish, then approximately 2–4 vials can be frozen from each dish.

Solution A (50% of final volume):	50% DMEM/F12	50% KSR
Solution B (50% of final volume):	80% DMEM/F12	20% DMSO

2. Remove any differentiation in the dishes by following Step 3.4.
3. Passage the cells off the cell culture dishes (up to Steps 3.3.1.G or 3.3.2.H).
4. Centrifuge the cells for 4 min at $200 \times g$, and then slowly resuspend the cell pellet in solution A (50% of the total volume).
5. Triturate to the desired colony size, which should be 150–200 cells per piece.
6. Add equal volume of solution B to the cells very slowly so as to not shock them. Do not break up the colonies any further.
7. Distribute the cells into cryopreservation vials at 1 mL/vial.
8. Store vials in a Mr. Frosty overnight at −70°C, then transfer to permanent storage in liquid nitrogen.

3.6. Thawing hESC

1. Equilibrate sufficient complete STEMPRO® hESC SFM for thawing into desired dish (refer to **Table 6.1**). Also, equilibrate at least 2 mL wash medium.
2. Recover one vial of hESC cells from liquid nitrogen storage and thaw quickly in 35–38°C water bath.
3. Bring the vial to the hood and disinfect with 70% isopropyl.
4. Using a 5-mL serological pipette, transfer the contents of the vial into a fresh 15-mL centrifuge tube.
5. Pipette 1 mL of the equilibrated wash medium and slowly add to vial.

6. Pipette the 1 mL out of the vial and slowly add it to the centrifuge tube, drop wise and against the side of the tube so as to not shock the cells (*see* **Note 7**).

7. Slowly add another 1 mL wash medium to the tube for a total of 3 mL.

8. Centrifuge the cells at $200 \times g$ for 2 min at room temperature.

9. During centrifugation of cells, prepare the new dish by removing excess CELLstart™ or Geltrex™ hESC Qualified from the dish.

10. After the cells are spun, pour off supernatant. Tap tube with finger to loosen the cell pellet from the bottom of the tube.

11. *Gently* resuspend the cell pellet in desired volume of prepared STEMPRO® hESC SFM. Titrate the cells one time to mix, being careful not to break up the colonies (*see* **Fig. 6.3**).

12. Transfer cells to the new culture dish and incubate at 37°C with 5% CO_2.

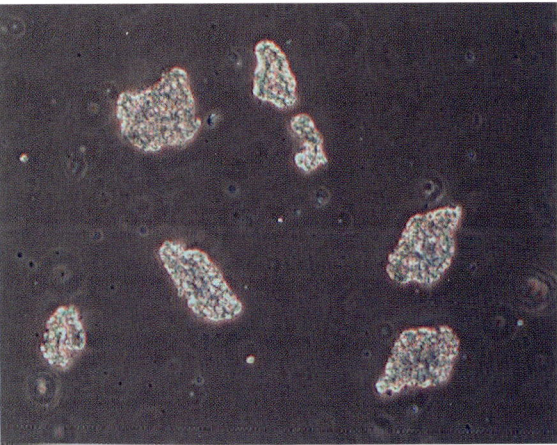

Fig. 6.3. hESC colony pieces just after a thaw, prior to placing in an incubator (40× magnification).

3.7. Embryoid Bodies (EBs)

1. EBs can be set up at a passage, where instead of placing a ratio of cells into a fresh culture dish, the entire dish of cells is transferred into a nontissue culture-treated 100-mm dish. This way, the cells would not stick. Usually a 60-mm dish of attached cells can be transferred into a 100-mm dish. Usually, larger colony pieces are better to use at this step.

2. Prepare 100 mL growth medium by adding 78 mL DMEM/F12 with GlutaMAX™ into a 100-mL container.

3. Add 20 mL KNOCKOUT™ Serum Replacement.

4. Add 1 mL nonessential amino acids.

5. Add 1 mL GlutaMAX™-1 Supplement.

6. Aliquot appropriate volume needed for the day to a centrifuge tube (refer to Table 6.1) and add 2-mercaptoethanol at a 1:550 dilution, giving a final concentration of 0.1 mM (*see* **Note 8**).

7. Change the medium every other day by allowing cells to settle in a 15-mL tube. Then, remove supernatant and replace with fresh medium, place cells back onto the same 100 mm.

8. After 4 days, transfer the cells equally into 2×100 mm tissue culture-treated dishes to allow for attachment.

9. Keep the same schedule for fluid changes, however as the cells expand the volume per dish may need to be increased to 15 mL, then to 20 mL if needed.

10. Allow the cells to expand for 14 days and 21 days. The entire 100-mm dish can be harvested for molecular analysis.

4. Notes

1. StemPro® hESC SFM is supplied as a three-part kit – 500 mL DMEM/F12 with GlutaMAX™ basal medium (Cat. No. 10565-018), 40 mL 25% bovine serum albumin (kit part no. A10008-01), and 10 mL StemPro® hESC SFM supplement (kit part no. A10006-01). Prepared medium is stable for 1 week when stored at 2–8°C.

2. An optimal dilution of Geltrex™ hESC Qualified may need to be determined for each cell line. Try various dilutions from 1:30 to 1:100.

3. An optimal dilution of CELLstart™ may need to be determined for each cell line. Try various dilutions from 1:50 to 1:100.

4. The 10-cc syringe is to be used as a handle for the needle, so any comfortable size will be suitable.

5. Wash medium consists of 0.1% BSA in DMEM/F12 with GlutaMAX. Alternatively, the normal culture medium can be used as a wash medium.

6. Do not remove the culture medium before rolling the dish.

7. Cells will not be a single cell suspension. They will be in clumps, which is an acceptable visual observation.

8. For EB medium, all but the 2-mercaptoethanol can be prepared and stored at 2–8°C.

Acknowledgments

The authors thank Mary Lynn Tilkins for her helpful comments and suggestions.

References

1. Thomson, J.A., Itskovitz-Eldor, J., Shapiro, S.S., Waknitz, M.A., Swiergiel, J.J., Marshall, V.S., Jones, J.M. (1998). *Embryonic stem cell lines derived from human blastocysts.* Science. **282**(5391): 1145–7.
2. Bryja, V., Bonilla, S., Cajanek, L., Parish, C.L., Schwartz, C.M., Luo, Y., Rao, M.S., Arenas, E. (2006). *An efficient method for the derivation of mouse embryonic stem cells.* Stem Cells. **24**(4): 844–9.
3. Vemuri, M.C., Schimmel, T., Colls, P., Munne, S., Cohen, J. (2007). *Derivation of human embryonic stem cells in xeno-free conditions*, in *Stem Cell Assays*, M.C. Vemuri, Editor. Humana Press: Totowa, NJ. pp. 1–10.
4. Ellerstrom, C., Strehl, R., Moya, K., Andersson, K., Bergh, C., Lundin, K., Hyllner, J., Semb, H. (2006) *Derivation of a xeno-free human embryonic stem cell line.* Stem Cells. **24**(10): 2170–6.
5. Wang, L., Schulz, T.C., Sherrer, E.S., Dauphin, D.S., Shin, S., Nelson, A.M., Ware, C.B., Zhan, M., Song, C., Chen, X., Brimble, S.N., McLean, A., Galeano, M.J., Uhl, E.W., D'Amour, K.A., Chesnut, J.D., Rao, M.S., Blau, C.A., Robins, A.J. (2007) *Self-renewal of human embryonic stem cells requires insulin-like growth factor-1 receptor and ERBB2 receptor signaling.* Blood. **110**(12): 4111–19.
6. Yao, S., Chen, S., Clark, J., Hao, E., Beattie, G.M., Hayek, A., Ding, S. (2006) *Long-term self-renewal and directed differentiation of human embryonic stem cells in chemically defined conditions.* Proc Natl Acad Sci USA. **103**(18): 6907–12.
7. Xu, C., Inokuma, M.S., Denham, J., Golds, K., Kundu, P., Gold, J.D., Carpenter, M. (2001) *Feeder-free growth of undifferentiated human embryonic stem cells.* Nature Biotechnology. **19**(10): 971–4.
8. Levenstein, M.E., Ludwig, T.E., Xu, R., Llanas, R.A., VanDenHeuvel-Kramer, K., Manning, D., Thomson, J.A. (2006) *Basic fibroblast growth factor support of human embryonic stem cell self-renewal.* Stem Cells. **24**(3): p. 568–74.

Chapter 7

Single Cell Enzymatic Dissociation of Human Embryonic Stem Cells: A Straightforward, Robust, and Standardized Culture Method

Catharina Ellerström, Johan Hyllner, and Raimund Strehl

Abstract

The routine culture and expansion of human embryonic stem (hES) cells has been and is still posing a challenge to researchers wishing to take advantage of the cells' unique potential. In contrast to mouse embryonic stem cells, hES cells usually have to be expanded by tedious mechanical microdissection or by enzymatic dissociation to cell clusters of a very narrow size range.

It is essential to use a culture system that allows the robust and reproducible enzymatic dissociation of viable hES cell cultures to single cells to allow the scale-up of hES cell cultures as well as the application of hES cells in various experiments, such as FACS, electroporation, and clonal selection.

By the development of enzyme-based protocols, which are less labor intensive and less time consuming, much progress has been made over the recent years with regard to improved culture systems for hES cell. We have developed a culture system that is based on single cell enzymatic dissociation (SCED) in combination with a highly supportive feeder cell layer of human foreskin fibroblasts (hFFs). The culture system allows defined enzymatic propagation while maintaining the hES cell lines in an undifferentiated, pluripotent, and normal state.

In this chapter, we will show how hES cells, which have been derived and passaged by traditional mechanical dissection, can be rapidly adjusted to propagation by enzymatic dissociation to single cells. The protocols we describe are widely applicable and should therefore be of general use for the reliable mass cultivation of hES cells for various experiments.

Key words: Human embryonic stem cell, human feeders, single cells, enzymatic dissociation, expansion, robust culture system.

1. Introduction

Human embryonic stem (hES) cells have traditionally been maintained on a feeder layer of mouse embryonic fibroblasts (MEF) and have been passaged by manual microdissection and subsequent transfer of selected parts of the dissected colonies to new culture dishes to maintain a pluripotent and undifferentiated state in vitro (1–4). This is, without question, a well-established protocol, which can provide the user with hES cells of very good quality. The protocol is, however, also very labor-intensive and requires highly trained personnel.

Feeder-free culture conditions for hES cells, using MEF conditioned or more recently defined media, have been employed with success to propagate hES cells (5–12). These methods though, still require exact dissociation of the hES cell colonies to clusters of a very narrow size range, which causes problems in the standardization and automation of cell expansion. The requirement to passage hES cells as clusters also limits the applicability to several important experimental techniques, which are based on single cells, e.g., FACS, clonal selection, or electroporation (13–15).

The existing feeder-free culture protocols are not generally applicable with equal success and quality and the search for a robust and standardized feeder-free culture system is still ongoing, e.g., in the International stem cell Initiative (ISCI-2).

It is well known that hES cells depend on cell-cell and cell-substrate contact for survival (16, 17) and Rho-associated kinase (Rock) inhibitors have been shown to decrease apoptosis or anoikis of dissociated hES cells cultured on, e.g., MEF cells (18). The cellular microenvironment plays a very important role in the natural survival of single hES cells, which is why we employ a highly supportive culture environment comprised of a high-density human foreskin fibroblast feeder substrate and a growth factor-supplemented culture medium.

The single cell enzymatic dissociation (SCED) protocols we present in this chapter allow robust bulk expansion of undifferentiated hES cells while maintaining pluripotency and genetic stability over extended culture periods. The protocols are easy to learn and only require a laboratory equipped for standard cell culture. In addition we have found the protocols widely applicable to already established hES cell lines, and the transition of cultures from the traditional manual propagation method only demands a short period of time (19).

2. Materials

2.1. Culture and Handling of Feeder Cells

1. We strongly recommend using the commercially available human foreskin fibroblast feeders (hFFs) named CRL-2429, the American Type Culture Collection (ATCC, Manassas, VAUSA) (see **Note 1**).

2. The hFF medium: Iscove's modified Dulbecco's medium (IMDM) (Invitrogen, Paisley, U.K.) supplemented with 10% fetal bovine serum, (FBS, Invitrogen) and 1% penicillin-streptomyocin (Invitrogen).
3. 1X Trypsin-EDTA (Invitrogen).
4. 1X PBS -Ca^{2+}/-Mg^{2+} (Invitrogen).
5. Mitomycin C (Sigma-Aldrich, St. Louis); Final concentration should be 10 μg/ml in the medium (*see* **Note 2**).
6. Gelatin (Sigma-Aldrich); Final concentration should be 0.1% when diluted in cell culture grade water. Autoclave the solution prior use.
7. VitroHES™ medium (Vitrolife AB, Kungsbacka, Sweden).
8. Human recombinant basic fibroblast growth factor (hrbFGF, Invitrogen). Final concentration should be 10 ng/mL culture medium. Dilute in Vitro-PBS and store aliquots in −20°C.
9. Vitro-PBS (Vitrolife).
10. Appropriate sterile cell-culture plasticware.

2.2. Culture and Handling of hES Cells

1. Any human ES cell line cultured on, e.g., mouse embryonic fibroblast (MEF) feeders by manual microdissection (*see* **Note 3**).
2. VitroHES™ medium (Vitrolife AB, Kungsbacka, Sweden, http://www.vitrolife.com).
3. hrbFGF (Invitrogen). Final concentration should be 10 ng/mL in VitroHES™ medium.
4. 1X PBS -Ca^{2+}/-Mg^{2+} (Invitrogen).
5. 1X TrypLE™ Select (Invitrogen).
6. Appropriate sterile cell-culture plasticware.

3. Methods

3.1. Culture and Handling of Feeder Cells

3.1.1. Thawing of hFFs

1. Prewarm hFF medium to 37°C and add appropriate volume to the culture flask (For a T-75 flask, use 10 mL of prewarmed medium).
2. Bring a beaker with 37°C tap water.
3. Transfer the frozen vial(s) using forceps and submerse them directly into the 37°C water (*see* **Note 4**).
4. Before the thawed vial is brought into the hood (a laminar flow cabinet), it has to be wiped with 70% ethanol for disinfection.

5. Gently transfer the thawed cells to the hFF medium in the culture flask.

6. Move the flask into the incubator at 37°C, 5% CO_2, 95% humidity.

3.1.2. Culture of hFFs

1. When hFFs are cultured for expansion purposes, the cells should never be allowed to grow confluent. Instead they should be passaged before they reach 90% confluence. Recommended subcultivation ratio is 1:4 to 1:8.

2. Change medium every second or third day (*see* **Note 5**).

3. Before each medium change, equilibrate the required amount of hFF medium at least to room temperature, preferably to 37°C.

4. Place the hFF cells and medium in the LAF-bench, remove all medium from each flask, and discard the medium into an appropriate waste bottle.

5. Add the prewarmed medium into each flask and return the culture flask(s) to the incubator at 37°C, 5% CO_2, 95% humidity.

3.1.3. Gelatin Coating of Culture Dishes

1. Coat the cell-culture wells to be used with 0.1% gelatin solution. Make sure that the surface is completely covered with gelatin solution.

2. Place the cell culture units in the LAF-bench for 0.5–1 h (*see* **Note 6**).

3. Aspirate the excess gelatin solution just prior to the seeding of the mitomycin C-treated hFFs.

3.1.4. Mitomycin C Treatment of hFFs

1. Remove the culture medium from the cell-culture flasks with hFF to be treated with mitomycin C and determine the volume of medium.

2. Add mitomycin C stock solution to the medium to a final concentration of 10 µg/ml and return the medium to the hFF cells to be inactivated (*see* **Note 7**).

3. Place the cell-culture flasks with the hFF medium and mitomycin C solution in the incubator for 2–3 h at 37°C, 5% CO_2, 95% humidity.

3.1.5. Seeding of Mitomycin C-Treated hFFs

1. Remove the gelatin solution from the coated cell-culture wells (as described above).

2. Remove the mitomycin C solution from each culture flask of inactivated hFF feeders.

3. Thoroughly wash the inactivated hFF feeders twice with 1X PBS (e.g., 10 ml to a T-75 culture flask).

4. Remove the 1X PBS and add 1X Trypsin-EDTA and allow cells to detach from the flask. Incubate the hFFs for 4–6 min at 37°C in order to dissociate the feeder cells.
5. Transfer the cell suspension to a tube and centrifuge at 400 g for 5 min.
6. After the centrifugation step, remove the supernatant and resuspend the cells in VitroHES^{TM} medium supplemented with 10 ng rhbFGF/mL.
7. Count the cells in a hemocytometer and calculate the number of cells per ml.
8. Dilute the cells with rhbFGF-supplemented VitroHES^{TM} medium to the required volume (*see* **Note 8**) to be able to seed the inactivated hFF cells at a density of 70×10^3 cells/cm^2.
9. Add the prepared hFF cell suspension to the dishes, plates, or flasks and place them in the incubator at 37°C, 5% CO_2, 95% humidity.
10. Visually inspect the hFF cells 24 h after seeding in order to confirm that the hFF cells have attached and that they are evenly distributed within the culture vessel.

3.2. Transfer of hES Cells to SCED Culture

3.2.1. hES Cell Quality

1. Before transferring hES cells to SCED culture, control the quality of the hES cell colonies under a microscope in phase contrast.
2. Use only high-quality hES cell cultures, in which the colonies should be homogenously undifferentiated and no or only a few differentiated areas should be present (*see* **Note 9**).

3.2.2. Transfer Procedure

1. Preheat the cell-culture medium to be used to 37°C.
2. Equilibrate 1X TrypLE^{TM} Select to room temperature.
3. Place the cultures to be transferred in the LAF bench. Remove the medium from the culture dishes to prepare for dissociation. Do not prepare and dissociate more than two wells at a time.
4. Rinse the cultures by adding 1X PBS. The entire cell surface should be covered.
5. Remove the 1X PBS and add 1X TrypLE^{TM} Select, again, the entire cell surface should be covered.
6. Incubate for approximately 3 min and control that the cells have lost contact with each other and started to detach from the surface using a microscope. If not, incubate for 1–3 additional minutes on a 37°C heated stage (*see* **Note 10**).
7. Gently draw the suspension up and down a 1000-μl pipette about 3–5 times and then pool the cell suspension in a 15-ml centrifuge tube.

8. Rinse the culture vessel once with fresh VitroHES™ medium and transfer to the cell suspension in the centrifuge tube.
9. Centrifuge the cell suspension at $400 \times g$ for 5 min.
10. Discard the supernatant and resuspend the cells in fresh VitroHES™ medium supplemented with 10 ng rhbFGF (*see* **Note 11**).
11. Split the cell suspension in 1:2, 1:4, and 1:8 in 3 new cell-culture units containing mitomycin C-treated hFF cells (2–5 days old) (*see* **Note 12**).
12. Adjust the total volume of culture medium (*see* **Note 8**).
13. Label the cell-culture wells with the name of the hES cell line, date, and passage number as well as split ratio applied.
14. Control the cells using a microscope and remove larger cell clumps by using a 1000-µl pipette.
15. Return the cells into the incubator at 37°C, 5% CO_2, 95% humidity.

3.3. Medium Change of SCED Cultures

1. Completely change the culture medium every 2–3 days (*see* **Note 13**).
2. Preheat the volume of fresh VitroHES™ medium that will be used to 37°C and supplement the medium with 10 ng rhbFGF directly before use.
3. Place the culture vessels in the LAF bench.
4. Remove the used medium and add fresh preheated VitroHES™ medium + 10 ng rhbFGF/mL to each cell-culture unit.
5. Return the culture vessels to the incubator at 37°C, 5% CO_2, 95% humidity.
6. Discard any left over preheated media. Do not warm up media repeatedly.

3.4. SCED Passage of Adjusted hES Cells

3.4.1. Estimate Quality and Confluence

1. Estimate the quality of the hES cells and the grade of confluence using a phase contrast microscope. Confluence of hES cell colonies should be between 25 and 60%.
2. The optimal split ratio has to be determined individually for each hES cell line. The results obtained from the initial split series (1:2, 1:4, and 1:8) will give you a very good indication of the split ratios to be used further on for the specific hES cell line in use (*see* **Note 14**).

3.4.2. SCED Passage Procedure

1. Before you start, make sure that you have enough culture vessels prepared with fresh mitomycin C-treated feeders (*see* **Note 15**).

2. Preheat the amount of fresh VitroHES™ medium that will be used to 37°C and supplement the medium with 10 ng rhbFGF directly before use.

3. Equilibrate the amount of TrypLE™ Select to be used to room temperature.

4. Place the culture vessels to be passaged in the LAF bench. Do not handle more than 2 cell culture wells or flasks at a time!

5. Remove the used culture medium from the cell cultures.

6. Rinse the culture vessel once by adding 1X PBS. The entire cell surface should be covered.

7. Remove the 1X PBS and add TrypLE™ Select. Again, the entire cell surface should be covered.

8. Incubate for approximately 4–6 min on a 37°C heating plate and control that the cells have detached from the surface using a microscope. If not – incubate for 1–3 additional minutes on a 37°C heating plate (*see* **Note 16** and **Fig. 7.1**).

9. Gently draw the suspension up and down a 1000-μl pipette about 3–5 times and then pool the cell suspension in a 15-ml centrifuge tube.

10. Rinse the culture well/flask with VitroHES™ medium and transfer to the cell suspension in the centrifuge tube.

11. Centrifuge the cell suspension at $400 \times g$ for 5 min.

12. Discard the supernatant and resuspend the cells in fresh pre-heated VitroHES™ medium + 10 ng rhbFGF (*see* **Note 17**). Further dilution may be done if required. Remove larger cell aggregates using a 1000-μl pipette (*see* **Note 18**).

13. Calculate the volume of cell suspension to be added to new cell-culture wells/flasks containing mitomycin C-treated hFF cells (2–5 days old).

14. Adjust the total volume of culture medium (*see* **Note 8**).

15. Label the cell-culture wells with the name of the hES cell line, current date, and passage number as well as the split ratio applied.

16. Return the cells into the incubator at 37°C, 5% CO_2, 95% humidity.

17. Depending on the growth rate of the specific hES cell-line you choose to work with, there is a range in time for when you will be able to detect the first sign of tiny hES cell colonies in the microscope. Fast and robust lines will reveal their colonies 1–2 days after passage while more slow growing lines may require 1–3 additional days before the small colonies can be identified (**see Fig. 7.2**). Consequently, the passage interval for the human ES cells in the SCED system can range from 7 to 12 days depending on the growth rate of the specific hES cell-line used (**see Fig. 7.2**).

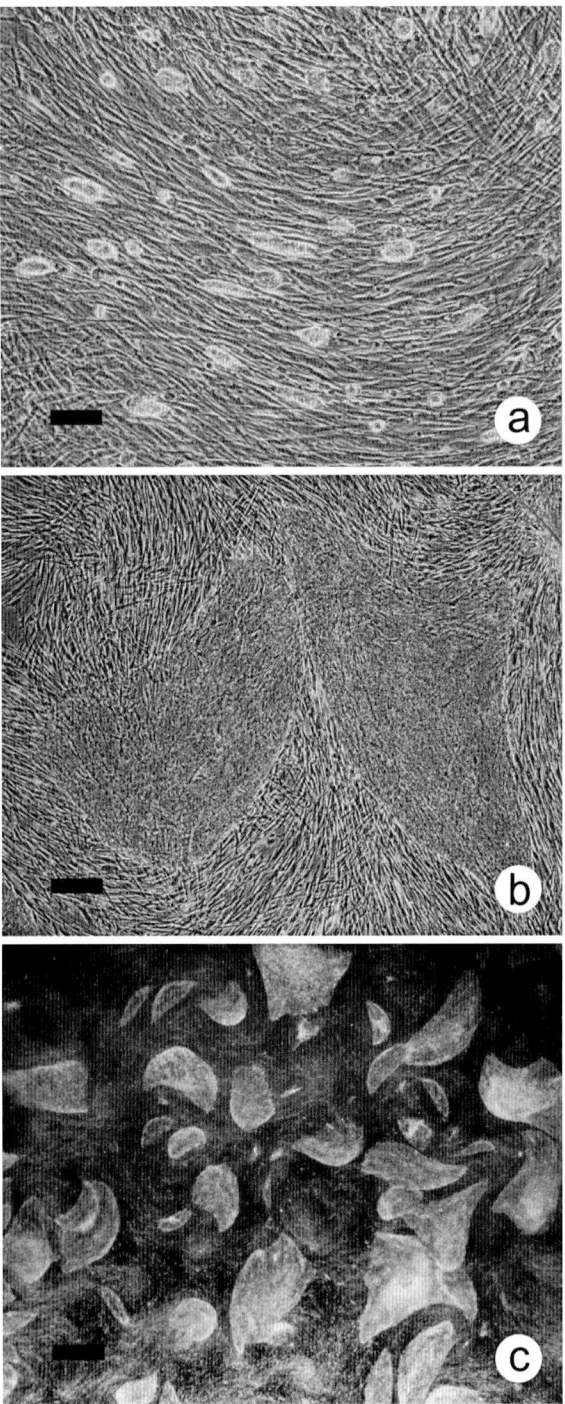

Fig. 7.1. Single-cell enzymatic dissociation of a hES cell colony. **(a)** hES cell colony just prior dissociation. **(b)** hES colony after 2–3 min of dissociation on a heated plate at 37°C. **(c)** hES colony after approximately 8 min of dissociation on a heated plate at 37°C. When the majority of the hES cells within the colonies have rounded up and when the edges of the colonies start to look uneven and rough, it is time to return the dish to the LAF bench and to start pipetting the cell suspension. Scale bars = 100 μm.

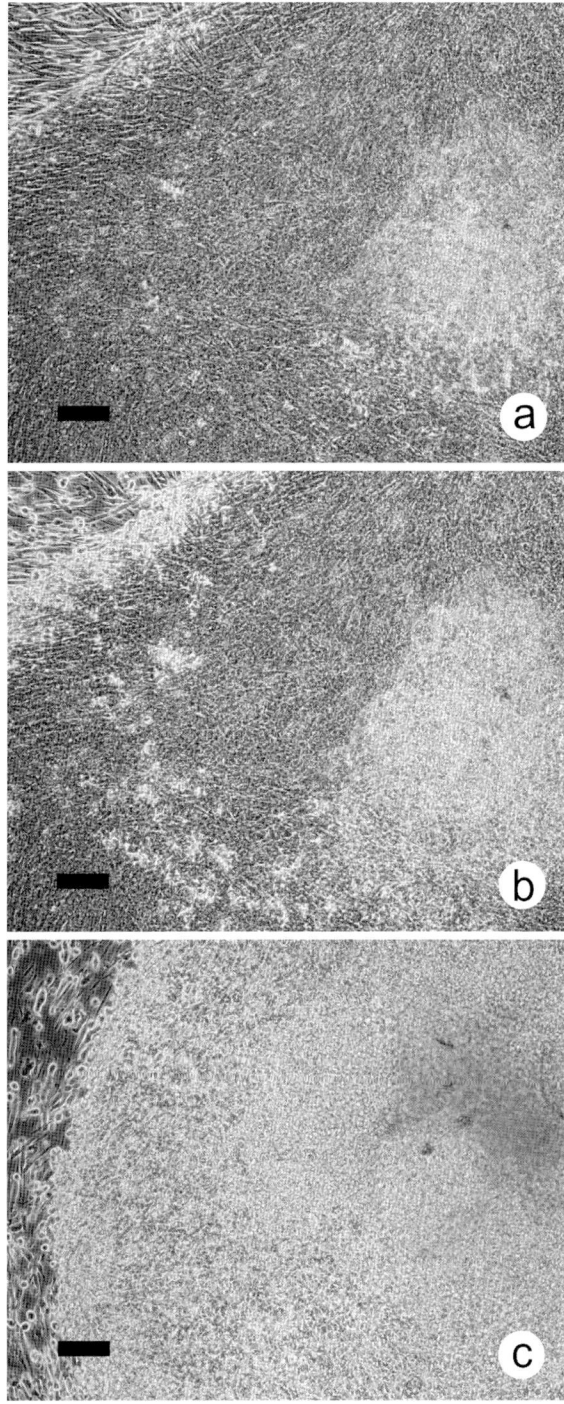

Fig. 7.2. Human ES cells cultured in the SCED culture system. **(a)** Adjusted hES cells 2 days after enzymatic dissociation. Scale bar = 100 μm. **(b)** Homogenous hES cell colonies growing on top of the hFF feeder cells. Scale bar = 250 μm. **(c)** Low magnification view of culture vessel with hES cell colonies ready to be passaged. Scale bar = 1 mm.

3.5. Characterization of SCED Cultured hES Cells

The morphological observation obtained from routine observation in the phase contrast microscope during culture, medium changes, and passages needs to be complemented with additional characterization of the hES cells. As with all hES cell culture work surveillance at regular intervals is highly recommended. If proper culture conditions are maintained, the majority of the colonies in the cultures should homogenously express the expected stem cell markers such as Oct-4, SSEA-3, SSEA-4, TRA1-60, TRA1-81, but remain negative for SSEA-1. In our laboratory, we routinely also use hES-Cellect™, a human ES cell-surface specific antibody that recognizes undifferentiated hES cells. Furthermore, genetic characterization should be performed at regular intervals, e.g., every 10 passages (*see* **Note 19**) (3, 6, 19–22).

4. Notes

1. Primary cultured human foreskin fibroblasts are capable of approximately 60 population doublings before the onset of senescence. hFFs obtained from ATCC usually have gone through a minimum of 16 doublings already at the time of shipping. Do not use them as feeders above 25 doublings (19, 23–25).

2. A commonly used alternative to mitomycin C treatment of the feeders is δ-irradiation. For details concerning the δ-irradiation of hFFs, please see (24).

3. In our experience, the adjustment of manually passaged hES cells to the SCED culture system is a general procedure and is not restricted to specific hES cell lines nor to any specific history of previous mechanical passages. In fact, hES cells appear to behave in a very similar way regardless of how many mechanical passages they had experienced prior to transfer to the SCED system (19).

4. Quickly submerging the vial will minimize the risk of explosion that can occur if liquid nitrogen has leaked into the vial.

5. One passage is equivalent to a medium change, so for one passage of the cells every week only one additional medium change is required.

6. It is important that the gelatin solution is not allowed to dry in prior to seeding of hFFs, since drying in may result in an unwanted uneven distribution of the hFFs within the culture dish.

7. It should be noted that mitomycin C is a biohazard and should be handled with care.

8. Suggested cell-culture medium volume for SCED-cultured hES cells in different culture formats:

Cell culture format	Total volume (ml)
Centre well dish	2
35-mm dish	3
60-mm dish	6–7
100-mm dish	10–15
6-well plate	5–7 per well
12-well plate	3–5 per well
24-well plate	1.5–2 per well
48-well plate	1–1.2 per well
96-well plate	0.1–0.2 per well
T25 bottle	5–8
T75 bottle	10–20
T150 bottle	20–40
T175 bottle	20–40

9. We usually start our SCED cultures by using one IVF dish of manually passaged hES cells cultured of a MEF feeder layer to transfer to three new hFF-containing IVF wells (with the hES cells diluted 1:2, 1:4, and 1:8). The better the starting material, the better the result of the transfer will be. Therefore choose the best starting material available. If optimal starting material is not available, start by mechanically removing the obviously differentiated colony areas before initiating the enzymatic dissociation. If the starting material is very poor, it may be necessary to increase the amount of starting material and to rigorously remove differentiated areas.

10. At the very first enzymatic dissociation when manually passaged hES cells on MEF feeders are used, the enzymatic dissociation may proceed fairly fast. It is therefore recommended that this step is monitored with extra care.

11. If the first enzymatic dissociation is based on the cells in an IVF dish (2.89 cm^2 culture area), the obtained cell pellet after centrifugation is tiny. This is expected.

12. At this very first enzymatic dissociation, it is imperative to seed the hES cells at different subcultivation ratios in order to establish optimal split ratio for the chosen hES cell line.

13. When the hES cells reach the log phase, they will most likely need more frequent media changes. The culture medium should never be allowed to be used up to a degree where an obvious pH change (indicated by yellow coloration) occurs. This exhausted and acidified medium is very unfavorable for the hES cells and can lead to inferior results especially when occurring directly prior to passage.

14. Subcultivation shall be performed when the majority of the hES cell colonies have spread out, but still are undifferentiated. When bigger colonies start to change density in the centre of the colony, passage should be considered (*see* **Fig. 7.2**). The subcultivation ratio depends on the confluence of the hES cells at the time of passage. We use the following criteria:

hES confluences (%)	Split ratio
<10	1:2
10–25	1:4
25–50	1:10
50–75	1:20
75–100	1:40

15. One advantage with employing hFFs as feeder cells is that they can be stored longer in the incubator before use than the MEFs. We have found that hFFs can be successfully used as feeders for hES cells for up to 5 days after mitomycin C treatment and seeding. If the inactivated hFFs in the culture dish to be used are more than 3 days old, perform a complete medium change prior to the addition of the newly dissociated hES cells. Make it a habit to always visually inspect the feeders prior to use and ensure that the density, attachment, and cell distribution within the well/flask is as required.

16. In contrast to hES cells cultured of MEF feeders, where the hES cell colonies appear to be pushing the feeder cell layer aside, hES cells cultured on hFF feeders actually grow on top of the feeder cells (*see* **Fig. 7.2**). As the hES cells are in direct cell–cell contact with the feeder cells in that case, the enzymatic dissociation steep will require longer time before the cells are sufficiently dissociated. The time range for dissociation may be 5–8 min typically. In our experience, slight enzymatic over-dissociation does not to have any significant impact on the survival of the hES cells (19, 24).

17. After removing the supernatant, the pellet should be clearly visible. The easiest way to resuspend the hES cell pellet is to give the bottom of the centrifuge tube one or two gentle sideways flicks with your finger before adding fresh culture medium. This will facilitate the resuspension.

18. Remove big aggregates of feeder cells and/or extracellular matrix that have not been sufficiently dissociated. If left in culture these aggregates can become the initial centers of differentiation.

19. Genetic changes in hES cells may arise under various growth conditions and the causes are still poorly understood (26–29). All hES cell cultures should therefore be genetically analyzed on a regular basis. In our hands, SCED cultured diploid normal hES cells have so far not revealed any of the genetic abnormalities reported earlier (21). But as in any cultured cells, genomic alterations over time cannot be ruled out (19).

Acknowledgements

The authors would like to thank Professor H. Semb for valuable scientific input and K. Noaksson for her technical assistance and advice.

References

1. Thomson J.A., Itskovitz-Eldor J., Shapiro S.S., Waknitz M.A., Swiergiel J.J., Marshall, V.S. et al. (1998) Embryonic stem cell lines derived from human blastocysts. *Science* **282**, 1145–1147.

2. Reubinoff B.E., Pera M.F., Fong C.Y, Trounson A., Bongso A. (2000) Embryonic stem cell lines from human blastocysts: somatic differentiation in vitro. *Nature Biotech* **18**, 399–404.

3. Heins N., Englund M.C.O., Sjöblom C., Dahl U., Tonning A., Bergh C. et al. (2004) Derivation, characterization, and differentiation of human embryonic stem cells. *Stem Cells* **22**, 367–376.

4. Zeng X., Miura T., Luo X., Bhattacharya B., Condie B., Chen J. et al. (2004) Properties of pluripotent human embryonic stem cell BG01 and BG02. *Stem Cells* **22**, 292–312.

5. Xu C., Inokuma M.S., Denham J., Golds K., Kundu P., Gold J.D. et al. (2001) Feeder-free growth of undifferentiated human embryonic stem cells. *Nature Biotech* **19**, 971–974.

6. Sjögren-Janson E., Zetterström M., Moya K., Lindqvist J., Strehl R., Eriksson P.S. (2005) Large-scale propagation of four undifferentiated human embryonic stem cell lines in a feeder-free culture system. *Dev Dyn* **233**, 1304–1314.

7. Rosler E.S., Fisk G.J., Ares X., Irving J., Miura T., Rao M.S., et al. (2004) Long-term culture of human embryonic stem cells in feeder-free conditions. *Dev Dyn* **229**, 259–274.

8. Carpenter M.K., Rosler E.S., Fisk G.J., Brandenberger R., Ares X., Miura T., et al. (2004) Properties of four human embryonic stem cell lines maintained in a feeder-free culture system. *Dev Dyn* **229**, 243–258.

9. Brimble S.N., Zeng X. Weiler D.A., Luo Y., Liu Y., Lyons I.G., et al. (2004) Karyotypic stability, genotyping, differentiation, feeder-free maintenance, and gene expression sampling in three human embryonic stem cell

lines derived prior to august 9. *Stem Cells Dev* **13**, 585–596.

10. Klimanskaya I., Chung Y., Meisner L., Johnson J., West M.D., Lanza R. (2005) Human embryonic stem cells derived without feeder cells. *Lancet* **365**, 1636–1641.

11. Ludwig T.E., Levenstein M.E., Jones J.M., Jones J.M., Berggren W.T., Mitchen E.R., et al. (2006) Derivation of human embryonic stem cells in defined conditions. *Nature Biotech* **24**, 185–187.

12. Wang L., Schulz T.C., Sherrer E.S., Dauphin D.S., Shin S., Nelson A.M., et al. (2007) Self-renewal of human embryonic stem cells requires insulin-like growth factor-1 receptor and ERBB2 receptor signalling. *Blood* **110**, 4111–4119.

13. Rohwedel J., Guan K., Hegert C., Wobus A.M. (2001) Embryonic stem cells as in vitro model for mutagenicity, cytotoxicity and embryotoxicity studies: present state and future prospects. *Toxicol In Vitro* **15**, 741–753.

14. Davila J.C., Cezar G.G., Thiede M., Strom S., Miki T., Trosko J. (2004) Use and application of stem cells in toxicology. *Toxicol Sci* **79**, 214–223.

15. Cowan C.A., Klimanskaya I., McMahon J., Atienza J., Witmyer J., Zucker J.P., et al. (2004) Derivation of embryonic stem-cell lines from human blastocysts. *N Eng J Med* **350**, 1353–1356.

16. Hasegawa K., Fujioka T., Nakamura Y., Nakatsuji N., Suemori H. A. (2006) A method for the selection of human embryonic stem cell sub-lines with high replating efficiency after single cell dissociation. *Stem Cells* **24**, 2649–2660.

17. Pyle A.D., Lock L.F., Donovan P.J. (2006) Neurotrophins mediate human embryonic stem cell survival. *Nature Biotech* **24**, 344–350.

18. Watanabe K., Ueno M., Kamiya D., Nishiyama A., Mattsumura M., Wataya T., et al. (2007) A ROCK inhibitor permits survival of dissociated human embryonic stem cells. *Nature Biotech* **25**, 681–686.

19. Ellerström C., Strehl R., Noaksson K., Hyllner J., Semb H. (2007) Facilitated expansion of human embryonic stem cells by single-cell enzymatic dissociation. *Stem Cells* **25**, 1690–1696.

20. Suemori H., Yasuchika K., Hasegawa K., Fujioka T., Tsuneyoshi N., Nakatsuji N. (2006) Efficient establishment of human embryonic stem cell lines and long-term maintenance with stable karyotype by enzymatic bulk passage. *Biochem Biophys Res Commun* **345**, 926–932.

21. Draper J.S., Smith K., Gokhale P., Moore H.D., Maltby E., Johnson J., et al. (2004) Recurrent gain of chromosomes 17q and 12 in cultured human embryonic stem cells. *Nature Biotech* **22**, 53–54.

22. Mitalipova M.M., Rao R.R., Hoyer D.M., Johnson J.A., Meisner L.F., Jones K.L., et al. (2005) Preserving the genetic integrity of human embryonic stem cells. *Nature Biotech* **23**, 19–20.

23. Amit M., Margulets V., Segev H., Shariki K., Laevsky I., Coleman R., et al. (2003) Human feeder layers for human embryonic stem cells. *Biol Reprod* **68**, 2150–2156.

24. Hovatta O., Mikkola M., Gertow K., Strömberg A-M., Inzunza J., Hreinsson J., et al. (2003) A culture system using human foreskin fibroblasts as feeder cells allows production of human embryonic stem cells. *Human Reprod* **18**, 1404–1409.

25. Richards M., Tan S., Fong C.Y., Biswas A., Chan W-K., Bongso A. (2003) Comparative evaluation of various human feeders for prolonged undifferentiated growth of human embryonic stem cells. *Stem Cells* **21**, 546–556

26. Maitra A., Arking D.E., Shivapurkar N., Ikeda M., Stastny V., Kassauei K., et al. (2005) Genomic alterations in cultured human embryonic stem cells. *Nature Genet* **37**, 1099–1103.

27. Caisander G., Park H., Frej K., Lindqvist J., Bergh C., Lundin K., et al. (2006) Chromosomal integrity maintained in five human embryonic stem cell lines after prolonged in vitro culture. *Chromosome Res* **14**, 131–137.

28. Buzzard J.J., Gough N.M., Crook J.M., Colman A. (2004) Karyotype of human ES cells during extended culture. *Nature Biotechnol* **22**, 381–382.

29. Pera M.F. (2004) Unnatural selection of cultured human ES cells? *Nat Biotechnol* **22**, 42–43.

Chapter 8

Monitoring Stemness in Long-Term hESC Cultures by Real-Time PCR

Amparo Galán and Carlos Simón

Abstract

Human embryonic stem cells (hESC) involve long-term cultures that must remain undifferentiated. The real-time PCR (RT-PCR) technique allows the relative quantification of target genes, including undifferentiation and differentiation markers when referred to a housekeeping control with the addition of a calibrator that serves as an internal control to compare different lots of reactions during the time. The main aspects will include a minimal number of cells to be analyzed, genes to be tested, and how to choose the appropriate calibrator sample and the reference gene. In this chapter, we present how to apply the RT-PCR technique, protocols for its performance, experimental set-up and software analysis, as of the gene expression of hESC lines in consecutive passages for long-term culture surveillance.

Key words: Real-time PCR, relative quantification, hESC, monitoring stemnness.

1. Introduction

The main objective of human embryonic stem cells (hESCs) is their future application in regenerative medicine (1–5). For that purpose, the main consideration to be taken into account is to establish standardized conditions to keep them in long-term cultures in an undifferentiated stage. Nowadays, most laboratories involved in hESC culture and maintenance have developed different systems to check stemness, although no real consensus exists regarding which markers should be used for routine survey. Therefore, it is important to establish standard quantitative monitoring systems that can alert of unwanted differentiation within our hESC cultures.

Over recent years, real-time PCR (RT-PCR) has become essential for nucleic acid quantification because it has unsurpassed sensitivity and a broad dynamic range. Before RT-PCR was available, conventional PCR methods based on end-point analysis were used for quantification of a specific target. Such methods have well-known limitations and disadvantages, i.e., reactions with a low initial copy number that can reach the same plateau as reactions starting with higher template concentrations and/or a different PCR efficiency. In contrast, RT-PCR provides accurate, kinetic quantification because it allows data analysis in the only phase (log-linear phase) where the amplification efficiency of each reaction is constant, as assessed by fluorescent emission (due to the incorporation of SYBR Green).

This chapter describes the use of RT-PCR to monitor seven genes involving undifferentiation and differentiation markers referring to a housekeeping gene as well as the selection of a proper calibrator. This monitoring approach involves the quantitative gene expression surveillance of typical gene markers of undifferentiation *OCT3/4* (*POU5F1*), *NANOG*, and *REX1* (*ZFP42*) (6–8). On the other hand, unwanted differentiation must also be discarded, thus rendering necessary the selection of early differentiation of ectoderm, mesoderm, and endoderm represented by Keratin (*KER*), Cartilage (*CMP*), Amylase (*AMY*) and α-fetoprotein (*AFP*), respectively (9). This monitoring system can be applied to not only compare different hESC lines, but also to survey a specific hESC line in long-term cultures. Although the LightCycler® (Roche) instrument is specified in this unit, the protocols may be utilized for any RT-PCR instrument.

The LightCycler® (Roche) system provides various experimental approaches to quantitative PCR, involving absolute and relative quantification. The most powerful is relative quantification, because the absolute values for most gene expression studies of the sample under investigation are not relevant. Various techniques express the target amount of an unknown sample in relation to another gene transcript, the so-called housekeeping gene, which is assumed to be constant. That identical concept can be achieved with the LightCycler®, which determines target gene levels in relation to the levels of a reference gene in the same sample (*see* **Table 8.1**)

Furthermore, if calibrator normalization is applied to all the result lots, two additional benefits can be derived: PCR efficiency correction, achieved by standard curves construction, and the comparison of the results in the different runs performed, i.e., to track changes over time. To correct PCR efficiency, standard curves are constructed for all the genes of interest from which RNA (cDNA) is measured, and linear regression analysis is applied to interpolate unknown sample values. The standard curve assay may be performed even if the PCR amplification

Table 8.1
Definitions to be used when performing relative quantification with LightCycler®

Term	Definition
Crossing point (Cp)	Crossing point is that where PCR amplification begins to become clearly positive above the background phase. Therefore, Cp is considered to be the most reliably proportional point to the initial concentration. Cp differs for each sample because different amounts of starting material need different numbers of cycles to reach this detection limit.
Housekeeping gene	A gene that is not only expressed constitutively, but also at the same level in all samples to be analyzed. It is used as a reference for mRNA quantification.
Calibrator	A positive sample for all genes analyzed to which all other samples are compared. It is used: – To normalize sample values within a run and between runs, and it must keep a constant ratio of the target gene expression to the reference gene expression. – To correct PCR efficiency.
Standard curves	A series of calibrator dilutions, which are used to generate standard curves for determining the efficiency of amplification in all the genes tested.
Negative control	A nontemplate control. It is necessary to be included for every lot of reactions to check for contamination and reaction artifacts.

efficiencies of the primer sets are not equal since the correction for unequal efficiencies is intrinsic to the linear regression formula.

2. Materials

2.1. Collection of Cells

1. Washing 30 colonies ($\cong 10^5$ cells) with PBS free of calcium^{2+} (Ca^{2+}) and magnesium^{2+} (Mg^{2+}) (Gibco/BRL).

2.2. RNA Extraction

1. Zymo extraction columns containing lysis buffer and washing buffer (Zymo Research).
2. RNase-free water for resuspension and quantification.
3. RNase Inhibitor (Clontech, BD, Epicentre).

2.3. Retrotranscription

1. Eco-RT MMLV Retrotranscription Enzyme (Ecogen), Oligo-dT, dNTPs, Buffer 5X (Clontech, BD), RNase Inhibitor (Clontech, BD, Epicentre).

2.4. Primers

1. Stock solutions are provided in 100-µM (Sigma) concentration.

2. Dilute to a working concentration of 10 μM with high purity water. Prepare small quantities to avoid several freeze-thaw cycles that can disturb the quality of the primers.

2.5. Real-Time Kit

1. LightCycler FastStart DNA MasterPLUS SYBR Green I® (Roche) Containing:
   a. Vial 1a (white cap) containing the Taq DNA polymerase.
   b. Vial 1b (green cap) reaction mix containing SYBR Green I®.
   c. Vial 2 containing H_2O of PCR grade.
   d. To prepare the Master Mix, pipette 14 μl from vial 1a into vial 1b. Each vial 1a contains enough enzyme for three vials of Reaction Mix. Mix gently by pipetting up and down. Do not vortex. Re-label vial 1b (green cap) to Master Mix. Always keep the Master Mix away from light. The Master Mix can be kept at −20°C in the dark for 1 week.

2.6. Agarose Electrophoresis

1. TAE 10x (Sigma), TAE 1x. Prepared with double distilled water.
2. Agarose 2% (to visualize small amplicons), prepared in TAE 1x.
3. Loading Buffer Blue Juice (10x) (Invitrogen, CA).
4. Ladder of 100 bps (Invitrogen, CA).

3. Methods

3.1. Material Preparation

3.1.1. hESC Cultures

hESC maintenance usually lasts over long periods, ranging from months to years of cultures. To assess their undifferentiation stage, samples must be obtained periodically. However, the minimum representative number of cells is a critical step. Common characterization processes in other cell cultures usually involve million-order figures. However, this is not a practical process to be applied routinely to hESC cultures, where conditions must be adjusted to take the minimum representative number of cells. In our laboratory practice, we have standardized 10^5 cells as an adjusted number to work with while performing RT-PCR for the established gene analysis. Monitoring is proposed every 4–5 passages, which could correspond to one month of passaging.

The determinant factors to be taken into account when routinely performing this technique involve a proper design of high-quality primers, setting up an appropriate calibrator, and a suitable housekeeping gene.

3.1.2. Preparation of RNA

RNA extraction is performed using the MiniRNA Isolation Kit (Zymo Research) that is based on a single-step RNA extraction/binding buffer combined with fast-spin column technology. This Kit allows the efficient isolation of total RNA from 1×10^1–1×10^5 cells. However, other similar procedures have been demonstrated to work well (*see* **Note 1**).

1. Collect cells at $400 \times g$ for 5 min at room temperature.
2. Wash cells with PBS free from Ca^{2+} and Mg^{2+}.
3. Allow the supernatant to dry as much as possible. At this stage, cells can be frozen at –80°C, or the protocol should be followed.
4. Add 200 μl RNA Extraction Buffer to the cell pellet. Resuspend the cells gently by vortexing.
5. Incubate the lysate on ice for 20 min with brief vortexing every 10 min.
6. Add one volume (200 μl) 95–100% ethanol to the lysate. Mix briefly and then incubate on ice for 10 min.
7. Transfer the mixture to a Zymo-Spin™ Column in a supplied collection tube.
8. Spin in a microcentrifuge at $\geq 9{,}300 \times g$ for 1 min and discard the flow-through.
9. Add 200 μl RNA wash buffer to the column. Centrifuge at $\geq 9{,}300 \times g$ for 1 min and discard the flow-through.
10. Repeat Step 9.
11. Transfer the column to a new RNase-free 1.5-ml microcentrifuge tube.
12. Add 10 μl RNAse free water directly to the column membrane. After 2 min, centrifuge at $\geq 9{,}300 \times g$ for 1 min to elute the RNA.
13. Quantify and assess the RNA Quality in a Nanodrop® Spectrophotometer (*see* **Note 2**).
14. The eluted RNA can be used immediately or stored at –80°C for future use after adding 0.5–1 μl RNase Inhibitor (*see* **Notes 3** and **4**).

3.1.3. Preparation of cDNA

1. Prepare a mix, containing:

Component	Volume
Total RNA	0.5–1 μg
Oligo(dT)$_{18}$	1 μl
Nuclease-free water	Up to 12 μl
Final volume	12 μl

2. Incubate the mix for 5 min at 70°C, and then chill on ice.
3. Add the following:

Component	Volume*
RNase inhibitor	0.5 μl
dNTP mix 100 mM	1 μl
5x Reaction buffer	4 μl
Water	up to 19.5 μl
Final volume	19.5 μl

* If more than 1 reaction is to be done, prepare a Master Mix by calculating volumes for $n + 1$, considering n the number of reactions.

4. Mix by pipetting.
5. Add in ice 0.5 μl of ECO-RT at 200 U/μl.
6. Incubate at 42°C for 60 min.
7. Stop the reaction by heating at 70°C for 10 min. Chill on ice.
8. Dilute the cDNA up to 100 μl, assess both quantity and quality in the Nanodrop®, and keep at –20°C or –70°C (*see* **Notes 2, 4** and **5**).

3.1.4. Primers Design

Real-time PCR results rely on the quality and the accuracy of the primers used, thus very strict parameters for their design and subsequent validation are required. Although some primers used for end-point PCR primers are also useful for RT-PCR, specific primer sets must be designed in most cases for these conditions. The concentrations of PCR reactions, such as Taq DNA polymerase, $MgCl_2$, other salts, and dNTPs, remain constant within the same chemistry (SYBR Green), thus making primers the only point of flexibility in performing assays.

Primer sets must reach a single peak in the melting curve as it is indicative of a single PCR product. Multiple peaks in this plot indicate that nonspecific products or primer dimers have formed. The formation of a single product can also be confirmed by running the PCR products on 2% agarose following the PCR run. The following steps must be taken into account for the proper design of primers.

1. Retrieve the sequence information from the appropriate source (e.g., GeneBank, or Ensembl)

2. Determine the locations of exon boundaries by aligning the mRNA sequence with its gene (i.e., GenAtlas). Some genes do not have introns, so this step may not be applied.

3. Copy the sequence into the design software (Genefisher, Primer3) for 130–300 bps expected amplicons.

4. Perform a BLAST (or equivalent) search of both the primers of the set together to verify that they will fully anneal to the correct sequence, and only to that sequence.

5. Validate the primer set according to the remaining steps of this protocol.

3.1.5. Selection of a Calibrator

The calibrator is a positive control sample with a constant ratio of the target gene to the reference gene. It plays a double role in real-time PCR experiments. Firstly, it can be used to check PCR efficiency with primers. Secondly, it is usually employed as an internal control between PCR runs thus allowing long-term studies when several LightCycler® runs must be compared over time.

In principle, any sample can be defined as the calibrator. However, it is highly recommended to choose a sample easy to obtain, since it needs to be included in every run. The calibrators suggested for hESC experiments include:

1. For undifferentiation genes: An hESC line.
2. For differentiation genes: Commercially available somatic tissues.
3. For housekeeping genes: Both the above-mentioned must be positive for reference genes.

3.1.6. Selection of Reference Genes

Housekeeping genes are a large group of genes that code for proteins whose activities are essential for the maintenance of the cell function. The detection of housekeeping gene mRNA is routinely used to control several variables that may affect RT-PCR. These endogenous controls are constitutively expressed in each experimental sample and, therefore, serve as perfect candidates for the normalization of the final result.

Commonly used housekeeping genes include *GAPDH*, albumin, actins, tubulins, cyclophilin, and ribosomal genes (10–13). However, the expression level of housekeeping genes may vary depending on the cell type analyzed, the extent of cell proliferation, differentiation, and the various experimental conditions. Specifically, *GAPDH* and *ACTB*, which are generally established as housekeeping markers, have been demonstrated to display variations that are unacceptable for reference genes as they change at different developmental stages (14–16). Thus ribosomal genes, like 18s or Rpl19, are better endogenous reference controls for RT-PCR experiments.

3.1.7. Selection of Target Genes

Consensus genes for the stemness monitoring of long-term hESC cultures are specified in **Table 8.2**, along with the designed primers and the amplified fragments.

Table 8.2
Proposed primers for differentiation and undifferentiation genes

Gene	Characteristic	Primers name	Primers sequence	Fragment (pbs)
OCT3/4	Undifferentiation	OC2FR OC2RR	CGAAAGAGAAAGCGAACCAG GCCGGTTACAGAACCACACT	157
REX1	Undifferentiation	REX-51 REX-31	GGCGGAAATAGAACCTGTCA CTTCCAGGATGGGTTGAGAA	152
NANOG	Undifferentiation	NAN-FR6 NAN-RR6	CCGTTTTTGGCTCTGTTTTG TCATCGAAACACTCGGTGAA	187
KER	Ectoderm	KER-FR KER-RR	CAGATGCTGTGTCCCTGTGT TTCAGATCCAGAAGGGGATG	130
CMP	Mesoderm	CMP-FR CMP-RR	CCGGAGCCAGGACTACATTA GGTCTTGAAGTCAGCCGTGT	160
AFP	Endoderm	AFP-FR AFP-RR	ACACAAAAAGCCCACTCCAG GGTGCATACAGGAAGGGATG	147
AMY	Endoderm	AMY-FR AMY-RR	CATCTGTTTGAATGGCGATG TTCCCACCAAGGTCTGAAAG	138
RPL19	Housekeeping	19-FR3 19-RR3	CGAATGCCAGAGAAGGTCAC CCATGAGAATCCGCTTGTTT	157

3.2. Experimental Set-Up

3.2.1. Establishing a Protocol

This procedure is available providing that a Lightcycler 2.0® (Roche) or similar equipments are used, as well as the elements included in the LightCycler FastStart DNA MasterPLUS SYBR Green I® (Roche).

3.2.1.1. LightCycler Protocol

Program the LightCycler Instrument before preparing the reaction mixes as indicated in **Table 8.3**.

3.2.1.2. Preparation of the PCR Mix

1. Place the required number of Lightcycler® Capillaries in pre-cooled centrifuge adapters or in a LightCycler® Centrifuge Bucket (*see* **Notes 6–10**).

Table 8.3
The PCR parameters that must be programmed for a normal LightCycler PCR run with the LightCycler FastStart DNA MasterPLUS SYBR Green I® Kit

Analysis mode	Cycles	Segment	Target temperature	Hold time	Acquisition mode
			Denaturation		
None	1		95°C	7 min	None
			Amplification		
Quantification	35–45	Denaturation	95°C	10 s	None
		Annealing	Primer dependent[a]	0–10 s[b]	None
		Extension	72°C	=amplicon (bp)/ 25 s	Single
			Melting curve		
Melting curves	1	Denaturation	95°C	0 s	None
		Annealing	65°C	15 s	None
		Melting	95°C Slope = 0.1°C/s[c]	0 s	Continuous
Cooling					
None	1		40°C	30 s	None

[a] For initial experiments, set the target temperature 5°C below the calculated primer Tm. Calculate the primer Tm according to the following formula: Tm 2°C (A+T) + 4°C (G+C).
[b] For typical primers, choose an incubation time of 0–10 s for the annealing step. To increase the specificity of primer binding, use an incubation time of <5 s.
[c] In all other cases, the slope will be 20°C/s.

2. In a 1.5-ml reaction tube on ice, prepare the PCR Mix for one or $n + 1$ (considering n = number of reactions) as follows:

Component	Volume
H$_2$O PCR grade (vial 2)	12 µl
Primer Forward 10 µM	1 µl
Primer Reverse 10 µM	1 µl
Master Mix, 5x conc (vial 1)	4 µl
Final volume	18 µl

3. Mix gently by pipetting up and down. Do not vortex.
   a. Pipette 18 μl PCR Mix into each LightCycler Capillary.
   b. Add 2 μl cDNA template (*see* **Note 11**).
   c. Seal each capillary with a stopper.
4. Place the adapters containing the capillaries into a standard benchtop microcentrifuge and centrifuge at $800 \times g$ for 5 s.
5. Transfer the capillaries into the sample carousel of the LightCycler® Instrument and follow the "LightCycler® Protocol".

3.2.1.3. Standard Curves. Correction of Efficiency

The accuracy of quantification depends solely on the differences in the amplification of the target and reference genes. To determine efficiency, we must construct a relative standard curve (log concentration of standard dilutions vs. cycle number). The slope of that curve can be directly converted into efficiency by the formula: $E=10^{-1/\text{slope}}$

A slope of −3.32 indicates the maximum possible reaction efficiency (2.0), which means that the amount of PCR product doubles during each cycle. However, amplification reactions operate at less than maximum efficiency because experimental conditions are not fully optimized and samples can contain PCR inhibitors. Thus, if efficiency is assumed to be 2 when PCR efficiency is less, an error is introduced that proportionally increases with the cycle number. For example, if there is a difference in the amplification efficiency of 0.05 between the target and reference genes, the calculated relative target value can be more than twofold (113%) off if it is calculated after 30 cycles. Therefore PCR efficiency correction should always be considered.

To assess the PCR efficiency correction, standard curves will not only be performed with the calibrator for all the target gene primers analyzed, but also with the reference primers (**Fig. 8.1**). For a statistically valid standard curve, the standard dilutions should span at least 3–5 orders of magnitude, and all dilutions should be within the same concentration range as the unknown ones to be analyzed. In addition, use a minimum of 4–5 serial dilutions (e.g., 1:10 dilutions) for each curve and prepare at least 3 replicates of each. Replicates are especially important for lower concentrations since the Cp distortion increases at low concentrations.

3.2.1.4. Experimental Design

Each experimental lot will include
1. Per gene analyzed:
   a. Sample/samples to be analyzed (x2)
   b. Calibrator (x2)
   c. Negative control (water x1)
2. Per passage analyzed:
   a. Undifferentiation gene markers: *OCT-4, REX-1, NANOG*
   b. Differentiation gene markers: *KER, CMP, AMY, AFP*
   c. Housekeeping gene marker: *RPL19*

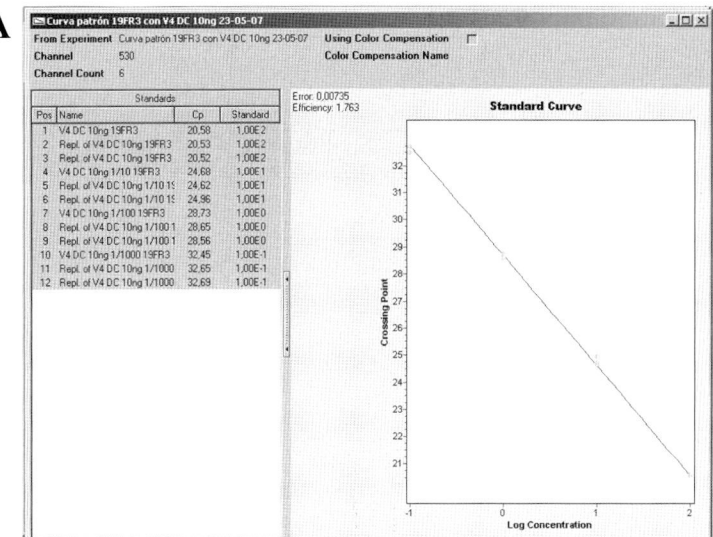

Fig. 8.1. Calibration curve with all the gene markers proposed to monitor the hESC lines. (**A**) Example of a calibration curve with the housekeeping gene *RPL19* using serial dilutions of differentiated VAL-4 (V4DC) as a template (5). (**B**) PCR efficiencies for all the target and reference genes using the calibrator as a template for both the undifferentiation (VAL-4 P42) and differentiation markers (VAL-4 DC).

GENE MARKER	CALIBRATOR	PCR EFFICIENCY
OCT-3/4	VAL-4 P42	1.622
NANOG	VAL-4 P42	1.735
REX	VAL-4 P42	1.728
KER	VAL-4 DC	1.763
CMP	VAL-4 DC	1.789
AFP	VAL-4 DC	1.982
AMY	VAL-4 DC	1.732
RPL19	VAL-4 DC	1.763
RPL19	VAL-4 P42	1.706

3.2.1.5. Data Analysis

Absolute Quantification

The LightCycler® program firstly allows to quantify by measuring the Cp of each sample (*see* **Note 12**). This value will be that which is to be normalized with both the reference and calibrator for the relative quantification study (**Fig. 8.2**).

Specificity Assessment

Melting Curves: Melting curves allow to check for amplicon specificity. Each fragment amplified has a unique and specific melting temperature (**Fig. 8.3**).

Agarose Electrophoresis. Apart from the melting curve, samples can be separated on an agarose gel (2%). If this is done immediately after the PCR run, no ethidium bromide is necessary as SYBR Green can also be visualized with UVA irradiation.

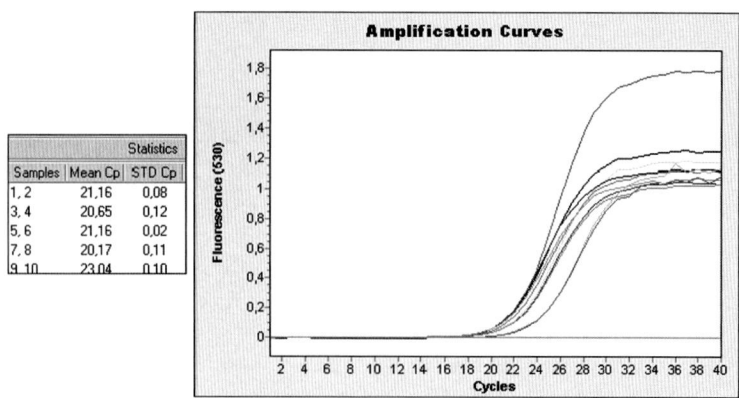

Fig. 8.2. Absolute quantification for the Oct4 experiment. Samples (1–8) and the calibrator VAL-4 P42 (9, 10) were analyzed per duplicate by the Crossing point (Cp) detection. The negative control was also included as sample 11 and corresponds to the *flat line* in the graph. All positive samples have a logarithmic exponential phase and a final plateau phase.

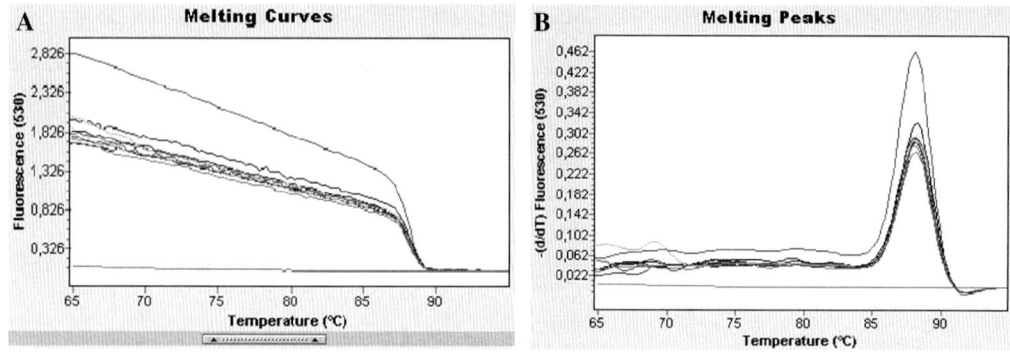

Fig. 8.3. (**A**) Melting curves of Oct4 primers with VAL-4 samples. (**B**) Positive samples have a single peak of 88°C. The negative control has no melting peak.

Relative Quantification

LightCycler® Software. The results are expressed as the target/reference ratio of each sample normalized by the target/reference ratio of each sample normalized by the target/reference ratio of the calibrator when analyzed for Relative Quantification Monocolor using the LightCycler® software (*see* **Notes 12** and **13**).

If replicates are included in the run, the software calculates and displays the median of the Cps from those replicates and the resulting Cp median values [Cp (median target) minus Cp (median of reference)]. The last two columns of the Results screen show the ratio concentration [(target concentration) divided by (reference concentration)] and the final result, the normalized ratio [(ratio concentration of sample divided by ratio concentration of calibrator)], which is the valid one (**Fig. 8.4**).

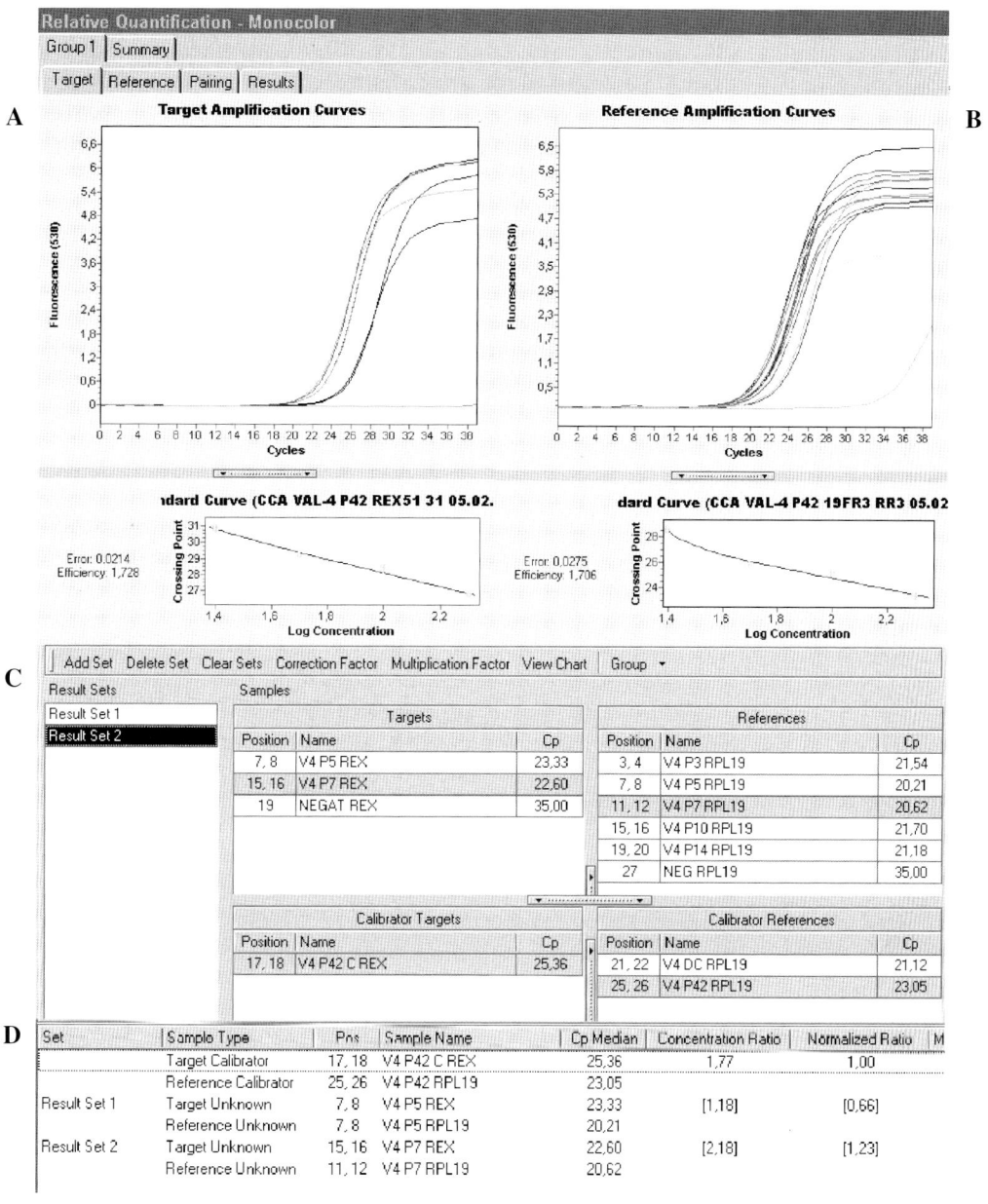

Fig. 8.4. Relative Quantification Monocolor, a four-step procedure. (**A**) The first step consists in introducing data related to Target Cp values and in importing the external standard curve performed with the calibrator. (**B**) The second step consists in introducing the corresponding reference data of the same samples in step (**A**), and also importing the external standard curve for the reference gene with the same calibrator. (**C**) The pair results of the median Cp of the target and reference samples and their corresponding calibrator data. (**D**) Normalized ratio results considering the PCR efficiency.

3.3. Interpretation of the Results

Following the assay, the resulting raw data are analyzed using second-party software, usually Microsoft Excel or equivalent. The data analyses are dependent on the type of assay performed and are outlined in detail as part of each protocol (*see* **Note 14**). RT-PCR allows the quantification of the gene expression over time when compared to a proper reference gene and calibrator. Undifferentiation markers must be present and in a similar ratio to that of the reference gene and must always be expressed constitutively, irrespectively of the culture conditions or cell cultures aging (**Fig. 8.5A**). Early differentiation markers must be absent in hESC cultures or at very low expression levels, thus showing that stemness still remains in long-term hESC cultures (**Fig. 8.5B**).

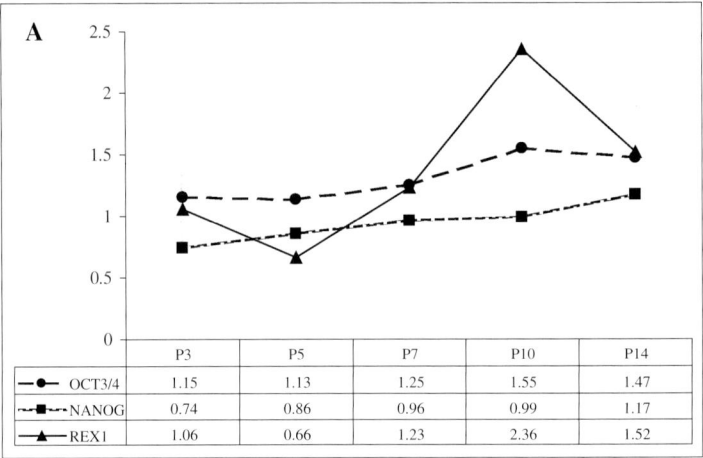

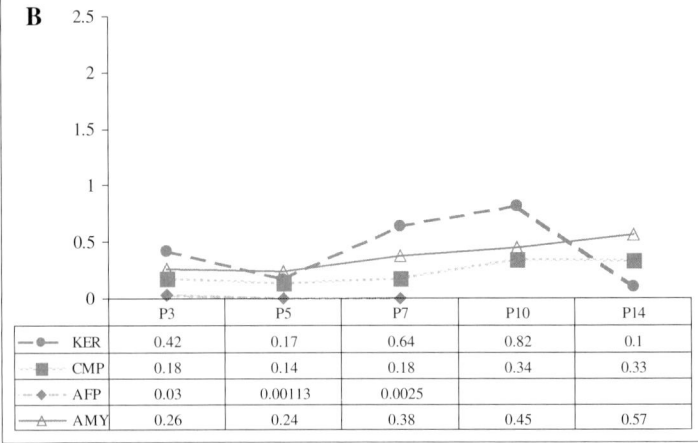

Fig. 8.5. Gene expression monitoring is the result of the relative quantification of selected undifferentiation (**A**) and differentiation (**B**) markers. The VAL-4 hESC line was analyzed in passages 3, 5, 7, 10, and 14 to monitor its undifferentiation and differentiation stages for more than a 2-month period. The data values indicate that the hESC line maintains its properties of stemness throughout the passages.

4. Notes

1. Unless stated otherwise, always use RNase-free water; treat the gel tank and cassette, the bench and other surfaces with RNAse ZAP (Sigma), use gloves and lab coat, and aerosol-resistant filter tips if possible.
2. Use high-quality RNA and cDNA. Whenever possible, use a ratio of absorbance at 260/280 of ~1.8 for DNA, and a ratio of ~2.0 for RNA. The 260/230 ratio may commonly be in the range of 1.8–2.2.
3. Aliquot RNA in the case of the calibrator and add RNase inhibitor to keep RNA at –80°C for long-term storage.
4. Follow the same RNA extraction procedure and cDNA synthesis for all samples analyzed, and in the same lot of experiments, if possible.
5. Perform a two-step reverse transcription PCR to obtain enough cDNA for all the experiments.
6. Aliquot primers to prevent contamination.
7. Keep the Reaction Mix, once prepared, for a maximum of 1 week at –20°C.
8. Do not touch the surface of the capillaries and always wear gloves when handling them.
9. Perform each step of the experiment in different areas to avoid cross-contamination. Use a hood to prepare real-time reactions. Run reactions on a separate bench and perform runs when electrophoresed in agarose separately from where real-time reactions are being prepared.
10. Sterilize the pipettes, the filter tips, and the bucket by UVA irradiation in a hood previous to use.
11. To fill real-time reactions, start from the negative control (water) followed by the most diluted template and finish with the most concentrated sample to avoid cross-contamination.
12. Since the result is expressed as a ratio, relative quantification assays are more or less independent of the initial amount of template. However, each sample must contain enough sample material to be detectable and to generate Cp values during the run.
13. The target and the reference genes are theoretically expressed at a constant ratio in the calibrator sample. However, this ratio may slightly fluctuate in different calibrator batches. For data consistency, it is very important to define a batch-specific correction factor to compensate for such batch-to-batch differences.

14. An enormous amount of data is generated when monitoring several passages, hESC lines, and genes. Therefore, an accurate evaluation by statistical analysis is necessary. Depending on the individual assay or system under investigation, various methods and/or programs may be used, including the classical standard analysis of variance (e.g., the Student's t-test), Poisson's error law, or other applied statistical systems available in the literature or on the Internet, which are suitable for each requirement.

References

1. Liew, C. G., Moore, H., Ruban, L., Shah, N., Cosgrove, K., Dunne, M., and Andrews P. (2005) Human embryonic stem cells: possibilities for human cell transplantation. *Ann Med* **37**, 521–32.
2. Yu, J., and Thomson, J. A. (2006) Embryonic stem cells. *In* Regenerative Medicine. NIH, eds. Terese Winslow, pp 1–12.
3. Guhr, A., Kurtz, A., Friedgen, K., and Loser, P. (2006) Current state of human embryonic stem cell research: an overview of cell lines and their use in experimental work. *Stem Cells* **24**, 2187–91.
4. The International Stem Cell Initiative: Characterization of human embryonic stem cell lines by the International Stem Cell Initiative. (2007) *Nat Biotechnol* **25**, 803–16.
5. Valbuena, D., Galán, A., Sánchez, E., Poo, M. E., Gómez, E., Sánchez-Luengo, S., Melguizo, D., Garcia, A., Ruiz, V., Moreno, R., Pellicer, A., and Simón, C. (2006) Derivation and characterization of three new Spanish human embryonic stem cell lines (VAL -3 -4 -5) on human feeder and in serum-free conditions. *Reprod Biomed Online* **13**, 875–86.
6. Schuldiner, M., Yanuka, O., Itskovitz-Eldor, J., Melton, D. A., and Benvenisty, N. (2000) Effects of eight growth factors on the differentiation of cells derived from human embryonic stem cells. *Proc Natl Acad Sci USA* **97**, 11307–12.
7. Henderson, J. K., Draper, J. S., Baillie, H. S., Fishel, S., Thomson, J. A., Moore, H., and Andrews, P. W. (2002) Preimplantation human embryos and embryonic stem cells show comparable expression of stage-specific embryonic antigens. *Stem Cells* **20**, 329–337.
8. Babaie, Y., Herwig, R., Greber, B., Brink, T. C., Wruck, W., Groth, D., Lehrach, H., Burdon, T., and Adjaye, J. (2007) Analysis of Oct4-dependent transcriptional networks regulating self-renewal and pluripotency in human embryonic stem cells. *Stem Cells* **25**, 500–10.
9. Xu, R. H., Peck, R. M., Li, D. S., Feng, X., Ludwig, T., and Thomson, J.A. (2005) Basic FGF and suppression of BMP signaling sustain undifferentiated proliferation of human ES cells. *Nat Methods* **2**, 185–190.
10. Eisenberg, E., and Levanon, E. Y. (2003) Human housekeeping genes are compact. *Trends Genet* **19**, 362–5.
11. Lee, P. D., Sladek, R., Greenwood, C. M., and Hudson, T. J. (2002) Control genes and variability: absence of ubiquitous reference transcripts in diverse mammalian expression studies. *Genome Res* **12**, 292–7.
12. Hsiao, L. L., Dangond, F., Yoshida, T., Hong, R., Jensen, R.V., Misra, J., Dillon, W., Lee, K. F., Clark, K. E., Haverty, P., Weng, Z., Mutter, G. L., Frosch, M. P., Macdonald, M. E., Milford, E. L., Crum, C. P., Bueno, R., Pratt, R. E., Mahadevappa, M., Warrington, J. A., Stephanopoulos, G., and Gullans, S. R. (2001) A compendium of gene expression in normal human tissues. *Physiol Genomics* **7**, 97–104.
13. Warrington, J. A., Nair, A., Mahadevappa, M., and Tsyganskaya, M. (2000) Comparison of human adult and fetal expression and identification of 535 housekeeping/maintenance genes. *Physiol Genomics* **2**, 143–7.
14. Synnergren, J., Giesler, T. L., Adak, S., Tandon, R., Noaksson, K., Lindahl, A., Nilsson, P., Nelson, D., Olsson, B., Englund, M. C., Abbot, S., and Sartipy, P. (2007) Differentiating human embryonic stem cells express a unique housekeeping gene signature. *Stem Cells* **25**, 473–80.
15. Al-Bader, M. D., and Al-Sarraf, H. A. (2005) Housekeeping gene expression during fetal brain development in the rat-validation by semi-quantitative RT-PCR. *Brain Res Dev Brain Res* **156**, 38–45.
16. Szabo, A., Perou, C. M., Karaca, M., Perreard, L., Quackenbush, J. F., and Bernard, P. S. (2004) Statistical modeling for selecting housekeeper genes. *Genome Biol* **5**, R59.

Chapter 9

Culture and Preparation of Human Embryonic Stem Cells for Proteomics-Based Applications

Charles C. King

Abstract

New challenges will arise as research into human embryonic stem (hES) cell differentiation moves from optimization and overcoming technical hurdles to mechanistic considerations. An immediate need will be to culture hES cells in the absence of contaminating feeder layers and allow for the preparation of purified DNA, RNA, and proteins to analyze changes in microRNA levels, gene expression, protein expression, and signal transduction. Purified, uniform populations of hES cells will allow researchers to better explore the biochemical mechanisms by which differentiation occurs.

Much recent work has focused upon genetic analysis of different stem cell populations. Expected variabilities between pluripotent hES cells, mesoderm, ectoderm, and definitive endoderm have been observed in microarray profiles (1–7). Interestingly, there also appears to be significant heterogeneity in mRNA expressed in different hES cell lines (8, 9). One approach to better understand how changes in mRNA levels in differentiating stem cells and individual hES cell lines relate to cell function is to study changes in signal transduction and global changes in protein expression. This chapter describes the methods routinely employed to prepare cells for analysis by traditional biochemistry (fractionation and western blotting) and proteomic analysis (2D electrophoresis/mass spectrometry and free-flow isoelectric focusing).

Key words: Human embryonic stem cells (hESCs), mouse embryonic fibroblasts (mEFs), definitive endoderm, cell fractionation, batch ion-exchange chromatography (BIEX), two dimensional (2D) electrophoresis, mass spectrometry, free-flow isoelectric focusing.

1. Introduction

Pluripotency of hES cells is maintained when cells are grown on mouse embryonic fibroblasts (mEFs), on laminin or matrigel supplemented with conditioned medium from mEFs, or on human feeder layers (10–13). Unlike mouse ES cells, signals received by hES cells from secreted chemicals and growth factors in feeder layers do not operate through the LIF/gp130 pathway (14, 15). Consequently, alternate signaling pathways, activated by the contact of hES cells to feeder layers and/or soluble factor(s) present in the conditioned media, mediate pluripotency. hES cell lines cultured on either mEFs or on laminin supplemented with conditioned medium from mEFs remained pluripotent and proliferative (15). Removal of conditioned medium resulted in the rapid loss of the pluripotency markers Oct-4, nanog, and TRA-1-60, indicating loss of "stemness". Therefore, it could be inferred that conditioned medium contains soluble factors secreted by the feeder layers that are instrumental in maintaining pluripotency. Previous work from our laboratory and others has demonstrated that some of the soluble factors secreted by mEFs can be recapitulated in the absence of mEFs (16, 17). Specifically, hES cells grown on laminin in the presence of activin A, nicotinamide (NIC), and keratinocyte growth factor (KGF) remain undifferentiated during continuous growth over 20 passages. This observation, combined with recent data indicating that hES cells cultured exclusively in the presence of defined human/recombinant chemicals and reagents can robustly differentiate into definitive endoderm (18), ectoderm (19), and mesoderm (20), provides the stem cell biologist with the initial tools to begin to examine the biochemistry of differentiation in a system free of contaminants.

2. Materials

2.1. hESC Culture and Passage

1. All water in these studies is generated using the MilliQ Plus water purification system (Millipore, Billerica, MA, USA) that has a resistance of 18.2 Ω-cm and less than 5 ppb contaminants. For simplicity, this is referred to as "water" throughout this chapter.
2. 0.1% porcine gelatin (Sigma-Aldrich, St Louis, MO, USA) diluted in water and filter sterilized (*see* **Note 1**).
3. CF1 mouse feeder layers, mitomycin-treated (ATCC, Manassas, VA, USA; cell line SCRC-1040.2).
4. Feeder layer media. Dulbecco's Modified Eagle's Medium (DMEM) containing 4 mM L-glutamine, 1.5 g/L sodium bicarbonate, and 4.5 g/L glucose (all from Invitrogen, Carlsbad, CA, USA), supplemented with 15% fetal bovine serum (FBS; HyClone, Logan, UT, USA) and filter sterilized.

5. DMEM serum replacement (DSR) medium. DMEM supplemented with Knockout Serum Replacement, glutamine, non-essential amino acids, 0.1 mM β-mercaptoethanol (all from Invitrogen), and FGF-2 (10 ng/ml) (Peprotech Inc., Rocky Hill, NJ, USA) and filter sterilized.
6. DSR + ANK (activin, nicotinamide, keratinocyte growth factor). DSR medium containing 50 ng/ml human recombinant activin A (Peprotech Inc.), 50 ng/ml human recombinant keratinocyte growth factor (KGF) (Peprotech Inc.), and 10 mM Nicotinamide (Sigma-Aldrich). Aliquoted stock solutions of activin, KGF, and Nic (1:1000) are frozen and stored at −20°C and added fresh to 50 ml of DSR medium and filtered using the Steriflip disposable vacuum filtration system (Millipore).
7. Collagenase IV (Invitrogen), 1 mg/ml in phosphate buffered saline pH 7.4 (PBS) (Sigma-Aldrich). Made fresh each passage and filter sterilized.
8. Laminin (Invirtrogen), 20 μg/ml diluted in PBS pH 7.4 and filter sterilized.

2.2. Generation of Definitive Endoderm

1. RPMI 1640 medium (Invitrogen) containing 100 ng/ml Activin A, but no serum.
2. RPMI containing 100 ng/ml Activin A in the presence of 0.2% human (Pel-Freeze, Rogers, AR, USA) or fetal bovine serum (HyClone).

2.3. Immunohistochemistry

1. Microscope cover glass slips (Fisher Scientific, Waltham, MA, USA).
2. 4% paraformaldehyde (16% stock solution from Electron Microscopy Sciences, Hatfield, PA, USA) in PBS, pH 7.4.
3. Permeabilization solution: 0.2% triton X-100 (w/v) in PBS, pH 7.4.
4. $NaBH_4$ (Sigma-Aldrich) (0.1% in PBS – FRESH) to quench auto-fluorescence and enhance antigenicity.
5. Antibody dilution buffer: 1% BSA diluted in PBS, pH 7.4 filter sterilized.
6. Secondary antibodies: AlexaFluor 488 goat anti-rabbit, AlexaFluor 546 goat anti-rabbit, AlexaFluor 488 goat anti-mouse, AlexaFluor 546 goat anti-mouse (Invitrogen).
7. Nuclear stain: Hoechst 33342 –10 mg/mL solution in water (Invitrogen).
8. Mounting media: ProLong Antifade kit (Invitrogen).

2.4. Cell Lysis and Basic Fractionation

1. Protease inhibitor cocktail. For routine cell lysis, the following final concentrations of protease inhibitor are added fresh to lysis buffer: 300 µM phenylmethylsulfonyl fluoride (PMSF), 1 mM vanadate, 50 µg/ml leupeptin, 1 µg/ml aprotinin, 1 µg/ml pepstatin, and 1 µM microcystin. The PMSF (Sigma-Aldrich) stock solution is 1 M diluted in EtOH. The vanadate (Sigma-Aldrich) stock solution is 100 mM diluted in water. The leupeptin (Sigma-Aldrich) stock solution is 5 mg/mg diluted in water. The pepstatin (Sigma-Aldrich) stock solution is 1 mg/ml diluted in 90% MeOH/10% acetic acid. The microcystin (Calbiochem; San Diego, CA, USA) stock solution is 1 mM in DMSO (Sigma-Aldrich). All protease inhibitors can be stored at –20°C for 1 year.

2. Cytoplasmic lysis buffer: 20 mM HEPES, pH 7.5, 10 mM KCl, 2 mM EDTA, 2 mM EGTA, 5 mM $MgCl_2$, and protease inhibitors. This buffer and the soluble protein lysis buffer can be directly used for batch ion exchange chromatography.

3. Soluble protein lysis buffer: The same composition as the cytoplasmic lysis buffer with the addition of 0.5% Triton X-100. This buffer and the cytoplasmic lysis buffer can be directly used for batch ion exchange chromatography.

4. Detergent insoluble protein lysis buffer: Dilute 4X sample buffer 1:4 in water. 4X sample buffer: 250 mM Tris-HCl, pH 6.8, 8% (w/v) SDS, 40% glycerol, 0.008 (w/v) bromophenol blue, 20% (v/v) β-mercaptoethanol (Sigma-Aldrich), and protease inhibitors.

5. Nuclear lysis buffer: 20 mM HEPES, pH 7.9, 420 mM NaCl, 0.1 mM EDTA, 1.5 mM $MgCl_2$, and 25% glycerol, protease inhibitors, and 150 U DNase 1 (Sigma-Aldrich).

6. Detergent insoluble lysis buffer: 5 M urea, 2 M thiourea, 2% CHAPS, 2% SB-3, 40 mM Tris-HCl, pH 7.5, 10% glycerol, and protease inhibitors.

7. Hypotonic lysis buffer: 20 mM HEPES, pH 7.5, 5 mM EDTA, and protease inhibitors.

8. P20-PM lysis buffer: 20 mM HEPES, pH 7.9, 150 mM NaCl, 5 mM EDTA, 1%v/v NP-40, and protease inhibitors.

2.5. Batch Ion Exchange Chromatography (BIEX) and Protein Desalting/Concentration

1. Binding buffer: 50 mM Tris-HCl, pH 7.5, 1 mM EDTA, 1 mM DTT (added fresh to each aliquot of buffer immediately before use), and 5 mM $MgCl_2$.

2. Elution buffer: 50 mM Tris-HCl, pH 7.5, 1 M NaCl, 1 mM EDTA, 1 mM DTT (added fresh to each aliquot of buffer immediately before use), and 5 mM $MgCl_2$. Stock solutions of buffers for BIEX (without the DTT) can be stored for at least 3 months at room temperature.

2. Sepharose Q (Amersham/Pharmacia, Uppsala, Sweden). Q Sepharose beads are supplied in a 50% EtOH solution designed to prevent microbial growth. For routine preparations of hESC (1–2 wells of a 6-well plate or 1 × 10 cm cell-culture dish); 0.15 ml of bead slurry is used to separate proteins.

3. For protein desalting/concentration, we follow the protocol of Wessel and Flugge (21). This requires reagent quality MeOH, chloroform (Sigma-Aldrich), and water.

2.6. SDS-PAGE and Western Blotting

1. Tris II (separating gel): 1.5 M Tris-HCl, pH 8.8, 0.4% SDS.
2. Tris III (stacking gel): 0.5 M Tris-HCl, pH 6.8, 0.4% SDS.
3. 30% acrylamide/bis solution (29:1), TEMED ($N,N,N,N,$'-Tetramethyl-ethylenediamie; Bio-Rad, Hercules, CA, USA), and 10% ammonium persulfate (Sigma-Aldrich).
4. Running buffer (10X): 247 mM Tris Base, 35 mM SDS, 1.9 M Glycine. Dilute 1 part in 9 parts water for use.
5. Mini Protean Tetra Cell (Bio-Rad) mini gel system.
5. Precision Plus Protein Dual Color Standards (Bio-Rad).
6. Trans-Blot Cell (Bio-Rad).
7. Transfer buffer: 25 mM Tris-HCl base, 192 mM glycine, 20% MeOH. Do not adjust the pH.
8. Immun-Blot PVDF/Filter Paper sandwiches (Bio-Rad).
9. Blocking buffer: 10 mM Hepes, pH 7.4, 500 mM NaCl, 3% bovine serum albumin (Calbiochem), 10% goat serum (Invitrogen).
10. Antibody diluent: 10 mM Hepes, pH 7.4, 500 mM NaCl, 0.2% Tween-20 (Calbiochem), 1% bovine serum albumin, 3% goat serum (Invitrogen).
11. Wash buffer: 10 mM Hepes, pH 7.4, 500 mM NaCl, 0.2% Tween-20.
12. Secondary antibodies: Goat anti-rabbit IgG H and L chain-specific peroxidase conjugate and Goat anti-mouse IgG H and L chain-specific peroxidase conjugate (Clabiochem).
13. Detection: SuperSignal West Pico Chemiluminescent Substrate and SuperSignal West Femto Maximum Sensitivity Substrate (Pierce, Thermo Fisher Scientific, Rockford, IL, USA).
14. Stripping buffer: 300 mM NaOH (Sigma-Aldrich; 1 M stock solution, diluted in water.)

2.7. 2D Electrophoresis

1. Ready Strip Immobilized pH Gradient (IPG) Strips (Bio-Rad). 7, 11, 18, or 24-cm strips can be used in a variety of pI ranges. For routine analysis, the 3–10NL (nonlinear) and 5–8 range IPG strips are recommended.

2. Protean IEF Cell (Bio-Rad).
3. 2D lysis buffer: 9 M urea, 4% CHAPS, with protease inhibitors.
4. 2D dilution buffer: 2 M thiourea, 5 M urea, 0.25% CHAPS, 0.25% Tween-20, 0.25% SB-3, 10% isopropanol, 12.5% water-saturated butanol, 0.01% bromophenol blue. 50 ml of this buffer is routinely prepared, aliquoted into 1-ml tubes, and frozen at −20°C for 6 months. Before use, 15-mg DTT (Sigma) and 0.5% ampholytes (Bio-Rad) are added to 1-mL aliquots. Tubes containing DTT/ampholytes can be frozen and reused once, but require additional 0.25% ampholytes. Depending on the pI range to be used in a specific experiment, a variety of different ampholytes can be used. A 1:3 mixture of 5–8 and 3–10 ampholytes is added to the 2D dilution buffer when separating proteins on a 3-10NL IPG strip and 5–8 ampholytes are added to the 2D dilution buffer when separating proteins on a 5–8 IPG strip.
5. Mineral Oil (Bio-Rad).
6. IPG gel equilibration buffer #1. To a 50-ml conical tube, add 9 g urea, 8.5 ml water, 1.25 ml 1 M Tris-HCl, pH 9.4 and dissolve at 37°C. Add 5 ml 10% SDS and 4 ml glycerol to give a total of 25 ml. Add 0.5 g DTT and mix. This will be enough buffer to cover 5 IPG strips.
7. IPG gel equilibration buffer #2. To a 50-ml conical tube, add 9 g urea, 8.5 ml water, 1.25 ml 1 M Tris-HCl pH 9.4 and dissolve at 37°C. Add 5 ml 10% SDS and 4 ml glycerol to give a total of 25 ml. Add 0.75 g iodoacetamide and mix. This will be enough buffer to cover 5 IPG strips. This recipe can be expanded to accommodate the number of IPG strips in the experiment.
8. Overlay Buffer – 0.5% agarose in 1X Sample Buffer (dilute 1:4 with water from a 4X Sample buffer stock; 100 ml–25 ml 1 M Tris-HCl, pH 6.8, 8 g SDS, 40 ml glycerol, 0.008 g bromophenol blue, 20 µl β-mercaptoethanol).
9. 12.5% Criteion Tris-HCl precast gels (Bio-Rad) and Criterion gel system. Use a single Criterion gel system for 1–2 gels and a Criterion Dodeca gel system for 3–12 gels. Precast gels are used for 2D gels because they greatly reduce the ariability that arises from pouring gels.

2.8. Staining

2.8.1. Silver Staining of Gels (Mass Spectrometry Compatible)

1. Fixative for silver staining – 40% methanol, 12.5% acetic acid.
2. Wash solution – 50% EtOH.
3. Buffer 1: For 1 L, add 0.13 g $Na_2S_2O_3$ anhydrous or 0.2 g/L $Na_2S_2O_3.5H_2O$ in water. This buffer is stable at room temperature for 3 weeks.

4. Buffer 2: 0.1 g/50 ml AgNO$_3$, 1 µL/ml (50 µL/50 mL) 37% formaldehyde in water. Make this buffer fresh within an hour of use.

5. Buffer 3: For 1 L, add 60 g Na$_2$CO$_3$, 1 mL 37% formaldehyde, and 30 mL Buffer 1 in water. This buffer is stable at room temperature for 3 weeks.

6. 10% acetic acid in water.

2.8.2. Colloidal Coomassie Gel Staining (Mass Spectrometry Compatible)

1. Fixative for Colloidal Coomassie – 50% methanol, 10% acetic acid.

2. Colloidal Coomassie. 17% ammonium sulfate (Fisher), 0.1% Coomassie G-250 (Bio-Rad), 3% phosphoric acid (Fisher), 34% MeOH.

    For solution #1, dissolve Coomassie Blue G250 in a minimum amount of methanol. Add phosphoric acid to the methanol/Coomassie blue solution and mix.

    For solution #2 (in a separate flask), dissolve ammonium sulfate in water. Slowly add methanol while stirring (a white precipitate will form). Gradually add solution #1 to solution #2, mix, and add immediately to the gels.

3. 10% acetic acid in water.

2.8.3. Flamingo Fluorescent Gel Staining (Mass Spectrometry Compatible)

1. Fixative for Flamingo – 40% ethanol, 10% acetic acid.
2. 10X Flamingo Fluorescent Gel Stain (Bio-Rad).
3. 0.1% (w/v) Tween-20 in water.
4. An imaging system capable of UV transillumination or laser-based detection. If a laser-based system is used to detect proteins, the sensitivity range will be 0.25–0.5 ng. If an UV-based system is used to detect proteins, the sensitivity range will be 0.5–2 ng.

2.8.4. Blot FastStain (Mass Spectrometry Compatible)

1. Blot FastStain kit (Chemicon; Temecula, CA, USA).
2. Dilute Fixer 1:10 with water. 40 ml is sufficient for one minigel.
3. Dilute Developer 1:7 with water. 35 ml is sufficient for one minigel.

2.9. Rotofor

1. Anode Buffer: 0.1 M H$_3$PO$_4$.
2. Cathode Buffer: 0.1 NaOH.
3. Protein dilution buffer: 2 M thiourea/7 M urea, 0.5% CHAPS, 0.25% Triton X-114 (digitonin at 0.5% also works well), 10% glycerol, 10 mM DTT, and 5% ampholytes (pI 3–10; Bio-Rad). This percentage of ampholytes is designed to work well with protein concentrations above 2.0 mg/ml.

If the protein concentration is between 1.0 and 2.0 mg/ml, use 2% ampholytes. For proteins with a concentration below 1.0 mg/ml, a 1.0% ampholyte solution is used. If protein precipitation occurs during focusing, the final ampholyte concentration can be increased to 8% without adversely affecting the results.

Note: If protein solubility remains a problem with the higher ampholyte concentration, this buffer can be adjusted further. Two easy and often effective changes are to replace the 2 M thiourea/7 M urea with 1–8 M urea or 0.1–2.0% glycine. Alternatively, the detergent concentrations can be adjusted empirically.

2.10. In-Gel Protein Digestion

1. Wash Solution 1: 75% 25 mM ammonium bicarbonate (Sigma-Aldrich)/25% acetonitrile (Sigma-Aldrich).
2. Wash Solution 2: 50% 25 mM ammonium bicarbonate/50% acetonitrile.
3. 25 mM ammonium bicarbonate (98.75 mg/50 ml water).
4. 1 M DTT in 25 mM ammonium bicarbonate.
5. 100 mM iodoacetamide (Sigma-Aldrich; 18 mg/ml in 25 mM ammonium bicarbonate).
6. 0.100 $CaCl_2$ (Sigma-Aldrich; 147 mg $CaCl_2$-dihydrate in 10 ml water).
7. Trypsin Proteomics Grade (Roche Diagnostics, Indianapolis, IN, USA) (20 ng/ul in 25 mM ammonium bicarbonate).

2.11. Peptide Desalting/Concentration

1. Uniplate 96 well/250 μl V-bottom plates (Whatman/GE Healthcare; Piscataway, NJ, USA).
2. 50% acetonitrile (Sigma)/0.1% trifluoroacetic acid (TFA; Sigma).
3. 5% acetonitrile/0.1% TFA.
4. Zip Tips (Millipore).
5. Matrix solution – 125 μl Alpha-Cyano-4-Hydroxycinnamic Acid (Agilent; New Castle, DE, USA), 256 μl 60% acetonitrile, 50 μl 100 mM diammonium citrate (or ammonium monobasic phosphate; both from Sigma), 5 μl 10% TFA.
6. Protein Lo-Bind tubes (Eppendorf; New York, NY, USA).

3. Methods

There are many different avenues of proteomic analysis one might undertake depending upon the interests of the laboratory. This section will focus upon generating cultures of stem cells that are free of contaminating feeder layers or other potentially contaminating animal-derived products used in the culture of hES cells. Subsequently, the outline of a variety of cell preparations will be described that can be used to probe hES cells for changes in signal transduction or protein expression. The focus will be on methods routinely used to fractionate cells for analysis of complex mixtures by 2D electrophoresis/mass spectrometry or free-flow isoelectric focusing (FF IEF). The goal is to provide a detailed roadmap to get the reader to the point where mass spectrometry analysis by MALDI-MS or MS/MS is possible. Because of the large variability in the types of mass spectrometers used in biological sciences today, the methods described here will only provide information about preparations of samples for analysis. Many other fantastic resources are available for mass spectrometry, including books edited by James, Pennington, and Sudizak.

3.1. hES Cell Culture (Feeder Layers)

1. Two NIH approved hES cell lines, HSF6 (NIH Code UC06) and H9 (NIH Code WA09) are routinely maintained on mitomycin C-treated CF-1 mouse embryonic fibroblasts (mEFs) at 37°C, 5% CO_2 in DSR medium in 35- or 60-mm dishes. The DSR medium is changed daily.

2. Cells are passaged every 3–4 days, depending on cell density.

3. The day before cells are to be passaged, treat the appropriate dishes with 0.1% gelatin for 1 h at 37°C. Wash the plates twice with PBS pH 7.4 to remove excess gelatin.

4. Quickly thaw the CF-1 mEFs in a 37°C bath, add the cells to 10 mL DSR medium in a 15-ml conical tube, and centrifuge for 3 min at 1000 rpm.

5. Aspirate the medium and resuspend cells in 8 ml fresh DSR medium and add to the gelatin-coated plates. Mitomycin-treated CF-1 fibroblasts at 5×10^6 cells will provide feeder layers for the following: 6×60-mm dishes, 2×6-well dishes, or 2×10-cm dishes.

6. To passage, first remove the conditioned DSR medium into a fresh 15-ml conical tube and wash the hES cells once with PBS pH 7.4.

7. Add filter sterilized collagenase IV (1 mg/ml) in PBS pH 7.4 and incubate for 5–7 min at 37°C.

8. Dislodge cells by pipetting and/or gentle scraping. Add the cells to conditioned media and centrifuge at $230 \times g$ for 5 min.

9. Aspirate the medium and resuspend in fresh DSR medium containing FGF-2 (*see* **Note 2**). Care should be taken to not excessively break the colonies into clusters that are too small as this will impede growth.

10. Aspirate the DSR medium from the mEFs plated the day before.

3.2. hES Cell Culture (Feeder-Layer Free)

11. Add the hES cells to the plate containing the mEFs, gently shake, and incubate at 37°C in 5% CO_2.

To change hES cells from mouse feeder layers to feeder-free conditions, cells grown under standard conditions must be switched to culture on plates treated with laminin and a different media. This process takes two steps for the change. The first step is the culture of hES cells in the absence of feeder layers, but with conditioned medium from mEFs. The second is the culture of hES cells in the presence of defined media. Depending on the samples to be analyzed, either of the two cell culture conditions can be sufficient for proteomics-based analysis. For routine analysis and mapping global protein expression in undifferentiated hES cells, culture is performed in absence of feeder layers, but with conditioned medium from mEFs. For analysis of signal transduction during differentiation and analysis of large-scale changes in protein expression that occurs during differentiation, hES cells are cultured in the presence of defined media. As with hES cells grown on mEFs, cells grown in feeder layer-free conditions are passaged every 3–4 days, depending on cell density.

1. Generate conditioned medium from mEF cells. Two days before switching hES cells to feeder layer-free conditions, treat 2 × 10 cm dishes with 0.1% gelatin for 1 h at 37°C. Wash the plates twice with PBS pH 7.4 to remove excess gelatin. Quickly thaw the CF-1 mEFs in a 37°C bath, add the cells to 10 mL DSR medium in a 15-ml conical tube, and centrifuge for 3 min at $230 \times g$.

2. Aspirate the medium and resuspend the cells in 15 ml of DSR and add to the gelatin-coated plates.

3. On the second day, remove the medium (now conditioned media) into a fresh 50-ml conical tube. Centrifuge at $230 \times g$ for 5 min to pellet any dead cells and remove the conditioned medium into a fresh tube and place at 4°C. The medium can be stored here for up to 4 days (*see* **Note 3**).

4. Add 8 ml of fresh DSR medium to the mEFs and incubate at 37°C for two days. Repeat the harvest of conditioned medium for a total of 8 days before new CF-1 mEFs need to be thawed.

5. (For either culture condition). The day of hES cell passage, incubate 35- or 60-mm dishes with filter sterilized laminin (20 µg/ml, Chemicon) in PBS pH 7.4 for 2 h at 37°C. Wash the plates twice with PBS pH 7.4 and once with DSR medium to remove excess laminin.

6. Aspirate the medium and resuspend in 8 ml fresh DSR medium and add to the gelatin-coated plates. Mitomycin-treated CF-1 fibroblasts at 5×10^6 cells will provide feeder layers for the following: 6 × 60-mm dishes, 2 × 6-well dishes, or 2 × 10-cm dishes.

7. To passage the hES cells, first remove the DSR medium into a fresh 15-ml conical tube and wash the hES cells once with PBS pH 7.4.

8. Add filter sterilized collagenase IV (1 mg/ml) in PBS pH 7.4 and incubate for 5–7 min at 37°C.

9. Dislodge cells by pipetting and/or gentle scraping. Add the cells to conditioned media and centrifuge at 1000 rpm for 5 min.

10. Aspirate the medium and resuspend in fresh mEF-conditioned medium containing 10 ng/ml basic fibroblast growth factor (FGF-2). Again, care should be taken to not excessively break the colonies into clusters that are too small as this will impede growth.

11. For routine culture, cells can be incubated in conditioned medium from mEFs (CM) supplemented with FGF-2. As with hESC grown on feeder layers, medium should be changed daily, and cell passaged as described above.

12. To prepare cells for differentiation, switch cells from mEF-conditioned medium supplemented with FGF-2 to DSR-ANK medium containing 50 ng/ml human recombinant activin A, 50 ng/ml human recombinant Keratinocyte Growth Factor (KGF), and 10 mM Nicotinamide.

13. Change the medium every other day for cells grown on laminin with DSR-ANK. hES cells cultured under these conditions are passaged weekly at 1:3 or 1:4 dilution. At least two passages under these conditions are performed before the differentiation process is initiated.

3.3. Differentiation Protocol

1. To generate definitive endoderm, hES cells grown in DSR-ANK for two passages are switched from DSR-ANK medium to RPMI medium containing 100-ng/ml Activin A in the absence of serum according to the protocol of D'Armour et al. (22, 23).

2. After 1 day, the medium should be changed to RPMI-containing 100 ng/ml Activin A and 0.2% serum and incubated for another 48 h.

3.4. Immunohistochemistry

Immunohistochemisrty of pluripotent hES cells is a rapid method to routinely determine whether cultured cell populations have maintained markers for pluripotency (Oct-4, nanog, Tra-1-60) or to quickly determine whether differentiation is successful. In addition to this technique, both RT-PCR and FACS analysis can be performed to determine the basic efficiency of the differentiation protocol.

1. During the passage of hES cells, split a portion of cells to 12-well plates containing glass coverslips coated with mEFs or laminin as described in the previous section.

2. Once grown to the appropriate conditions, cells should be rinsed twice with PBS, and then fixed with 4% paraformaldehyde for 15 min at room temperature.

3. Discard the paraformaldehyde and wash the cells once with PBS pH 7.4.

4. Lyse the cells in 0.2% triton X-100 in PBS pH 7.4 for 5 min at room temperature and wash once again with PBS pH 7.4.

5. Treat the cells for 10 min with fresh $NaBH_4$ in 0.1% PBS pH 7.4 to reduce background fluorescence and then wash again with PBS pH 7.4.

6. Add 1° antibody (Oct-4; Cell Signaling Technology, Danvers, MA, USA) in 1% BSA diluted in PBS pH 7.4 and rotate for at least 1 h at room temperature or overnight at 4°C. (*see* **Notes 4** and **5**).

7. Wash 3 times with PBS pH 7.4 to remove any unbound 1° antibody.

8. Add 2° antibody in 1% BSA diluted in PBS pH 7.4 and incubate for 1 h at room temperature covered in foil. Again, the dilutions for this step are low (1:400) and one can greatly reduce the amount of antibody used by incubating in 100-μl total volume and spotted onto a piece of Parafilm (*see* **Note 5**).

9. Wash away the unbound 2° antibody by washing 3 times with PBS pH 7.4 covered in foil.

10. If nuclear staining is to be performed, wash the glass coverslip 3 times in water. Add Hoechst stain for 20 min covered in foil.

11. Dry the coverslips slightly in air and place on 3–5 μl of Pro-Long Antifade mounting medium spotted onto a microscope slide. Adhere using clear commercially available fingernail polish and let dry in the dark.

12. The slides can be viewed using a fluorescence microscope using appropriate filters for up to 1 month if stored in the dark.

3.5. Cell Lysis and Basic Fractionation

Depending on the ultimate method used to analyze the proteins – western blots, 2D electrophoresis/mass spectrometry, free-flow IEF – different cell lysis conditions and fractionation protocols may be employed. Because the primary focus of this section is centered on analysis using proteomics-based applications, it is important that all cell preparations have as many contaminants removed prior to lysis as possible. A primary reason for switching to cells grown in the absence of feeder layers is to remove contaminating mouse cells that may influence results or confuse data interpretation. However, simply switching to a feeder-layer free system does not guarantee a clean preparation. Without proper washing, hES cells grown in conditioned medium from mEFs or in defined medium in laminin will have large contaminants from the medium (including albumin and soluble mouse proteins) and laminin itself. Therefore, it is highly recommended that before any lysis protocol is initiated, hES cells are adequately washed. The protocol is as follows:

1. To passage, first aspirate the conditioned DSR medium and wash the hES cells once with PBS.

2. Add filter-sterilized collagenase IV (1 mg/ml) in PBS and incubate for 5–7 min at 37°C.

3. Dislodge cells by pipetting or gentle scraping, add to a 15-ml conical tube containing 10 ml fresh PBS and centrifuge at $200 \times g$ for 5 min.

4. Aspirate the media, resuspend in PBS, and centrifuge at $200 \times g$ for 5 min. Repeat twice. The remaining cell pellet can now be lysed using the protocols described below.

3.5.1. Fractionation for SDS-PAGE and Western Blotting

At the appropriate time in the protocol, hES cells should be washed and prepared for lysis as described in **Section 3.5**.

1. Resuspend cell pellet in cytoplasmic lysis buffer, transfer to a fresh tube, and freeze at –80°C (*see* **Note 6**). For a 6-well plate, 100 µl of lysis buffer is used; for a 10-cm dish, 250 µl of lysis buffer is used. Alternatively, the cells can be sonicated briefly for 2×10 s using a Branson bath sonicator.

2. Centrifuge whole cell lysates at $100,000 \times g$ for 30 min at 4°C and remove the resulting supernatant containing cytoplasmic proteins into a fresh eppendorf tube.

3. Resuspend the pellet in 50 µl soluble lysis buffer containing 0.5% Triton X-100 and centrifuge at $100,000 \times g$ for 20 min at 4°C. Transfer the resulting supernatant containing the membrane-associated proteins (pellet) into a fresh eppendorf tube.

4. The remaining pellet containing insoluble proteins should be resuspended in 50 µl 1% SDS sample buffer and transferred to a new eppendorf tube.

3.5.2. Fractionation for Batch Ion-Exchange Chromatography

At the appropriate time in the protocol, hES cells should be washed and prepared for lysis as described in **Section 3.5**.

1. For samples that will be further be fractionated using batch ion-exchange chromatography (BIEX), resuspend the cell pellet in 100 µl cytoplasmic lysis buffer, transfer to a fresh eppendorpf tube, and freeze at –80°C.

2. Thaw at 4°C and pellet the insoluble material in a benchtop centrifuge set to $16,100 \times g$ for 10 min.

3. Remove the supernatant containing the soluble proteins into a fresh tube.

4. Add 50 µl nuclear lysis buffer and let sit at 4°C for 15 min, mixing occasionally.
5. Pellet the insoluble material at 4°C in a benchtop centrifuge set to 16,100×*g* for 10 min. Remove the soluble fraction that contains primarily nuclear proteins. It is recommended that a chloroform/methanol precipitation is performed on this sample before further analysis to remove the high amount of salt.
6. Resuspend the insoluble fraction in 50 µl detergent insoluble lysis buffer or whole cell lysis buffer.

3.5.3. Whole Cell Lysis for 2D Electrophoresis

At the appropriate time in the protocol, hES cells should be washed and prepared for lysis as described in **Section 3.5**. Although it is more common to fractionate proteins for 2D electrophoresis as described in the previous two sections, sometimes whole cell lysates are generated for specific purposes. When this is the case, the following protocol is employed.

1. Resuspend cell pellet in whole cell lysis buffer, transfer to an eppendorpf tube, and sonicate briefly for 5 × 10 s with a Branson bath sonicator.
2. Pellet the insoluble material at 4°C in a benchtop centrifuge set to 16,100×*g* for 10 min.
3. Remove the soluble proteins into a fresh tube and determine protein concentration using the BCA protein assay kit according to the manufacturer's specifications (Pierce).

3.5.4. Whole Cell Lysis for Free-Flow Isoelectric Focusing

A significant drawback to free-flow isoelectric focusing is the amount of sample needed for this type of experiment. Although microgram quantities of protein can be accommodated in the MicroRotofor instrument, experience has found that higher quantities of starting material yield much better results. Routinely, over 1 mg of protein is loaded per run keeping the concentration above 2.0 µg/µl. In our laboratory, free-flow isoelectric focusing is used primarily as a method to separate membrane proteins using a two-step fractionation. Traditional isoelectric focusing using IPG strips has limited ability to resolve membrane-associated proteins, although significant advances have recently been made in the field (24–28). These proteins are notorious for crashing out of solution at their isoelectric point or not entering the second dimension gel because of the electroendosmotic effect. Therefore, to analyze these proteins, separation in the first dimension is performed in solution to mainly eliminate these problems. As with the other lysis and fractionation methods described here, hES cells should first be washed and prepared for lysis as described in **Section 3.5**.

1. Resuspend cell pellet first in 500–1000 µl of hypotonic lysis buffer, transfer to a 2 ml eppendorpf tube, and freeze at −80°C.
2. Thaw at 4°C, and pellet the insoluble material by centrifugation at $100,000 \times g$ for 30 min.
3. Remove the soluble proteins (can be used for other applications) and resuspend the pellet in 250 µl P20-PM buffer. Mix well and centrifuge at $100,000 \times g$ for 30 min.
4. Remove the soluble proteins into a fresh tube and determine protein concentration using the BCA protein assay kit according to the manufacturer's specifications (Pierce).

3.6. Batch Ion-Exchange

To reduce sample complexity, routinely cytoplasmic proteins are further separated into four fractions by adsorption onto slurries of Q Sepharose High Performance beads.

1. Wash a 1:1 slurry of beads in an excess of binding buffer and centrifuge at $1,700 \times g$ for 2 min. Repeat. For 1×10-cm plate of cells, 150 µl of the stock slurry of beads in EtOH is used.
2. Add approximately 0.5 ml of hES cell cytoplasmic proteins (3–4 µg/µl) to the washed beads and mix often at 4°C for 5 min.
3. Pellet the beads by gentle centrifugation at $1,700 \times g$ for 2 min, remove the resulting supernatant to an eppendorpf tube, and label Flow Through, FT.
4. Resuspend the beads in 300 µl Binding Buffer containing 200 mM NaCl (an 80:20 mixture of binding buffer and elution buffer), mix at 4°C for 5 min, and the pellet the beads by centrifugation at $1,700 \times g$ for 2 min.
5. Remove the resulting supernatant to an eppendorpf tube and label 200-mM fraction.
6. Resuspend the beads in 300 µl Binding Buffer containing 400 mM NaCl (a 60:40 mixture of binding buffer and elution buffer), mix at 4°C for 5 min, and the pellet the beads by centrifugation at $1,700 \times g$ for 2 min.
7. Remove the resulting supernatant to an eppendorpf tube and label 400-mM fraction.
8. Resuspend the beads in 300 µl Binding Buffer containing 650 mM NaCl (a 35:65 mixture of binding buffer and elution buffer), mix at 4°C for 5 min, and the pellet the beads by centrifugation at $1,700 \times g$ for 2 min.
9. Remove the resulting supernatant to an eppendorpf tube and label 650 mM fraction (*see* **Note 7**).
10. Resuspend the beads in 300 µl Binding Buffer containing 1 M NaCl, mix at 4°C for 5 min, and the pellet the beads by centrifugation at $1,700 \times g$ for 2 min.

11. Remove the resulting supernatant to an eppendorpf tube and label 1-M fraction.

3.7. Protein De-salting, and Concentration

The following is a slight modification of the protocol of Wessel and Flugge (21).

1. In a 2-ml eppendorpf tube, add 100–300 µl of sample to be desalted.
2. Add 400 µl methanol and vortex.
3. Add 100 µl chloroform and vortex.
4. Add 300 µl water and vortex.
5. Centrifuge for 1 min at $16,100 \times g$.
6. Aspirate the top aqueous layer making sure not to disturb to protein between the two layers.
7. Add 400 µl methanol and vortex,
8. Centrifuge for 2 min at $16,100 \times g$.
9. Remove as much MeOH as possible without disturbing pellet. Air dry for 5 min and resuspend in a minimal amount of 2X sample buffer for SDS-PAGE/Western or whole cell lysis buffer for 2D electrophoresis (*see* **Note 8**).

3.8. SDS-PAGE

For routine analysis, the Bio-Rad Tetra Cell mini gel system is used with 1.5-mm spacing plates and assembled according to the manufacturer's protocol.

1. In a casting stand, prepare two 10% stacking gels, mix 9.56 ml water with 10 ml Tris II, 10 ml 30%acrylamide/bis (29:1), 50 µl TEMED, and 120 µl ammonium persulfate. Add to the gel plates. Leave approximately 2 cm for a stacking gel and overlay with 100% EtOH. The gel will polymerize in 10–15 min.
2. During the polymerization, make the stacking gel by adding 1.2 ml Tris III to 3.2 ml water, 0.7 ml 30%acrylamide/bis (29:1), and 20 µl TEMED.
3. Once the separating gel has polymerized, wash away the EtOh with ddH$_2$O and dry.
4. Add 60 µl ammonium persulfate to begin the polymerization step. Have the comb with the desired number of wells (10 or 15 well combs) ready, the stacking gel polymerizes quickly. Remove from the casting stand and assemble the gel(s) according to the manufacturer's instructions.
5. Add 1X running buffer to the inner and outer chambers of the gel box.

6. 4X sample buffer is added to each sample to give a final concentration of 1X. The samples are boiled for 5 min at 100°C and loaded into the individual wells along with 5 μl of the Precision Plus Dual Color Standards.

7. The gel is run at 200 V for 55 min.

3.9. Western Blotting

1. During the last 10 min of SDS-PAGE, the Immune Blot PVDF/Filter paper sandwiches are prepared. The PVDF membrane is incubated in MeOH for 30 s and then placed in transfer buffer until the SDS-PAGE run is finished.

2. The gel containing proteins separated by SDS-PAGE is placed into transfer buffer and the assembly Trans-Blot Cell is dissembled. The gel is laid upon the PVDF membrane, which is sandwiched between two pieces of filter paper and two fiber pads in a gel holder cassette.

3. The Trans-Blot Cell is filled with transfer buffer, and the cassette is placed in the tank such that the PVDF membrane lies between the gel and the anode.

4. The tank, containing a magnetic stir bar, is moved to a cold room and placed on stir plate and run at 260 mAmp for 4 h or overnight at 40 V.

5. The membrane is removed and efficiency of transfer is checked by the movement of the prestained markers from the gel onto the PVDF.

6. Place the PVDF in blocking buffer for 1 h or overnight. Longer times in a cold room have no effect on western blotting.

7. Rinse the blot once with wash buffer and add primary antibody in antibody diluent. Check the technical specifications for the proper dilution and incubation time.

8. Rinse the blot thrice with wash buffer for 5 min each time and then add the secondary antibody in antibody diluent. Both the goat-anti-mouse and goat-anti-rabbit antibodies should be diluted to 1:10,000.

9. Rinse the blot four times with wash buffer for 5 min each time. During the washing, prepare the SuperSignal West Pico Chemiluminescent Substrate developing solutions. Mix 5 ml of Reagent A and B together during the last wash. Place the PVDF in the Pico solution and incubate for 1–2 min. Blot dry, wrap in Saran Wrap, place in an X-ray exposure cassette, and expose to film for 5 sec to 5 min (*see* **Note 9**).

10. PVDF membranes can be reprobed multiple times. It is recommended that the lowest abundance protein is probed first because stripping/reprobing does remove some protein.

11. Place the PVDF membrane in stripping buffer for 5 min gently rocking at room temperature, wash once with wash buffer and begin the blotting process by blocking the membrane.

3.10. Two-Dimensional Electrophoresis

For standard experiments, the 11-cm Bio-Rad Protean IEF system is used. These conditions can be adjusted for other sizes of IEF (7 cm, 18 cm, 24 cm), or for other 2D systems.

1. Dilute samples containing 75–200 µg protein in 2D dilution buffer to give a final volume of 200 µl in an 11-cm tray. It should be noted that the more concentrated the protein solution, the better. The 2D dilution buffer works best when the urea/thiourea solution is high. Mix well and add to the focusing tray. Remove the backing on the IPG strip and place gel side down in the chamber. The printed pI range should be on the user's left and the gel should cover the electrodes located on either side of the focusing tray. Use a pair of tweezers to gently remove air bubbles, close the lid, and press the "rehydrate" button, then select "passive rehydration."

2. After 4 h of passive rehydration at room temperature, cover the gel with mineral oil and again remove any air bubbles with a pair of tweezers.

3. Set the instrument with the following parameters: active rehydration at 50 V for 8–10 h, no pause after rehydration, rapid voltage increase, S1=250 V for 15 min, S5=8000 V for 2.5 h, separate for 35,000–50,000 Volt-hours, and a final 500 V hold. Normally, the passive rehydration is started in the early afternoon and the main run started late in the afternoon. This allows the samples to focus overnight and be ready for the second dimension early the next morning.

4. To equilibrate the gels and prepare for the running of the second dimension, remove excess mineral oil on the IPG strips by gently blotting on paper towels. If desired, the gels may be frozen at $-80°C$ at this step.

5. Add each IPG strip to a disposable focusing tray, and overlay with 4 ml of equilibration buffer #1 for 15 min.

6. Wash briefly in SDS-PAGE running buffer for a few seconds, then overlay with 4 ml of equilibration buffer 2 for 15 min. Wash again briefly in SDS-PAGE running buffer.

7. During the incubation in equilibration buffer #2, prepare the precast Criterion Tris-HCl gels (IPG + 1 Well) for use by removing the comb and washing out the IPG well with SDS-PAGE running buffer.

8. Briefly heat the overlay buffer.

9. Using tweezers, add the IPG strip to the Criterion gel. Again, the printing on the IPG strip should be on the left side of the gel by the 1 well and should be legible. Add 8 μl of the Precision Plus Dual Color Standards to the 1 well, cover everything with overlay buffer, add SDS-PAGE running buffer (*see* **Note 10**).
10. Run the gel at 200 V for 55 min.

3.11. Staining

3.11.1. Silver Staining

1. Place gel in fixative 30–60 min. If necessary, the gel can remain in this solution overnight.
2. Wash the gel (with gentle shaking) with 50% EtOH five times for 10–15 min per wash. The stacking will become opaque during this process (*see* **Note 11**).
3. Briefly wash the gel in water (5 s) followed by incubation in buffer #1 for 60–90 s.
4. Briefly wash the gel in water (5 s), and then add freshly prepared silver stain for 15–20 min.
5. Wash twice with water for not more than 20 s per wash (*see* **Note 12**).
6. Add buffer #3 and watch for the appearance of spots. Normally, this takes between 0.5 and 10 min.
7. It is important to anticipate stopping the staining process a few seconds before optimal spot density is reached. This is by nature an empirical process. To stop developing, quickly wash the gel thrice with water (5 s each), followed by the addition of 10% acetic acid for 30 min to lower the pH and stop the staining.
8. Gels can be kept covered in 1% acetic acid for up to 1 month. If mass spectrometry is to be performed on the gels, it is recommended that the gel is scanned, analyzed, cut, destained, and frozen within 4 h of silver staining.

3.11.2. Colloidal Coomassie Staining

1. Place gel in fixative for at least 2 h. If necessary, the gel can remain in this solution overnight.
2. Add the colloidal solution to the gel and gently rock 4 h to overnight.
3. Destain using multiple washes with water and a Kem-Wipe.

3.11.3. Flamingo Staining

1. Place gel in fixative for at least 2 h. If necessary, the gel can remain in this solution overnight.
2. Dilute the 10X Flamingo solution 1:9 with water and add to the gel. Gently rock overnight covered with aluminum foil.
3. Add a 0.1% (w/v) Tween-20 solution to the gel for 10 min at room temperature.

4. Wash three times with water. Gel is stable covered in aluminum foil at 4°C for 4 months.

3.11.4. Nitrocellulose Staining

1. Place the unblocked PVDF or nitrocellulose membrane in Fixer for 2–3 min with gentle shaking.

2. Transfer the blot to the Developer solution and shake gently for 1 min. Normally, this is sufficient for bands to appear. The membrane can be refrigerated overnight for better results.

3.12. Free-Flow Isoelectric Focusing

Like traditional isoelectric focusing, free-flow isoelectric focusing (FF-IEF) isolates proteins by generating a linear pH gradient in the presence of an electrical field. In FF-IEF, proteins migrate in solution rather than a solid support. Charged proteins migrate in the electrical field until they reach a pH that matches the pI. From there, discrete protein fractions are collected and can subsequently be separated by SDS-PAGE. Membrane-associated proteins are often insoluble at their ioselectric point, making traditional 2D electrophoresis unreliable and selective depletion of membrane and larger proteins (27). FF-IEF followed by traditional SDS-PAGE provides a nontraditional separation technique for membrane-associated proteins. Correct assembly of the Micro-Rotofor apparatus is absolutely necessary for optimal results. The directions below give a detailed description of the sample preparation, but assume the reader is familiar with the correct assembly of the unit. For information on assembly of the Bio-Rad unit, please see the excellent video available from Bio-Rad (Catalog #10008759; MicroRotofor Cell Tutorial DVD).

1. Equilibrate the ion exchange membranes overnight at room temperature. The anode (cation exchange membrane; red) should be equilibrated in 0.1 M H_3PO_4, and the cathode (anion exchange membrane; black) should be equilibrated in 0.1 M NaOH.

2. Lyse 2 × 10-cm dishes of hES cells in 2 ml hypotonic lysis buffer and centrifuge at $100,000 \times g$ for 30 min at 4°C. Remove the supernatant containing the soluble proteins for other experiments if desired.

3. Resuspend the pellet in 500 µl P20-PM buffer and centrifuge at $100,000 \times g$ for 30 min at 4°C. Remove the supernatant containing the membrane-associated proteins and determine the concentration using the BCA assay.

4. Dilute the protein to 3.3 ml with Protein Dilution Buffer and load in the chamber according to the manufacturer's specifications (*see* **Note 13**).

5. Add 6 ml of 0.1 M H_3PO_4 to the anode assembly and 6 ml 0.1 M NaOH to the cathode assembly.

6. Attach the assembly to a vacuum hose, set the cooling switch to setting II (20°C), and run at 1 W set to constant power for 3–4 h (until the voltage stabilizes) (see **Note 14**).

7. Harvest the fractions according to the manufacturer's specifications and desalt/concentrate/remove ampholytes from the proteins using the chloroform-methanol precipitation.

8. Resuspend the samples in 30 µl 2X SDS sample buffer and separate using SDS-PAGE (**Section 3.8**). The samples can be subsequently stained for mass spectrometry or western blotted. Data obtained using this protocol to detect endogenous Insulin Receptor are shown in **Fig. 9.1**.

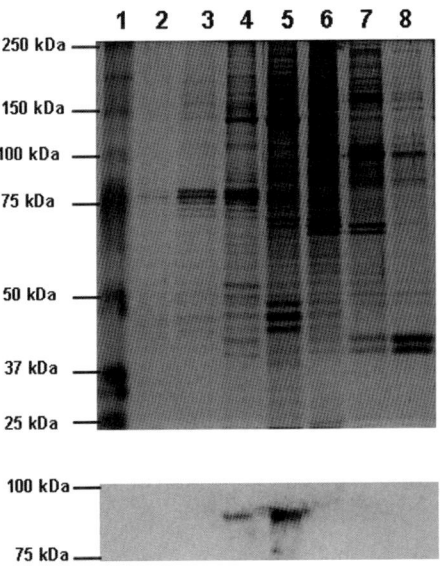

Fig. 9.1. Detection of insulin receptor in pluripotent hES cells by free-flow isoelectric focusing. Membrane preparations of hES cells were generated and fractionated on a MicroRotofor for 3 h at 1 W. Individual fractions from seven regions of the chamber were harvested and desalted/concentrated. Identical samples were separated by electrophoretic mobility using traditional SDS-PAGE. One set (*top panel*) was silver stained and the other set (*bottom panel*) was transferred to PVDF and blotted for endogenous insulin receptor using a rabbit monoclonal antibody against insulin receptor β (Cell Signaling Technologies Danvers, MA; http://www.cellsignal.com/index.jsp). The fractionation of the proteins shown on the top, as well as the detection of endogenous insulin receptor demonstrates the power of this technology.

3.13. Analysis and Quantification

The following section assumes quantitative data on changes in protein expression from a defined fraction of hES cells during differentiation is available. Depending on the end user's needs, a number of different analysis steps can be taken and is empirical for each laboratory's specific needs. For less rigorous studies, isolated proteins from 2D gels or bands from traditional SDS-PAGE gels can be selected and excised using an automated spot cutter or by hand (*see* **Section 3.14**).

1. Gels is performed using the PDQuest software version 7.4.0 (Bio-Rad).

2. Selected spots from individual gel images are first matched to at least two other replicate gels for a given time point of differentiation state using the "classic gel match" algorithm function. Refined maps for individual time points are subsequently created through use of landmarking and manual matching.

3. Once a master gel has been created for each time point, higher analysis set matching is performed to identify qualitative changes.

4. Quantitative changes are determined by normalization of all valid spots followed by matching triplicate samples from individual time points to other time points.

5. For each analysis, the 2D gel from an undifferentiated hES cell sample is set as the baseline standard. From this point, "Analysis Sets" for each of the individual time points during the generation of definitive endoderm can be created for representative gels using the normalized parameters to identify qualitative changes and a 2-fold increase/decrease threshold to identify quantitative changes.

3.14. Spot Cutting

An automated spot cutter has a number of benefits over the hand excision of spots, including tracking of samples, precise cutting with little or no contamination of other proteins, elimination of human-introduced keratins, and speed (600 spots cut/hr). However, the instruments are expensive and therefore not recommended for the casual user.

1. (If using an integrated spot cutter). Selected spots from the analysis were excised from gels using an integrated ExQuest Spot cutter (Bio-Rad) according to the manufacturer's specifications.

2. (To excise by hand). Powder-free gloves should be worn at all times during this process to reduce the potential for contamination from keratins.

3. For SDS-PAGE gel or 2D gel. Use a fresh razor blade or hole punch to cut out the band(s) of interest, making sure to cut as close to the stained portion of the gel as possible.

4. Place gel piece in Eppendorph Lo-Bind tubes or Uniplate 96-well/250-µl V-bottom plates.

5. For nitrocellulose. Use a fresh razor blade to cut out the band(s) of interest, making sure to cut as close to the stained portion of the gel as possible.

6. Place gel piece in Eppendorph Lo-Bind tubes or Uniplate 96-well/250-μl V-bottom plates.

3.15. In-Gel Protein Digestion

1. Wash each gel piece three times with wash solution 1 for 10 min in a 55°C oven. Samples in eppendorph tubes can be placed directly in a 55°C water bath.

2. Wash each gel piece once in wash solution 2 for 10 min at 55°C. The gel pieces should be colorless by the end of this step.

3. For proteins on nitrocellulose, add PVP-360 (Sigma-Aldrich; diluted in 100 mM acetic acid) to each sample and incubate for 30 min at 37°C, followed by 5 water washes and one wash in wash solution 1 (see **Notes 15–18** for optional preparative steps).

4. Wash gel pieces or nitrocellulose 3 times in 25 mM ammonium bicarbonate buffer.

5. Add trypsin diluted in 25 mM ammonium bicarbonate buffer to the protein gel pieces and incubate at 37°C for 12–18 h.

3.16. Peptide Desalting/Concentration

1. For sample preparation in 96-well plates, we use an eight-channel mulitpipetor.

2. Charge the tip with 3 × 30 μl washes in Wetting Solution.

3. Wash away the acetonitrile with 3 × 30 μl washes in Equilibration Solution.

4. Bind peptides by 7–10 aspirations/dispenses of the pipette.

5. Wash 3 × 30 μl washes in a separate vial of Wetting Solution.

6. Elute peptides into 5–8 μl of 50% acetonitrile (for MS/MS) or into 2 μl of matrix solution for MALDI-MS.

7. For routine analysis, we perform experiments on an Applied Biosystems Q-Star XL hybrid mass spectrometer that yields a MALDI-TOF fingerprint and MS/MS of the most abundant 4 spectra.

8. MALDI-TOF fingerprint data are analyzed using the online database at Rockefeller University (http://prowl.rockefeller.edu) and MS/MS data are analyzed with the Mascot online database (http://www.matrixscience.com). An example of hES cells fractionated by BIEX, separated by 2D electrophoresis, and identified by mass spectrometry is shown in **Fig. 9.2**.

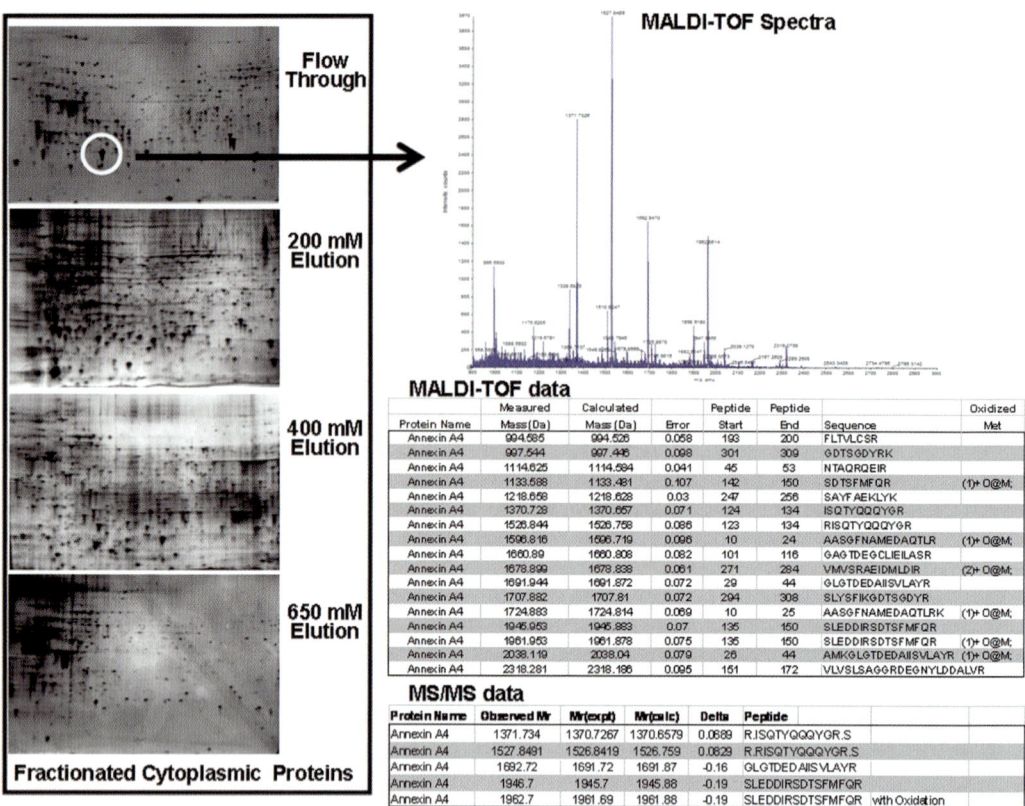

Fig. 9.2. Identification of Annexin A4 after cell fractionation, batch ion-exchange chromatography, 2D electrophoresis separation, trypsin digestion, and mass spectrometry analysis. This representative data from pluripotent hES cells demonstrated the preparative steps and expected results described in this chapter. On the left are four different 2D gels from the BIEX fractionation of soluble proteins from hES cells lysed in cytoplasmic lysis buffer. After determining the protein concentration, the samples were prepared for 2D electrophoresis by addition of 2D dilution buffer, separated, and stained with silver. Selected proteins, including the ∼36-kDa protein in the flowthrough fraction (*white circle*), were excised with an automated spot cutter, digested with trypsin, and prepared for mass spectrometry. Analysis was performed on Applied Biosystems Q-Star XL hybrid mass spectrometer that provides both a MALDI-TOF fingerprint (spectra and top table) and MS/MS of the most abundant 5 spectra. Peptide data from the MALDI-TOF run were analyzed using the online database at Rockefeller University (http://prowl.rockefeller.edu) and MS/MS data were analyzed with the Mascot online database (http://www.matrixscience.com). The results from both independent data streams suggest that Annexin A4 is the most abundant protein in the spot.

4. Notes

1. Unless otherwise stated, all cell culture reagents are filter sterilized through a 0.22-μm filter (Millipore).
2. FGF-2, Activin A, Nicotinamide, and KGF are aliquoted into 50-μl aliquots and stored at −20°C for up to 6 months.
3. Alternatively, we have found that conditioned medium can be filtered through a 0.22-μm filter and be stable for up to 7 days at 4°C without any adverse effects.

4. There is no blocking step in this protocol. Addition of antibody in 1% BSA is sufficient to block nonspecific binding in our experiments. An additional step of blocking cells in 3% filter-sterilized BSA in PBS can be added between Steps 5 and 6 in this protocol if desired.

5. Because most antibodies used for immunofluorescense are used at low dilutions (1:50–1:100), we have devised a method to save money on this potentially expensive step. The antibody is made in 100 μl total volume and spotted onto a piece of Parafilm (Fisher). The glass coverslip is incubated on the Parafilm for the appropriate time (2 hours to overnight) and then transferred back to the 12-well plate for washes.

6. At each step during the lysis and fractionation protocol, Protein Lo-Bind Tubes from Eppendorf are used unless otherwise noted. For simplicity "tube" or "eppendorf tube" will be used to refer to this specific brand of microfuge tubes throughout the chapter. Use of these tubes dramatically decreases protein loss from adhesion to the tube and prevents the leaching of chemicals into the protein mixture that can adversely effect mass spectrometry analysis.

7. Optional – for complex samples. Rarely is this portion used with hES cells. Only a small portion of samples (~5%) of the proteins are not eluted by the 650-mM wash. Additionally, multiple rounds of de-salting/concentration are required before 2D electrophoresis can be performed.

8. Samples can be dried completely using a Speed-Vac. However, we have found that a completely dried sample is much harder to resuspend.

9. If no bands appear, Pierce offers a SuperSignal West Femto Maximum Sensitivity Substrate that can be used. This is much more sensitive than the Pico kit and should gradually be titrated into the existing Pico mixture.

10. It is important to make certain that the IPG strip is placed directly in contact with the top of the Criterion gel. If gaps containing air bubbles or large amounts of overlay buffer come between the IPG strip and the gel, the efficiency of transfer of proteins between the IPG strip and gel will be dramatically decreased.

11. Alternatively, if time is short, the gel (in 50% EtOH) can be placed in the microwave for 30–40 s followed by shaking for 3–5 min in between.

12. This is an important step in the process. If longer washes are performed, the silver is washed off the gel and less intense spots will be detected.

13. Special care should be taken to remove all air bubbles from the chamber prior to the run. Failure to do so will result in erratic fluctuations in voltage that may result in improper running and poor separation.

14. Depending on power source used, it may be necessary to deactivate the Rapid Resistance Change Detection (RRCd) setting on the powerpack because resistance load detected often changes by more than 25% over brief time-periods and cause many units to shut down.

15. *(Optional)*. Remove final wash solutions from gel piece and dry for 5 minutes on speedvac. Add 100 μl of 100 mM DTT solution to the gel pieces. Vortex well and incubate at room temperature for 30 min at 55°C. This step is only for samples cut from a traditional 1D SDS-PAGE gel. The oxidation of proteins occurs between dimension 1 and 2 for 2D gels.

16. *(Optional)*. Remove from heat. Remove DTT solution and add 100 μl of 100 mM iodoacetamide. Incubate at room temperature for 45 min in the dark. This step is only for samples cut from a traditional 1D SDS-PAGE gel. The alkylation of proteins occurs between dimension 1 and 2 for 2D gels.

17. *(Optional)*. Decant all iodoacetamide solution and dry on speedvac for 5 min.

18. Add 89 μl of 25 mM ammonium bicarbonate, 1 μl of 100 mM $CaCl_2$ solution, and 10 μl of trypsin stock (400 ng).

Acknowledgements

The authors would like to thank Benjamin Lu for help developing methods for free-flow isoelectric focusing. This work was supported by the Larry L. Hillblom Foundation.

References

1. Beqqali, A., et al. (2006) *Genome-wide transcriptional profiling of human embryonic stem cells differentiating to cardiomyocytes.* Stem Cells. **24**(8): 1956–67.
2. Liu, Y., et al. (2006) *Genome wide profiling of human embryonic stem cells (hESCs), their derivatives and embryonal carcinoma cells to develop base profiles of U.S. Federal government approved hESC lines.* BMC Dev Biol. **6**: 20.
3. Zeng, X., et al. (2004) *Properties of pluripotent human embryonic stem cells BG01 and BG02.* Stem Cells. **22**(3): 292–312.
4. Rao, R.R., and Stice, S.L. (2004) *Gene expression profiling of embryonic stem cells leads to greater understanding of pluripotency and early developmental events.* Biol Reprod. **71**(6): 1772–8.
5. Rosler, E.S., et al. (2004) *Long-term culture of human embryonic stem cells in feeder-free conditions.* Dev Dyn. **229**(2): 259–74.
6. Carpenter, M.K., et al. (2004) *Properties of four human embryonic stem cell lines maintained in a feeder-free culture system.* Dev Dyn. **229**(2): 243–58.

7. Sperger, J.M., et al. (2003) *Gene expression patterns in human embryonic stem cells and human pluripotent germ cell tumors.* Proc Natl Acad Sci USA. **100**(23): 13350–5.

8. Player, A., et al. (2006) *Comparisons between transcriptional regulation and RNA expression in human embryonic stem cell lines.* Stem Cells Dev. **15**(3): 315–23.

9. Kim, C.G., et al. (2006) *Profiling of differentially expressed genes in human stem cells by cDNA microarray.* Mol Cells. **21**(3): 343–55.

10. Thomson, J.A., et al. (1998) *Embryonic stem cell lines derived from human blastocysts.* Science. **282**(5391): 1145–7.

11. Reubinoff, B.E., et al. (2000) *Embryonic stem cell lines from human blastocysts: somatic differentiation in vitro.* Nat Biotechnol. **18**(4): 399–404.

12. Xu, C., et al. (2001) *Feeder-free growth of undifferentiated human embryonic stem cells.* Nat Biotechnol. **19**(10): 971–4.

13. Richards, M., et al. (2002) *Human feeders support prolonged undifferentiated growth of human inner cell masses and embryonic stem cells.* Nat Biotechnol. **20**(9): 933–6.

14. Sato, N., et al. (2004) *Maintenance of pluripotency in human and mouse embryonic stem cells through activation of Wnt signaling by a pharmacological GSK-3-specific inhibitor.* Nat Med. **10**(1): 55–63.

15. Humphrey, R., Beattie, G.M., Lopez, A.D., King, C.C., Bucay, N., Hayek, A. (2004) *Maintenance of pluripotency in human embryonic stem cells is STAT3 independent.* Stem Cells. **22**(4): 522–30.

16. Xiao, L., Yuan, X., and Sharkis, S.J. (2006) *Activin A maintains self-renewal and regulates fibroblast growth factor, Wnt, and bone morphogenic protein pathways in human embryonic stem cells.* Stem Cells. **24**(6): 1476–86.

17. Beattie, G.M., et al. (2005) *Activin A maintains pluripotency of human embryonic stem cells in the absence of feeder layers.* Stem Cells. **23**(4): 489–95.

18. King, C.C., et al. (2008) *Generation of definitive endoderm from human embryonic stem cells cultured in feeder layer-free conditions.* Regen Med. **3**(2): 175–80.

19. Nasonkin, I.O., and Koliatsos, V.E. (2006) *Nonhuman sialic acid Neu5Gc is very low in human embryonic stem cell-derived neural precursors differentiated with B27/N2 and noggin: implications for transplantation.* Exp Neurol. **201**(2): 525–9.

20. Heiskanen, A., et al. (2007) *N-glycolylneuraminic Acid xenoantigen contamination of human embryonic and mesenchymal stem cells is substantially reversible.* Stem Cells. **25**(1): 197–202.

21. Wessel, D., and Flugge, U.I. (1984) *A method for the quantitative recovery of protein in dilute solution in the presence of detergents and lipids.* Anal Biochem **138**(1): 141–3.

22. D'Amour, K.A., et al. (2005) *Efficient differentiation of human embryonic stem cells to definitive endoderm.* Nat Biotechnol. **23**(12): 1534–41.

23. D'Amour, K.A., et al. (2006) *Production of pancreatic hormone-expressing endocrine cells from human embryonic stem cells.* Nat Biotechnol. **24**(11): 1392–401.

24. Zahedi, R.P., Moebius, J., and Sickmann, A. (2007) *Two-dimensional BAC/SDS-PAGE for membrane proteomics.* Subcell Biochem. **43**: 13–20.

25. Braun, R.J., et al. (2007) *Two-dimensional electrophoresis of membrane proteins.* Anal Bioanal Chem. **389**(4): 1033–45.

26. Molloy, M.P. (2000) *Two-dimensional electrophoresis of membrane proteins using immobilized pH gradients.* Anal Biochem. **280**(1): 1–10.

27. Santoni, V., Molloy, M., and Rabilloud, T. (2000) *Membrane proteins and proteomics: un amour impossible?* Electrophoresis. **21**(6): 1054–70.

28. Herbert, B. (1999) *Advances in protein solubilisation for two-dimensional electrophoresis.* Electrophoresis. **20**(4–5): 660–3.

Chapter 10

A Two- and Three-Dimensional Approach for Visualizing Human Embryonic Stem Cell Differentiation

Christian B. Brøchner, Peter S. Vestentoft, Niels Lynnerup, Claus Yding Andersen, and Kjeld Møllgård

Abstract

Undifferentiated human embryonic stem cells are characterized by expression of specific cell markers like the transcription factors OCT4, SOX2, and NANOG, the stage-specific embryonic antigen SSEA4, and the tumor-related antigens TRA-1-60 and TRA-1-81 and by their ability to differentiate under proper conditions into cells of the three germ layers and later into derivatives of these germ layers. Recent studies suggest a certain micro-heterogeneity of the expression of hESC markers, which demonstrates that not all cells in a hESC colony of apparently undifferentiated cells express all the expected markers. We describe a technique allowing paraffin embedding an entire hESC colony (e.g., ∼ 150 μm thick) and prepare 2-μm thick serial sections. Different staining procedures applied to individual sections produce a 2D survey of the developing hESC colony. Furthermore, a new and useful visualization of this 2D-expression pattern can be created by developing a 3D-model of the culture, based on serial paraffin sections. Individual sections are stained using individual markers. Using 3D image processing software such as Mimics or 3D-Doctor, the actual 3D-rendering of an entire colony can be accomplished. An extended version of this technique even allows for a high-magnification 3D-reconstruction of an area of interest (AOI), e.g., the developing hepatic stem cells. These techniques allow both a 2D and a 3D visualization of hESC colonies and lead to new insights into and information about the interaction of stem cells.

Key words: hESC, pluripotency markers, 3D-reconstruction, Mimics, 3D-Doctor.

1. Introduction

Pluripotent human embryonic stem cells (hESCs) were first successfully derived by Thomson and colleagues in 1998 (1) using the inner cell mass of preimplantation blastocysts. The potential use of such cells differentiated to specialized cell types for therapeutic purposes and regenerative medicine has created a huge interest in understanding how controlled differentiation occurs and what cues determine specialization and cell function.

Undifferentiated human embryonic stem cells are characterized by expression of specific cell markers like the transcription factors OCT4, SOX2, and NANOG, the surface markers stage-specific embryonic antigen SSEA4, the tumor-related antigens TRA-1-60 and TRA-1-81 and lack of expression of SSEA1 and by the ability to differentiate under proper conditions into cells of the three germ layers and later into derivatives hereof. Traditionally, characterization is done by the "one marker – one colony" – approach in which an entire colony is used for the detection of a single marker. However, recent studies suggest a certain micro-heterogeneity of the expression of hESC markers (2, 3), which demonstrates that not all cells in a hESC colony of apparently undifferentiated cells express all the expected markers. In fact, neighboring cells may express markers characteristic of totally different cell types (2), which at the macroscopic levels may reflect the different shapes and pattern of hESC colonies. In order to characterize the differentiation processes that occur in colonies of hESCs leading to more specialized cell types, an understanding of the starting point and the growth and development of the hESC colonies is needed. The present chapter describes approaches to detect expression of a number of different markers of both undifferentiated and differentiated cells in individual colonies of hESCs.

A technique is described in which an entire hESC colony (e.g., ~150 μm thick) can be paraffin embedded and serial sectioned. Different immunohistochemical staining procedures can be applied to individual sections and produce a 2D survey of the developing hESC colony. Furthermore, this 2D-expression pattern can be used to create a 3D model of the culture. Photos of individual immunohistochemical-stained sections can be aligned using an image editing program (for instance Adobe® Photoshop®). The actual 3D image can be created with software such as Mimics or 3D-Doctor allowing for new information about interaction of individual cells within a given stem cell colony.

2. Materials

Generation of specific 3-dimensional reconstructions requires technical equipment which is described in detail in **Sections 2.3, 2.4,** and **2.5**.

2.1. Cell Culture and Derivation

Legislation of the use of hESCs may vary from country to country, but hESCs can be obtained from stem cell banks or other public sources or may be generated via human embryos obtained from couples undergoing IVF-treatment, after informed consent.

Culture of hESCs requires access to sterile culture facilities including flowbenches, microscopes, and incubators, and can normally be performed in a standard equipped culture laboratory. The hESC culture medium consisted of: KO-DMEM, 15% KnockOut serum replacement, 2 mM Glutamax, 1 x nonessential amino acids, 50 IU/ml penicillin, 50 μg/ml streptomycin, 0.1 mM β-mercaptoethanol (Invitrogen), 5 mg/ml human serum albumin (State Serum Institute, Copenhagen), and 4 ng/ml basic fibroblast growth factor (bFGF; RD Systems, Minneapolis, MN, USA). Cells were maintained in humidified incubator at 37°C with an atmosphere consisting of either 6% CO_2, 7% O_2, and 87% N_2 or 5% CO_2 in air.

2.2. Paraffin Embedding and Immunohistochemistry

Only standard technical equipment, including microtome and microwave oven, of a histochemistry laboratory is needed. Graded alcohols, xylene, Histowax with melting point 56–58′, standard buffers, and an antigen retrieval system are also required. Details about antibodies used are listed in **Table 10.1**.

Table 10.1
Details on commercially available antibodies and detection

Antigen	Manufacturer	Code number	Antibody-species	Dilution	
OCT 3/4	Santa Cruz	Sc-8629	Goat		1:100
OCT4	Abcam	Ab 19857	Rabbit	TEG	1:250
NANOG	R&D	AF 1997	Goat	Citrate pH 6	1:50
TRA-1-60	Chemicon	MAB 4360	Mouse		1:100
SSEA4	Chemicon	MAB 4304	Mouse		1:75
SSEA1/CD15	BD-PharMingen	555400	Mouse		1:150
HNF-3β	Santa Cruz	Sc-6554	Goat	TEG	1:100
CD 34	DakoCytomation	M 7165	Mouse		1:25
p63	Oncogene	OP 132	Mouse	TEG	1:100
SOX2	R&D	MAB 2018	Mouse	TEG	1:70
nestin	Chemicon	MAB 5326	Mouse		1:300
PAX6	Chemicon	Ab 5409	Rabbit		1:8000
CD68	DakoCytomation	M 0814	Mouse		1:400

2.3. Alignment

1. Paraffin sections with similar slice thickness, e.g., 3 μm, preferably immunohistochemically stained – it is however possible to use unstained sections for the 3D-model.
2. Light microscopic photographs of sections at desired magnification (i.e., low for entire colony, higher for AOI).
3. Image editing software such as Adobe® Photoshop®, which is used in the present study.

2.4. Three-Dimensional Reconstruction of Entire Colony Using Mimics

1. Pictures aligned as described in **Section 2.3** must be saved in a BMP-format and all images must have an equal size (e.g., 4.933 KB).
2. Advanced 3D image processing software such as Mimics, which is used in the present study (Materialise Headquarters, Leuven, Belgium).
3. Preferably a pen tablet and pen as input-device instead of an ordinary mouse (e.g., a Wacom Board, Wacom Europe, Krefeld, Germany).

2.5. Three-Dimensional Reconstruction of AOI Using 3D-Doctor

1. Pictures aligned as described in **Section 2.3** must be saved in a BMP-format and all images must have an equal size (e.g., 4.933 KB).
2. Advanced 3D image processing software such as 3D-Doctor which is used in the present study (Able Software Corp., Lexington, MA, USA).
3. Preferably a pen tablet and pen as input-device instead of an ordinary mouse (e.g., a Wacom Board, Wacom Europe, Krefeld, Germany).

3. Methods

A special paraffin-embedding technique is required for the 2D survey. The procedure combines cutting of 2-μm thick slices of the entire colony, with immunohistochemistry, and enables the visualization of the regional distribution and colocalization of both embryonic stem cell and germ layer markers (an OCT4-staining is shown in **Fig. 10.1**). Undifferentiated hESCs are often characterized by staining for markers such as OCT4, SOX2, TRA-1-60, and SSEA4 (1, 4–8) while more differentiated germ layers may be characterized using antibodies against markers like HNF-3β, CD 34, p63, and nestin (endoderm, mesoderm, ectoderm, and neuroectoderm) (9–13). The 3D reconstructions of either an entire colony or an AOI require image editing such as alignment as well as special 3D-rendering software. This process also requires plotting of the used markers on the created figure in order to visualize the distribution of the different cell types (**Fig. 10.2**).

Fig. 10.1. Survey of a flat colony, OCT4-stained. Certain areas are specific for the OCT4-staining while other areas are found negative. Neighboring sections showed different groupings of cells based on the markers used, i.e., p63-areas were found, often in close connection to the OCT4-cells. Several times, there was a cell-type change from OCT4-positive to p63-positive on a one to one cell basis. Section number 29 out of a total of 70, colony age is 32 days. Scale bar: 1 mm.

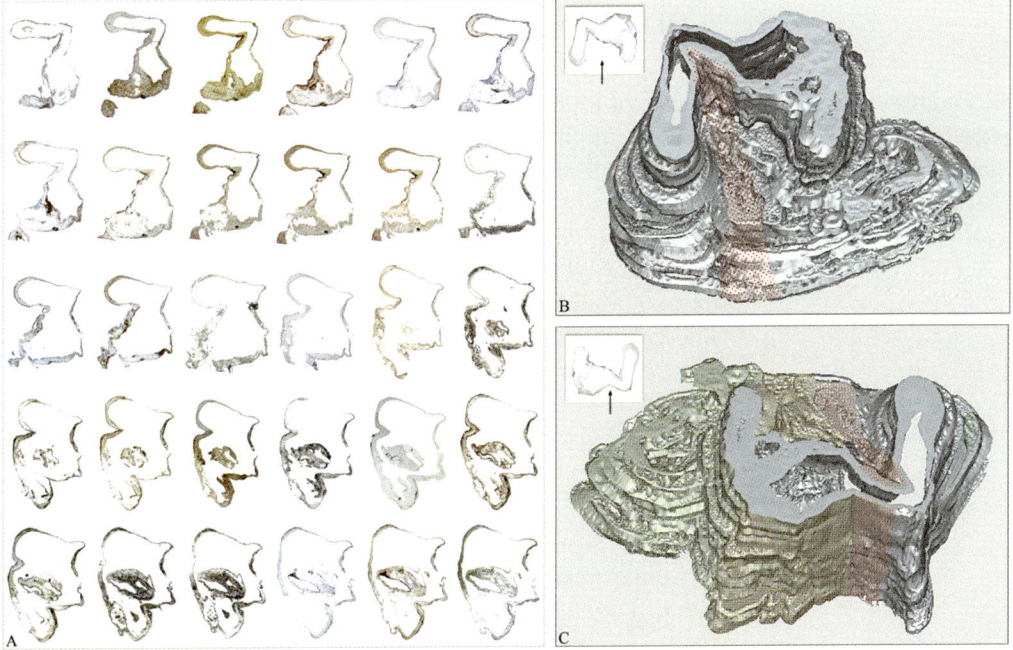

Fig. 10.2. Overview of sections used for generating the 3D model (**A**). Reconstruction of entire colony, seen from two sides (**B, C**). Inserts show direction of view and section-level. *Red dots* correspond to human embryonic stem cells positive for markers for undifferentiated embryonic stem cells, such as OCT4, NANOG, and TRA-1-60 (the "stem cell belt") while *yellow dots* correspond to a more differentiated area, positive for several known germ layer markers such as p63 (ectoderm), CD 34 (mesoderm), and HNF-3β (endoderm).

3.1. Cell Culture and Derivation

Danish legislation allows surplus human embryos from couples undergoing IVF treatment to be donated for hESC derivation.

1. Embryos were grown to the blastocyst stage and the inner cell mass was manually isolated using 27G needles attached on 1-ml syringes under a stereomicroscope.

2. Culture dishes were prepared by coating the bottom with 0.1% gelatin solution on which feeder cells (i.e., mitotically inactivated human foreskin fibroblasts (hFF) or mouse embryonic fibroblasts (MEF)) were allowed to attach at least one day prior to plating out each individual inner cell mass.

3. Colonies of outgrowing hESCs were manually isolated with hypodermic needles and passaged on fresh hFF once every 7–10 days. Areas with just a single confluent layer of cells that displayed the typical appearance of undifferentiated hESC as determined under the inverted microscope were chosen for passage.

4. 1-ml syringes attached with 27G needles were used to cut out clumps of cells containing around 100–200 cells. From four to eight clumps of cells were normally passaged from one culture dish to the new containing fresh feeder cells or they could alternatively be cryopreserved and stored in liquid nitrogen for later use. The passage number increases with one for every passage and the day of transfer is designated "the starting point" or day 1 for that particular dish.

3.2. Paraffin Embedding and Immunohistochemistry

Colonies of four well-characterized hESC lines (LRB01-04) were grown on mouse embryonic feeder cells for 4, 11, 21, 28, and 30 days in hESC culture medium (*see* Laursen et al. for detailed information (2)).

In an attempt to preserve the 2D structure of flattened colonies and the 3D structure of the more differentiated – and thus bulging – colonies, individual samples were fixed and embedded using two different strategies depending on their appearance (**A** and **B**).

1. **A**: Flat colonies were fixed in Bouin's fixative in situ in the culture dish. After 1–2 h of fixation, the fixative was replaced with 70% ethanol, and 24–48 h later the 70% ethanol was replaced with 90% ethanol.

2. Following overnight dehydration in 90% ethanol, the samples were exposed to 99% ethanol for 12 h.

3. Then the colonies were gently dissected free from the bottom using a Cell Scraper (Nunc) starting from the periphery. Colonies were then lifted carefully from the bottom of the culture dish to the small metal-embedding mould into which xylene was pipetted.

4. After 1 hour's exposure to xylene, paraffin was gently added to the embedding mould.

5. **B**: Prior to fixation the more differentiated, bulging colonies intended for 3D analysis were carefully dissected free and lifted from the bottom of the culture dish using hypodermic needles.

6. Individual colonies were placed between two cover-glasses and fixed in Bouin's fixative for 1–2 h followed by dehydration in graded alcohols.

7. Finally the colonies were cleared for 1 h in xylene and embedded in paraffin.

8. All specimens were cut in 3-μm thick serial sections. Neighboring sections were stained with different stem cell and differentiation markers to study regional distribution and colocalization of these markers in the differentiating colonies. In the present study, the panel of markers included antibodies against OCT4, NANOG, TRA-1-60, SSEA4, and SSEA1 as well as antibodies against HNF-3β, CD 34, p63, SOX2, nestin, PAX6, and CD68, but in theory any antibody against a marker may be used. (**Table 10.1**).

3.3. Alignment Using Adobe® Photoshop®

1. In order to align the pictures on the digital canvas, editing of the pictures of the paraffin sections is required using an image editing program, for instance Adobe® Photoshop®. Basing the alignment on serial sections closer to each other leads to a more accurate alignment compared to using sections far from each other (i.e., sections #1, #4, #7, #10, etc. leads to a more accurate alignment than using sections #1, #10, #20, #30, etc.) (*see* **Note 1**).

2. Initially, it is recommended that all pictures intended for the 3D-reconstruction are purified of undesired background colors using the eraser and lasso-tool. This should leave a picture only showing the cell colony.

3. Adjust canvas dimensions on all pictures to that of the largest picture.

4. Save the first slide in a BMP-format, it is recommended to use filenames showing their numerical order (e.g., 001.bmp, 002.bmp, 003.bmp).

5. Set background as white and create a new image with these newly set dimensions. Create a new layer. This image works as a background picture.

6. Select the cleansed figure on slide #1 and copy this into the new layer on the background picture.

7. Select the next figure on the following slide and copy this selection into a new layer on the background picture.

8. Set transparency on this new layer to 60%, which allows both layers to be visible at the same time. Position slide #2 exactly on top of slide #1, in order to recreate the cells' position before paraffin sectioning. Then reset transparency level to 100%. Using a selected fix point on the images may help this repositioning (it is mandatory that the fix point remains consistent throughout the entire colony).

9. Delete the primary layer, flatten image, and save this new image in numerical order, i.e., 002.bmp, if the previous image was saved as 001.bmp. Now restart from 7, until this is done with all images in the reconstruction library (*see* **Note 2**).

3.4. Three-Dimensional Reconstruction of Entire Colony Using Mimics

This description is based on Mimics version 10.01.

1. In order to create a 3D reconstruction of consecutive immunohistochemically colored paraffin-embedded sections, pictures of the sections intended for the 3D-model are saved as BMP files in the same folder and named in a numerical order such as 001, 002, and 003.

2. Open Mimics, click on File and Import Images.

3. Find your saved images and select all the images intended for the 3D-figure. Mark the "Manuel Import" box in the bottom of the screen and click Next.

4. The "Manuel Conversion Options" screen now appears. Make sure that the desired images are shown on the left-hand screen (Multiple Files), and sorted in numerical order. This can be changed using the "Num" button. Set the image size to the correct number of pixels (information you can find either by right-clicking the image, and then Preferences or by getting image information using Photoshop). The next step is setting slice distance. The exact distance between sections can be very small, e.g., 3 µm, which will lead to a very flat 3D-figure (depending on the total number of sections used). Hence, it is recommended to select a much higher value, e.g., 500 µm, leading to a 3D-visualization which is more easily viewed. When desired slice distance is selected, click OK.

5. Make sure all images are selected on the next screen and press the Convert button.

6. A Mimics project is now created and opened, showing the Change Orientation window. Here the image orientation is set (Anterior/Posterior, Top/Bottom, Left/Right). Click OK when adjusted.

7. The current project is now ready for modifications before the actual 3D calculation. First contrast needs to be set as well as threshold. These values are applied to all images and by scrolling through them, the effect of the threshold value can

be viewed on the individual images. Normally, the best pictures (with less holes or blank/white areas) intended for the reconstruction are obtained by selecting a low value for the lower threshold value and a high value for the upper threshold value.

8. Additional useful image editing can be done using the Segmentation menu. Here, tools such as Thresholding, Region Growing, Edit Masks, and Calculate 3D can be found. The Region Growing tool makes it possible to split the segmentation created by thresholding into several objects and to remove floating pixels. Click the Region Growing button, and select Source and Target mask. By clicking on one point in the source area (the cell culture), a segmentation calculation is commenced. All points in the current segmentation that are connected to the selected point will be used to form a new mask, given a new color (*see* **Note 3**).

9. To make this new mask active, deselect the original mask (by clicking on the glasses next to it) and enable the new mask, by clicking on the glasses. Fine editing can now be done using this new layer. By reviewing the individual images, the effects of this new mask can be seen, and where needed further fine-tuning can be done before building the 3D-representation.

10. All manual editing functions are performed on the active mask. Using the Edit Masks tool, it is possible to draw, erase, or restore the image with a certain threshold value. The tool can be set to either a circle or a rectangle and can have different sizes. On the newly created mask, adjustments can now be made (cells deselected during Region Growing can be redrawn into the mask, and floating pixels can be deleted). This action should be performed on all images, leaving a final result of all colored cells.

11. On the mask tab, the selected mask is highlighted. Make sure this is the mask containing the colored cells from the previous steps. Now, click on the Calculate 3D button either in the Segmentation Menu or in the Mask Tab. A Calculate 3D dialog will pop up, giving the choice of selecting quality. High quality will give long rendering time but a better 3D-recreation. Press Calculate to start calculating the 3D. When done, the 3D recreation will automatically appear on the screen.

12. Depending on the number of markers used, these can now be added to the figure. This is done by adding a new mask (rename it like the marker used), and on this new mask only highlight the cells positive for the marker (or markers, if it is a general undifferentiated stem cell area). The Edit Masks menu is useful for this purpose, and it should be carried out on all images involved.

13. Now calculate the 3D on this new layer. When done, both 3D-figures are shown on the screen. By clicking the glasses next to either of these, they will be selected or deselected (shown/not shown).

14. Using the Transparency tool in the 3D menu, it is possible to view cells deeper within the culture, which may well be positive for certain markers used. This can be toggled on or off as well as the 3D-figure can be rotated in all directions for further visualization.

15. Continue this procedure with as many layers as desired.

3.5. Three-Dimensional Reconstruction of Area of Interest Using 3D-Doctor

This description is based on 3D-Doctor version 4.0.061212.

1. In order to create a 3D reconstruction of consecutive immunohistochemically stained paraffin-embedded sections, pictures of the AOI are saved as TIFF or BMP files in the same folder and named in a numerical order such as 001, 002, and 003.

2. Start 3D-Doctor and press "File→New Stack" in case the "Create 3D Image Stack" window has not automatically appeared.

3. Press the "Add Files" button in the "Create 3D Image Stack" window and choose all the photos to be used.

4. Press "Open" and the photos will be added to the "Create 3D Image Stack Window". Then press "Sort by Name" securing that the photos will be arranged in the numerical order.

5. Press "Save List" and choose a directory to save the chosen photos as a file list (.lst). Press "Save". In the "Create 3D Image Window" press the "Open" button. This will open the main window enabling photo editing.

6. In order to commence the editing process, the photos have to be converted to grayscale photos. This is done by pressing "Image" → "Conversion" →"24 RGB → GrayScale". Choose a name (erase the ".tiff" already written in the name field) and a directory where to save the converted photos and then press "Save". In the dialogue box "Would you like to open image file?" that appears choose "Yes".

7. A new window then appears containing a tile of the grayscale photos on the right hand and an enlarged photo on the left-hand side. By double-clicking on a photo from the tile, it will be enlarged and become editable on the left-hand side.

8. If the structure to be 3D rendered is not perfectly aligned from photo to photo, an automatic or a manual alignment may be necessary. In order to do an automatic alignment of the photos it is preferable to choose a region of interest that the alignment will be based upon. Choosing a region of

interest is not a "must," but it does specify what structure the alignment should be based upon and it lowers the processing time. Press "Edit" → "Region of Interest (ROI)" → "ROI Tool". Encircle the structure that the alignment is to be based on by "left-clicking" around the structure on the enlarged photo. "Double-clicking" ends the procedure. Make sure that the desired area is included on each photo by double-clicking and inspecting each photo on the tile. If the region of interest requires modification, right click on the enlarged photo and choose the desired modification mode. When satisfied, right click on the enlarged photo and press "Done."

9. The automatic alignment is executed by pressing "Image" → "Auto Alignment". In the "Auto Image Alignment Dialog" that appears choose the range of photos to align, whether the alignment should based on matching the photos to the current enlarged one, or between neighbors and how many pixels the program is allowed to move the photos. Press "Save Image As" and choose a directory to save the aligned photos in. Enter a filename for the aligned photos in the name field and press "Save". Finally, press "OK" and the software will compute an optimal alignment based on the ROI. Press "Yes" in the "3D-Doctor" dialog box that eventually appears. This will open the aligned file (*see* **Note 4**).

10. The next procedure is to decide which structures that should appear in the 3D reconstruction. Click "3D Rendering"→ "Interactive Segment…" and then click "Set Object" in the "Interactive Segmentation" dialogue box that appears. Choose a color that will be given to the desired object and press "Ok." Click on the word "Default" to give the current object a specific name, for instance the name of the cell type being 3D rendered. Pressing the tab "Add" gives the option to add new objects with specific colors to the 3D structure. On the right-hand side, the objects can be chosen to be "On," "Off," and "Current." "On" enables a desired object to be visualized, "Off" hides the object, and "Current" dictates what object is currently being modified by the user. Click "Ok" and you will return to the "Interactive Segmentation" dialogue box.

11. Choose an image threshold range either by entering values or by dragging the vertical bar. The chosen threshold should give as much stain of the desired object and as little background as possible. Choose the boundary type for the object, whether the same object on each consecutive photo should be connected or not and finally press "Segment Plane" (*see* **Note 5**).

12. In order to move to another image, press "Prev Plane" or "Next Plane" and carry out the segmentation process for the objects of choice again. When the segmentation process has been carried out, press "Finish" in the "Interactive Segmentation" dialogue box.

    The segmentation process is likely to have segmented other structures than the ones intended. Therefore some editing has to be done. First make sure that the object that is to be modified is set to "Current" and the remaining objects are set to "On" or preferably "Off." This is done by right clicking on the enlarged photo and choosing "Object Settings." Having decided which object to modify press "Ok" in the "Object Management" dialogue box. Once again right click on the enlarged photo and choose "Boundary Editor" in the pop-up box. This enables the user to do a wide range of useful editing procedures such as "Trace Boundary" and "Delete Boundary." The "Trace Boundary" function enables the user to draw and thereby add new boundaries to the chosen object. With this function enabled, clicking and holding down the left mouse button will draw a boundary following the path of the mouse cursor. To complete the boundary, double click the left mouse button. With the powerful tool "Snap to Edges" enabled, the boundary will not only follow the line dictated by the path that the mouse-cursor moves but at the same time it will adhere to color gradients in the photo. This can give more accurate boundaries.

13. To delete boundaries, right click on the enlarged photo and choose "Delete Boundary." In this mode, left clicking on a boundary will delete it. If boundaries in larger areas are to be deleted, the process can be speeded up by pressing and holding the "Shift" key on the keyboard while at the same time clicking and holding the left button on the mouse while the cursor is on the enlarged photo. By moving the mouse, a rectangle will be made on the enlarged photo with the size of the rectangle dictated by the mouse movements. By right clicking on the enlarged photo and pressing "Delete Boundary" all of the boundaries within the rectangle will be deleted. Errors can be undone by pressing and holding down the control key "Ctrl" while pressing the "Z" key on the keyboard. This shortcut undoes the last modifications made to the enlarged photo.

14. To modify another photo, double click on the chosen one on the tile so that it will appear enlarged on the left-hand side. Repeat the modification process for each object on each photo. When finished, right click on the enlarged photo and press "Done."

15. Press "Edit" → "Calibrations" and enter the values required for the 3D structure.

    The actual 3D rendering is done by pressing "3D Rendering"→"Surface Rendering"→"Complex Surface" or "Simple Surface." In the "Complex/Simple Surface Rendering" dialogue box that appears, enter values of your choice or keep the recommended values and press "Ok." The software will then calculate the 3D structure.

16. Save the 3D-reconstruction as a .SUF file by pressing "File"→"Save Model...." In the "Save Surface Model File" dialogue box, choose where to save the 3D model or accept the proposed directory and press "Ok."

17. Close the "3D-Surface Model" model box and save the entire project as a "Project File" (*prj). This is done by pressing "File"→"Save"→"Save Project As." In the "Save As" dialogue box that then appears, choose a directory to save the project in and name the project. Press "Save." It is generally recommended to save the entire project at frequent intervals during the entire modeling process.

4. Notes

1. In case the entire colony should be recreated, and the light microscopic camera cannot contain the entire colony in one picture, it is possible to divide the colony into multiple pictures and then recombine them using the Photomerge-function in Photoshop®. When doing this, it is important to retain overlapping regions in the pictures in order to recombine these correctly.

2. It is advisable to save one image containing layers. When alignment is done with the newly added layer, flatten image and save this as the next BMP-image. Then undo the flatten-image action, delete the older of the two layers, and then add the next layer. This should facilitate the alignment process.

3. Region growing alone may not be sufficient if all cells are not all connected to each other. Instead, an area around the cells can be selected and a region growing can be performed using this selection.

4. Alternatively, the photo alignment can also be done manually. Rather than choosing a region of interest, a manual alignment is based upon giving structures appearing on each photo a certain number with the same number appearing on the same structure on each photo. To do this press "Image" →

"Alignment" → "Define Markers." Double-click on a photo on the tile so it appears enlarged on the left-hand side. Left clicking on a desired spot on the enlarged photo will assign the spot the number "0." Next clicking at another spot will give it the number "1" and so on. Double click on the next photo on the tile and carry out the previous procedure in the exact same order so in the end each consecutive photo has the same number on the structures alike. Right clicking on an enlarged photo will give the option to modify the markers. The alignment is then carried out be pressing "Image"→"Alignment"→"Align Image…." Choose a directory and a file name for the alignment file, press "Save" and an alignment will be computed.

5. If you wish to segment the same image but for another structure press "Set Object," choose the desired object in the "Object Management" dialogue box by setting it to "Current." Press "Ok" and repeat the segmentation process for the chosen object.

Acknowledgments

The authors would like to thank Sussi Forchhammer, Hanne Hadberg, Pernille Froh, and Ha Nguyen for the expert technical assistance. Keld B. Ottesen for the final layout of the photos.

References

1. Thomson JA, Itskovitz-Eldor J, Shapiro SS et al. (1998) Embryonic stem cell lines derived from human blastocysts. Science, 282:1145–1147.
2. Laursen SB, Møllgård K, Olesen C et al. (2007) Regional differences in expression of specific markers for human embryonic stem cells. Reprod Biomed Online 15:89–98.
3. Stewart MH, Bosse M, Chadwick K, Menendez P, Bendall SC, Bhatia M. (2006) Clonal isolation of hESCs reveals heterogeneity within the pluripotent stem cell compartment. Nat Methods, 3:807–815.
4. Boiani M, Schöler HR. (2005) Regulatory networks in embryo-derived pluripotent stem cells. Nat Rev Mol Cell Biol, 6:872–884.
5. Brivanlou AH, Gage FH, Jaenisch R, Jessell T, Melton D, Rossant J. (2003) Stem cells. Setting standards for human embryonic stem cells. Science, 300:913–916.
6. Heins N, Englund MC, Sjoblom C et al. (2004) Derivation, characterization, and differentiation of human embryonic stem cells. Stem Cells, 22:367–376.
7. Hoffman LM, Carpenter MK. (2005) Characterization and culture of human embryonic stem cells. Nat Biotechnol, 23:699–708.
8. Trounson AO. (2006) The production and directed differentiation of human embryonic stem cells. Endocr Rev, 27:208–219.
9. Brunner HG, Hamel BC, Bokhoven HH. (2002) P63 gene mutations and human developmental syndromes. Am J Med Genet, 112:284–290.
10. Department of Health and Human Services. Stem Cells. (2008) Scientific Progress and Future Research Directions. Appendix E: Stem Cell Markers. (Accessed April 12th

2008 at http://stemcells.nih.gov/info/scireport/appendixE.asp).

11. Mandal A, Tipnis S, Pal R et al. (2006) Characterization and in vitro differentiation potential of a new human embryonic stem cell line, ReliCellhES1. Differentiation, 74:81–90.

12. Mikkola M, Olsson C, Palgi J et al. (2006) Distinct differentiation characteristics of individual human embryonic stem cell lines. BMC Dev Biol, 6:40.

13. Yang A, McKeon F. (2000) P63 and P73: P53 mimics, menaces and more. Nat Rev Mol Cell Biol, 1:199–207.

Chapter 11

Immunoflourescence and mRNA Analysis of Human Embryonic Stem Cells (hESCs) Grown Under Feeder-Free Conditions

Aashir Awan, Roberto S. Oliveri, Pernille L. Jensen, Søren T. Christensen, and Claus Yding Andersen

Abstract

This chapter describes the procedures in order to do immunofluorescence (IF) microscopy and quantitative PCR (qPCR) analysis of human embryonic stem cells (hESCs) grown specifically under feeder-free conditions. A detailed protocol outlining the steps from initially growing the cells, passaging onto 16-well glass chambers, and continuing with the general IF and qPCR steps will be provided. The techniques will be illustrated with new results on cellular localization of transcriptional factors and components of the Hedgehog, Wnt, and PDGF signaling pathways to primary cilia in stem cell maintenance and differentiation. Furthermore, a sample qPCR experiment will be shown illustrating that these techniques can be important tools in answering basic questions about hESC biology.

Key words: hESCs, immunofluorescence microscopy, qPCR, feeder-free, primary cilia.

1. Introduction

Stem cell research is a new research field whose potential benefits include understanding some of the molecular mechanisms behind a wide variety of human pathologies in the hopes of obtaining therapeutical tools in dealing with these conditions. These pluripotent embryonic stem cells can give rise to all three germinal layers and can differentiate to form specific cell types dependent on the cell-culturing conditions present. They are seen as a cure for many human diseases including Alzheimer's disease and different types of cancer (1). However, the molecular and cellular functions leading to proliferation and differentiation still have yet to be characterized completely.

One set of proteins that are examined specifically for human embryonic stem cells (hESC) include the transcription factors, OCT4, SOX2, and NANOG. This triumvirate of proteins is important for stem cells whose interplay is thought to be responsible for the maintenance of the undifferentiated state. For example, it is thought there are some 17,000 transcriptional targets just for the OCT4 protein alone (2). Therefore, it is critical to examine the roles of these transcription factors in how they may help to regulate pluripotency of stem cells. One way of doing this is by examining their temporal and spatial locations and by quantifying their relative levels under different experimental conditions. The present chapter employs immunofluorescence microscopy and RNA analysis to illustrate the importance of these two techniques in analyzing hESCs.

We also examined the appearance of primary cilia in hESCs and the localization of signaling pathways, which may impinge on the mechanisms that control stem cell maintenance, differentiation, and self-renewal. Primary cilia are microtubule-based, nonmotile organelles that emerge as solitary organelles during growth arrest on many eukaryotic cells, including most cells of the human body (3) to coordinate a plethora of signaling pathways in cell cycle control, survival, migration, and differentiation (4–6). Consequently, defects in assembly or function of the primary cilia lead to a plethora of developmental defects, diseases, and disorders, including polycystic kidney and liver disease, Bardet Biedl syndrome, obesity and diabetes, heterotaxia, pulmonary dysfunction, holoprosencephaly, skeletal defects, retinal degeneration, and cancer. Also, hESCs were recently shown to possess primary cilia and with Hedgehog signaling that may control cell differentiation (7, 8). Therefore, studies in hESC primary cilia in feeder-free cultures may help understanding the complex signaling mechanisms that control early differentiation of hESCs, which have great potential for use in research and regenerative medicine.

2. Materials

2.1. Growing hESCs Under Feeder-Free Conditions

1. 2 X 16-well Lab-Tek Chamber Slide (Nalge Nunc, Rochester, NY, USA, Cat # 178599)
2. Petri dish (BD Bioscience, Franklin Lakes, USA, Cat # 353003)
3. 35-mm dish with confluent layer of mitotically inactivated human foreskin fibroblasts (hFF) (American Type Culture Center (ATCC), Manassas, VA, USA Line # CCD-1112Sk)

4. Trypsin-resistant hESC line
5. 0.1% gelatin diluted in PBS
6. Sterile PBS (prewarmed at 37°C)
7. IMDM medium (IMDM (Gibco, Carlsbad, CA, USA, Cat # 21980) + 10 % FBS (Invitrogen, Cat # 10108-165) + Penicillin/streptomycin (prewarmed at 37°C))
8. hESC medium (Knockout DMEM (Gibco, Cat # 10829) + 15 % Knockout Serum replacement + bFGF (4 ng/mL) (R&D, Minneapolis, MN, USA, Cat # 234-FSE)+ Pen/Strep + 1X nonessential amino acids (Invitrogen, Cat # 11140) + 1X Glutamax (Invitrogen, Cat # 35050), 0.1 mM β-mercaptoethanol (Invitrogen, Cat # 31359-010))
9. hFF medium (irradiated hFF grown in 175-cm^2 flask in a confluent layer (ca. 2×10^6 cells) with 25 mL IMDM medium and harvested every 9 days and filter sterilized)
10. Conditioned medium (hFF medium: hESC medium = 1:1)

2.2. IF Analysis

1. 4 % paraformaldehyde
2. PBS
3. 0.1% Triton-X 100 (in PBS)
4. 4 % FBS (in PBS)
5. Tweezers
6. Alexa Fluor 568 Donkey anti-mouse (1:600 PBS) (Invitrogen, Cat # A10037)
7. Alexa Fluor 488 Donkey anti-goat (1:600 PBS) (Invitrogen, Cat # A10055)
8. Alexa Fluor 488 Donkey anti-rabbit (1:600 PBS) (Invitrogen, Cat # A21206)
9. DAPI (14.4 mM) (1:600 PBS)
10. Primary antibodies (**Table 11.1**)
11. Nail polish
12. Coverslip (Nalge Nunc, Cat # 171080)
13. Mounting medium 0.1 g n-propyll gallate (Sigma Aldrich, ST. Louis, USA, Cat #P3130) + 0.5 ml 10X PBS + 4.5 ml glycerol

2.3. Real-Time RT-PCR

2.3.1. Total RNA Extraction (Including DNase Treatment)

1. RNeasy Mini Kit (Qiagen, Venlo, The Netherlands, Cat # 74104)
2. QIAshredder homogenizer (Qiagen, Cat # 79654)
3. Turbo DNase Kit (Ambion, Austin, TX, USA, Cat # AM2238)

Table 11.1
Primary antibodies used for IF protocol

Primary antibodies Name	Species	Company	Catalog number	Dilution
Sox2	Mouse	R & D Systems	MAB2018	1:100
Sox2	Goat	R & D Systems	AF2018	1:100
Oct4	Goat	Santa Cruz	SC8629	1:100
Oct4	Rabbit	AbCam	AB19857	1:100
Nanog	Goat	R&D Systems	AF1997	1:100
Acetylated α-tubulin	Mouse	Sigma Aldrich	T6793	1:5000
Detyrosinated α-tubulin	Rabbit	Abcam	AB48389	1:500
Dvl1	Mouse	Santa Cruz	SC8025	1:100
β-catenin	Rabbit	Cell Signaling	#1247S	1:200
Phospho β-catenin	Rabbit	Cell Signaling	#9561S	1:200
PDGFRα	Rabbit	Santa Cruz	SC12910-R	1:100
PDGFRβ	Rabbit	Santa Cruz	SC432	1:100
Centrin	Goat	Santa Cruz	SC8719	1:100
Pericentrin	Goat	Santa Cruz	SC28145	1:100
Smo	Rabbit	MBL International	A2668	1:200
Ptc	Rabbit	Santa Cruz	SC6149	1:200

4. Nuclease-free pipette tips with filter
5. 96–100% ethanol
6. 70% ethanol
7. Microcentrifuge
8. Waterbath or heating block

2.3.2. cDNA Synthesis

1. DNase-treated template RNA (from **Section 2.3.1**)
2. Nuclease-free pipette tips with filter
3. 200-μl PCR tubes (Abgene, Surrey, UK, Cat # AB-0620)
4. Nuclease-free water (Fermentas, Burlington, ON, Canada, Cat # R0582)
5. 200 ng/μl random hexamer primers (Fermentas, Cat # S0142)

6. 100 μM oligo (dT)$_{18}$ primers (Fermentas, Cat # SO131)
7. 10 mM dNTP mix (Invitrogen, Cat # 18427-013)
8. SuperScript III Reverse Transcriptase kit (Invitrogen, Cat # 18080-093)
9. RiboLock recombinant ribonuclease inhibitor (Fermentas, Cat # EO0311)
10. Thermal block or waterbath
11. Thermal cycler (MJ Research, Waltham, MA, USA, Cat # PTC-200)

2.3.3. Real-Time PCR

1. cDNA
2. LightCycler instrument and software (Roche Applied Science, Basel, Switzerland)
3. Polycarbonate capillaries with rubber caps (Genaxxon, Birkenfeld, Germany, Cat # I2250.0480)
4. Modified carousel for polycarbonate capillaries (Genaxxon, Cat # I2250.0001)
5. Transfer pin (Genaxxon, Cat # I2250.0003)
6. Capillary metal adapters and cooling block (store in 4–8°C)
7. SYBR Premix Ex Taq kit (Takara, Osaka, Japan, Cat # RR041A)
8. Nuclease-free pipette tips with filter
9. 10 μM primer working solutions for genes *NANOG*, *POU5F1* (i.e., the gene for OCT4), *GABRB3*, *TDGF1*, *GDF3*, *DNMT3B*, *TBP*, *GAPDH*, and *ACTB*. For primer details, *see* **Table 11.2**.

3. Methods

Conventional growth techniques for most hESC lines rely on a feeder layer of mitotically inactivated mouse embryonic fibroblasts (MEFs) or hFFs usually (for general hESC morphology, *see* **Fig. 11.1A**). This introduces contamination and ambiguity problems when analyzing hESCs for microscopy and/or RNA work. Therefore, it is advantageous to grow them under feeder-free conditions (**Fig. 11.1B–D**).

As such, when using trypsin-resistant hESC lines, it is important to have a supply of mitotically inactivated feeder layers from which to harvest the supernatant. Presumably, the supernatant contains factors

Table 11.2
Details for primers for six hESC genes (*NANOG, POU5F1, GABRB3, TDGF1, GDF3, DNMT3B*) and three reference genes (*ACTB, GAPDH, TBP*)

Gene symbol	Product name	Primer sequence (5'→3')	Amplicon size (bp)
NANOG	Nanog	F: CAAAGGCAAACAACCCACTT R: CTGGATGTTCTGGGTCTGGT	426
POU5F1	Oct4	F: GACAACAATGAAAATCTTCAGGAGA R: TTCTGGCGCCGGTTACAGAACCA	218
GABRB3	GABA A receptor 3β	F: CAAGCTGTTGAAAGGCTACGA R: ACTTCGGAAACCATGTCGATG	108
TDGF1	Teratoma-derived growth factor 1 (Cripto)	F: AGCACAGTAAGGAGCTAAACA R: CAGTTCCGTCCGTAGAAGGAG	101
GDF3	Growth differentiation factor 3	F: GTACTTCGCTTTCTCCCAGAC R: GCCAATGTCAACTGTTCCCTT	131
DNMT3B	DNA methyltransferase 3β	F: AGCCACCTCTGACTACTG R: GACAAACAGCCATCTTCCA	149
ACTB	β-actin	F: CCTGGCACCCAGCACAAT R: GGGCCGGACTCGTCATAC	144
GAPDH	Glyceraldehyde-3-phosphate dehydrogenase	F: TCGGAGTCAACGGATTTGGT R: TTGCCATGGGTGGAATCATA	148
TBP	TATA-sequence binding protein	F: CCCGAAACGCCGAATATAATC R: GACTGTCTTCACTCTTGGCT	130

Primers were selected from PrimerBank (17) or designed with PerlPrimer (18). All primer pairs have a melting temperature within the range of 60–62°C. Abbreviations: F = forward primer; R = reverse primer; bp = base pairs

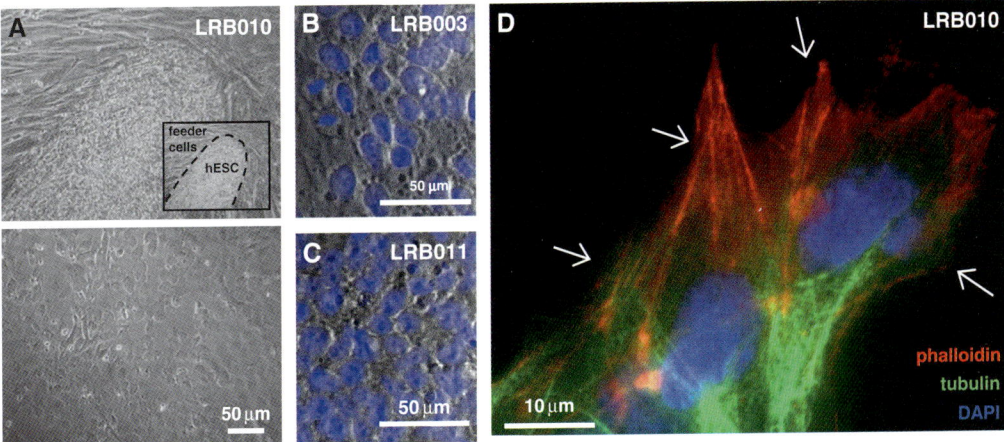

Fig. 11.1. Morphology of hESCs grown under feeder-free conditions. **A**. Light microscopy analysis of hESC line LRB010 grown in the presence (*upper panel*) and the absence of (*lower panel*) feeder cells (human foreskin fibroblasts, hff). **B** and **C**. Morphology of hESC cell lines LRB003 and LRB011 grown in feeder-free cultures. Nuclei were stained with DAPI (*blue*). IF analysis of hESC (line LRB010) grown in feeder-free cultures and stained for F-actin (phalloidin, *red*), tubulin (antitubulin, *green*). and DNA (DAPI, nucleus, *blue*). The arrows indicate migrating cells at the border of the hESC colony.

which help the hESCs grow independently of feeders (9). Finally, in growing the hESCs without feeders, it is feasible only for experimental analysis and not for continuing with passaging of cell lines. Therefore, the line should be maintained independently on feeder dishes and only passaged onto feeder-free conditions for analysis.

Colocalization studies using IF can generate valuable information about stem cell biology, especially in terms of the differentiation modules involved. By using antibodies to the traditional stem cell markers (OCT4, SOX2, NANOG), we see the expected nuclear localization which denote the undifferentiated state (**Fig. 11.2A–C**) which are absent in differentiated cells such as hFFs (**Fig. 11.2D,E**). However, recently, we have shown for the first time that hESCs possess primary cilia that coordinate Hedgehog (Hh) signaling with Patched (Ptc) and Smoothened (Smo) trafficking in and out of the cilium in response to Hh pathway activation (4) (**Figs. 11.2F and 11.3A**). Because Hh signaling facilitates the differentiation and proliferation of hESCs (10), this suggests that the primary cilia may play a critical role in modulating the signals for differentiation and/or proliferation in hESCs with components of the Hh pathway present (**Fig. 11.3A, B**). This suggests that the primary cilia may play a critical role in modulating the extracellular signals for differentiation and/or proliferation in stem cells (4). Using an antibody to acetylated α-tubulin to mark the primary cilia as well as to the stem cell markers, we also unexpectedly observed localization of OCT4, SOX2, and NANOG along the length of the cilium in a subpopulation of the stem cells (**Fig. 11.2G, H**). The localization of these

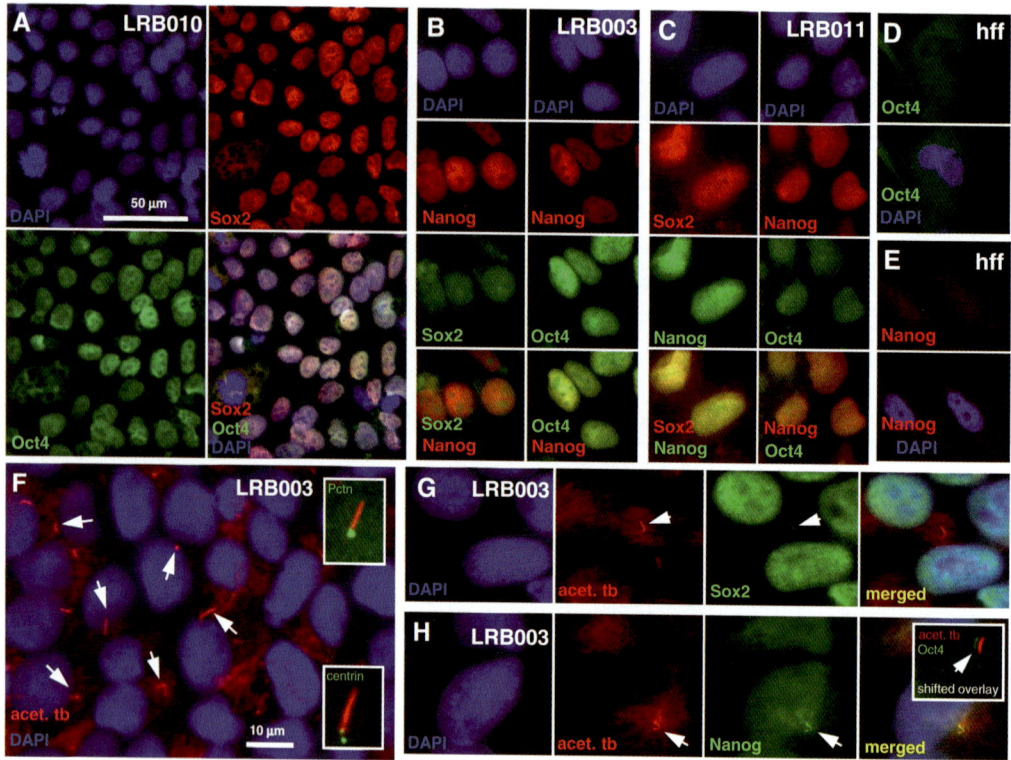

Fig. 11.2. IF analysis on the localization of stem cell markers in hESCs in feeder-free cultures. **A**. Colocalization of Sox2 and Oct4 to the nucleus of hESC line LRB010. **B** and **C**. Colocalization of Sox2, Nanog, and Oct4 to the nucleus of hESC lines LRB003 and LRB011. **D** and **E**. Absence of Oct4 and Nanog in nuclei of feeder cells (hff: human foreskin fibroblasts). **H**. Presence of primary cilia (antiacetylated α-tubulin, acet. tb, arrows, *red*) in confluent cultures of hESC line LRB003 that has entered growth arrest. *Inserts*: *Upper panel*: antipericentrin (Pctn, *green*) marks the centrosome at the base of the primary cilium. *Lower panel*: Thecilium (acet-tb, *red*) in LRB003 hESC specifically emerges from one of the centrosomal centrioles marked with anticentrin (centrin, *green*). From Kiprilov et al. (8) with permission, with permission, courtesy of J. Cell Biology. **I** and **J**. Localization of Sox2 (I) and Nanog (J) (*green*) to the primary cilium (acet. tb, *arrows*, *red*) in hESC line LRB003. *Insert* (**J**). Shifted overlay showing localization of Oct4 (*green*) to the primary cilium acet. tb, *arrows*, *red* in hESC line LRB003. Nuclei were stained with DAPI (*blue*).

transcription factors as well as components of the Hh machinery to this organelle suggests that the cilium may play an important role in coordinating the differentiation and/or self-renewal profile of stem cells via translocation of these and other transcription factors through the cilium. Use of IF analysis with feeder-free cultures resulted in the localization of other signaling pathways in stem cells, including Wnt signaling that maintains the self-renewal (humans:10; mice:11) and Platelet Derived Growth Factor (PDGF)-receptor (PDGFR) signaling that helps maintaining hESCs in an undifferentiated state (11). As shown in **Fig. 11.3C** and **D**, Dishevelled-1 (Dvl-1) and the phosphorylated form of β-catenin in the Wnt pathway localize to the hESC primary cilium or at the base of the cilium, whereas PDGFRα (but not PDGFRβ)

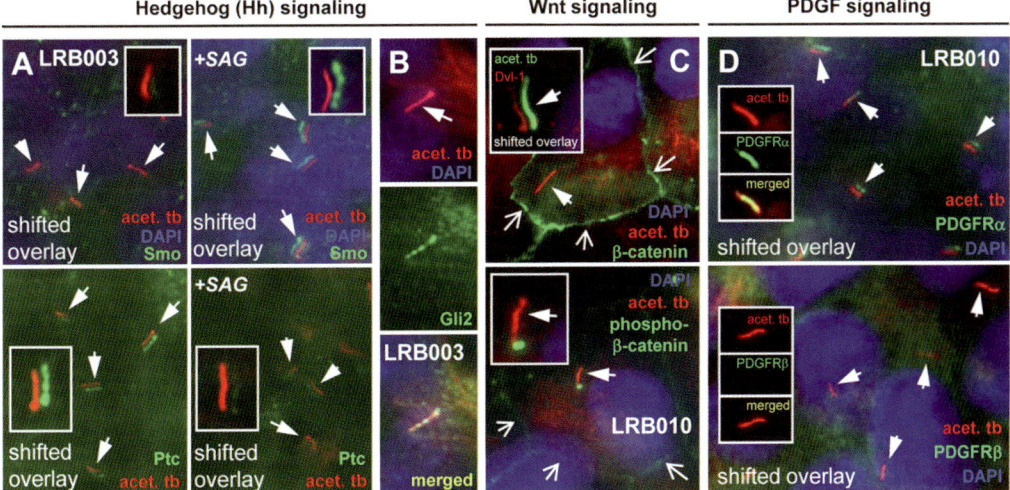

Fig. 11.3. IF analysis on the localization of signaling components to the primary cilium (acet. tb, *bold arrows*) in hESCs grown in feeder-free cultures. **A** and **B**. Localization of Hedgehog signaling components to the primary cilium of hESC line LRB003. In the absence of Hh pathway stimulation, the Hedgehog receptor patched (Ptc, *green*) colocalizes with the cilium all along the ciliary membrane (**A**, *lower left panel*). Red and green channels are displaced in the images to define colocalization more clearly. Upon stimulation with SAG, as part of the signaling cascade, Ptc leaves the cilium (**A**, *lower right panel*) and smoothened (Smo) enters to activate the Hedgehog signaling cascade (**A**, *upper right panel*). **B**. The Hedgehog transcription factor, Gli2 (*green*), localizes to the primary cilium. From Kiprilov et al. (8), with permission, courtesy of J. Cell Biology. **C**. Localization of Wnt signaling components to the primary cilium in hESC line LRB010. β-catenin (*green*) predominantly localizes to the cell membrane (*upper panel*), whereas the phosphorylated form of β-catenin predominantly localizes to the base of the primary cilium (*lower panel*). *Insert upper panel*: Shifted overlay to show that Dishevelled-1 (Dvl-1, *red*) localizes to the primary cilium (acet. tb, *bold arrow*, *green*). *Open arrows* mark the cell membrane. **D**. Localization of PDGFR isoforms to the primary cilium in hESC line LRB010. Shifted overlay of PDGFRα (*green*) localizing to the primary cilium (acet. tb, *bold arrows*, *red*) (*upper panel*), and shifted overlay of PDGFRβ (*green*) not localizing to the primary cilium acet. tb, *bold arrows*, *red*) (*lower panel*). *Inserts* show merged images of PDGFR isoforms and the primary cilium. Nuclei are stained with DAPI (*blue*).

localizes along the entire length of the primary cilium, similar to that observed in fibroblasts (12). These observations favor the conclusion that the hESC primary cilium is an important sensory organelle both in maintenance of stem cell pluripotentiality and in regulation of early differentiation and proliferation.

Characterization of hESCs also involves RNA analysis and especially the markers used for hESCs have a few pitfalls which need to be addressed in order to obtain valid results. Extracted mRNA needs to undergo complete degradation of genomic DNA in connection with the RNA-extraction procedure due to presence of a substantial number of processed pseudogenes of *POU5F1* and *NANOG* in the human genome (13, 14). Processed pseudogenes are "transcriptional remnants," which have been incorporated into the genome by retrotransposition of the mRNA in early evolutionary history. Accordingly, they consist of a DNA sequence complementary to the same gene's mRNA (including untranslated regions). Therefore, the designing of primers with overlapping exon-intron boundaries will not evade the amplification of such pseudogenes, since the amplified genomic fragment will have the same amplicon size as the targeted gene sequence leading to false positive results.

In characterizing hESCs, we used six transcripts (*NANOG, POU5F1, GABRB3, TDGF1, GDF3,* and *DNMT3B*) previously thought to form a core group of ubiquitous hESC markers (15). These six transcripts were evaluated in our hESC lines and an abundant concomitant expression was observed (**Fig. 11.4**). Furthermore, all six transcripts became significantly down-

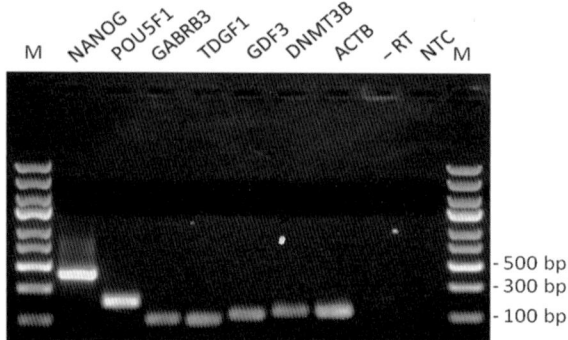

Fig. 11.4. Agarose gel electrophoresis of six proposed hESC markers and one reference gene (*ACTB*) evaluated in the LRB003 line (19). The band sizes correspond to the predicted amplicon sizes (*see* **Table 11.2**). The minus reverse transcriptase control was checked using *POU5F1* primers: The lack of a distinct band confirms the complete degradation of genomic DNA and hence POU5F1 pseudogenes during the RNA extraction procedure. Abbreviations: M = ladder; –RT = minus reverse transcriptase control; NTC = no template control; bp = base pairs.

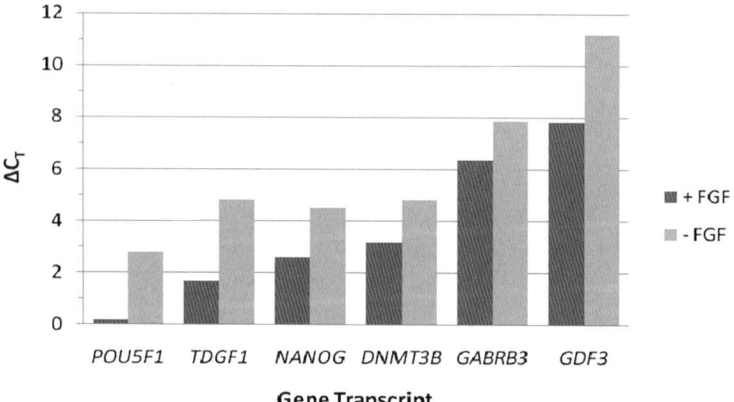

Fig. 11.5. ΔC_T values for a core group of six markers evaluated in the LRB010 hESC line. The smaller the ΔC_T value, the larger the amount of initial mRNA. ΔC_T values for each gene transcript were calculated by subtracting the geometric mean C_T value of three reference genes (*GAPDH, ACTB, TBP*) from the C_T (threshold cycle number) of each individual gene transcript. The increase in ΔC_T value for all six markers upon removal of FGF (mirroring incipient cell differentiation) reflects transcriptional down-regulation (3- to 11-fold). A very similar pattern was observed for line LRB003, whereas the markers were completely absent in human foreskin fibroblasts (hFFs).

regulated upon removal of FGF from the culture medium as determined by quantitative RT-PCR (**Fig. 11.5**). Thus, the use of this core group for hESC characterization on the mRNA level is validated. Primer pairs for all six transcripts (and three reference genes) are listed in **Table 11.2.**

3.1. Growth of hESCs

3.1.1. Day 1

1. Place a 16-well slide into a petri dish (helps prevent contamination) and remove cover from wells.
2. Add 200 μl gelatin to each well.
3. Place in 37°C incubator overnight to allow gelatin to adhere to glass slide.

3.1.2. Day 2

1. Trypsinize 35-mm dish grown with a near confluent layer of hESCs (which were plated on irradiated hFFs 5–7 days prior) by incubating in 1 mL 1X trypsin (prewarmed) for approximately 5 min. Pipette up and down gently to dissolve clumps.
2. To stop the reaction, pipette the hESC-containing solution (1 mL) into 9 mL IMDM (discard any remaining clumps that have not dissolved).
3. Mix and centrifuge (300g for 5 min).
4. Resuspend pellet in 7 ml conditioned media.
5. Remove the 200 μl gelatin solution in each well and replace with 200 μl media containing hESCs.
6. The remainder of hESCs solution can be used for at continuous passage of the cells by plating them in culture dishes containing irradiated hFF cells or onto additional 16-wells.
7. Incubate in 37°C incubator.
8. Replace half the media every 2 days until the cells form nice confluent layers.

3.2. IF Analysis

3.2.1. Day 1

1. Thaw the paraformaldehyde from the freezer.
2. Aspirate media from wells and wash with 200 μl PBS once. Be careful not to leave the wells too dry for too long.
3. Add 200 μl paraformaldehyde and incubate for 20 min at room temperature.
4. Wash 3 × 5 min with PBS.
5. Add 200 μl triton solution and incubate for 20 min at room temperature.
6. Wash with PBS once and wait 5 min.
7. Repeat above step twice more.
8. Add 200 μl 4% FBS for 45 min.
9. Make a humidified chamber by putting wet paper in the petri dish and place slide inside.

10. Remove the wells gently (they snap off with a little pressure from the sides) and remove the silcone rings with forceps by gently placing the forceps tip under rings and pulling to other side slowly. If the rings tear apart, go back and repeat until all of the silcon is removed from the slides.

11. Use suction pipette tip (attached to vacuum) to dry the areas between wells. Place 25 µL PBS to avoid dehydrating in each of the 16 spots where the well used to be. Make sure each well is distinct from the others so ultimately there will be 16 small pools of PBS per slide.

12. Dilute the primary antibodies in PBS (for complete list of antibodies, see **Table 11.1**).

13. Add 25–50 µl (less may be used as long as it covers the well) of the desired antibody solution to the appropriate wells. Make sure the pattern is recorded.

14. Incubate at 4°C overnight (alternatively, the primary incubation can be 60–90 min at room temperature).

3.2.2. Day 2

1. Wear gloves.
2. With vacuum suction, do 5 quick on/off washes per well with PBS.
3. Repeat with 3 quick on/off washes.
4. Suck off PBS and turn off the lights. Add 25 µl of the correct secondary antibody mix to the appropriate wells.
5. Let the slide incubate on the table with cover on top of petri dish, under foil for 30–60 min.
6. Wash 5 times per well with PBS.
7. Suck off PBS from the top row and put 30 µl DAPI per well (should stay on wells ca. 15 s). Suck off and replace with PBS on the entire top row once.
8. Wash the top row again 2 times per well with PBS.
9. Repeat above 2 steps for the bottom row.
10. Dry the slide with the suctions and remove uneven surfaces (i.e., silcone pieces) on the glass (this is important so that there are no air bubbles left after the slide is sealed).
11. Place slide on napkin. Add drops of mounting media (ca. 200 µl in total for a 16-well) on the coverslip (make sure no air bubbles) and gently place a coverslip on top. Gently press on the coverslip to remove air bubbles. Afterwards, take end of the napkin, fold over, and press firmly with the palm of your hand over the slide for 15 s to remove any

remaining air bubbles. Wipe off excess mounting media with the napkin (very criticial to make a tight seal with the nail polish).

12. Seal the coverslip with nail polish around all 4 edges a few times and store it in a dark box at 4°C for at least 20 min before viewing in microscope.

3.3. Real-Time RT-PCR

3.3.1. Total RNA Extraction (Including DNase Treatment)

1. Harvest cells, sediment, and completely aspirate cell-culture medium. Loosen the cell pellet thoroughly by flicking the tube.
2. Add 350 μl Buffer RTL and mix briefly by pipetting.
3. Pipette the lysate directly into a QIAshredder spin column placed in a 2-ml collection tube, and centrifuge for 2 min at full speed.
4. Follow Handbook for Qiagen Rneasy Mini Kit at this point.
5. Split each eluate of 30 μl into two 1.5-ml tubes and add the following to each sample of 15 μl RNA:
   - 2 μl 10× Turbo DNase buffer
   - 1 μl nuclease-free water
   - 2 μl Turbo DNase (2 U/μl)
6. Incubate at 37°C for 30 min in waterbath or heating block.
7. Add 80 μl nuclease-free water to each 20-μl sample to inactivate the DNase.
8. For each sample, add 350 μl Buffer RLT, and mix well.
9. Add 250 μl 96–100% ethanol to the diluted RNA and mix well by pipetting. Do not centrifuge.
10. At this point, follow Qiagen RNeasy Mini Kit protocol again to purify DNase-treated RNA.
11. Pool all identical samples and aliquot 3 μl for determination of RNA yield and quality on a spectrophotometer.
12. Store immediately the RNA at –80°C (NOTE: Under these storage conditions, no degradation of RNA is detectable for at least 1 year).

3.3.2. cDNA Synthesis

NOTE: set up all reactions on ice.
1. For each individual sample, prepare a master mix 1:
   - 1 μl dNTP mix
   - 1 μl random hexamer primers
   - 1 μl oligo $(dT)_{18}$ primers
   - 2 μl RNA (corresponding to 10 pg–250 ng)
   - 8 μl nuclease-free water

2. Heat samples to 70°C for 5 min and place immediately on ice for 3 min (NOTE: this facilitates unfolding of secondary RNA structures).

3. In the meantime, prepare for each sample a master mix 2 in a 200-μl PCR tube (NOTE: refreeze DTT and first-strand buffer immediately):
   - 4 μl first-strand buffer (5×)
   - 1 μl DTT
   - 1 μl RNase inhibitor
   - 1 μl SuperScript III reverse transcriptase

4. Collect the contents of master mix 1 by brief centrifugation and distribute the 7 μl into each PCR tube containing master mix 2 (i.e., 20 μl total).

5. Mix by pipetting gently and incubate at room temperature for 5 min.

6. In a thermal cycler, incubate samples at 55°C for 1 h followed by inactivation of reverse transcriptase at 70°C for 15 min.

7. Store cDNA at −20°C or use immediately for PCR.

3.3.3. Real-Time PCR

1. Thaw 2× SYBR Premix Ex Taq mix, template cDNA, gene specific primers, and nuclease-free water. Mix the individual solutions by inversion and brief spin (NOTE: SYBR mix should at all times be protected from light!)

2. Prepare a master mix for each reaction (or multiples hereof):
   - 5 μl SYBR Premix Ex Taq mix
   - 2 μl nuclease-free water

3. Mix the master mix thoroughly and dispense 7 μl master mix into each polycarbonate capillary placed in metal adaptor and precooled cooling block. Reach for the funnel-shaped part of the capillary.

4. Add 0.5 μl of forward primer (10 μM) and 0.5 μl of reverse primer (10 μM) to appropriate capillary (end concentration 0.5 μM each).

5. Add 2 μl of template cDNA (< 100 ng) making a total of 10 μl.

6. Place rubber caps using the transfer pin.

7. Spin down the adapters with capillaries briefly and replace in cooling block.

8. Turn on the LightCycler instrument and load samples.

9. Run a PCR program on the Lightcycler software in accordance with primer melting temperatures (**Table 11.3**).

Table 11.3
LightCycler program for the real-time quantification of hESC transcripts and reference genes using SYBR Green I dye detection chemistry

Program	Analysis Mode	Temperature (°C)	Time (s)	Ramp (°C/s)	Acquisition mode	Cycles
1. Initial denaturation	None	95	10	20	NONE	1
		95	5	20	NONE	
2. PCR	Quantification	55	20	20	NONE	37
		72	10	20/	SINGLE	
		95	0	20	NONE	1
3. Melting curve	Melting curve	65	15	20	NONE	1
		95	0	0.1	CONT	
4. Cooling	None	40	30	20	NONE	1

4. Notes

1. During trypsinization of stem cells, observe under the microscope when the cells de-adhere and be careful not to over-trypsinize. For each cell line and condition, times may vary accordingly.

2. When working with fluorophores conjugated to secondary antibodies or the SYBR mix, one has to work in the dark since they are light-sensitive.

3. During cDNA synthesis, the short room temperature incubation facilitates primer annealing and gives the reverse transcriptase an opportunity to extend the primers for greater thermostability when the incubation temperature is increased in the following step leading to longer cDNA molecules.

4. After cDNA synthesis, efficiency is around 50–100% depending on the polymerase. Check with manufacturer.

5. It is recommended to dilute the cDNA (e.g., 100×) before performing real-time PCR

6. The thermocycler program listed has an annealing temperature of 55°C during PCR, which is approximately 5°C below the T_m of the primers listed in **Table 11.2**.

7. For subsequent data analysis, we calculate the geometric C_T mean of the three reference genes *TBP*, *GAPDH*, and *ACTB*. Next, geometric reference C_T mean is subtracted from the C_T (threshold cycle) value for each target gene yielding a ΔC_T value. The ΔC_T values can be compared between separate cell populations or a fold change may be calculated using the comparative $\Delta\Delta C_T$ method (16).

References

1. Guillaume DJ, Zhang SC (2008) Human embryonic stem cells: a potential source of transplantable neural progenitor cells. Neurosurg Focus 24:E3. Review.
2. Shulz WA, Hoffmann MJ (2007) Transcription factor networks in embryonic stem cells and testicular cancer and the definition of epigenetics. Epigenetics 2:37–42. Review.
3. Satir P, Christensen ST (2007) Overview of structure and function of mammalian cilia. Annu Rev Physiol 69:377–400.
4. Badano JL, Mitsuma N, Beles PL, Katsanis N (2006) The ciliopathies: an emerging class of human genetic disorders. Annu Rev Renomics Hum Genet 7:125-148
5. Fliegauf M, Benzing T Omran H (2007) When Cilia go bad: cilia defects and ciliopathies. Nat Rev Mol Cell Biol 8:880–893.
6. Christensen ST, Pedersen LB, Schneider L, Satir P (2007) Sensory cilia and integration of signal transduction in human health and disease. Traffic 8:97–109.
7. Corbit KC, Aanstad P, Singla V, Norman AR, Stainier DY, Reiter JF (2006) Vertebrate Smoothened functions at the primary cilium. Nature 437:1018–1021.
8. Kiprilov EN, Awan A, Desprat R, Velho M, Clement CA, Byskov AG, Andersen CY, Satir P, Bouhassira EE, Christensen ST, Hirsch RE (2008) Human embryonic stem cells in culture possess primary cilia with hedgehog signaling machinery. J Cell Biol 180:897–904.
9. Chin AC, Fong WJ, Goh LT, Philip R, OH SK, Choo AB (2007) Identification of proteins from feeder conditioned medium that support human embryonic stem cells. J Biotechnol 130:320–328.
10. Rho JY, Yu K, Han JS, Chae JI, Koo DB, Yoon HS, Moon SY, Lee KK, Han YM (2006) Transcriptional profiling of the developmentally important signalling pathways in human embryonic stem cells. Hum Reprod 21:405–412.
11. Pebay A, Wong RC, Pitson SM, Wolvetang EJ, Peh GS, Filipczyk A, Koh KL, Tellis I, Nguyen LT, Pera MF (2005) Essential roles of sphingosine-1-phosphate and platelet-derived growth factor in the maintenance of human embryonic stem cells. Stem Cells 23:1541–1548.
12. Schneider L, Clement CA, Teilmann SC, Pazour GJ, Hoffmann EK, Satir P, Christensen ST (2005) PDGFRalphaalpha signaling is regulated through the primary cilium in fibroblasts. Curr Biol 15:1861–1866.
13. Takeda J, Seino S, Bell GI (1992) Human Oct3 gene family: cDNA sequences, alternative splicing, gene organization, chromosomal location, and expression at low levels in adult tissues. Nucleic Acids Res 20:4613–4620.
14. Booth HA, Holland PW (2004) Eleven daughters of NANOG. Genomics 84: 229–238.
15. Adewumi O, Aflatoonian B, Ahrlund-Richter L, Amit M, Andrews PW, Beighton G, et al. (2007) Characterization of human embryonic stem cell lines by the International Stem Cell Initiative. Nat Biotechnol 25:803–816.
16. Pfaffl MW (2001) A new mathematical model for relative quantification in real-time RT-PCR. Nucleic Acids Res 2:e45.
17. Wang X, Seed B (2003) A PCR primer bank for quantitative gene expression analysis. Nucleic Acids Res 31:e154.
18. Marshall O (2007) Graphical design of primers with PerlPrimer. Methods Mol Biol 402:403–414.
19. Laursen SB, Møllgaard K, Olesen C, Oliveri RS, Bøchner CB, Byskov AG, et al. (2007) Regional differences in expression of specific markers for human embryonic stem cells. Reprod Biomed Online 15:89–98.

Chapter 12

Study of Gap Junctions in Human Embryonic Stem Cells

Raymond C.B. Wong and Alice Pébay

Abstract

Gap junctional intercellular communication (GJIC) has been described in different cell types including stem cells and has been involved in different biological events. GJIC is required for mouse embryonic stem cell maintenance and proliferation, and various studies suggest that functional GJIC is a common characteristic of human embryonic stem cells (hESC) maintained in different culture conditions. This chapter introduces methods to study gap junctions in hESC, from expression of gap junction proteins to functional study of GJIC in hESC proliferation, apoptosis, colony growth, and pluripotency.

Key words: Connexin, gap junction, gap junctional intercellular communication, human embryonic stem cells.

1. Introduction

Gap junctions are intercellular channels consisting of two connexons localized in the membrane of adjacent cells and have been described in stem cells (1–10). Each connexon consists of six membrane proteins, termed connexins (Cx) (11, 12). Numerous molecules can diffuse through gap junction channels, including small ions, second messengers, amino acids, metabolites, short interfering RNA, and peptides involved in cross-presentation of major histocompatibility complex class I molecules (13–16). Such intercellular coupling is termed gap junctional intercellular communication (GJIC). GJIC is involved in various cellular mechanisms, including control of cell migration, proliferation, differentiation, metabolism, apoptosis, and carcinogenesis (14, 17–21). In human embryonic stem cells (hESC), various connexins have been found to be expressed: Cx25, Cx26, Cx30, Cx30.2,

Cx30.3, Cx31, Cx31.1, Cx31.9, Cx32, Cx36, Cx37, Cx40, Cx43, Cx45, Cx46, Cx47, Cx59, and Cx62 (1–3, 22). Among these connexins, Cx43 and Cx45 are highly enriched in undifferentiated hESC compared to its differentiated counterpart in many microarray studies, and the two connexins were identified as "undifferentiated hESC markers" (23). Moreover, Cx43 was identified as a downstream target of key pluripotent transcription factors Oct4, Sox2, and Nanog (24). Undifferentiated hESC can communicate through functional GJIC, as determined by dye coupling (1–3) and ionic coupling (22). Finally, functional GJIC is a common characteristic of undifferentiated hESC maintained in different culture conditions, suggesting an understudied role of gap junctions in mediating hESC maintenance (2, 3). Studies of intercellular communication through gap junctions can potentially lead to novel methods to improve clonal survival and maintenance of hESC, which is fundamental to realize the therapeutic potential of these cells.

2. Materials

2.1. General (see Note 1)

1. hESC were cultivated in different formats depending on the experiments to perform. In all cases, we culture hESC with a feeder layer of mouse embryonic fibroblasts (MEF) supplemented with 20% fetal calf serum or 20% Knockout serum replacement (Invitrogen, #10828-028) plus 4 ng/ml bFGF (R&D, #233-FB-025/CF).
2. For RT-PCR, Western blot, cell proliferation assay, cell pluripotency assay, and colony growth assay, we culture hESC in centre-well organ culture dishes, 60×15-mm style (35 mm culture dishes, Falcon, #353037).
3. Alternatively for Western blot, we culture hESC with MEF in 6-well plates (Falcon, #353046).
4. For immunocytochemistry and SL/DT, we culture hESC in Lab-Tek Chamber slide w/cover, permanox slide sterile (8-well chamber slides, Lab-Tek, #177445).
5. TrypLE Express (Invitrogen, #12604).
6. Phosphate buffered saline (PBS) 10X: 1.37 M NaCl, 27 mM KCl, 100 mM Na_2HPO_4, 18 mM KH_2PO_4, pH 7.4.
7. Dispase (Invitrogen, #17105-041).
8. Ethanol and methanol (Merck).
9. Tris base (Merck).

2.2. RT-PCR

1. RNase-free microtubes and pipettes tips
2. Dynabeads® Oligo (dT)$_{25}$ (Invitrogen, Dynal, #610.02, 610.05, 610.50) and Magnetic Particle Concentrator (Invitrogen, Dynal, #120.20D)
3. Lysis/binding buffer: 20 mM Tris-HCl (pH7.5), 1 M LiCl, 2 mM EDTA
4. Washing buffer A: 10 mM Tris-HCl (pH7.5), 0.15 M LiCl, 1 mM EDTA, 0.1 % LiDS
5. Washing buffer B: 10 mM Tris-HCl (pH7.5), 0.15 M LiCl, 1 mM EDTA
6. Superscript II Rnase H$^-$ Reverse Transcriptase (Invitrogen, #18064-014)
7. TAQ DNA polymerase (Biotech International, #TAQ-1)
8. dNTP mix
9. Sense and anti-sense primers for Cx43 and Cx45: Cx 43, sense 5′-ATGAGCAGTCTGCCTTTCGT-3′, antisense 5′-TCTGCTTCAAGTGCATGTCC-3′; Cx 45, sense 5′-GGAAGATGGGCTCATGAAAA-3′, antisense 5′-GCAAAGGCCTGTAACACCAT-3′
10. Agarose molecular biology grade (Scientifix, #9010B)
11. TAE buffer: 20 mM Tris-Cl, pH 7.8, 10 mM sodium acetate and 0.5 mM EDTA
12. Ethidium bromide (Sigma, #E1510)
13. MinElute Gel Extraction Kit (Qiagen, #28604)

2.3. Western Blot Analysis

1. Mini-protean 3 system (Biorad, #165-3301, 165-3302, 170-3930, 170-3935)
2. 30% Acrylamide/Bisacrylamide solution 29:1 3.3% C (w/v) (Biorad, #161-056): toxic product that should be used with care under a fume hood.
3. Resolving gel buffer: 1.5 M Tris-HCl, pH 8.8.
4. Stacking gel buffer: 0.5 M Tris-HCl, pH 6.8.
5. Ammonium persulfate (APS, Biorad, #161-0700). Prepare fresh solution of APS 10% in water.
6. TEMED (Biorad, #161-0800).
7. Laemmli sample buffer (Biorad, #161-0737) or reducing sample buffer: 0.1 M Tris-HCl (pH 6.8), 41.6% (v/v) glycerol, 3.3% (w/v) SDS, 0.02% (w/v) bromophenol blue. Add 0.46 mM β-mercaptoethanol (Sigma, #M7154), 1 mM sodium orthovanadate (Sigma, #S6508), and 1 mM phenylmethylsulphonyl fluoride (PMSF, Sigma, #P7626) to the sample buffer prior to cell lysis.
8. Kaleidoscope prestained ladder (Biorad, #161-0324).

9. Running buffer 1X: 25 mM Tris base, 192 mM glycine, 0.1% SDS (w/v), pH 8.3. Mix 3.03 g Tris base, 14.4 g glycine, 1 g SDS and bring final volume to 1 L with distilled water, check pH.
10. Transfer buffer 1X: 25 mM Tris base, 192 mM glycine, 20% methanol, 0.1% SDS (w/v), pH 8.3. Mix 3.03 g Tris base, 14.4 g glycine, 200 ml methanol, 1 g SDS and bring final volume to 1 L with distilled water, check pH.
11. Hybond-P PVDF membrane (Amersham Pharmacia Biotech #RPN 303F) or PVDF membrane (Biorad #162-0177) cut to the size of the gel or precut (ready gel Blotting Sandwiches, Biorad, #162-0219).
12. Bovine serum albumin fraction 5, min 96% (BSA, Sigma, #A4503).
13. Tris-buffered saline with Tween 20 (TBST) pH 7.6: 20 mM Tris base, 137 mM sodium chloride, 3.8 mM HCl, 0.05% Tween-20 (Biorad, #170-6531). Mix 2.42 g Tris base, 3.8 ml HCl 1 M, 0.5 ml Tween-20 and bring final volume to 1 L with distilled water, check pH.
14. Rabbit anti-mouse Connexin 43 affinity-purified polyclonal antibody (Chemicon, #AB1728).
15. Negative control Rabbit Immunoglobulin Fraction, Solid-Phase Absorbed (Dako, #X0936).
16. Mouse monoclonal anti-β-tubulin 1 (Sigma #T7816).
17. Polyclonal Goat Anti-Rabbit Immunoglobulins/Horseradish peroxidise (HRP) (Dako, #P0448).
18. ECL Plus (Amersham Pharmacia Biotech #RPN2132, RPN2133) or Pierce ECL western blot substrate (chemiluminescent detection reagent, Pierce #32209).
19. Hyperfilm (Amersham Biosciences) or Gel-Doc system (Biorad).
20. Western Blot Stripping Buffer (Pierce, #21059).
21. Ponceau S (Sigma, #P-3504).

2.4. Immunocytochemistry

1. Glass coverslips 22×60 mm.
2. Wax pen/liquid blocker.
3. Rabbit anti-mouse Connexin 43 affinity-purified polyclonal antibody (Chemicon, #AB1728).
4. Rabbit anti-Connexin 45 polyclonal antibody (Chemicon, #AB1745)/
5. 6-diamidino-2-phenylindole (DAPI, Sigma, #D9542) or Bis-benzimide H 33342 (Hoechst-33342, Sigma, #382065). Prepare a fresh solution at 1 µg/ml in water.

6. Mouse GCTM-2 antibody (gift from Prof M. Pera, University of Southern California).
7. Mouse anti-Oct-3/4 C-10 antibody (Santa Cruz Biotechnology, Inc. #sc-5279).
8. Mouse TG-30 antibody, recognizing CD9 (gift from Prof M. Pera, University of Southern California).
9. TRA-1-60 antibody (gift from Prof P. Andrews, University of Sheffield)
10. Negative control rabbit immunoglobulin fraction, solid-phase absorbed (Dako, #X0936).
11. Negative control mouse IgM antibody (Dako, #X0942).
12. FITC-conjugated swine antirabbit immunoglobulins (Dako, #F0205) or Alexa Fluor® 488 goat antirabbit IgG (H+L) (Invitrogen, #A11008).
13. FITC-conjugated rabbit antimouse immunoglobulins (Dako, #F0261) or Alexa Fluor® 568 goat antimouse IgG (H+L) (Invitrogen, #A11004).
14. Vectashield (Vector Laboratories, # H-1000).
15. Nail varnish.

2.5. Scrape Loading/ Dye Transfer Assay

1. Ca^{2+} Mg^{2+}-PBS buffer (Buffer 1): 140 mM NaCl, 5.5 mM KCl, 1.8 mM $CaCl_2$, 1 mM $MgCl_2$, 10 mM glucose, 10 mM Hepes, pH 7.35.
2. Lucifer Yellow CH dipotassium salt (Sigma, #L0144) to freshly dilute in Buffer 1 (1 mg/ml).
3. Dextran, tetramethylrhodamine (Rhodamine dextran, Molecular Probes, #D1868) to freshly dilute in Buffer 1 (1 mg/ml).
4. Surgical blades (Swann-Morton, #0201).

2.6. Chemical Closure of Gap Junctions

1. Phorbol 12-Myristate 13-acetate (PMA, Sigma, #P8139)
2. U0126 (Promega, #V112A)
3. α-glycyrrhetinic acid (Sigma, #G8503)
4. Recombinant human BMP-4 (R&D, #314-BP-010)

2.7. Cell Apoptosis Assay

1. In situ Cell Death Detection Kit, Fluorescein (Roche, #11684795910) containing TUNEL reaction mixture.
2. 2% (w/v) paraformaldehyde (PFA) in PBS. Stock can be aliquoted and stored at −20°C. Once thawed, aliquots should be used on the day and discarded.
3. 0.1% Triton X-100 in PBS.

2.8. Cell Proliferation Assay

1. In situ Cell Proliferation Kit, FLUOS (Roche, #1810740): containing 10 mM BrdU, mouse anti-BrdU antibody conjugated to fluorescein
2. 4 M HCl solution
3. 70% ethanol
4. Propidium iodide (Sigma, #70335)

2.9. Cell Pluripotency Assay

1. TRA-1-60 antibody (Santa Cruz, #sc-21705)
2. TRA-1-81 antibody (Santa Cruz, #sc-21706)
3. Goat serum (Invitrogen, #16210-046)
4. Goat antimouse IgM antibody conjugated to Alexa fluor 488 (Invitrogen, #A21042)
5. Negative control mouse IgM antibody (Dako, #X0942)

2.10. Image and Data Analysis

1. Inverted fluorescence microscope
2. Western blot analysis: Scion image software (NIH) or Geldoc system (Biorad)
3. Statistical analysis: Graphpad Prism
4. Cell proliferation, apoptosis, pluripotency assay: Flow cytometer (Moflo, DIVA, or FC500)
5. Colony growth assay: cell^B analysis software (Olympus Software Imaging Solutions) or Adobe Photoshop

3. Methods

3.1. Reverse Transcriptase-Polymerase Chain Reaction (RT-PCR) for Connexin mRNA Expression

1. Prewarm dispase (10 mg/ml of culture medium) to 37°C.
2. Harvest 10–12 day-8 hESC colonies by dispase treatment (10 min, 37°C, *see* **Note 2**) and transfer cells to an eppendorf tube.
3. Wash colonies 4 times with PBS. Centrifuge at 300g for 2 min to spin down cells.
4. Isolate Poly-A+ mRNA using Dynabeads Oligo (dT)$_{25}$ adapted from the supplier's instructions (*see* **Note 3**). Using the magnetic particle concentrator, all Dynabead-binded mRNA are captured to one end of the tube, and the mRNA-free supernatant is discarded.
5. Add 300 μl of the lysis/binding buffer to hESC extracts in an eppendorf tube and homogenize with a pipette until complete lysis.

6. Transfer 20 µl of homogenized Dynabeads into an eppendorf tube, place onto the magnetic particle concentrator for 2 min. Remove supernatant, resuspend the beads in 200 µl of lysis/binding buffer, place onto the magnetic particle concentrator for 2 min and remove supernatant.

7. Mix lysed hESC with Dynabeads and incubate for 10 min at room temperature, then place onto the magnetic particle concentrator for 2 min and remove supernatant.

8. Mix 400 µl of washing buffer A with the beads, place onto the magnetic particle concentrator for 2 min and remove supernatant.

9. To enhance the purity of the mRNA preparation, repeat Step 8 first with 400 µl of washing buffer A, then twice with 200 µl of washing buffer B.

10. Add 200 µl of washing buffer B to the beads, mix thoroughly and separate into 2 eppendorfs, each containing 100 µl of washing buffer B and beads. Place onto the magnetic particle concentrator for 2 min and remove supernatant. Label one tube as +RT and the other tube as −RT

11. Perform RT using Superscript II Rnase H⁻ Reverse Transcriptase, following the supplier's protocol. A negative control without the addition of reverse transcriptase (RT⁻) must be performed for each RNA sample to check the absence of contaminating genomic DNA. A 20-µl reaction volume is prepared as follows (**Table 12.1**) and added to the +RT and −RT tubes.

12. Incubate for 1 h at 42°C, then place on ice, mix solution, place onto the magnetic particle concentrator for 2 min, and remove supernatant. Resuspend in 20 µl distilled/RNase free water.

Table 12.1
Reverse transcription buffer for cDNA synthesis

Reagents	RT^a volume	RT⁻ volume	Final concentration
5× first-strand buffer (Invitrogen)	4 µl	4 µl	1×
Dithiothreitol (DTT)	2 µl	2 µl	0.01 M
dNTP mix^a	1 µl	1 µl	0.5 mM
Superscript II reverse transcriptase	1 µl		200 Units
MilliQ	12 µl	13 µl	

^{a}dNTP mix contained dATP, dCTP, dGTP, dTTP

13. Perform PCR experiments using Taq DNA polymerase. A negative control (−RT) must be performed for each cDNA sample. A water control (no cDNA) should also be included. For each reaction, a total volume of 25 μl is prepared as shown in **Table 12.2**. For the cDNA samples, homogenize the preparation (cDNA and beads) before pipetting. It is recommended to perform PCR immediately after RT.

14. PCR reaction is performed with the following conditions using the specific primers for Cx43 and Cx45: initial denaturation at 94°C for 5 min, 35 cycles of "denaturation at 94°C for 30 sec, annealing at 55°C for 2 min, extension at 74°C for 2 min", ending with a final incubation at 74°C for 7 min.

15. Amplicons are sized by electrophoresis on 2% (w/v) agarose gel stained with 0.001% (v/v) ethidium bromide in TAE buffer.

16. Optional: Confirmation of the identity of amplicons: Excise DNA fragments of interest from the gels using a scalpel with the aid of a UV transilluminator; purify the amplified products using the QIAquick gel extraction kit following the supplier's instructions and confirm the identity of the purified amplicons by DNA sequencing.

Table 12.2
PCR reaction mix

Reagents	cDNA volume	−RT volume	Water control volume	Final concentration
10 × Buffer	2.5 μl	2.5 μl	2.5 μl	1×
dNTP mix	0.5 μl	0.5 μl	0.5 μl	0.2 mM
Taq polymerase	0.25 μl	0.25 μl	0.25 μl	0.25 Units
cDNA	3 μl	0 μl	0 μl	
−RT	0 μl	3 μl	0 μl	
Primer mix	2 μl	2 μl	2 μl	2 μM
MgCl$_2$	1.5 μl	1.5 μl	1.5 μl	1.5 μl
MilliQ	Make up to 25 μl	Make up to 25 μl	Make up to 25 μl	

3.2. Western-Blot Analysis of Connexion 43 Phosphorylation States: SDS-Polyacrylamide Gel Electrophoresis (SDS-PAGE, see Note 4)

3.2.1. SDS-PAGE Gel Preparation

1. Western blot analysis is carried out using the Mini-protean 3 system, following the supplier's specifications.
2. Prepare a 10% resolving gel and a 4% stacking gel (**Table 12.3**). TEMED and freshly prepared 10% (w/v) APS are to be added immediately prior to pouring the gel in order to catalyze polymerization.
   a. Resolving gel: 50 µl 10% (w/v) APS and 5 µl TEMED
   b. Stacking gel: 50 µl 10% (w/v) APS and 10 µl TEMED
3. Pour the 10% resolving gel into a prepared gel cassette, leaving space for the stacking gel, overlay with water and allow polymerization for ~40 min at room temperature.
4. Pour the 4% stacking gel, add combs, and allow polymerization for ~40 min at room temperature.

Table 12.3
SDS-polyacrylamide gel formulations

Percent gel	MilliQ H$_2$O (ml)	30% Acrylamide/Bis (ml)	Gel Buffer (ml)	10% w/v SDS (ml)
4	6.1	1.3	2.5[a]	0.1
10	4.1	3.3	2.5[a]	0.1

[a]Resolving gel buffer = 1.5 M Tris-HCl, pH 8.8; Stacking gel buffer = 0.5 M Tris-HCl, pH 6.8

3.2.2. Preparation of Cell Lysates

1. Harvest Day 8-hESC colonies using dispase (10 min, 37°C, see **Note 2**). Transfer cells to an eppendorf tube.
2. Dilute 1:1 with commercially available Laemmli sample buffer or with a reducing sample buffer containing β-mercaptoethanol supplemented with 1 mM sodium orthovanadate and 1 mM PMSF.
3. Boil samples for 4 min and centrifuge for 10 min at 16,110×*g*.
4. Samples should be kept at −80°C for long-term storage. Avoid freeze-thawing samples, as this might impact on the dephosphorylation of the samples. Instead, aliquot samples.

3.2.3. Gel Electrophoresis and Transfer

1. Protein extracts (25 µl/well) are resolved on a SDS-PAGE gel in running buffer at 200 V for ~35 min, using a kaleidoscope prestained ladder (8 µl/well) to estimate the size of the resultant bands.
2. After separation, remove stacking gel with a surgical blade and transfer proteins in the resolving gel to a PVDF membrane. Do not touch the PVDF membranes with hands, use forceps.

3. Prior to use, the PVDF membrane must be activated with 100% methanol for 30 s followed by a rinse in distilled water.
4. For optimum transfer, pre-equilibrate gels and the PVDF membrane in transfer buffer for at least 10 min prior to transfer.
5. Prepare a "gel sandwich" with prewet filter papers, the gel and the membrane in between, according to the manufacturer's instructions. Make sure there is no air space (bubbles) between gel and membrane to ensure a good transfer.
6. Allow transfer for 1 h at 100 V in transfer buffer.
7. After transfer, remove the membrane and orientate it.

3.2.4. Immunoblotting and Protein Detection

1. All steps are performed on a rocking platform.
2. Block the PVDF membrane with 1% BSA (*see* **Note 5**) in TBST either overnight at 4°C or 1 h at room temperature.
3. Incubate the membranes with the primary rabbit polyclonal antibody against Cx43 (0.5 µg/ml in TBST) for 2 h or overnight. Negative control membranes should be incubated with the appropriate immunoglobulin negative fraction at the same concentration.
4. Wash membranes 3 times in TBST (15 min each).
5. Incubate the membranes with the HRP-conjugated secondary antibody for 1 h at room temperature (0.15 µg/ml in TBST).
6. Wash membranes 3 times in TBST (15 min each).
7. Incubate the membranes with the chemiluminescent detection reagent for 5 min and expose to Hyperfilm. Hyperfilm exposure time must be optimized for highest signal to noise ratio. Alternatively, chemiluminescence can be detected using the Gel-Doc system (Biorad). Cx43 appears as a triplet of band at approximately 43 kDa.

3.2.5. Membrane stripping

1. Antibodies on the membrane can be stripped by incubation with the Western Blot Stripping Buffer (15–30 min at room temperature).
2. Following TBST washes, the membrane can be block and re-probe membranes with another antibody of interest, as described above.
3. Beta-tubulin antibody (1/10,000 in TBST for 1 h) followed by HRP-conjugated secondary antibody can be used as a lysate loading control, molecular weight 55 kDa (*see* **Note 6**).

3.3. Immunocytochemistry for the Expression of Cx43, Cx45, and Pluripotency markers

1. Wash cells in PBS.
2. Fix cells in cold 100% ethanol for 10 min at room temperature. Allow air-drying of the samples.
3. With a wax pen, delimit each individual well (to limit risks of "cross contamination" of reagents).
4. Wash cells 3 times in PBS.
5. Samples should be kept at –80°C for long-term storage.
6. Block sample with 1% serum for 1 h at room temperature (*see* **Note 7**).
7. Incubate cells with the following primary antibodies: Cx43 (20 µg/ml, 1/50–1/100 in PBS containing 0.1% serum), Cx45 (1/50–1/100 in PBS containing 0.1% serum) for 30–60 min at room temperature. Negative control should be performed using sample incubated with the appropriate antibody isotype control.
8. Wash samples three times in PBS.
9. Incubate the samples with the appropriate secondary antibody conjugated with FITC (20 µg/ml, 1/40) or Alexa Fluor® 488 (6.67 µg/ml) for 30 min at room temperature.
10. Optional: To assess pluripotency, double staining can be performed using specific hESC markers such as the following antibodies: GCTM-2 (undiluted hybridoma supernatant), Oct-4 (4 µg/ml, 1/50 in PBS), TG-30/CD9 (undiluted hybridoma supernatant), TRA-1-60 (undiluted) for 30 min at room temperature, followed by incubation with the appropriate secondary antibody conjugated with Alexa Fluor® 568 (6.67 µg/ml) for 30 min at room temperature.
11. Wash samples three times in PBS.
12. Counterstain nuclei with DAPI or Hoechst-33342 (1 µg/ml in water) for 5 min at room temperature.
13. After PBS washes, mount samples in Vectashield to enhance visualization of the immunostaining.
14. Cover the slide with a glass coverslip and seal with nail varnish.
15. Cx43 staining appears as a dotted staining at the membrane of cells, while we observed Cx45 staining to be at the membrane and the cytoplasm of hESC. Specificity is verified by the absence of immunostaining in the antibody isotype controls.

3.4. Scrape Loading/Dye Transfer Assay (SL/DT Assay, see Notes 8 and 9)

1. GJIC in hESC is determined using the SL/DT assay as described in (25, 26).
2. In all experiments, hESC must be kept moisturized in buffers at all time to prevent dehydration.

3. Although GJIC can be modulated by Ca^{2+} in different cell types, we previously demonstrated that Ca^{2+} does not modify GJIC in hESC. Thus, for an easier handling of the cells, we suggest to perform SL/DT in the presence of Ca^{2+}.

4. Wash hESC colonies 3 times in a prewarmed $Ca^{2+}Mg^{2+}$-PBS buffer (Buffer 1, **Note 10**).

5. Remove the plastic chambers of the slide and if necessary, individualize each well with a wax pen.

6. Cut hESC colonies with a surgical blade followed by 5 min of incubation with Lucifer yellow (1 mg/ml) and rhodamine-dextran (1 mg/ml) diluted in Buffer 1 (*see* **Note 11**).

7. After further washes with Buffer 1, live colonies are viewed under a fluorescence microscope.

8. Control colonies incubated with both biochemical dyes without scraping should demonstrate no uptake or dye transfer of Lucifer yellow or rhodamine-dextran, confirming that the Lucifer yellow transfer is solely due to gap junction coupling rather than a leaky membrane.

9. If experiments are performed in order to assess effect of specific acute treatments, the drugs used must be incubated at each step of the experiments (i.e., in buffers and Lucifer yellow/rhodamine dextran solutions).

3.5. Chemical Closure of Gap Junctions

We previously found that a number of specific inhibitors and ligands can induce chemical closure of gap junction in hESC, such as PMA (1 µM, 60 min to activate protein kinase C), U0126 (60 µM, 60 min to inhibit MEK phosphorylation), glycyrrhetinic acid (α-GA, 10 µM, 24 h), and BMP-4 (10 ng/ml, 30 min). Using these gap junction blockers, one can readily study the effect of gap junction closure on cell apoptosis, proliferation, pluripotency, and colony growth of hESC (*see* **Note 12**).

3.6. Cell Apoptosis Assay

1. Cell apoptosis is quantified using the In situ Cell Death Detection Kit by flow cytometry analysis.

2. hESC were cultured with or without α-GA (10 µM) for 24 h.

3. Collect floating apoptotic bodies in the media. Wash cells with PBS, centrifuge at 1300*g* for 2 min to collect cells.

4. Harvest hESC colonies using dispase (10 min, 37°C) and transfer hESC colonies to an eppendorf tube.

5. Incubate with TrypLE Express (*see* **Note 13**) for 5 min in 37°C. Wash cells with PBS, centrifuge at 300*g* for 2 min to collect cells. Gently pipette the cells up and down to break clumps to achieve single-cell suspensions. Mix the apoptotic bodies to the hESC samples. Use no more than 2×10^7 cells/ml.

6. Fix cells with 2% PFA for 1 h at room temperature.
7. Wash fixed cells twice with PBS. Centrifuge at 1300g for 2 min to collect samples.
8. Permeabilize the samples with 0.1% Triton X-100 in PBS for 2 min on ice.
9. Wash fixed cells twice with PBS. Centrifuge at 1300g to collect samples.
10. Incubate cells with 50 µl of "TUNEL reaction mixture" for 60 min at 37°C. Negative controls are performed by incubating the cells in "label solution" only for 60 min in an incubator at 37°C. An unstained sample should be prepared as a control to determine autofluorescence background.
11. Wash samples twice with PBS. Centrifuge at 1300g to collect samples.
12. Samples are analyzed by a flow cytometer. Negative control samples were used to set the gate and determine the background due to the secondary antibody, run the samples with voltages set so that the majority of the cells are in the left quadrant. All other subsequent samples should be run with the same voltage to ensure consistency. Collect at least 50,000 cells for analysis.

3.7. Cell Proliferation Assay

1. Cell proliferation in hESC can be quantified using the In situ Cell Proliferation Kit by flow cytometry analysis.
2. Add BrdU (10 µM final) in the culture medium for 2 h in an incubator (37°C) prior to harvesting the hESC. Negative control is performed with hESC without incubation with BrdU.
3. Harvest hESC by dispase treatment (10 min, 37°C) and transfer cells to an eppendorf tube.
4. Incubate with TrypLE Express (**Note 13**) for 5 min in 37°C. Wash cells with PBS, centrifuge at 300g for 2 min to collect cells. Gently pipette the cells up and down to break clumps to achieve single-cell suspensions.
5. Rinse with PBS. Centrifuge at 300g for 2 min to collect samples.
6. Inject the cell suspension into 70% ethanol and incubate for 30 min at −20°C. Do not resuspend the cell pellet in 70% ethanol to avoid cell aggregation.
7. Wash cells with PBS. Centrifuge at 300g for 2 min to collect samples.
8. Denature DNA with HCl 4 M for 10 min. Wash cells with PBS until pH > 6.5 (~3–4 times). Centrifuge at 1300g for 2 min to collect samples.

9. Incubate cells with 50 µl mouse anti-BrdU antibody conjugated with fluorescein for 45 min at 37°C.
10. Rinse samples twice with PBS. Centrifuge at 1300g for 2 min to collect samples.
11. Incubate with propidium iodide (20 µg/ml) for 15 min at room temperature. Do not wash away propidium iodide. Take samples directly to flow cytometry analysis.
12. Samples are analyzed by a flow cytometer. Negative control samples were used to set the gate and determine the background due to the antibody; run the samples with voltages set so that the majority of the cells are in the left lower quadrant. All other subsequent samples should be run with the same voltage to ensure consistency. Collect at least 50,000 cells for analysis.

3.8. Cell Pluripotency Assay

1. hESC pluripotency can be quantified by flow cytometry using stem cell markers TRA181 or TRA160.
2. Harvest hESC by dispases treatment (10 min, 37°C) and transfer cells to an eppendorf tube.
3. Incubate with TrypLE Express (**Note 13**) for 5 min at 37°C. Wash cells with PBS, centrifuge at 300g for 2 min to collect cells. Gently pipette the cells up and down to break clumps to achieve single-cell suspensions.
4. Block with 1% goat serum in hESC culture medium for 30 min on ice. *See* **Note 7**.
5. Incubate samples with TRA-1-60 or TRA-1-81 antibodies (1 µg/1 million cells, diluted in PBS with 1% goat serum) for 30 min on ice. Negative control should be performed with hESC incubated with the appropriate concentration of mouse IgM antibody. An unstained sample should be prepared as a control to determine autofluorescence background.
6. Wash samples twice with PBS. Centrifuge at 300g for 2 min to collect samples.
7. Incubate with goat antimouse IgM antibody conjugated with Alexa Fluor® 488 (4 µg/ml, diluted in PBS with 1% goat serum) for 30 min on ice.
8. Wash samples twice with PBS. Centrifuge at 300g for 2 min to collect samples.
9. Samples were analyzed by a flow cytometer. Negative control samples were used to set the gate and determine the background due to the antibody; run the samples with voltages set so that the majority of the cells are in the left lower quadrant. All other subsequent samples should be run with the same voltage to ensure consistency. Collect at least 50,000 cells for analysis.

3.9. Colony Growth Assay

1. We previously found that the chemical inhibition of GJIC was accompanied by cell death and decrease of colony growth in serum-free medium, but not in culture medium containing serum.

2. Incubate hESC colonies for 5–7 days in serum-free medium in the presence or absence of α-GA (10 μM), change medium every 2 days.

3. Capture phase-contrast images of the morphology for at least 16 hESC colonies every day for 5–7 days.

4. Colony growth can be assessed by measuring hESC colony area using cell^B analysis software (Olympus Software Imaging Solutions). Alternatively, the colony diameter can be measured with reference to scale bar using Adobe Photoshop. The colony diameter is recorded as the average of the longest and shortest diameter of the colony.

3.10. Statistical Analysis

All experiments must be performed at least 3 times to ensure consistent results. Statistical analysis on raw data is performed using Graphpad Prism software. Different statistical tests may be used for the different experiments performed, such as the two-tailed t-test or one- and two-way ANOVA followed by Bonferroni or Tukey tests. Statistical significance is established at *$p<0.05$, **$p<0.01$, and ***$p<0.001$.

4. Notes

1. Formats of culture are reflective of what we found to be the easiest handling format to perform experimental work. These are only indicative and can be modified according to specific needs.

2. Dispase is a protease that cleaves adhesion molecules, thus allowing extraction of hESC from MEF and the plastic surface of the dish. hESC colonies can be readily sucked off using a pipette, leaving the MEF layer intact on the dish.

3. The system uses polyT magnetic beads to hybridize at high efficiency with the $3v$ polyA tail of mRNA, thus yielding highly purified mRNA.

4. This technique allows for the study of the level of phosphorylation of Cx43 in hESC. Indeed, Cx43 can be present as unphosphorylated, or phosphorylated once or twice. There is no strict correlation between the phosphorylation states of connexin proteins and the degree of functional coupling. Indeed, phosphorylation of Cx43 can influence GJIC both positively and negatively depending on the cell type (27, 28).

The antibody used in this protocol recognizes respectively 3 forms of Cx43 (unphosphorylated, phosphorylated once and twice).

5. Blocking the membrane with BSA rather than with dried milk gives a better resolution for the detection of phosphorylation.

6. If no detection is observed, the membrane can be incubated with 0.1% (w/v) PonceauS staining solution (0.1% (w/v) PonceauS, 5% (v/v) acetic acid: mix 50 mg Ponceau S, 2.5 ml acetic acid and bring to a final volume of 50 ml with distilled water) for 10–20 min with agitation to check for successful protein transfer. Membrane can then be washed in distilled water until clean.

7. The serum used for blocking should correspond to the species in which the secondary antibodies were raised. A combination of serum from different species can be used together.

8. hESC communicate through functional and opened gap junctions. The SL/DT assay allows for a quick, cheap, and reliable study of GJIC in cells. Controls: hESC in serum (GJIC are opened); PMA (1 μM) for 60 min (GJIC are closed).

9. For an easier handling of cells during these experiments, it is advisable to cultivate hESC cells on 8-well chamber slides. Generally we use day-5 colonies as the colonies are large enough to handle and not yet started to spontaneously differentiate.

10. $Ca^{2+}Mg^{2+}$-free PBS buffer (Buffer 2: 140 mM NaCl, 5.5 mM KCl, 10 mM glucose, 10 mM Hepes, 2 mM EGTA, pH 7.35) can be used to determine the effect of exogenous Ca^{2+} and Mg^{2+} in modulating gap junctions in hESC.

11. Due to its low molecular weight (522 Da), Lucifer yellow is able to diffuse from cell to cell through functional gap junctions. On the other hand, rhodamine–dextran (10 kDa) is too large to diffuse through gap junctions, thus serves as a negative control. Time of incubation with Lucifer yellow can be modified depending on the size of the colonies.

12. Other techniques not used in the laboratory are available for the study of gap junctions in hESC. In particular, siRNA can be used to down-regulate specific connexin proteins instead or in complement of a chemical inhibition of GJIC. Furthermore, other potent inhibitors of GJIC in different cell types, but not yet used in hESC include heptanol, octanol, and halothane.

13. TrypLE Express is a trypsin-like enzyme used to dissociate cells into single cells. In our experience, TrypLE Express is more gentle to hESC than trypsin.

Acknowledgements

This study was supported by The University of Melbourne and the National Health and Medical Research Council of Australia (NHMRC 454723). R.C. Wong is currently supported by the California Institute of Regenerative Medicine Grant RC1-00110-1.

References

1. Carpenter M. K., Rosler E. S., Fisk G. J., et al. (2004) Properties of four human embryonic stem cell lines maintained in a feeder-free culture system. *Dev Dyn.* **229**, 243–258.
2. Wong R. C., Dottori M., Koh K. L., Nguyen L. T., Pera M. F., and Pebay A. (2006) Gap junctions modulate apoptosis and colony growth of human embryonic stem cells maintained in a serum-free system. *Biochem Biophys Res Commun.* **344**, 181–188.
3. Wong R. C., Pebay A., Nguyen L. T., Koh K. L., and Pera M. F. (2004) Presence of functional gap junctions in human embryonic stem cells. *Stem Cells* **22**, 883–889.
4. Duval N., Gomes D., Calaora V., Calabrese A., Meda P., and Bruzzone R. (2002) Cell coupling and Cx43 expression in embryonic mouse neural progenitor cells. *J Cell Sci.* **115**, 3241–3251.
5. Cheng A., Tang H., Cai J., et al. (2004) Gap junctional communication is required to maintain mouse cortical neural progenitor cells in a proliferative state. *Dev Biol* **272**, 203–216.
6. Potapova I., Plotnikov A., Lu Z., et al. (2004) Human mesenchymal stem cells as a gene delivery system to create cardiac pacemakers. *Circ Res.* **94**, 952–959.
7. Valiunas V., Doronin S., Valiuniene L., et al. (2004) Human mesenchymal stem cells make cardiac connexins and form functional gap junctions. *J Physiol.* **555**, 617–626.
8. Todorova M. G., Soria B., and Quesada I. (2008) Gap junctional intercellular communication is required to maintain embryonic stem cells in a non-differentiated and proliferative state. *J Cell Physiol.* **214**, 354–362.
9. Oyamada Y., Komatsu K., Kimura H., Mori M., and Oyamada M. (1996) Differential regulation of gap junction protein (connexin) genes during cardiomyocytic differentiation of mouse embryonic stem cells in vitro. *Exp Cell Res.* **229**, 318–326.
10. Worsdorfer P., Maxeiner S., Markopoulos C., et al. (2008) Connexin expression and functional analysis of gap junctional communication in mouse embryonic stem cells. *Stem Cells* **26**, 431–439.
11. Kumar N. M., and Gilula N. B. (1996) The gap junction communication channel. *Cell* **84**, 381–388.
12. Revel J. P., and Karnovsky M. J. (1967) Hexagonal array of subunits in intercellular junctions of the mouse heart and liver. *J Cell Biol.* **33**, C7–C12.
13. Alexander D. B., and Goldberg G. S. (2003) Transfer of biologically important molecules between cells through gap junction channels. *Curr Med Chem.* **10**, 2045–2058.
14. Krysko D. V., Leybaert L., Vandenabeele P., and D'Herde K. (2005) Gap junctions and the propagation of cell survival and cell death signals. *Apoptosis* **10**, 459–469.
15. Valiunas V., Bukauskas F. F., and Weingart R. (1997) Conductances and selective permeability of connexin43 gap junction channels examined in neonatal rat heart cells. *Circ Res.* **80**, 708–719.
16. Neijssen J., Herberts C., Drijfhout J. W., Reits E., Janssen L., and Neefjes J. (2005) Cross-presentation by intercellular peptide transfer through gap junctions. *Nature* **434**, 83–88.
17. Sheardown S. A., and Hooper M. L. (1992) A relationship between gap junction-mediated intercellular communication and the in vitro developmental capacity of murine embryonic stem cells. *Exp Cell Res.* **198**, 276–282.
18. De Maio A., Vega V., and Contreras J. (2002) Gap junctions, homeostasis, and injury. *J Cell Physiol.* **191**, 269–282.
19. Mesnil M., Crespin S., Avanzo J. L., and Zaidan-Dagli M. L. (2005) Defective gap junctional intercellular communication in the carcinogenic process. *Biochim Biophys Acta* **1719**, 125–145.

20. Trosko J. E., and Chang C. C. (2003) Isolation and characterization of normal adult human epithelial pluripotent stem cells. *Oncol Res.* **13**, 353–357.

21. Yamasaki H., Krutovskikh V., Mesnil M., Tanaka T., Zaidan-Dagli M. L., and Omori Y. (1999) Role of connexin (gap junction) genes in cell growth control and carcinogenesis. *C R Acad Sci III* **322**, 151–159.

22. Huettner J. E., Lu A., Qu Y., Wu Y., Kim M., and McDonald J. W. (2006) Gap junctions and connexon hemichannels in human embryonic stem cells. *Stem Cells* **24**, 1654–1667.

23. Assou S., Lecarrour T., Tondeur S., et al. (2007) A meta-analysis of human embryonic stem cells transcriptome integrated into a web-based expression atlas. *Stem Cells* **25**, 961–973.

24. Boyer L. A., Lee T. I., Cole M. F., et al. (2005) Core transcriptional regulatory circuitry in human embryonic stem cells. *Cell* **122**, 947–956.

25. El-Fouly M. H., Trosko J. E., and Chang C. C. (1987) Scrape-loading and dye transfer. A rapid and simple technique to study gap junctional intercellular communication. *Exp Cell Res.* **168**, 422–430.

26. Venance L., Piomelli D., Glowinski J., and Giaume C. (1995) Inhibition by anandamide of gap junctions and intercellular calcium signalling in striatal astrocytes. *Nature* **376**, 590–594.

27. Lampe P. D., and Lau A. F. (2000) Regulation of gap junctions by phosphorylation of connexins. *Arch Biochem Biophys.* **384**, 205–215.

28. Lampe P. D., and Lau A. F. (2004) The effects of connexin phosphorylation on gap junctional communication. *Int J Biochem Cell Biol.* **36**, 1171–1186.

Chapter 13

hESC Engineering by Integrase-Mediated Chromosomal Targeting

Ying Liu, Uma Lakshmipathy, Ali Ozgenc, Bhaskar Thyagarajan, Pauline Lieu, Andrew Fontes, Haipeng Xue, Kelly Scheyhing, Chad MacArthur, and Jonathan D. Chesnut

Abstract

Bacteriophage recombinases can target specific loci in human embryonic stem cells (hESCs) at high efficiency allowing for long-term expression of transgenes. In this chapter, we describe a retargeting system where phiC31 integrase is used to deliver a chromosomal target for a second integrase, R4. The engineered hESC line can be adapted for complex element assembly using Multisite Gateway technology. Retargeted clones show sustained expression and appropriate regulation of the transgenes over long-term culture and upon differentiation. The system described here represents a method to rapidly assemble complex plasmid-based assay systems, controllably insert them into the hESC genome, and have them actively express in pluripotent as well as in differentiated lineages there from.

Key words: Embryonic stem cells, molecular engineering, phiC31 integrase, R4 integrase, Gateway, stem cell culture.

1. Introduction

R4 human embryonic stem cells (hESCs) are engineered to contain a chromosomal target for R4 Integrase for efficient and site-specific insertion of multiple genetic elements (e.g., promoter-reporter pairs) in a retargeting construct that has been adapted for complex element assembly using MultiSite Gateway® Technology. Retargeted clones exhibit sustained expression and appropriate regulation of the transgenes over long-term undifferentiated culture, as well as upon their random differentiation and directed induction into various cell lineages (1, 2).

hESC/R4 platform cells are created by using an integration vector containing the R4 *attP* target sequence and the Hygromycin-resistance gene, and phiC31-mediated recombination to stably integrate the target site into the genome of hES cells (1) (**Fig.13.1**). This results in cell lines containing an R4 *attB* site placed upstream of a selectable marker lacking a promoter (Zeocin-resistance gene, *sh ble*) at a specific chromosomal locus. The Hygromycin-resistant colonies are tested extensively to make sure that they contain a single copy of the target site, maintained parental karyotype, and retain hESC properties. Further validation of these clones confirm that they are pluripotent and able to differentiate into representatives of all three primary germ layers.

Verstility is added to this system by the ability to assemble multigene expression constructs easily using MultiSite Gateway® for simultaneous cloning of DNA fragments to generate a retargeting construct. Based on the Gateway® Technology (3–5), the MultiSite Gateway® Technology uses site-specific recombinational cloning to allow simultaneous cloning of multiple DNA fragments in a defined order and orientation.

This chapter provides an overview of hESC/R4 hESC platform lines and offers instructions and guidelines for:

- Maintaining the hESC/R4 (*h*uman *e*mbryonic *s*tem *c*ell) culture in StemPro® hESC SFM (*s*erum- and *f*eeder-free *m*edium), as well as in MEF-CM (*m*ouse *e*mbryonic *f*ibroblast *c*onditioned *m*edium) and on MEF feeders.

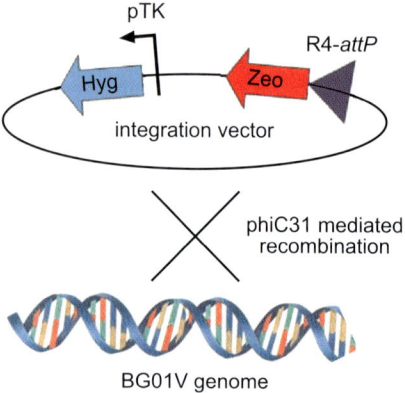

Fig. 13.1. PhiC31 integrase-mediated insertion of a target site into the cellular genome. A target plasmid (containing a wild type R4 integrase recognition (attP) site along with an activatable antibiotic resistance gene, a phiC31 integrase attB site, and a constitutively expressed antibiotic gene) is used to transfect hESCs in combination with a plasmid directing the expression of phiC31 integrase. This enzyme directs integration of the target plasmid at phiC31 "pseudo sites" within the cellular genome. Cells are allowed to recover and then selected by virtue of their antibiotic resistance. Isolated colonies are evaluated for proper integration, copy number, and site of integration, then banked for use as "platform lines" in secondary targeting.

- Characterization and quality control of hESC/R4 platform lines before and after retargeting.
- Retargeting the hESC/R4 platform lines with the retargeting construct, and subsequent selection and expansion of transformants.

For more information about the MultiSite Gateway® Technology, refer to the MultiSite Gateway® Pro manual (25-0942). For more information on the BG01V/R4 hESC or other platform lines, see www.invitrogen.comor (1, 2). For more information on culturing hESC using StemPro® hESC SFM, refer to www.invitrogen.com.

2. Materials

2.1. Thawing hESC/R4 Cells in StemPro® hESC SFM

1. BG01v/R4 hES cells (store frozen cells in liquid nitrogen until ready to use).
2. StemPro® hESC SFM prewarmed to 37°C. (Invitrogen).
3. Disposable, sterile 15-ml conical tubes.
4. 37°C water bath.
5. Geltrex™ (Invitrogen) coated 35-mm dishes.
6. *Incubator:* 36–38°C, humidified atmosphere of 4–6 % CO_2 in air.
7. *Culture Conditions:* Adherent; ensure proper gas exchange and minimize exposure to light.
8. *Recommended Culture Vessels:* 35- or 60-mm dishes.

2.2. Propagating hESC/R4 Cells in StemPro® hESC SFM

1. bFGF (10 μg/ml, Invitrogen): Prepare 10 μg/ml bFGF working solution in D-MEM/F-12 with 0.1% BSA, aliquot 80 μl per tube, and store frozen at −20°C. Thaw at 37°C water bath immediately before use.
2. StemPro® hESC SFM complete medium, prewarmed to 37°C.
3. Collagenase Type IV: Dissolve 10 mg/ml collagenase in D-MEM/F-12, filter to sterilize, and freeze in aliquots. Thaw immediately before use.
4. Geltrex™ Reduced Growth Factor Basement Membrane Matrix (Invitrogen): Thaw the Geltrex™ bottle at 4°C overnight to prevent polymerization. Next day, dilute Geltrex™ 1:2 with D-MEM/F-12 at 4°C to make 100X stock solution, using an ice bucket to keep the bottles cold. Quickly prepare 0.5 ml aliquots in 50-ml conical tubes (prechilled on ice) and store at −20°C. Thaw at 4°C before use.

5. Dulbecco's Phosphate Buffered Saline (D-PBS) (1X).
6. D-MEM/F-12 (Invitrogen).
7. 37°C water bath.
8. Geltrex™ coated 35-mm culture plates.

2.3. Freezing hESC/R4 Cells Cultured on StemPro® hESC SFM

1. Plates with hESC/R4 cells on StemPro® hESC SFM.
2. BSA Wash Medium.
3. D-MEM/F-12 with GlutaMAX™ (Invitrogen).
4. Collagenase Type IV working solution (10 mg/ml) in D-MEM/F-12.
5. Knockout™ Serum Replacement (KSR, Invitrogen).
6. DMSO (use a bottle set aside for cell culture; open only in a laminar flow hood).
7. Disposable, sterile 15-ml conical tubes.
8. Sterile freezing vials.

2.4. Preparing a MEF Feeder Cell Layer

1. Mitotically inactivated, hygromycin-resistant human feeder cells (Invitrogen), hygromycin-resistant mouse embryonic fibroblasts (Millipore or ATCC), or generate them as described in *Generating Mitomycin C-Treated MEFs*.
2. Dulbecco's Modified Eagle Medium (D-MEM), high glucose with L-glutamine and sodium pyruvate.
3. Fetal Bovine Serum (FBS), ES Cell-Qualified (Invitrogen).
4. MEM Non-Essential Amino Acids Solution 10 mM (100X) (NEAA).
5. 2-Mercaptoethanol (1,000X).
6. D-MEM/F-12 with GlutaMAX™ (2 mM).
7. Knockout™ Serum Replacement (KSR).
8. bFGF: Reconstitute lyophilized human bFGF in sterile D-MEM/F-12 containing 0.1% BSA to 10 μg/ml. Divide stock solution into working aliquots and store at $\leq -20°C$.
9. Porcine skin gelatin (Sigma). Prepare 0.1% (w/v) porcine skin gelatin in sterile, distilled water and sterilize by filtration using a 0.2-μm filter.

    Store up to 1 year at 4°C.
10. 37°C incubator with a humidified atmosphere of 5% CO_2.

2.5. Establishing and Maintaining Your hESC/R4 Cells on MEF Feeders

1. Plates with hESC/R4 cells.
2. hESC medium prewarmed to 37°C.
3. D-MEM/F-12 with GlutaMAX™.
4. Feeder layer plates with mitotically inactivated MEFs – prepare at least two days in advance.

5. Collagenase Type IV.
6. Hygromycin B.
7. Disposable, sterile 15-ml tubes.
8. A 37°C incubator with a humidified atmosphere of 5% CO_2.

2.6. Freezing hESC/R4 Cells Cultured MEF Feeders

1. Plates with hESC/R4 cells on MEF feeders.
2. hESC Medium.
3. Collagenase Type IV working solution (1 mg/ml) in D-MEM/F-12.
4. Fetal Bovine Serum, ES Cell-Qualified.
5. DMSO (use a bottle set aside for cell culture; open only in a laminar flow hood).
6. Disposable, sterile 15-ml conical tubes.
7. Sterile freezing vials.

2.7. Quality Control of hESC

1. Parafomaldehyde (Electron Microscopy Services)
2. BSA (Sigma)
3. Triton-X (Sigma)
4. DAPI (Invitrogen)
5. LIVE/DEAD Cell Vitatlity Assay Kit (Invitrogen)

2.8. Characterization of hESC/R4 Platform Line

1. hESC/R4 cells
2. Phosphate buffered saline (PBS)
3. CellsDirect Resuspension and Lysis Buffers (Invitrogen)
4. PCR primers (see below for primer sequences)
5. AccuPrime™ *Taq* DNA Polymerase High Fidelity
6. Thermocycler
7. Water bath or heat block at 75°C

2.9. Transfection Using the Neon™ Transfection System

1. Retargeting expression construct generated in pJTI™-R4 DEST using MultiSite Gateway® Technology and pJTI™-R4Int plasmid-encoding R4 Integrase.
2. 60-mm culture dish containing mitotically inactivated MEF feeders in a 37°C incubator.
3. MEF-CM prewarmed to 37°C.
4. TrypLE™ express dissociation enzyme without phenol red.
5. D-PBS (Ca- and Mg-free).
6. Neon™ transfection system. Refer to www.invitrogen.com for ordering information.

7. 100-μl gold tips, 20 microporation tubes, 20 ml of resuspension buffer, and 300 ml of electrolytic buffer E2) (available from Invitrogen; search for Neon transfection systems at www.invitrogen.com for ordering information).
8. LIVE/DEAD® Cell Vitality Assay Kit or Trypan Blue Stain for exclusion counting.

2.10. Characterization of Retargeted hESC/R4 Line

1. Retargeted hESC/R4 line
2. Phosphate buffered saline (PBS)
3. CellsDirect Resuspension and Lysis Buffers (Invitrogen, Cat. No. 11739-010)
4. PCR primers (see below for primer sequences)
5. AccuPrime™ Taq DNA Polymerase High Fidelity
6. Thermocycler
7. Water bath or heat block at 75°C

3. Methods

Follow the general guidelines below to grow and maintain the hESC/R4 platform lines:
1. All solutions and equipment that come in contact with the cells must be sterile. Always use proper sterile technique and work in a laminar flow hood.
2. When subculturing hESC/R4 cells on mouse embryonic fibroblast feeder layers (MEF feeders), always use mitotically inactivated MEFs to prevent overgrowth of hESCs. Make sure to start preparing the feeder layer two days before culturing hESC/R4 cells.
3. Before starting experiments, be sure to have your cells established (at least 5 passages) and also have some frozen stocks on hand. We recommend using early-passage cells for your experiments (below 30 passages).
4. For general maintenance of cells, pass hESC/R4 cells before colonies start contacting each other.
5. When thawing or subculturing cells, transfer cells into pre-warmed medium.
6. 10 ml/L of antibiotic-antimycotic containing penicillin, streptomycin, and amphotericin B may be used if required.

3.1. Thawing hESC/R4 Cells in StemPro® hESC SFM

Follow the protocol below to thaw hESC/R4 cells to initiate cell culture in StemPro® hESC SFM. The BG01VR4 hESC line is supplied in a vial containing 1 ml of cells at 3×10^6 cells/ml in freezing medium. For harvesting and freezing these and other hESC/R4 cells in StemPro® hESC SFM, see below. *See* **Note 1** (*Important*).

Store frozen cells in liquid nitrogen until ready to use. To thaw and establish hESC/R4 cells in StemPro® hESC SFM:

1. Remove the cryovial of cells from the liquid nitrogen and thaw quickly in a 37°C water bath (to prevent crystal formation).

2. When thawed, immediately transfer cells into a 15-ml conical tube.

3. Add 1 ml of prewarmed StemPro® hESC SFM to the 15-ml tube with thawed cells in a *dropwise* fashion (*see* **Note 2**). Repeat 2–3 times.

4. Centrifuge the 15-ml tube with cells at 1000 rpm ($200 \times g$) for 2 min.

5. Aspirate medium, resuspend cells in warm StemPro® hESC SFM, and plate onto Geltrex™-coated dish (2 ml for a 35-mm dish).

6. Grow cells in a 37°C incubator with a humidified atmosphere of 5% CO_2. Change the medium every day.

3.2. Human Embryonic Stem Cell Culture on StemPro® hESC SFM

Traditional hESC culture methods require the use of mouse or human fibroblast feeder layers, which are labor intensive and hard to scale. In addition, the undefined conditions on feeder cultures make it difficult to maintain hESCs in an undifferentiated state. StemPro® hESC SFM enables culture of hESCs in a serum-free medium (SFM) without feeder cells. Go to www.invitrogen.com/stempro/hesc for an instructional video on how to use StemPro® hESC SFM (*see* **Note 3**).

To prevent differentiation and/or slow growth of hESC/R4 cells grown in StemPro® hESC SFM, *see* **Notes 4–11**.

Regarding the physical conditions for hESC culture: *Medium*: StemPro® hESC SFM (Invitrogen, Cat. No. A10007-01) contains D-MEM/F-12 with GlutaMAX™, StemPro® hESC Supplement, and bovine serum albumin 25% (BSA). Refer to the manual supplied with StemPro® hESC SFM for storage and handling.

3.2.1. Coating Plates with Geltrex™

1. Thaw 1 tube of Geltrex™ (0.5 ml, aliquoted as above) slowly at 4°C and add 49.5 ml of cold D-MEM/F-12 (1:100 dilution). Mix gently.

2. Cover the whole surface of each culture plate with the Geltrex™ solution (1.5 ml for a 35-mm dish, 3 ml for a 60-mm dish).

3. Seal each dish with parafilm to prevent drying and incubate for 1 h at room temperature in a laminar flow hood.

4. You may store the Geltrex™-treated dish at 4°C for up to 1 month.

5. Before plating cells, tip the plate slightly and aspirate the Geltrex™ solution. Immediately plate cells in pre-equilibrated complete medium.

3.2.1.1. Medium Preparation

Follow the instructions below for preparing the various medium required to maintain your hESC culture in StemPro® hESC SFM. *See* **Notes 12–14** for Storage and Handling of StemPro® hESC SFM Supplement

3.2.1.2. BSA Wash Medium

Add BSA 25% (supplied with the StemPro® hESC SFM kit, Cat. No. A10007-01) at a final concentration of 0.1% to D-MEM/F-12 containing GlutaMAX™.

3.2.1.3. StemPro® hESC SFM Complete Medium

Thaw the StemPro® hESC SFM Supplement in 37°C water bath (minimize dwell time) and prepare the StemPro® hESC SFM Complete Medium according to **Table 13.1**.

Table 13.1
StemPro hESC-SFM media components

Component	Final concentration	For 500 ml	For 100 ml
D-MEM/F-12 with GlutaMAX™	1X	454 ml	90.8 ml
StemPro® hESC SFM Supplement (50X)	1X	10 ml	2 ml
BSA 25%	1.8%	36 ml	7.2 ml
bFGF (10 µg/ml)	8 ng/ml	400 µl	80 µl
2-Mercaptoethanol (55 mM)	0.1 mM	909 µl	182 µl

3.2.2. Collagenase Preparation

1. Prepare 1 and 10 mg/ml aliquots of collagenase IV in D-MEM/F-12.

2. Filter to sterilize and freeze at –20°C (*see* **Note 15**).

3.2.3. Passaging Using Collagenase

1. Warm appropriate amounts of 10 mg/ml Collagenase IV solution (~2000 U/ml), StemPro® hESC SFM Complete Medium, and BSA wash medium to 37°C in a water bath. Minimize dwell time.

2. Set up hESC plate on a dissecting microscope in a bio-safety cabinet or laminar flow hood to comfortably observe colonies.

3. Cut out and remove any overtly differentiated colonies with a 21½-gauge needle.

4. Aspirate the medium and gently add 1–2 ml of collagenase.

5. Leave for 3 min to loosen the cells in between colonies and to round up colony edges.

6. Gently tap the sides of the dish to dislodge cells.

7. Remove collagenase, rinse with D-PBS, and then add 3 ml of BSA wash medium.

8. Gently scrape dish using a sterile 1000-µL pipette tip.

9. Gently transfer clumps using a 5-ml pipette and place into a 15-ml tube.

10. Wash plate with 3 ml of BSA wash medium and add to tube.

11. Centrifuge cells at $200 \times g$ for 2 min at room temperature.

12. Gently aspirate medium and flick tube to loosen cells from the bottom.

13. Gently resuspend the cells in pre-equilibrated complete medium using a 1- or 5-ml serological pipette.

14. Remove a Geltrex^{TM}-coated plate from 2–8°C and tip slightly to aspirate the Geltrex^{TM} solution. Immediately plate the cells. Do not allow the surface to dry out before plating.

15. Mix plates gently to evenly spread out the clumps and place the plate into an incubator set at 37°C with 5% CO_2 in air. See the next page for an image of the BG01v/R4 platform line immediately after passaging using collagenase.

16. Gently change medium the next day to remove excess cells and provide fresh nutrients, and every day thereafter.

17. Observe cells every day and passage by the above protocol whenever required (approximately every 5–7 days) (*see* **Note 16**).

3.2.3.1. Images of the BG01v/R4 Platform Line

Below are representative images of the BG01v/R4 platform line 3 days after thawing on StemPro® hESC SFM (**Fig. 13.2**), and right after passaging using collagenase (**Fig. 13.3**). The images are provided to display examples of typical BG01v/R4 hESC colony morphology.

3.2.4. Preparing SFM Freezing Medium

Prepare SFM Freezing Medium 1 and 2 immediately before use. Discard any unused medium.

3.2.4.1. SFM Freezing Medium 1

For every 1 ml, mix together the following in a separate sterile 15-ml tube:
D-MEM/F-12 with GlutaMAX^{TM} (0.5 ml), Knockout^{TM} Serum Replacement (KSR)(0.5 ml)

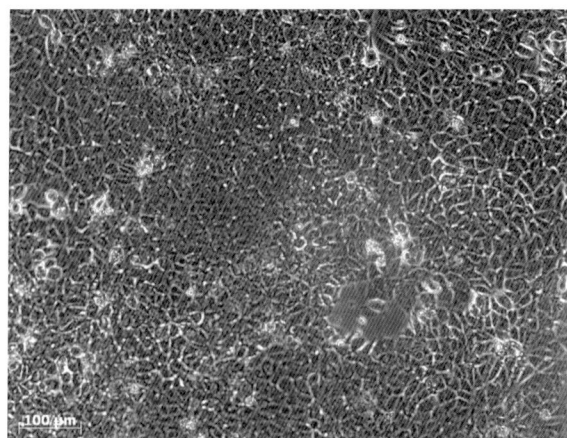

Fig. 13.2. hESC/R4 cells 3 days after being thawed on StemPro® hESC SFM.

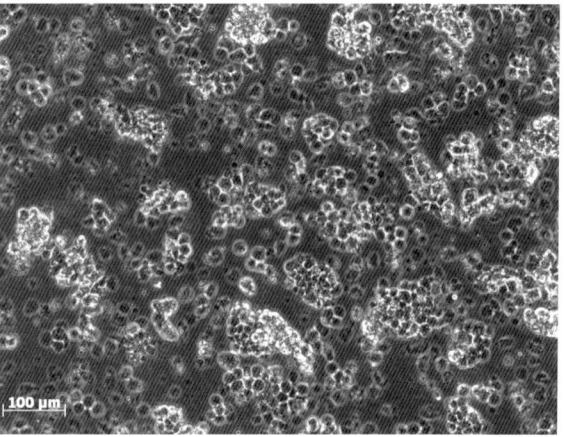

Fig. 13.3. Clumps of BG01v/R4 platform line immediately after passaging.

3.2.4.2. SFM Freezing Medium 2

For every 0.5 ml, mix together the following in a separate sterile 15-ml tube:

DMEM/F12 with GlutaMAX™ (0.8 ml), DMSO (0.2 ml)

Keep SFM Freezing Medium 1 at room temperature and place tube with SFM Freezing Medium 2 on ice.

3.2.5. Freezing Protocol for hESCs Cultured on StemPro® hESC SFM

1. Aspirate serum-free culture medium from the cells and gently add 1–2 ml of 10 mg/ml collagenase solution. (Alternatively, use the StemPro® EZPassage™ Disposable Stem Cell Passaging Tool to cut the cell colonies into pieces; follow the protocol provided with the tool and then proceed to Step 4.)
2. Leave for 3 min to dislodge cell colonies from the substrate.
3. Remove collagenase and rinse with D-PBS.
4. Add 3 ml of BSA wash medium.

5. Gently scrape the dish using a sterile 1000-μl pipette tip.

6. Gently transfer the cell clumps using a 5-ml pipette and place into a 15-ml tube.

7. Wash plate with 3 ml of wash medium and add to the tube.

8. Spin cells down for 2 min at $200 \times g$ at room temperature.

9. Gently aspirate medium and resuspend the BG01v/R4 hES cells in SFM Freezing Medium 1 at room temperature, using 1 ml of freezing medium per cells from one 60-mm dish.

10. Add the same volume of cold SFM Freezing Medium 2 to cells in a *dropwise* manner, swirling the tube after each drop.

11. Resuspend the cells by gently pipetting 2–3 times. Aliquot 1 ml of the cell suspension to each freezing vial and store at −80°C overnight in isopropanol chamber.

12. Transfer frozen vials to liquid nitrogen tank for long-term storage (*see* **Notes 17** and **18**).

3.3. Culturing hESC/R4 Cells on MEF Feeders

You must maintain your hESC/R4 cells on mitotically inactivated Hygromycin-resistant mouse embryonic fibroblast (MEF) feeder cells (MEF feeders) under Hygromycin selection for at least two weeks and in MEF-conditioned medium (MEF-CM) as a feeder-free culture for at least one passage prior to transfection with your retargeting construct. After retargeting, you must plate your transfected cells on MEF feeders to allow them to recover, before starting selection in MEF-CM with Zeocin™. The cells should be expanded in MEF-CM on Geltrex™-coated dishes under selection before their transfer into StemPro® hESC SFM under selection.

This section provides instructions and guidelines for preparing the MEF feeders, thawing and maintaining hESC/R4 cells (frozen in StemPro® hESC SFM) on MEF feeders, and freezing the hESCs cultured on MEF feeders.

In order to prepare a MEF feeder cell layer, follow the protocol below to prepare the matrix (feeder layer) for thawing and establishing hESC/R4 cells. hESCs frozen in StemPro® hESC SFM may directly be thawed onto the MEF feeder layer. Use mitotically inactivated MEFs as your feeder cell layer to prevent overgrowth of the hESCs. Both mitomycin C treatment and irradiation methods can be used to mitotically inactivate your MEFs.

3.3.1. MEF Medium Preparation

To prepare 500 ml of MEF medium, mix the reagents as seen in **Table1 3.2**. Filter through a 0.22-μm filtration unit to sterilize. Preheat the medium to 37°C before use.

3.3.2. hESC Medium Preparation

To prepare 500 ml of hESC medium, mix the reagents as seen in **Table 13.3**. If stored at 4°C, hESC medium can be kept for up to 1 week. Preheat the medium to 37°C before use.

Table 13.2
MEF medium composition

Component	Volume	Final concentration
D-MEM	445 ml	1X
FBS	50 ml	10%
NEAA (10 mM)	5 ml	0.1 mM
2-Mercaptoethanol, 1,000X (55 mM)	500 µl	55 µM

Table 13.3
hESC KSR medium composition

Component	Volume	Final concentration
D-MEM/F-12 with GlutaMAX™	79 ml	1X
Knockout™ Serum Replacement (KSR)	20 ml	20%
NEAA (10 mM)	1 ml	0.1 mM
2-Mercaptoethanol, 1,000X (55 mM)	100 µl	55 µM
bFGF (10 µg/ml)	40 µl	4 ng/ml

3.3.3. Preparation of Gelatin-Coated Plates

Coat culture plates for 20–60 min at room temperature with sterile 0.1% (w/v) porcine skin gelatin in dH$_2$O.

3.3.4. Generating Mitomycin C-Treated MEFs

Prepare 0.1% (w/v) porcine skin gelatin (Sigma, Cat. No. G1890) in sterile, distilled water and sterilize by filtration using a 0.2-µm filter. Store up to 1 year at 4°C.

3.3.4.1. Preparing Gelatin-Coated Plates

Coat plates for 20–60 min at room temperature with 0.1% gelatin in distilled water.

3.3.4.2. Preparing Mitomycin C (see Note 19)

Prepare 10 µg/ml mitomycin C in MEF medium; filter sterilize and store at −20°C in the dark until use. Mitomycin C can also be kept at 4°C in the dark for up to 2 weeks.

3.3.4.3. Mitomycin C Inactivation

Use the procedure below to generate mitotically inactivated MEFs in T175 culture flasks. Make sure that the MEFs to be treated with mitomycin C are 90–95% confluent in T175 flasks 3 days after the initial thawing. Observe each flask individually under the microscope to ensure cell growth and culture sterility.

1. Culture MEFs in MEF medium (*see* **Section 3.3.1**).
2. In a biosafety cabinet, aspirate the medium from T175 flasks and add 16 ml of mitomycin C solution (10 µg/ml).

3. Incubate MEFs treated with 10 µg/ml mitomycin C in the flasks for 2–3 h at 37°C, 5% CO_2. Work in sets of no more than six flasks at a time.

4. After 2–3 h of incubation, aspirate off the mitomycin C solution and neutralize the waste with bleach (see **Note 20**).

5. Wash cells five times with Dulbecco's phosphate-buffered saline (D-PBS) containing Mg^{2+} and Ca^{2+}.

6. Aspirate D-PBS and wash cells with 20 ml D-PBS that is Mg^{2+}- and Ca^{2+}-free.

7. Add 3 ml of 0.05% Trypsin-EDTA solution per flask to trypsinize cells. At room temperature, monitor the degree of cell detachment, while gently rocking and tapping the flask (see **Note 21**).

8. When cells are sufficiently detached from the flask, add 5 ml of MEF medium to each flask, rock to disperse, and pool cell suspensions from 1–6 flasks into 2×50-ml conical tubes.

9. Add 15 ml of MEF medium to the first flask to rinse out the cells. Rinse the subsequent flask using the same 15 ml MEF medium, and pool with cell suspension. Discard the flasks.

10. Adjust the volume in each tube to 50 ml with MEF medium and centrifuge cells at $200 \times g$ for 4 min at room temperature.

11. Resuspend cell pellets with MEF medium and pool into one 50-ml tube, using a maximum of 12× T175 flasks of cells per 50-ml tube.

12. Centrifuge cells at $200 \times g$ for 4 min at room temperature.

13. Resuspend the cell pellet in 40 ml of MEF medium, using a 10-ml serological pipette and ensuring that the cells are resuspended fully. Adjust the volume to 50 ml with MEF medium.

14. Centrifuge cells at $200 \times g$ for 4 min at room temperature. At this stage, the cells will have been washed a total of 9 times: 6 times before trypsin, once at trypsinization, and twice post-trypsinization.

15. Resuspend the cell pellet in 10 ml of MEF medium and then bring to a final volume of 40 ml with MEF medium, mixing vigorously before counting cells with trypan blue. Mixing is critical to get an accurate cell count.

16. Plate MEFs at a density of 3×10^4 cells/cm^2 of culture surface area in MEF medium with 2.5 ml per well of a gelatin-coated 6-well dish.

17. Freeze the cells for later use or use within 2–5 days after plating for hESC cell culture. The medium should be changed every other day if they are not used immediately.

3.3.5. Plating MEF Feeder Cell Layer

If you are using commercially available mitotically inactivated MEF, follow the procedure below. For mitotically inactivating your own MEF culture using mitomycin C *see* **Section 3.3.4**.

1. Two days before hESC coculture, thaw frozen vial of mitotically inactivated, Hygromycin-resistant MEFs in a 37°C water bath.
2. Add contents of vial to 5 ml of MEF medium in a 15-ml tube.
3. Centrifuge for 4 min at $200 \times g$.
4. Aspirate medium and resuspend MEFs in an appropriate volume of MEF medium.
5. Plate MEFs on a 0.1% gelatin-coated culture plate at a density of $30,000/cm^2$.
6. One day before hESC coculture, replace medium with hESC medium.
7. Next day, the feeder layer is ready to be seeded with hESC/R4 in fresh hESC medium.

3.3.6. Thawing hESC/R4 Cells on MEF Feeders

You may thaw your hESC/R4 cells frozen in StemPro® hESC SFM directly onto the established MEF feeder layer in either 35- or 60-mm culture dishes for conditioning prior to transfection. To thaw and establish hESC/R4 cells on MEF feeders:

1. Remove the cryovial of cells from the liquid nitrogen and thaw quickly in a 37°C water bath (to prevent crystal formation).
2. When thawed, immediately transfer cells into a 50-ml tube and add warm hESC medium *dropwise* up to 10 ml (*see* **Note 22**).
3. Spin cells down for 4 min at $200 \times g$ (1000 rpm).
4. Aspirate supernatant.
5. Resuspend cells in hESC medium (2 ml for a 35-mm dish).
6. Aspirate feeder layer plates, and plate resuspended BG01v/R4 cells on the prepared MEFs.
7. Grow cells in a 37°C incubator with a humidified atmosphere of 5% CO_2. Change the medium every day.
8. When colonies become visible (usually 5–10 days after thawing), passage cells onto Hygromycin-resistant MEF feeders using collagenase.

3.3.7. Passaging hESC/R4 Cells on Hyg-Resistant MEF Feeders

The hESC/R4 cells must be maintained on Hygromycin B selection on MEF feeders for at least two weeks prior to transfection.

1. Aspirate culture medium and add 1 ml of 10 mg/ml collagenase solution for every 10 cm^2 of culture vessel surface area. (Alternatively, you may use the StemPro® EZPassage™ Disposable Stem Cell Passaging Tool to cut the cell colonies into pieces; follow the protocol provided with the tool and then proceed to Step 4.)

2. Incubate in a 37°C incubator until the edges of colonies curl up (usually less than an hour).

3. Aspirate collagenase solution.

4. Add hESC medium or 0.1% BSA in D-MEM/F-12.

5. Gently scrape dish using a 5-ml serological pipette and transfer clumps into a 15-ml tube. Do not make the clumps too small; there should be >100 cells per clump.

6. Spin cells down for 2 min at $200 \times g$ at room temperature.

7. Gently aspirate medium and resuspend the hESC/R4 cells in hESC medium.

8. Aspirate feeder layer plates and plate resuspended hESC/R4 cells on the prepared MEFs (passage ratio 1:3 or 1:4).

9. Add a final concentration of 50 μg/ml Hygromycin B (1:1,000 dilution of 50 mg/ml Hygromycin B stock).

10. Grow cells in a 37°C incubator with a humidified atmosphere of 5% CO_2. Change the medium everyday.

11. Feed cells every day and passage by the above protocol whenever required (before colonies start contacting each other; typically every 4–7 days).

3.3.8. Freezing hESC/R4 Cells Cultured on MEF Feeders

When freezing hESC/R4 cells that are cultured on MEF feeder cells, we recommend the following:

1. Freeze cells at a density of $2–3 \times 10^6$ viable cells/ml.

2. For every 20 cm^2 of cells (one 60-mm dish), prepare 1 ml of MEF Freezing Medium A and 1 ml of MEF Freezing Medium B.

3. Bring hESC/R4 cells into freezing medium in two steps, as described in this section.

3.3.8.1. Preparing MEF Freezing Medium

Prepare MEF Freezing Medium 1 and 2 immediately before use. Discard any unused medium.

3.3.8.2. MEF Freezing Medium 1

For every 1 ml, mix together the following in a separate sterile 15-ml tube:

hESC Medium	0.5 ml
Fetal Bovine Serum, ES Cell-Qualified	0.5 ml

3.3.8.3. MEF Freezing Medium 2

For every 0.5 ml, mix together the following in a separate sterile 15-ml tube:

hESC Medium	0.8 ml
DMSO	0.2 ml

Keep MEF Freezing Medium 1 at room temperature and place tube with MEF Freezing Medium 2 on ice.

3.3.8.4. Freezing Protocol for hESC/R4 Cells Cultured on MEF Feeders

1. Aspirate culture medium from the cells and add 1 ml of 1 mg/ml collagenase solution for every 10 cm^2 of culture vessel surface area. (Alternatively, use the StemPro® EZPassage™ Disposable Stem Cell Passaging Tool to cut the cell colonies into pieces; follow the protocol provided with the tool and then proceed to Step 4.)
2. Incubate in a 37°C incubator until the edges of colonies curl up (usually less than an hour).
3. Aspirate collagenase solution.
4. Add hESC medium or 0.1% BSA in D-MEM/F-12.
5. Gently scrape dish using a 5-ml serological pipette and transfer clumps into a 15-ml tube. Try not to make the clumps too small; there should be > 100 cells per clump.
6. Spin cells down for 2 min at $200 \times g$ at room temperature.
7. Gently aspirate medium and resuspend hESC/R4 cells in Freezing Medium A (e.g., resuspend cells from one 60-mm dish in 1 ml of freezing medium).
8. Add the same volume of Freezing Medium B to cells in a dropwise manner, swirling the tube after each drop.
9. Resuspend the cells by gently pipetting 2–3 times. Aliquot 1 ml of the cell suspension to each freezing vial and store overnight at −80°C in isopropanol chamber.
10. Transfer frozen vials to liquid nitrogen tank for long-term storage (see **Note 23**).

3.4. hESC Culture in Feeder-Free Conditions

You must maintain your hESC/R4 cells in MEF-conditioned medium (MEF-CM) as a feeder-free culture for at least one passage prior to transfection with your retargeting construct, as well as for the duration of selection and expansion after transfection.

3.4.1. Preparation of MEF-Conditioned Medium

1. Plate mitotically inactivated MEF cells as described in **Section 3.3.5**.
2. Change the MEF medium to hESC medium after 24 h of incubation.
3. Collect the hESC medium, now MEF-conditioned medium, from culture dishes every 24 h and supplement with bFGF to a final concentration of 4 ng/ml before using for hESCs.

3.4.2. Passage of hESCs on Geltrex™ Coated Dishes

1. Prepare Geltrex™-coated dishes as described above.
2. Aspirate medium from hESCs and add 1 ml collagenase IV per well of a 6-well plate.

3. Incubate for 5–20 min at 37°C. Incubation time will vary among different batches of collagenase, therefore you need to determine the appropriate incubation time by examining the colonies. Stop incubation when the edges of the colonies are starting to pull away from the plate.

4. Aspirate the collagenase and add 2 ml of MEF-CM into each well.

5. Gently scrape cells using a cell scraper or a 10-ml pipette to collect most of the cells on the well and transfer cells into a 15-ml tube.

6. Gently dissociate cells into small clusters (50–500 cells) by gentle pipetting. Do not take cells to a single-cell suspension.

7. Remove the Geltrex^{TM} from the Geltrex^{TM}-coated plates and wash once with D-MEM/F-12.

8. Seed the cells into each well of Geltrex-coated plates. The final volume of medium should be 2–3 ml per well.

9. Return the plate to the 37°C incubator. Make sure to obtain an even distribution of cells by gently shaking the plate left to right and back to front.

10. The day after seeding, undifferentiated cells should be visible as small colonies. Single cells in between the colonies will begin to differentiate. As the cells proliferate, the colonies will become large and compact, representing the majority of surface area of the culture dish. In this system, the hESCs are maintained at high density with 300,000–500,000 cells/cm^2 at confluence. In our experiments, we found the optimal split ratio to be 1:3 to 1:4. Using these ratios, the seeding density is approximately 50,000–150,000 cells/cm^2.

3.4.3. Daily Maintenance of Feeder-Free hESC Culture

1. Collect CM from feeders and add bFGF to a final concentration of 4 ng/ml.

2. Feed hESCs with 2–3 ml MEF-CM supplemented with bFGF for each well of 6-well plates every day.

3. Passage when cells are 100% confluent. At this time, the undifferentiated cells should represent at least 90% of the surface area.

3.5. Quality Control of hESC

We recommend that you perform quality control assays on your hESC/R4 cells prior to retargeting to confirm the karyotype of the strain, to assess the pluripotency of the culture, and to measure the viability of the cells. Guidelines and instructions for quality control of hESC/R4 cells are provided below.

3.5.1. Karyotype Analysis

1. Passage cells onto a Geltrex^{TM}-coated T-25 cell culture flask in a feeder-free fashion using MEF-CM as you would under normal conditions.

2. Feed the cells with MEF-CM until they are 50% confluent (i.e., the culture is in log phase).

3. Proceed with karyotype analysis using established protocols or send the live culture by overnight delivery service to a karyotype analysis service vendor.

4. If sending to a vendor, fill the culture flask to the top with conditioned medium, tighten the cap, and seal with parafilm (*see* **Note 24**).

3.5.2. Immunocytochemistry

3.5.2.1. Paraformaldehyde Solution

To prepare 20% paraformaldehyde (PFA) stock solution:

1. Add PBS to 20 g of EM grade paraformaldehyde and bring the volume up to 100 ml.

2. Add 0.25 ml of 10 N NaOH and heat at 60°C using a magnetic stirrer until completely dissolved.

3. Filter through 0.22-µm filter and cool on ice. Make sure the pH is 7.5–8.0.

4. Aliquot 2 ml in 15-ml tubes, freeze on dry ice, and store at −20°C.

5. To prepare 4% PFA for fixing:

6. Add 8 ml PBS into each 15-ml tube containing 2 ml of 20% PFA and thaw in a 37°C water bath.

7. Once dissolved, cool on ice.

3.5.2.2. Fixing Cells

1. Remove culture medium and gently rinse 3X with D-PBS without dislodging the cells.

2. Fix the cells with 4% fresh *Paraformaldehyde Fixing Solution* at room temperature for 15 min. Rinse 3X with D-PBS.

3. Check for presence of cells and autofluorescence after fixing

4. Proceed to staining below. You may also store slides for up to 3–4 weeks in D-PBS at 4°C. *Do not* allow slides to dry.

3.5.2.3. Staining Cells

1. Incubate cells for 30–60 min in blocking buffer (5% serum of the secondary antibody host species, 1% BSA, 0.1% Triton-X in PBS).

2. Remove the blocking buffer and incubate cells overnight at 4°C with primary antibody diluted in 0.1% BSA in PBS. Ensure that the cell surfaces are covered uniformly with the antibody solution.

3. Rinse the wells with PBS.

4. Wash the cells for 5 min with blocking buffer (if using a slide, use a staining dish with a magnetic stirrer).

5. Incubate the cells with fluorescence-labeled secondary antibody *in the dark* at 37°C for 30–45 min.

6. Wash the cells 3X with PBS and in the last wash counter stain with DAPI (1:1000 diluted) for 5 min and rinse with PBS.

7. If desired, mount with 3 drops of ProLong® Gold antifade reagent per slide and seal with the cover slip. You may store the slides *in the dark* at 4°C.

3.5.2.4. Flow Cytometry

We recommend using the LIVE/DEAD® Cell Vitality Assay Kit, available separately from Invitrogen, to assess the vitality of your hESCs by flow cytometry. For more information on how to distinguish metabolically active cells from cells that are dead or injured, refer to the manual provided with the LIVE/DEAD® Cell Vitality Assay Kit (Cat. No. L34951).

3.5.2.5. LIVE/DEAD® Cell Vitality Assay

The assay has been optimized using Jurkat cells. Some modifications may be required for use with other cell types. A negative control for necrosis should be prepared by incubating cells with 2 mM hydrogen peroxide for 4 h at 37°C. Untreated cells should be used as a positive control for C_{12}-resazurin staining.

1. Prepare a 1 mM stock solution of C12-resazurin. Dissolve the contents of the vial of C_{12}-resazurin (Component A) in 100 μL of DMSO (Component C). It may be necessary to agitate the solution in an ultrasonic water bath to fully dissolve the C_{12}-resazurin. The C_{12}-resazurin stock solution should be stable for 3 months if stored at $\leq -20°C$, protected from light. Prepare a fresh 50 μM working solution of C_{12}-resazurin by diluting 1 μL of the 1 mM C_{12}-resazurin stock solution in 19 μL of DMSO.

2. Prepare a 1 μM working solution of SYTOX Green stain. For example, dilute 5 μL of the 10 μM SYTOX Green stain stock solution (Component B) in 45 μL of DMSO (Component C). The unused portion of this working solution may be stored at £ –20°C for up to 1 month.

3. Prepare a 1X phosphate-buffered saline (PBS) solution. For example, for about 20 assays, add 2 ml of 10X PBS (Component D) to 18 ml of deionized water (dH_2O). Pass the 1X PBS through a 0.2-μm filter before use.

4. Harvest the cells and dilute as necessary to about 1×10^6 cells/ml using the 1X PBS. The cells may be washed with 1X PBS if desired.

5. Add the dyes to the cell suspension. Add 1 μL of the 50 μM C_{12}-resazurin working solution (prepared in step 1) and 1 μL of the 1 μM SYTOX Green stain working solution (prepared in step 2) to each 100 μL of cell suspension (final concentrations of 500 nM C_{12}-resazurin and 10 nM SYTOX Green dye) (*see* **Note 25**).

6. Incubate the cells at 37°C in an atmosphere of 5% CO_2 for 15 min.

248   Liu et al.

7. Dilute the cell suspension. After the incubation period, add 400 μL of the 1X PBS, mix gently, and keep the samples on ice.

8. Analyze the cell sample. As soon as possible, analyze the stained cells by flow cytometry, exciting at 488 nm and measuring the fluorescence emission at 530 and 575 nm. The population should separate into two groups: live cells with a low level of green and a high level of orange fluorescence and necrotic cells with a high level of green fluorescence and a low level of orange fluorescence. Confirm the flow cytometry results by viewing the cells with a fluorescence microscope, using filters appropriate for fluorescein (FITC) and tetramethylrhodamine (TRITC).

3.6. Characterization of BG01v/R4 hESC Platform Line

The BG01v/R4 hESC platform line was created by the phiC31-mediated site-specific integration of the R4 retargeting sequences into the chromosome 13 of BG01V hESCs. Follow the guidelines below for a nested PCR to confirm the integration of the R4 retargeting site into the correct chromosomal locus on the BG01v/R4 genome.

3.6.1. Preparation of hESC/R4 Cells for PCR

1. Pellet 10,000–30,000 cells total.
2. Wash the cells with 500 μl PBS.
3. Centrifuge cells down and remove PBS.
4. Resuspend cell pellet in a mixture of 20 μl of Resuspension Buffer and 2 μl of Lysis Solution.
5. Incubate the cell suspension at 75°C for 10 min.
6. Quick spin down for 1 min.

3.6.2. PCR to Screen for Chromosome 13 Integration

Set up a PCR with the primers and conditions found in **Tables 13.4, 13.5,** and **13.6,** using AccuPrime™ Taq DNA Polymerase High Fidelity. Use of nested PCR with primary and secondary reactions is required to eliminate the high background observed with only the primary PCR.

Table 13.4
Platform line PCR diagnostic PCR reaction mix

10X AccuPrime™ PCR Buffer II	5 μl
Forward PCR primer (10 μM stock)	1 μl
Reverse PCR primer (10 μM stock)	1 μl
AccuPrime™ TaqDNA Polymerase High Fidelity (5 U/μl)	1 μl
Cell lysate (from step 6 above)	3 μl
Sterile, distilled water	39 μl

Table 13.5
Platform line PCR diagnostic PCR conditions

Step	Time	Temperature	Cycles
Initial denaturation	2 min	94°C	1X
Denaturation	30 s	94°C	See table below
Annealing	30 s	See table below	
Extension	1 min	72°C	
Final extension	7 min	72°C	1X

Table 13.6
Platform line diagnostic PCR reaction primer sequences

PCR target	Primer name	Sequence	Size	PCR conditions	Chr 13	
Hyg^R gene	HygPF	ATGAAAAAGCCTGAACTCACC	430 bp	52°C, 35–40	Hyg control	Primary
	HygPR	ATTGACCGATTCCTTGCG				
	HygSF	AAAAGTTCGACAGCGTCTCC	205 bp	52°C, 30 cycles		Secondary
	HygSR	TCGCTGAATTCCCCAATG				
Target site	13PL1PF	AGTTAAGCCAGCCCCGACAC	592 bp	58°C, 35 cycles	Plus strand, AttL	Primary
	13PL1PR	TTTTGGCTACCAGTACTAGGCAGG				
	13PL1SF	CTTGTCTGCTCCCGGCATCC	467 bp	55°C, 30 cycles		Secondary
	13PL1SR	CCCAGCAATGGAGCCTGATT				

3.6.3. Southern Blot Analysis (Optional)

Nested PCR is usually sufficient to confirm the presence of the retargeting sequences in the BG01v/R4 chromosome 13; however, you may also perform a Southern blot analysis as an additional check to verify that a single copy of the R4 retargeting sequence is integrated into the BG01v/R4 genome. Use the Southern blot protocol of your choice with a radiolabeled lacZ probe from pcDNA3.1/lacZ plasmid digested previously with

250 Liu et al.

EcoRV (~2 kb). We recommend using the DNAzol® Reagent (Cat. No. 10503-027) to isolate the genomic DNA from the platform cell line.

3.7. Constructing the Retargeting Expression Vector with the MultiSite Gateway® Pro 2.0 Vector Module

For generating the retargeting construct using MultiSite Gateway® Technology, follow the protocol as outlined in the MultiSite Gateway® Pro manual (25-0942) supplied with the kit. "This section does not provide instructions for generating the retargeting construct but provides additional comments and suggestions to help you obtain the best results in multi-fragment vector construction." Note that the assembly of more than 3 fragments is an inefficient process, and following the suggestions below will help maximize (but still not guarantee) the chances of getting the right clone. For more information on the MultiSite Gateway® Technology, visit www.invitrogen.com.

3.7.1. Experimental Outline

Two PCR products flanked by specific *att*B or *att*Br sites and two MultiSite Gateway® Pro Donor vectors, pDONR™ 221 P1-P5r and pDONR™ 221 P5-P2, are used in separate BP recombination reactions to generate two entry clones (*see* **Fig. 13.4**). The two entry clones and the pJTI™R4 DEST destination vector are used together in a MultiSite Gateway® Pro LR recombination reaction to create your retargeting expression clone containing two DNA elements. Refer to the MultiSite Gateway® Pro manual (25-0942) supplied with the kit for detailed instructions.

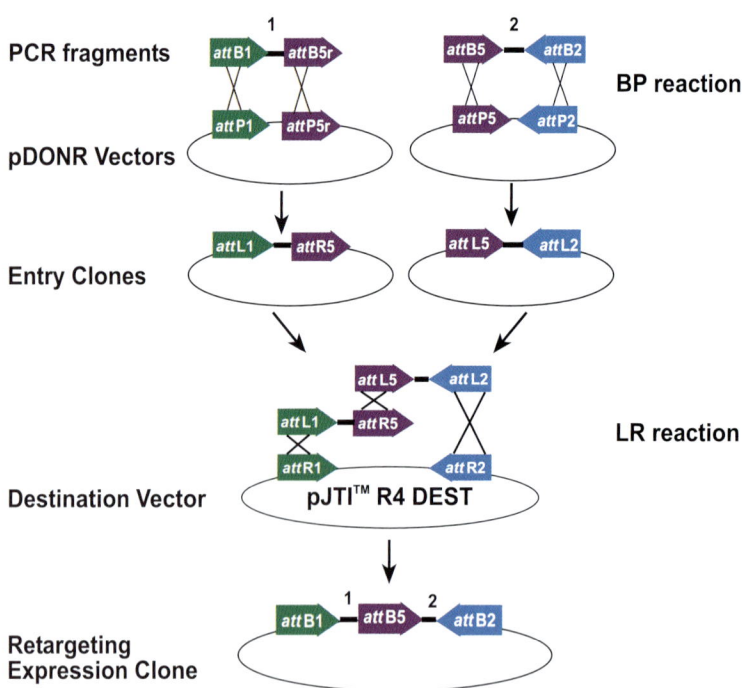

Fig. 13.4. Two-fragment Multisite Gateway reaction.

3.7.2. MultiSite Gateway® Pro Donor Vectors

The pDONR™ 221 P1-P5r and pDONR™ 221 P5-P2 donor vectors, included in the MultiSite Gatewayâ Pro 2.0 Vector Module, are used to clone *att*B- or *att*Br-flanked PCR products to generate entry clones to facilitate the generation of a retargeting expression construct containing multiple DNA elements. For a map and a description of the features of each MultiSite Gatewayâ Pro donor vector, refer to the MultiSite Gateway® Pro manual (25-0942) supplied with the kit (*see* **Note 26**).

3.7.3. pJTI™R4 DEST Destination Vector

The pJTI™R4 DEST vector is designed specifically to be used in a MultiSite Gateway® Pro LR recombination reaction to create your retargeting expression clone to site-specifically integrate your multiple DNA elements into the BG01v/R4 genome (*see* **Scheme 13.1**). The pJTITMR4 DEST vector (5583 bp) contains the λ Integrase *att*R1 and *att*R2 sites for the transfer of DNA elements of interest from pDONR entry clones to generate the retargeting expression clone, the R4 *att*B site for site-specific integration of the DNA elements into the BG01v/R4 genome, and the human EF1α promoter for constitutive expression of Zeocin™ resistance

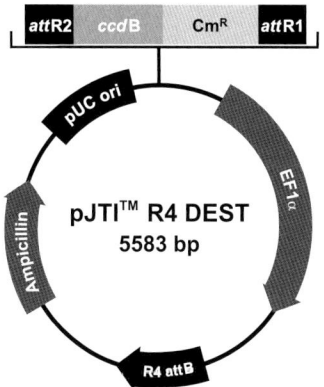

Features of pJTI™ R4 DEST
5583 nucleotides

EF1α: bases 1-1179
R4 *att*B: bases 1258-1552
Ampicillin resistance gene (ORF): bases 1696-2553
pUC origin: bases 2701-3374
*att*R2 recombination site: bases 3777-3901
*ccd*B gene: bases 3942-4247 (complementary strand)
Chloramphenicol resistance gene: bases 4567-5247 (complementary strand)
*att*R1 recombination site: bases 5356-5480 (complementary strand)

Scheme 13.1. Map of pJTI™R4 DEST. The pJTI™R4 DEST vector (5583 bp) contains the λ Integrase *att*R1 and *att*R2 sites for the transfer of DNA elements of interest from pDONR entry clones to generate the retargeting expression clone, the R4 *att*B site for site-specific integration of the DNA elements into the BG01v/R4 genome, and the human EF1α promoter for constitutive expression of Zeocin™ resistance upon successful integration. The complete sequence of pJTI™R4 DEST is available from www.invitrogen.comor by contacting Technical Support.

upon successful integration. "The complete sequence of pJTI™R4 DEST is available from www.invitrogen.com or by contacting Technical Support."

3.7.4. Recombination Region of pJTI^{TM}R4 DEST

The recombination region of the retargeting expression clone resulting from pJTI™R4 DEST/pDONR entry clone is shown in **Fig. 13.5**. Shaded regions correspond to those DNA sequences transferred from the entry clone into pJTI™R4 DEST by recombination. Nonshaded regions are derived from the pJTI™R4 DEST vector.

"The complete sequence of pJTI™R4 DEST is available at www.invitrogen.com."

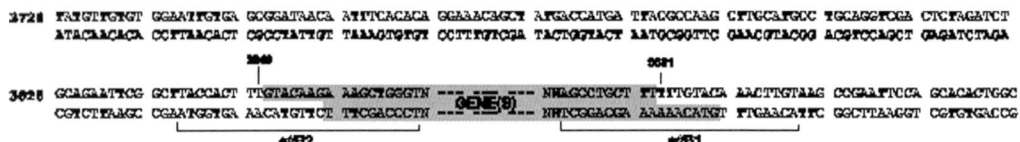

Fig. 13.5. Flanking att site sequence in pJTI™ R4 DEST.

3.7.5. Preparation of Plasmid DNA

For retargeting experiments, it is essential that the plasmid DNA used for transfection is of very high quality. Typically, best results have been obtained using plasmid DNA that has very low levels of endotoxins. If using large quantities of DNA, the recommendation would be to have the plasmid DNA commercially prepared. If smaller quantities are required, use a commercial kit that delivers pure DNA that is free of endotoxins. Follow the manufacturer's recommended protocol for DNA preparation.

3.7.6. Generation of Entry Clones

1. Ensure that primers used for PCR amplification are of good quality. Since these primers are generally ~45 bases in length, the possibility of mutations is greater. This in turn may lead to inefficient recombination with the pDONR vectors.

2. As far as possible, avoid using a plasmid containing the kanamycin-resistance gene as the template for PCR.

3. If the fragment of interest is longer than ~3 kb, longer incubation times will help. Incubate the BP reaction at 16°C overnight instead of 1 h at room temperature.

4. When picking colonies for analysis, replica plate them on kanamycin and the drug resistance of the template for PCR to reduce background from inadvertently purified template. The colony should only grow on Kanamycin.

5. Generation of retargeting expression clones. Produce clean DNA preparations of the entry clones to be used in the LR reaction. DNA from "minipreps" will suffice for the assembly of up to two fragments. For assembly of 3 or more fragments, "midiprep" or "maxiprep" quality DNA is essential.

6. Sequence the entry clones with appropriate primers to ensure the *att* sites do not have mutations.

7. Dilute the DNA to a convenient concentration for the reactions. Since the MultiSite Gateway® Pro manual recommends 20 femtomoles of the DEST vector and 10 femtomoles of each of the entry vectors per reaction, we recommend maintaining a working concentration of 20 fmoles/μl for the DEST vector and 10 fmoles/μl for each of the entry vectors to allow the addition of 1 μl of each vector to the recombination reaction. The vector aliquots should be stored at –20°C.

8. While it may be tempting to use a "master mix" when setting up multiple LR reactions, this does not give the best results. The LR clonase should always be added at the end. Add the DNA first, briefly centrifuge the tubes, and then add the enzyme to the liquid phase at the bottom.

9. Longer incubation times are essential for the assembly of greater than 2 fragments. Generally, overnight incubation at either room temperature or at 16°C should work.

10. Performing multiple transformations is more efficient than performing one large transformation. For a 5-fragment assembly, it may be necessary to transform the complete reaction volume to get enough colonies for analysis. Five transformations of 2 μl each will yield more colonies than two transformations of 5 μl each.

11. Replica plate the colonies obtained on ampicillin and kanamycin plates. True recombinant clones will only grow on ampicillin plates.

3.7.7. Retargeting Construct Assembly from >2 DNA Fragments

The StemPro® TARGET™ hESC BG01V Kit contains the MultiSite Gateway® Pro 2.0 Vector Module and the pJTI™R4 DEST vector, which together allow the construction of a 2-fragment retargeting expression construct. For generating expression constructs from up to five individual DNA elements, Invitrogen offers additional MultiSite Gateway® Pro Kits. For more information, visit our website at www.invitrogen.com or contact Technical Support.

3.7.8. Schematic for Assembly of Multiple Fragments

The MultiSite Gateway® Pro Kits combined with the pJTI™R4 DEST vector facilitate rapid and highly efficient construction of retargeting constructs containing your choice of two, three, or

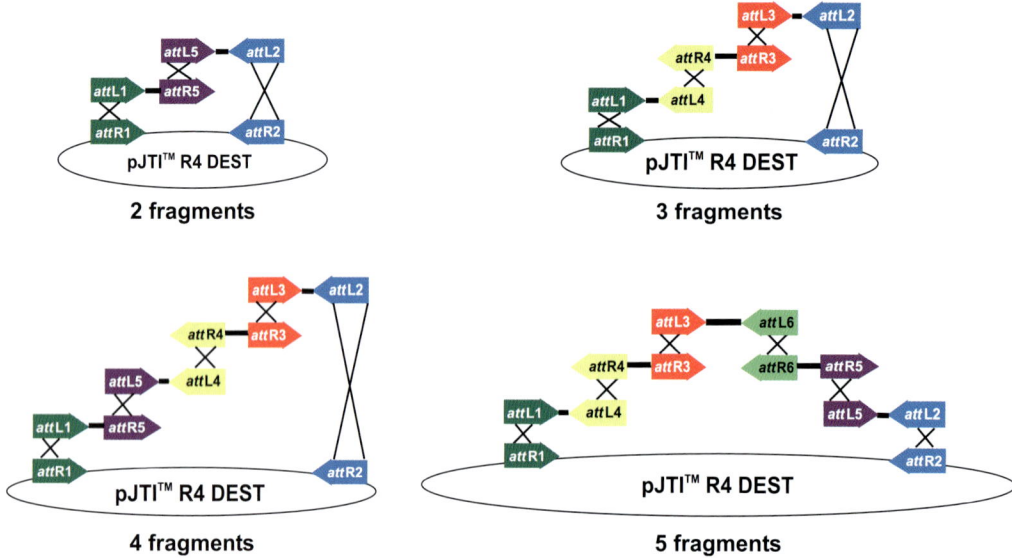

Fig. 13.6. Various Multisite Gateway assemblies possible with pJTI™ R4 DEST.

Table 13.7
Donor vectors for Multisite Gateway assembly of various complexity

Number of fragments	Donor vectors required
1	pDONOR201 or pDONOR221
2	pDONOR221 P1P5r and pDONOR221 P5P2
3	pDONOR221 P1P4, pDONOR221 P4rP3r, and pDONOR221 P3P2
4	pDONOR221 P1P5r, pDONOR221 P5P4, pDONOR221 P4rP3r, and pDONOR221 P3P2
5	pDONOR221 P1P4, pDONOR221 P4rP3r, pDONOR221 P3P6, pDONOR201 P6rP5r, and pDONOR221 P5P2

four separate DNA elements in a defined order and orientation (**Fig.13.6**). **Table 13.7** depicts the strategies for the multifragment assembly of retargeting constructs with the available pDONR vectors.

3.8. Retargeting hESC/R4 Cells

The hESC/R4 platform lines are designed for site-specific retargeting mediated by R4 Integrase and show sustained expression and appropriate regulation of the transgenes over long-term culture and random and directed differentiation into a variety of lineages. The MultiSite Gateway® Pro 2.0 vector module, supplied with the kit, allows the assembly of complex genetic elements from two DNA fragments that can efficiently be inserted into the target site on the hESC/R4 genome. This section provides instructions and guidelines on:

1. Preparing your hESC/R4 platform line for retargeting.
2. Transfection procedure for site-specific integration of genetic elements of interest into the hESC/R4 genome.
3. Selection, expansion, and characterization of the retargeted clones.

3.8.1. Growth of hESC/R4 Cells Prior to Transfection

1. Thaw hESC/R4 cells onto MEF feeders in either 35- or 60-mm dishes.
2. Passage hESC/R4 cells onto Hyg-resistant MEF feeders in 60-mm dishes using collagenase (1 mg/ml) when the colonies become visible, usually 5–10 days after thawing.
3. Start Hyg selection (10 μg/ml) and keep selection for at least two weeks.
4. Passage cells onto MEF-CM in Geltrex™-coated dishes and *stop* Hygromycin selection.
5. Passage hESC/R4 cells once more as a feeder-free culture on MEF-CM and Geltrex™ and proceed to transfection for retargeting.

Important: The following factors are important for a successful retargeting event:

- *Cells*: Cells that are 80–90% confluent are ideal for transfection. A higher confluency often results in a higher proportion of dead cells in culture. Carry out a live/dead assay using either FACS (LIVE/DEAD® Cell Vitality Assay Kit) or Trypan Blue exclusion counting.

- *DNA*: The quality and the concentration of DNA used play a central role for the efficiency of transfection. It is crucial that the DNA is free of endotoxins. If using large quantities of DNA, we recommend commercially prepared plasmid DNA. For smaller quantities, use a commercial kit that delivers pure DNA that is free of endotoxins. *Do not* precipitate DNA with ethanol to concentrate because it reduces efficiency and viability due to the salt contamination.

You must use a *high-efficiency transfection method* such as microporation or electroporation to retarget the hESC/R4 cells. The following pages contain two alternative transfection conditions for your convenience.

3.8.2. Transfection Protocol Using the ProCell™ Transfection System

1. Turn on ProCell™ instrument; adjust the parameters to 850 V, 30 ms, 1 pulse.
   Mix 5 μg of retargeting expression vector and 5 μg of pJTI™-R4Int plasmid in a 1.5-ml microcentrifuge tube for a total of 10 μg DNA per 1×10^6 cells for each transfection. The pJTI™-R4Int vector (5647 bp) contains the gene for R4 Integrase from the *Steptomyces* phiC31 phage (*see* **Scheme 13.2**). The R4 Integrase allows the site-specific integration of DNA elements into the HESC/R4 genome from the pJTITMR4 DEST retargeting expression construct upon cotransfection of the platform line with both vectors. "The complete sequence of pJTI™R4 DEST is available from www.invitrogen.com or by contacting Technical Support."

2. Add 5 ml of prewarmed MEF-CM to a 60-mm culture dish containing mitotically inactivated MEF feeder layer for each transfection and put back into 37°C incubator.

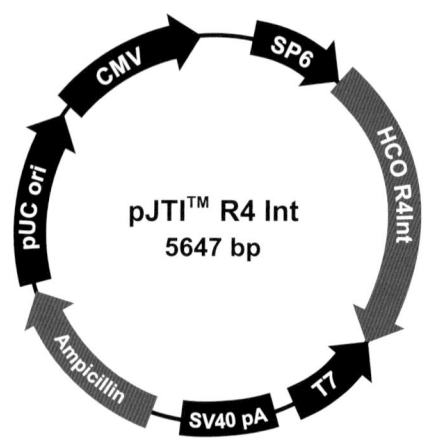

Features of pJTI™ R4 Int
5647 nucleotides

SP6 promoter: bases 15-34
HCO R4Int: bases 51-1460
T7 promoter: bases 1483-1502 (complementary strand)
SV40 promoter and origin: bases 1889-1970
Ampicillin resistance gene (ORF): bases 2650-3510
pUC origin: bases 3655-4328
CMV promoter: bases 5039-5639

Scheme 13.2. Map of pJTI™-R4Int. The pJTI™-R4Int vector (5647 bp) contains the gene for R4 Integrase from the *Steptomyces* phiC31 phage. The R4 Integrase allows the site-specific integration of DNA elements into the HESC/R4 genome from the pJTITMR4 DEST retargeting expression construct upon cotransfection of the platform line with both vectors. The complete sequence of pJTI™R4 DEST is available from www.invitrogen.com or by contacting Technical Support.

3. To harvest target hESC lines, rinse the cells with 2 ml of D-PBS (Ca- and Mg-free) and add 1 ml of TrypLE^{TM} per 60-mm dish. After 4 min of incubation at room temperature, gently dislodge cells and transfer to a 15-ml conical tube. Wash the culture dish with at least 2 ml of D-PBS and transfer the wash to the same tube.

4. Set aside a small aliquot of cells for QC and centrifuge the rest at $200 \times g$ for 4 min. During centrifugation, count the cells in the aliquot to determine the resuspension volume.

5. Discard the supernatant and resuspend the cells in Resuspension Buffer R or D-PBS, using 100 μl buffer per 1×10^6 cells. A 60-mm dish should yield about $3–5 \times 10^6$ hES cells.

6. Combine the DNA (10 μg total, step 2) and 100 μl of resuspended cells (1×10^6 cells) in the same 1.5-ml microcentrifuge tube that contains the DNA.

7. Add 3 ml of Electrolytic Buffer E2 in the microporation tube, pipette 100 μl of the DNA-cell mixture into Gold Tip, and insert Gold Tip into the station.

8. Press the start button, and then the high voltage button to microporate.

9. Immediately transfer microporated cells to MEF-CM/MEF feeder dish from step 3, prewarmed to 37°C.

3.8.3. Alternative Transfection Procedure Using BTX or Bio-Rad Electroporator

1. Culture feeder-free hESC/R4 cells to 80–90% confluency. Two 60-mm culture dishes usually provide enough cells for one electroporation.

2. Prepare 30 μg total DNA for each electroporation in a 1.5-ml microcentrifuge tube by mixing 15 μg each of the retargeting expression vector and the pJTI^{TM}-R4Int plasmid.

3. For each transformation, add 3 ml of MEF-CM at 37°C to a Geltrex^{TM}-coated dish and put back into 37°C incubator.

4. For each electroporation, add 3 ml of prewarmed MEF-CM into a 15-ml conical tube and place in a 37°C water bath.

5. To harvest target hESC lines, rinse the cells with 2 ml of D-PBS (Ca- and Mg-free) and add 1 ml of TrypLE^{TM} per 60-mm dish. After 4 min of incubation at room temperature, gently dislodge cells and transfer to a 15-ml conical tube.

6. Wash the culture dish with at least 2 ml of D-PBS and transfer the wash to the same tube. Centrifuge to pellet cells.

7. Resuspend the cells in D-PBS to wash and recentrifuge.

8. Remove the D-PBS and resuspend the cell in 800 μl of OptiPro^{TM} SFM.

9. Combine the DNA (10 µg total, Step 2) and the resuspended cells (Step 8) in the same 1.5-ml microcentrifuge tube that contains the DNA and transfer the mixture to an electroporation cuvette with a gap of 0.4 cm.
10. Electroporate the cells once using the BTX ECM630 electroporator with the following parameters: 200 V, 10 ms, 2 pulses. If you are using the Bio-Rad Gene pulser II, the parameters are: 500 V, 250 µF.
11. Immediately transfer the cells into the 15-ml conical tube (prepared in Step 4) using a Pasteur pipette and incubate at room temperature for 5 min.
12. Transfer the mixture from Step 11 into the 60-mm dish (prepared in Step 3) and incubate in the 37°C incubator (*see* **Note 27**).

3.9. Selection and Expansion of Retargeted hESC/R4 Clones

1. Replace the MEF-CM in your culture dishes with fresh MEF-CM everyday.
2. Each clone recovers at a different rate. Monitor morphology and size of the colonies.

3.9.1. Selecting Retargeted Clones

3. When your retargeted hESC/R4 cells have recovered from transfection and the colonies are well defined, start drug selection with 2.5 µg/ml Zeocin™. If you have used microporation for retargeting and have been recovering your cells on MEF feeders supplemented with MEF-CM, you should also start Hygromycin B selection to prevent overgrowth of your colonies by MEFs (*see* **Note 28**).
4. Change MEF-CM containing drug every day. Colonies start appearing as early as day 5 of drug selection. Mark the colonies and observe them for an additional week (total of 12–14 days under selection).

3.9.2. Picking Retargeted Clones

1. *Optional*: Before picking colonies, add collagenase (1 mg/ml) to the culture dishes containing the transfectants and incubate for 5–15 min. This loosens the colonies and makes the subsequent mechanical picking easier.
2. Using a StemPro® EZPassage™ Disposable Stem Cell Passaging Tool or a sharp knife made from a Pasteur pipette, chop the colonies into several pieces, and transfer to a well of a 24- or a 12-well plate containing mitotically inactivated MEF feeder layers and MEF-CM.
3. Check the cells and change the MEF-CM the next day.
4. Start drug selection with Zeocin™ (2.5 µg/ml) and Hygromycin B (10 µg/ml) on the second day after the colonies are picked and maintain selection for future passages.

3.9.3. Expanding Retargeted Clones

1. Continue to feed cells using MEF-CM with drug selection. When colonies grown on a 24- or 12-well plate become visible and show growth and expansion (usually after 7–14 days), cells from one well can be passaged onto a 35-mm dish (or a well of a 6-well plate). Passaging should follow the protocol as described in **Section 3.3.6**.

2. Freeze at least 4 vials of cells for each clone before conducting any further experiments. Follow the freezing protocol as described in **Section 3.3.8**.

3. Once the retargeted clones become established on 35-mm MEF feeder dishes supplemented with MEF-CM and under drug selection, you may passage them onto 60-mm MEF feeder dishes and switch from MEF-CM to hESC medium.

4. To adapt cells from hESC medium to StemPro® hESC SFM, you *must* first passage cells onto Geltrex™-coated dishes and grow in feeder-free fashion on MEF-CM for at least one day (refer to **Section 3.4**). The confluency of cells at this stage should be between 30 and 70% (i.e., 1–3 million cells/dish). Next time you change the medium, you may directly replace the MEF-CM with StemPro® hESC SFM and start passaging the hESC/R4 cells in StemPro® hESC SFM as described above.

3.10. PCR to Screen Retargeting Events

Upon retargeting the hESC/R4 platform line, follow the guidelines below to PCR screen for successful retargeting events. Successful retargeting of the hESC/R4 genome introduces the human EF1α promoter upstream of the *Sh ble* gene on chromosome 13, resulting in Zeocin™ resistance of successfully retargeted clones. PCR amplification of the EF1α promoter region in successfully retargeted clones will result in a 534-bp product.

Table 13.8
Secondary targeted cell line diagnostic PCR primer sequences

Target	Primer name	Sequence	Size	Chr 13
EF1α-Zeo	ELZP Frw	GCCTCAGACAGTGGTTCAAAGTTT	534 bp	EF1α promoter
	ELZP Rev	TGATGAACAGGGTCACGTCGT		
Hyg^R gene	HygPF	ATGAAAAAGCCTGAACTCACC	430 bp	Hyg control
	HygPR	ATTGACCGATTCCTTGCG		

Use the primer sequences found in **Table 13.8** for the PCR reaction. Primers for the Hygromycin-resistance gene are given to use in a positive control reaction and have the same sequences as those used in the primary PCR for the initial characterization of the hESC/R4 platform line.

3.10.1. Preparation of Retargeted hESC/R4 Cells for PCR

1. Pellet 10,000–30,000 cells total.
2. Wash the cells with 500 µl PBS.
3. Centrifuge cells to pellet and remove PBS.
4. Resuspend cell pellet in a mixture of 20 µl of Resuspension Buffer and 2 µl of Lysis Solution.
5. Incubate the cell suspension at 75°C for 10 min.
6. Centrifuge for 1 min to pellet cell debris.

3.10.2. PCR to Screen for Chromosome 13 Integration

Set up a PCR with the primers listed in **Table 13.8** and the below conditions, using AccuPrime™ Taq DNA Polymerase High Fidelity. Use of nested PCR with primary and secondary reactions is required to eliminate the high background observed with only the primary PCR (*see* **Table 13.9**).

10X AccuPrime™ PCR Buffer II	5 µl
Forward PCR primer (10 µM stock)	1 µl
Reverse PCR primer (10 µM stock)	1 µl
AccuPrime™ Taq DNA Polymerase High Fidelity (5 U/µl)	1 µl
Cell lysate (from step 6, previous page)	3 µl
Sterile, distilled water	39 µl

3.10.3. Southern Blot Analysis (Optional)

PCR is usually sufficient to confirm the presence of the retargeted sequences in the BG01v/R4 chromosome 13 after transfection. However, you may also perform a Southern blot analysis as an additional check to screen for a single copy number. Use the

Table 13.9
Secondary targeted line diagnostic PCR conditions

PCR conditions	Step	Time	Temperature	Cycles
	Initial denaturation	2 min	94°C	1X
	Denaturation	30 s	94°C	40X
	Annealing	30 s	55°C	
	Extension	1 min	72°C	
	Final extension	7 min	72°C	1X

Southern blot protocol of your choice with a radiolabeled lacZ probe from pcDNA3.1/lacZ plasmid digested previously with EcoRV (~2 kb). We recommend using the DNAzol® Reagent (Cat. No. 10503-027) to isolate the genomic DNA from the platform cell line.

3.11. Sustained Expression and Appropriate Regulation of Transgenes in Retargeted Clones

The images in **Fig. 13.7** are included to provide an example of the sustained expression and appropriate regulation of transgenes over long-term culture, upon random differentiation, as well as directed induction into specific lineages. For more information on the long-term behavior of BG01v/R4 platform line upon retargeting, and the subsequent active expression of the transgenes in pluripotent as well as in differentiated lineages, refer to Liu et al. (1).

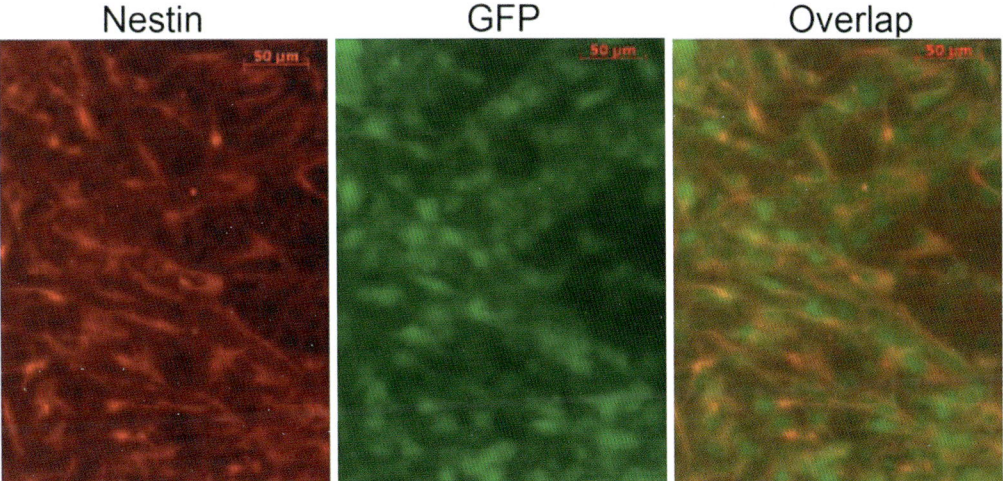

Fig. 13.7. BG01v/R4 platform line was retargeted using a GFP cassette driven by a constitutive promoter EF1α. Retargeted cells continue to express GFP after they are differentiated into Nestin+ cells.

4. Notes

1. If you have frozen hESCs harvested from MEF-CM or MEF feeder layers, you must thaw your cells into either MEF-CM or MEF feeder layers to achieve high plating/survival of your hESC culture.

2. Dropwise addition of prewarmed StemPro® hESC SFM during the thawing procedure is crucial.

3. Before starting experiments, we recommend that you first prepare ample frozen stocks of the BG01v/R4 hES cells in SFM Freezing Medium, as described in **Section 2.3**.

4. *Starter culture:* This must be a high-quality culture, with a high density of cells, and primarily undifferentiated. The starter culture should be hESCs maintained on Geltrex™ in Mouse Embryonic Fibroblast-Conditioned Medium (MEF-CM).

    The cells should not be maintained on MEF feeder cells.

5. You may also maintain your starter culture on Matrigel™ MEF-CM or on CELLstart™ without MEF-CM.

6. *Passaging:* Passaging the cells is *the* most likely point of difficulty. It is critical to achieve high plating/survival of colony pieces. The pieces must be smaller than typical collagenase passaging on Geltrex™/MEF-CM.

    We recommend using the StemPro® EZPassage™ Disposable Stem Cell Passaging Tool for reproducible and optimal passaging.

7. Some *cell death* at passaging is normal; however, wide-scale cell death (i.e., <20% survival) indicates poor passaging.

8. *Timing of passaging is a critical factor.* Do *not* passage the cells too early as they will plate poorly and differentiate. The cultures need to grow to near-confluence before they are harvested, i.e., a day or two after the colonies are just touching. This usually results in a harvest of 5–8 million cells per 60-mm culture dish.

9. *Exposure to collagenase.* hESCs in StemPro® hESC SFM are very sensitive to overexposure to collagenase, which causes poor plating and differentiation. Do *not* expose cells to collagenase for longer than 3 min. Do *not* use lower concentrations of collagenase or treat for longer periods.

10. *Density.* The cultures must be maintained at a high density (200+ colonies in a 60-mm dish). A lower cell density will cause the culture to deteriorate, slowing proliferation and causing differentiation. If this happens, allow the culture to proliferate to near confluence before splitting.

11. hESCs grown in culture are always under pressure between *proliferation* and *differentiation*. The cultures should be fed every day; do not exhaust medium by not feeding. Scrape clearly differentiated areas out with a 21½-gauge needle.

12. StemPro® hESC SFM supplement is supplied as a frozen sample. Thaw supplement prior to use, refreeze in desired volumes, and store it immediately at −20°C.

13. *Avoid multiple freeze thaw cycles* of supplement. Thawed StemPro® hESC SFM Growth Supplement must be stored at 2–8°C (Stable up to 1 week).

14. StemPro® hESC SFM Complete Medium can be stored at 2–8°C *in dark* for up to 7 days. *Add 2-Mercaptoethanol daily during storage*, at volumes listed in the table above.

15. One mg/ml aliquots of collagenase IV are usually used with hESCs maintained on MEF feeders, while the 10 mg/ml aliquots are used with hECSs cultured on StemPro® hESC SFM.

16. If you are transferring your cells into StemPro® hESC SFM from MEF feeder layers or from MEF-conditioned medium (MEF-CM), prepare a frozen stock of the cells as a precaution. You may check the viability and recovery of frozen cells 24 h after storing cryovials in liquid nitrogen.

17. *Important:* Before starting experiments, we recommend that you first expand and bank at least 10–20 vials of frozen hESC/R4 cells in SFM Freezing Medium, as described above.

18. *Important:* You may thaw hESCs frozen in StemPro® hESC SFM directly onto MEF feeders or MEF-CM; however, "if your hESCs were frozen from cultures grown on MEF feeder cells, they must first be thawed and established on MEF feeders." After your hESCs have sufficiently recovered and established on 60-mm culture dishes, you may adapt the cells from hESC medium to StemPro® hESC SFM by first passaging them on Geltrex™-coated plates in a feeder-free fashion, then transitioning them to StemPro® hESC SFM as described above.

19. *Caution:* Mitomycin C is highly toxic. Read and understand the MSDS and handle accordingly.

20. Used mitomycin C *must* be neutralized by addition of 15 ml bleach (Clorox) per 500 ml mitomycin C solution. Swirl to mix, incubate for 15 min, and discard.

21. MEFs are trypsin sensitive. 1–2 min of incubation is sufficient to detach cells. *Do not* overexpose.

22. *Dropwise addition* of prewarmed hESC medium during the thawing procedure is crucial.

23. You may check the viability and recovery of frozen cells 24 h after storing cryovials in liquid nitrogen.

24. We recommend Cell Line Genetics as a karyotype analysis service vendor. *Cell Line Genetics,* Suite 254, 510 Charmany Drive, Madison, WI 53719 Phone: (608) 441-8163; Fax: (608) 441-8162; http://www.clgenetics.com/

25. If the fluorescence intensity of the SYTOX Green dye is too low, the final dye concentration can be increased up to 50 nM.

26. pDONR™ 221 is provided as a positive control for the BP recombination reaction, but should not be used to generate multifragment entry clones.
27. After transfection, you must allow the retargeted hESC/R4 cells to recover and the colonies to become well defined (usually 5 days postmicroporation or 2 days postelectroporation) before starting selection with Zeocin™
28. Generally, the cells will have recovered fully 5 days postmicroporation or 2 days postelectroporation; however, some microporated clones may take longer to recover.
29. Troubleshooting: **Table 13.10** lists some potential problems and possible solutions to help you troubleshoot your hESC cell culture studies retargeting experiments. For troubleshooting any potential problems that might arise when generating your retargeting expression construct, refer to the MultiSite Gateway® Pro manual (25-0942) supplied with the kit.

Table 13.10
Troubleshooting

Problem	Cause	Solution
No viable cells after thawing stock	Stock not stored correctly	Order new stock and store in liquid nitrogen. Keep in liquid nitrogen until thawing.
	Home-made stock not viable	Freeze cells at a density of 2–3 × 10^6 viable cells/ml.
		Use low-passage cells to make your own stocks.
		Follow the freezing procedure for your type of cell culture exactly. Slow freezing and fast thawing are crucial. Add the cold freezing medium in a dropwise manner (slowly), swirling the tube after each drop. At the time of thawing, thaw quickly and do not expose vial to the air, but quickly change from nitrogen tank to 37°C water bath.
		Obtain new BG01v/R4 cells.
	Thawing medium not correct	Use specified medium.
	Cells too diluted	Generally, we recommend thawing one vial in a 35-mm dish. If you need to concentrate cells, spin down the

(continued)

Table 13.10 (continued)

Problem	Cause	Solution
		culture for 4 min at 200×g at room temperature and dilute the cells at higher density.
	MEFs suboptimal and do not support recovery of BG01v/R4 cells (if thawed on MEF feeders)	Purchase or make a new batch of mitotically inactivated MEFs (see p. 59).
MEFs overgrow plate	MEFs not inactivated	Inactivate mitosis in MEFs as described above, or purchase inactivated MEFs.
Cells grow slowly	Growth medium not correct	Use correct growth medium.
	bFGF inactive	bFGF is not stable when frequently warmed and cooled. Add bFGF to medium just before use or store medium with bFGF in aliquots at −20°C.
	Cells too old	Use healthy BG01v/R4 cells, under passage 30; do not overgrow.
	Cells too diluted	Spin down cells for 4 min 200×g at room temperature; aspirate medium and dilute cells at higher density.
	Clump size is too small and differentiated	Be gentle at time of passage so the clumps of cells do not get too small.
	Mycoplasma contamination	Discard cells, medium, and reagents and use early stock of cells with fresh medium and reagents.
Cells differentiated	Cells not thawed and established on StemPro® hESC SFM	Thaw and culture a fresh vial of BG01v/R4 cells. Make sure to thaw on StemPro® hESC SFM as described above.
	Suboptimal quality of feeder layer (if cells are maintained on feeder layers)	Check the concentration of feeder cells used. Purchase or make new batch of mitotically inactivated MEFs, if necessary. Use Hygromycin-resistant MEFs.
	Culture conditions not correct	Thaw and culture fresh vial of BG01v/R4 cells. Follow thawing instructions and subculture/maintenance procedures exactly.

(continued)

Table 13.10 (continued)

Problem	Cause	Solution
	Cells overexposed to collagenase	hESCs are very sensitive to collagenase overexposure. Avoid exposing cells to collagenase for more than 3 min. Do *not* use lower concentrations of collagenase and treat for longer periods.
	Cells passaged too early	Passaging cells too early causes poor plating and differentiation. Grow to cells to near-confluence, i.e., a day or two longer than when the colonies are just touching.
No growth after transfection	Incorrect amount of Zeocin^{TM} and/or Hygromycin B is used for selection.	Use 10 µg/ml Hygromycin B and 2.5 µg/ml Zeocin^{TM} for selection.

Retargeting hESC/R4 cells: The table below lists some potential problems and solutions that help you troubleshoot your problems during transfection for retargeting.

Problem	Cause	Solution
Low-survival rate after transfection	Poor DNA quality	The quality of the retargeting construct DNA strongly influences the results of transfection experiments. Use endotoxin-free DNA for all transfections. Make sure that the A260:A280 ratio of the DNA is between 1.8 and 2.0. *Do not* use phenol:chloroform extraction or ethanol precipitation.
	Cells are cultured in suboptimal conditions	Cells that are 80–90% confluent are ideal for transfection. A higher confluency often results in a higher proportion of dead cells in culture. Avoid excessive cell densities of high confluency.
	Cells are harvested from StemPro® hESC SFM plates prior to transfection	After two weeks of Hygromycin B selection on MEF feeders, BG01v/R4 cells must be passaged at least once as a feeder-free culture on MEF-CM and Geltrex^{TM} without drug selection.
	Cells are damaged during harvesting and subsequent handling prior to transfection	Avoid damaging cells conditions during harvesting. Centrifuge cells at lower speeds (150–200×g). Avoid

(continued)

Table 13.10 (continued)

Problem	Cause	Solution
		overexposure to TrypLE™, trypsin, or accutase. Pipette cells gently.
	Cells remained too long in electroporation cuvette or the Gold-Tip.	Immediately after electroporation/microporation, transfer cells into prewarmed medium at 37°C to prevent damage.
	Multiple use of Gold-Tip (if MT-100 MicroPorator is used for transfection)	Maximum recommended use of Gold-Tip is between 1 and 3 times, because the electric pulses that are applied drastically reduce its quality and impair its physical integrity.
Low transfection efficiencies	Poor optimization of transfection parameters	Optimize transfection parameters following electroportaor/microporator manufacturers' recommendations.
	Amount of DNA too low	Use the correct amount of DNA for the transfection method of choice as described above.
	Cell density too low or too high	Too low or too high cell densities could drastically reduce the transfection efficiency. Use 1×10^6 cells per microporation, or $0.6–1.0 \times 10^7$ cells per electroporation as described above.
	Poor DNA quality	Use endotoxin-free DNA for all transfections. Make sure that the A260:A280 ratio of the DNA is between 1.8 and 2.0. *Do not* use phenol:chloroform extraction or ethanol precipitation.
	Cells are contaminated with Mycoplasma	Test cultures for Mycoplasma contamination. Start a new culture from a fresh stock.

References

1. Liu, Y., Thyagarajan, B., Lakshmipathy, U., Xue, H., Lieu, P., Fontes, A., MacArthur, C. C., Scheyhing, K., Rao, M. S., and Chesnut, J. D. (2009), *Generation of a platform human embryonic stem cell line that allows efficient targeting at a predetermined genomic location.* Stem Cells Dev, 2009 Apr 8. [Epub ahead of print].

2. Thyagarajan, B., et al., (2008) *Creation of engineered human embryonic stem cell lines using phiC31 integrase.* Stem Cells, **26**(1): 119–26.

3. Hartley, J.L., Temple, G.F., and Brasch, M.A. (2000) *DNA cloning using in vitro site-specific recombination.* Genome Research, **10**: 1788–95.

4. Sasaki, Y., et al., (2008) *Multi-gene gateway clone design for expression of multiple heterologous genes in living cells: eukaryotic clones containing two and three ORF multi-gene cassettes expressed from a single promoter.* J Biotechnol, **136**: 103–12.

5. Sasaki, Y., et al., (2004) *Evidence for high specificity and efficiency of multiple recombination signals in mixed DNA cloning by the Multisite Gateway system.* J. Biotechnol, **107**: 233–43.

Chapter 14

Human Embryonic Stem Cell Differentiation on Periodontal Ligament Fibroblasts

Y. Murat Elçin, Bülend İnanç, and A. Eser Elçin

Abstract

Human embryonic stem cells' (hESCs) unlimited proliferative potential and differentiation capability to all somatic cell types made them potential cell source in different cell-based tissue engineering strategies as well as various experimental applications in fields such as developmental biology, pharmacokinetics, toxicology, and genetics. Periodontal tissue engineering aims to improve the outcome of regenerative therapies which have variable success rates when contemporary techniques are used. Cell-based therapies may offer potential advantage in overcoming the inherent limitations associated with guided tissue-regeneration procedures, such as dependency on defect type and size and the pool and capacity of progenitor cells resident in the wound area. Elucidation of developmental mechanisms of different periodontal tissues may also contribute to valuable knowledge based upon which the future therapies can be designed. Prior to the realization of such a potential, protocols for the differentiation of pluripotent hESCs into periodontal ligament fibroblastic cells (PDLF) as common progenitors for ligament, cementum, and alveolar bone tissue need to be developed. The present protocol describes methods associated with the guided differentiation of hESCs by the use of coculture with adult PDLFs, and the resulting change of morphotype and phenotype of the pluripotent embryonic stem cells toward fibroblastic and osteoblastic lineages.

Key words: Human embryonic stem cell, periodontal ligament fibroblastic cell, directed differentiation, osteogenic induction, cell coculture, periodontal tissue engineering.

1. Introduction

Pluripotent nature of human embryonic stem cells (hESCs) renders them the ability to differentiate into all somatic and germ line representatives, and their virtually unlimited proliferative potential has implications in experimental elucidation of developmental mechanisms, testing the effects of newly developed drugs at the cellular level, revealing potential teratogenic and toxic effects of existing drugs on cellular differentiation, and also as unlimited cell source for prospective tissue engineering applications (1–3). These general utilities apply to a variety of tissues, including tooth supporting structures. Periodontal apparatus consists of three different but related tissues, namely cementum on the tooth root surface, alveolar bone of the socket wall, and periodontal ligament between them (4). All of these are descendants of ectomesenchyme, and progenitor cells in the periodontal ligament contribute to the regeneration of the lost structures to some extent with guided tissue regeneration procedures (5). In order to achieve the regeneration of the whole periodontium with tissue engineering, the cells to be used should bear the potential to differentiate into cells of all three tissue types, like periodontal progenitors (6). Studying periodontal development and ectomesenchymal differentiation in its early phases with cells at initial stages of lineage development could substantially contribute to the understanding of tooth development as well as periodontal regeneration processes (7). Human embryonic stem cells may become a useful cell source with unlimited supply; however, the differentiation to periodontal progenitors is a prerequisite for their successful utilization. Overcoming immune rejection and demonstration of the efficacy and safety of the specifically designed tissue-engineering structures are other challenges for using the cells in clinical therapies (8). Our group has established protocols for initial differentiation toward periodontal lineages using coculture with adult human periodontal ligament fibroblasts (hPDLFs) (*see* **Tables** 14.1 and 14.2) (9, 10). Here, the

Table 14.1
Fibrogenic differentiation of hESCs on hPDLFs for 3 weeks[a]

	% Staining						
	SSEA-4	Col-I	Col-III	FBN	FSP	Vimentin	Pan-CK
Weeks	colony/ SC(2-D)	colony/ SC(2-D)	colony/ SC(2-D)	colony/ SC(2-D)	colony/ SC(2-D)	colony/ SC(2-D)	colony/ SC(2-D)
1	18.2±3.48/ 6.9±1.31	12.3±1.53/ 3.57±1.40	NA/ NA	7.67±0.58/ 4.0±1.28	4.33±2.52/ 3.13±0.80	2.33±0.58/ NA	3.77±1.47/ NA
2	4.8±1.85/ NA	12.3±2.08/ 5.27±1.16	NA/ 11.1±4.78	7.67±2.08/ 9.0±2.93	11.3±3.51/ 6.43±2.0	19.7±7.77/ 16.3±2.84	10.5±4.52/ 14.4±1.71
3	NA/ NA	15.3±1.15/ 15.1±2.49	NA/ 21.8±3.73	18.7±2.52/ 20.2±3.08	13.3±3.79/ 9.27±1.23	19.3±4.51/ 16.5±1.15	10.2±2.44/ 23.7±3.52

[a] Immunohistomorphometric data representing mean ± standard deviation of staining percent for specific undifferentiation-related stem cell, mesenchymal, fibrogenic, and epithelial markers at indicated time points. SC, single cell; 2-D, monolayer colony areas; colony, three-dimensional colony areas; SSEA-4, antistage-specific embryonic antigen-4; Col-I, anticollagen type I; Col-III, anticollagen type III; FBN, antifibronectin; FSP, antifibroblast surface protein; Vimentin, antivimentin; Pan-CK, antipan-cytokeratin; NA, not available.

Table 14.2
Osteogenic differentiation of hESCs on hPDLFs for 4 weeks[a]

Days/Area	% Staining				
	ALP	OSN	OSP	BSP	OSC
1/2-D	19.5±4.05	NA	NA	NA	NA
28/2-D	9.07±2.25	NA	3±0.954	16±2.51	8.3±2.91
28/3-D	19.9±2.36	7.87±1.39	14.1±2.11	25.1±2.4	10±2.26

[a] Immunohistomorphometric data representing mean ± standard deviation of staining percent for specific osteogenic differentiation markers at indicated time points. 2-D, monolayer colony areas; 3-D, three-dimensional colony areas; ALP, antialkaline phosphatase; OSN, antiosteonectin; OSP, antiosteopontin; BSP, antibone sialoprotein; OSC, antiosteocalcin; NA, not available.

culture expansion of hPDLFs and human embryonic stem cells are described, followed by the coculture procedures and immunohistochemical and RT-PCR detection of differentiation marker expression as well as microscopic demonstration of morphological changes during the experiments (*see* **Fig. 14.1**). The methodology represents initial steps in obtaining definitive periodontal progenitors, which could be used in experimental and prospective clinical applications in the field of periodontology.

2. Materials

2.1. Culture of hPDLFs

1. Expansion medium (PDLF-EM) was prepared with Dulbecco's Modified Eagle's Medium (DMEM; Gibco, Invitrogen, Paisley, UK, Cat. No. 12800-082) supplemented with: fetal bovine serum, 10–15% (FBS; Sigma, St. Louis, MO, USA, Cat. No. F6178), nonessential amino acid stock solution, 1% (NEAA; Gibco, Cat. No.11140-035), L-glutamine, 1 mM (Gibco, Cat. No. 25030-032), penicillin-streptomycin, 100 u/ml-100 μg/ml (Sigma, Cat. No. A5955).

2. Solution of trypsin, 0.05% and ethylenediamine tetraacetic acid, 0.53 mM (EDTA; Gibco, Cat. No. 25300054).

3. Mitomycin C (Sigma, Cat. No. M4287), dissolved in PBS at 1 mg/ml and stored in single-use aliquots of 100 μl at −20°C.

4. Tissue culture flasks, 25 and 75 cm^2 (Corning, Shiphol Rijk, The Netherlands, Cat. Nos. 430168 and 430641, respectively).

5. Centrifuge tubes, 50 ml (Corning, Cat. No. 430921).

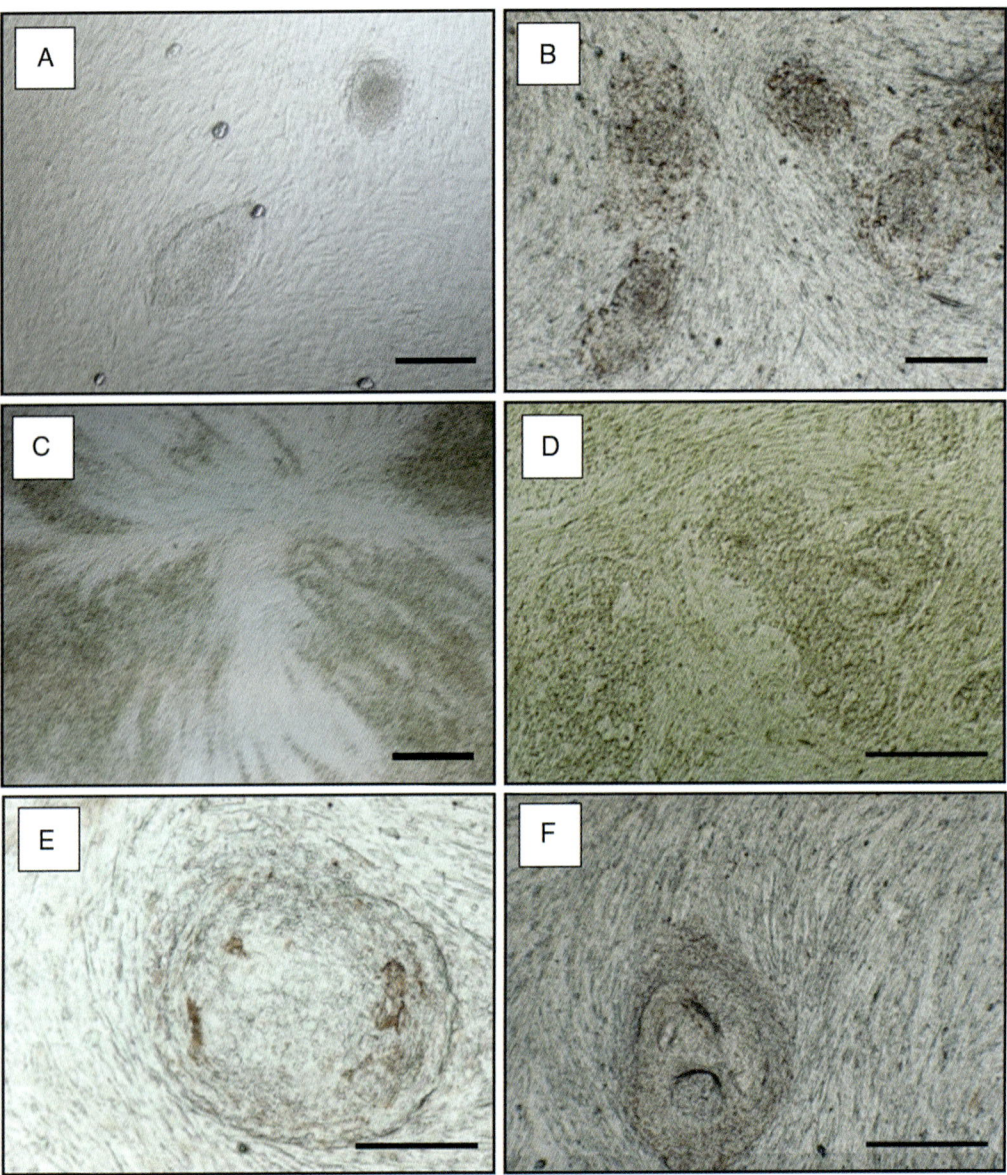

Fig. 14.1. Undifferentiated and differentiated human embryonic stem cells. (**A**) Undifferentiated hESC colonies on MEFs at day 5 after passage. (**B**) Colonies stain positively for SSEA-4. (**C**) After differentiation for 21 days in coculture with PDLFs, embryonic stem cells spread in 2-D and intermingle with adult fibroblastic cells. (**D**) Differentiation in osteogenic medium for 21 days results with more compact and rounded cell clusters having 3-D areas and rough surface appearance. (**E**) Round colonies differentiated under osteogenic induction conditions exhibit staining foci for BSP at day 28. (**F**) Slight staining for osteocalcin at day 28 indicates early phase of osteogenic differentiation. Scale bars = 200 μm.

2.2. Culture of hESCs

1. Human embryonic stem cell line (HUES-9, HUES Facility, Melton Laboratory, HHMI, Harvard University, Cambridge, MA, USA).

2. Expansion medium (ES-EM) was prepared with Knock-Out™ Dulbecco's Modified Eagle's Medium (KO™-DMEM; Gibco, Cat. No. 10829-018), supplemented with: KO™-Serum Replacer, 20% (KO™-SR, Gibco, Cat. No. 10828), nonessential amino acid stock solution, 1% (NEAA; Gibco, Cat. No. 11140-035), L-glutamine, 1 mM (Gibco, Cat. No. 25030-032), penicillin-streptomycin, 100 u/ml-100 μg/ml (Sigma, Cat. No. A5955), basic fibroblast growth factor, 8 ng/ml (bFGF, Sigma, Cat. No. F0291). bFGF is dissolved in KO-DMEM™ at 1 mg/ml concentration and stored in single use aliquots at –40°C.

3. Six-well culture plates (Corning, Cat. No. 3516).

4. Collagenase type IV, 1 mg/ml, (Gibco, Cat. No.17104-019).

2.3. hESC Transfection

1. Green fluorescent protein (GFP)-encoding plasmid pQBI pgk (Quantum Biotech Inc., Q-Biogene, Montreal, Canada, Cat. No. AFP2043).

2. FuGENE6 transfection reagent (Roche Diagnostics, Basel, Switzerland).

3. Vials, DNAse-free, 0.65 ml (Corning, Cat. No. 3208).

4. G418 (Sigma, Cat. No. A1720).

2.4. Fibrogenic Differentiation

1. Thermanox™ coverslips, tissue culture treated, 13 and 25 mm (Nalge Nunc Int., Rochester, NY, USA, Cat. Nos. 174950 and 174985).

2. Six-well culture plates (Corning, Cat. No. 3516), 15-ml centrifuge tubes (Corning, Cat. No. 430766), and micropipette tips 1 ml (Corning, Cat. No. 4809).

3. Differentiation medium (ES DM): same as ES-EM except that KO™-SR was replaced with 20% FCS and no bFGF added.

4. Soybean trypsin inhibitor, 1 mg/ml (Gibco, Cat. No. 17075-029).

2.5. Osteogenic Induction

1. Osteogenic induction (ES-OI) medium was prepared by supplementing ES-DM medium with: β-glycerophosphate, 10 mM (Sigma, Cat. No. G9891), ascorbic acid, 50 μg/ml (Sigma, Cat. No. A-4544), and dexamethasone, 10^{-7} M (Sigma, Cat. No. D4902).

2.6. Immunohistochemistry

1. Methanol (Riedel-de Haën, Fluka, Buchs, Switzerland, Cat. No. 24229).

2. Permeabilization solution: Triton X, 0.01% (v/v) (Sigma, Cat. No. T8787).

3. Endogenous hydrogen peroxidase block (Lab Vision, Fremont, CA, USA, Cat. No. TA-125-HP).
4. Nonspecific binding block (Lab Vision, Cat. No. TA-125-UB).
5. Antibody dilution buffer: bovine serum albumin, 2% (w/v) (BSA; Sigma, Cat. No. A4919) in PBS.
6. Primary antibodies:
    a. Anti-SSEA-4, goat polyclonal IgG (Santa Cruz Biotech., Santa Cruz, CA, USA, Cat. No. SC-21704).
    b. Antifibroblast surface protein, mouse monoclonal IgM (Sigma, Cat. No. F4771).
    c. Antifibronectin, mouse monoclonal IgM (Sigma, Cat. No. F6140).
    d. Anticollagen type I, goat polyclonal IgG (Santa Cruz Biotech.; Cat. No. SC-8786).
    e. 6.5. Anticollagen type III, mouse monoclonal IgG1 (Sigma, Cat. No. C7805).
    f. Antivimentin, mouse monoclonal IgM (Sigma, Cat. No. V6630).
    g. Anti-Pan-cytokeratin, mouse monoclonal IgG1 (Santa Cruz Biotech.; Cat. No. SC8018).
    h. Antiosteonectin, mouse monoclonal IgG (Alexis, San Diego, CA, USA; Cat. No. 804-317-C100).
    i. Antiosteopontin, goat polyclonal IgG (Santa Cruz Biotech.; Cat. No. SC-21742).
    j. Antibone sialoprotein, rabbit polyclonal IgG (Alexis, Cat. No. 210-312-R100).
    k. Antiosteocalcin, goat polyclonal IgG (Santa Cruz Biotech.; Cat. No. SC-18319).
7. Secondary Antibodies: biotinylated goat antimouse IgM (Cat. No. 62-6840), goat antimouse IgG (Cat. No. 81-6511), mouse antigoat IgG (Cat. No. 81-6540), and goat antirabbit IgG (Cat. No. 65-6140), all from Zymed, San Francisco, CA, USA.
8. Detection system: Horseradish peroxidase enzyme system and AEC chromogen (Lab Vision, Cat. No. TA-125-HA).

2.7. SEM

1. Fixing solution: glutaraldehyde, 2.5% (v/v) (Sigma-Aldrich, Cat. No. G7651) in 0.1 M sodium cacodylate buffer (Sigma, Cat. No. C0250).
2. Ethanol (Riedel-de Haën, Cat. No. 32221).

2.8. RT-PCR

1. mRNA extraction: Direct mRNA minipreparation kit (GenElute™, Sigma, Cat. No. DMN-70).

2. cDNA synthesis (SuperScript™ first-strand synthesis system, Invitrogen, Cat. No. 11904-018).

3. Reverse transcription mixture: $MgCl_2$ (25 mM), DTT (0.1 M), random hexamer (50 ng/ml), dATP, dCTP, dGTP, and dTTP (10 mM each), reverse transcriptase (50 U/ml, SuperScript™ II), and RT buffer (10X), all from Invitrogen.

4. PCR mixture: *Taq* polymerase (5 U/µl), dNTP mixture (10 mM each), specific primers (10 pM each), $MgCl_2$ (50 mM), cDNA, dH_2O, and PCR buffer (10X), all from Invitrogen.

5. Gel: Agarose (Sigma, Cat. No. A5093).

6. Other chemicals: Trizma R base (Cat. No. T-1503), boric acid (Cat. No. B-6768), EDTA (Cat. No. ED2SS), all from Sigma.

3. Methods

3.1. Culture of hPDLFs

1. Scrape the periodontal ligament tissue from the middle third of a healthy tooth root (extracted for orthodontic treatment) with sterile surgical blades.

2. Transfer the explants into a 15-ml centrifuge tube in PDLF-EM containing 10X antibiotics.

3. Discard the medium and wash the explants three times with sterile PBS at room temperature in a laminar air flow cabinet.

4. After final wash, add 2 ml of PDLF-EM into the tube and pipette back the explants, then transfer them to a 25-cm^2 flask containing 8 ml PDLF-EM, and place inside an incubator at 37°C, 5% CO_2.

5. Change the medium every 2–3 days, removing 4.0 ml and adding 4.0 ml of newly prepared PDLF-EM.

6. Continue to culture for 3–6 weeks until proliferating fibroblasts from explants become confluent on the flask surface.

7. Discard the medium, wash the cells three times with sterile PBS, add dropwise trypsin/EDTA at room temperature until just covering the surface, and keep the flask for 5 min inside the incubator.

8. Remove the flask, tap rigorously with hand to completely dissociate the rounded fibroblastic cells, then add 7.0 ml of PDLF-EM.

9. Pipette out the cell suspension and transfer to a 15-ml centrifuge tube, wash the remaining cells (*see* **Note 1**) with additional 5.0 ml medium, and add to the tube.

10. Centrifuge the cell suspension for 5 min at 800 rpm, 25°C.
11. Discard the supernatant, add 3.0 ml of PDLF-EM to the remaining cell pellet.
12. Gently pipette up and down to resuspend the cells, then transfer the suspension into a 75-cm^2 flask containing 20.0 ml PDLF-EM, and designate as passage 1 (P-1).
13. Continue to culture for 2–3 days until cells reach confluence.
14. Change the medium 1 day before the next passage.
15. Continue the expansion by splitting cells 1:2 or 1:3, until obtaining desired cell number for the experiments.

3.2. Culture of hESCs

1. Remove the vial containing frozen cell line from the liquid nitrogen tank and quickly immerse in water at 37°C.
2. Thaw the cells with gentle shaking until little ice crystal remains, then transfer the vial to the laminar air flow cabinet.
3. Pipette the cell suspension with a micropipette, and add dropwise (see **Note 2**) into 15-ml centrifuge tube containing 12.0 ml ES-EM at 25°C.
4. Centrifuge the cell suspension for 5 min at 650 rpm, 25°C.
5. Remove the supernatant (and DMSO-containing medium), add 3.0 ml of ES-EM.
6. Pipette up and down very gently several times to barely dissociate cell clumps from one another.
7. Transfer the cells on 6-well plates covered with mitotically inactivated primary mouse embryonic fibroblast (MEF) feeder cells.
8. Add ES-EM until reaching 3.0 ml/well, then place the plate inside the incubator at 37°C, 5% CO_2.
9. Change the medium daily removing 2.5 ml, and adding 2.5 ml.
10. Maintain the culture for 8–10 days, until ESC colonies with undifferentiated morphology become prevalent.
11. Select the colonies demonstrating the undifferentiated morphology under inverted microscope (see **Note 3**) and pick them up mechanically with a micropipette tip (see **Note 4**).
12. Colonies are dissociated into several particles during mechanical selection; transfer the clumps into ES-EM-containing 35-mm dish or 15-ml centrifuge tube.
13. Pipette the collected hES cell clumps several times, only to dissociate from one another (see **Note 5**), then transfer the cells to 6-well tissue culture plates with newly prepared MEFs.
14. Add ES-EM to reach 3.0 ml/well and shake the plate to ensure even distribution of hES cells before placing in the incubator.

15. Designate as Pn+1 where "n" is the passage number at which line is received.
16. Change medium daily during expansion, and perform the next passage after 5 days, again selecting the colonies with undifferentiated morphology.

3.3. Transfection of hESCs

1. Remove the hESC-containing 6-well culture plate at the third day following passage from the incubator and discard the medium.
2. Wash twice with KO^{TM}-DMEM, then add 400 μl of KO^{TM}-DMEM or ES-EM (see **Note 6**) and keep for 15 min in the incubator.
3. Add 97 μl of KO^{TM}-DMEM in sterile DNase-free 0.6-ml vial, then add 6 μl of FuGENE, stir gently to mix the components for 5 minutes at room temperature.
4. Add 1 μg of plasmid DNA in 20 μl of carrier to the mixture and mix; then keep for 20 min at room temperature.
5. Add 20 μl from the mix to every well containing 400 μl of KO^{TM}-DMEM or ES-EM, then keep the plate in the incubator for 6 h.
6. Complement the medium to 3.0-ml/well adding ES-EM.
7. Change the whole medium after 24 h.
8. Observe the cells under fluorescent microscope at 24, 48, and 72 h posttransfection.
9. To determine the transfection percent, count the GFP+ colonies under fluorescent microscope and all colonies under inverted light microscope.
10. To achieve stable transfection, select GFP+ colonies, pick them up with a micropipette tip, and transfer to newly prepared MEFs.
11. Culture in the presence of 600 μg/ml G-418 for 14 days, with passages every other day in nonresistant MEFs, and every 5 days when gentamycin-resistant MEFs are used.
12. Select the GFP+ colonies to be used for experiments.

3.4. Differentiation of hESCs

1. Seed PDLFs at a density of ~50,000 cells/cm^2 on Thermanox^{TM} coverslips in 6-well plates.
2. Incubate the cells in PDLF-EM until reaching confluence (2–3 days).
3. Mitotically inactivate the PDLFs through incubation with PDLF-EM containing 10 μg/ml mitomycin C for 2–3 h (see **Note 7**).

4. Wash several times with PDLF-EM to completely eliminate the antibiotic, then incubate for 24 h in PDLF-EM.

5. Remove the hESC-containing plate from incubator at day 5 after the last passage and mechanically select the colonies with undifferentiated morphology.

6. Remove the colony particles with micropipette tip under inverted microscope in laminar air flow cabinet and transfer the clumps into a 15-ml centrifuge tube containing 3.0 ml ES-EM.

7. Collect the hES cells from the same passage in the tube, then pipette gently to dissociate the clumps.

8. Alternatively, add 2.0 ml of trypsin/EDTA, incubate for 5 min, and then dissociate the cells into single-cell suspension by gentle pipetting with a micropipette.

9. Quench the trypsin activity by adding soybean trypsin inhibitor.

10. Transfer the hESC colony particles (or single-cell suspension) to 6-well plates on Thermanox™ coverslips covered with confluent layer of PDLF cells.

11. Add ES-EM until reaching 3.0 ml/well.

12. Place the cells inside the incubator at 37°C, 5% CO_2.

13. After 24 h, the hESC cell clumps will be largely attached to the PDLF surface (designate as day 1).

14. Discard the ES-EM medium and add the same amount of ES-DM or ES-OI medium for fibrogenic or osteogenic differentiation, respectively.

15. Daily change the medium during 28 days of differentiation experiments.

3.5. Immunohistochemistry

1. Remove the selected culture plates at predetermined time points from the incubator, discard the medium, and wash three times with PBS at room temperature.

2. Add 1.0 ml/well of ice-cold methanol, keep for 45 min at room temperature, then discard the excess methanol and leave the plate to air-dry.

3. Permeabilize the fixed cells with 0.01% (v/v) Triton X for 30 min, wash three times with PBS.

4. Quench the endogenous peroxidase activity by incubating at room temperature with peroxide block for 10 min, wash three times with PBS.

5. Block the nonspecific binding by incubating at room temperature with blocking solution for 10 min, wash three times with PBS.

6. Incubate with primary antibody at predetermined concentration (by titration) for 1 h at room temperature or overnight at 4°C.

7. Incubate with species and isotype-specific secondary antibody at predetermined concentration (by titration) for 30 min at room temperature, wash three times with PBS.

8. Incubate for 20 min with avidin-biotin horseradish peroxidase enzyme system, wash three times with PBS.

9. Incubate for 5–15 min at room temperature with AEC chromogen until red staining develops.

10. Wash three times with distilled water, mount, and visualize under the microscope.

3.6. SEM

1. Remove the selected culture plates (coverslips) at predetermined time points from the incubator, discard the medium, and wash three times with PBS at room temperature.

2. Fix with 2.5% (v/v) glutaraldehyde in cacodylate buffer for 30 min at room temperature or overnight at 4°C.

3. Dehydrate with graded ethanol series (50, 70, 95, and 100%).

4. Dry in air.

5. Place samples on stubs.

6. Sputter-coat with gold and visualize under scanning electron microscope.

3.7. RT-PCR

1. Prepare cell culture for mRNA isolation. Put the cell suspension in a sterile 1.5-ml microfuge tube.

2. Centrifuge the tubes at maximum speed for 2 min at room temperature in a microfuge.

3. Discard all but 25 µl of each supernatant.

4. Use the direct mRNA miniprep kit's procedure.

5. Keep the mRNA in aqueous solution on ice.

6. Add the mRNA to reverse transcription mixture in 0.65-ml tubes and follow the procedure.

7. Amplify cDNA in aqueous solution. Transfer the cDNA to PCR tubes and then add the PCR mixture. End volume of the mix solution is 50 µl (*see* **Note 8**).

8. Preheat the lid to 105°C. Amplify the nucleic acids using denaturation (1 min at 95°C), annealing (1 min at 55–60°C), and polymerization cycles (1 min at 72°C).

9. Amplify for 25–35 cycles.

10. Take samples (10 μl each) from the test reaction mixture and the four control reactions, run them by agarose electrophoresis, and stain the gel with ethidium bromide. Analyze the gel using UV visualization and documentation system.

4. Notes

1. Multiple explant particles may remain undissociated on the flask surface at this first passage, and further PDL fibroblasts can be expanded by continuing their culture.
2. Adding thawed cell suspension dropwise into freshly prepared media aims at reducing the osmotic shock that affects cell survival rate and is a general cell culture thawing rule.
3. The microscope should be located inside the laminar air flow cabinet, and the culture plate should be kept on a metal plate at 37°C, minimizing temperature fluctuations which affect differentiation of hESCs.
4. The hESC colonies could be collected after enzymatic incubation with collagenase type IV for 5 min; however, this diminishes the selection ability of undifferentiated colonies, albeit shortens the time needed for passaging procedure.
5. Rigorous pipetting may result with increased dissociation into single cells, which subsequently differentiate spontaneously during expansion culture.
6. KO^{TM}-SR did not adversely affect the transfection efficiency in our experiments, therefore ES-EM could be used instead of KO^{TM}-DMEM in the first 6 h of transfection.
7. The PDLFs may or may not be mitotically inactivated, for they proliferate slower than MEFs and contact inhibition at confluence further diminishes their expansion.
8. Mineral oil should be used to cover the PCR mixture.

References

1. Menendez, P., Wang, L., and Bhatia, M. (2005) Genetic manipulation of human embryonic stem cells: a system to study early human development and potential therapeutic applications. *Curr. Gene Ther.* **5**(4), 375–85.
2. Davila, J.C., Cezar, G.G., Thiede, M., Strom, S., Miki, T., and Trosko, J. (2004) Use and application of stem cells in toxicology. *Toxicol. Sci.* **79**(2), 214–23.
3. Hyslop, L.A., Armstrong, L., Stojkovic, M., and Lako, M. (2005) Human embryonic stem cells: biology and clinical implications. *Expert Rev. Mol. Med.* **7**(19), 1–21.
4. Lekic, P., and McCulloch, C.A. (1996) Periodontal ligament cell population: the central role of fibroblasts in creating a unique tissue. *Anat. Rec.* **245**(2), 327–41.

5. Pitaru, S., McCulloch, C.A., and Narayanan, S.A. (1994) Cellular origins and differentiation control mechanisms during periodontal development and wound healing. *J. Periodontal Res.* **29**(2), 81–94.

6. Taba, M. Jr., Jin, Q., Sugai, J.V., and Giannobile, W.V. (2005) Current concepts in periodontal bioengineering. *Orthod. Craniofac. Res.* **8**(4), 292–302.

7. Jernvall, J., and Thesleff, I. (2000) Reiterative signaling and patterning during mammalian tooth morphogenesis. *Mech. Dev.* **92**(1), 19–29.

8. Hentze, H., Graichen, R., and Colman, A. (2007) Cell therapy and the safety of the embryonic stem cell-derived grafts. *Trends Biotechnol.* **25**(1), 24–32.

9. Inanc, B., Elcin, A.E., Unsal, E., Balos, K., Parlar, A., and Elcin, Y.M. (2008) Differentiation of human embryonic stem cells on periodontal ligament fibroblasts *in vitro*. *Artif. Organs.* **32**(2), 100–9.

10. Inanc, B., Elcin, A.E., and Elcin, Y.M. (2007) Effect of osteogenic induction on the *in vitro* differentiation of human embryonic stem cells cocultured with periodontal ligament fibroblasts. *Artif. Organs.* **31**(11), 792–800.

Chapter 15

Generation of Neural Crest Cells and Peripheral Sensory Neurons from Human Embryonic Stem Cells

Ronald S. Goldstein, Oz Pomp, Irina Brokhman, and Lina Ziegler

Abstract

Peripheral somatic sensory neurons (PSNs) are responsible for the critical function of transmitting multiple modalities of information from the outside world, including heat, touch, and pain, as well as the position of muscles required for coordinated voluntary movement to the central nervous system. Many peripheral neuropathies exist, including hereditary neurodegeneration in Familial Dysautonomia, infections of PSNs by viruses such as *Varicella zoster* and damage to PSNs and/or their process resulting from other disease conditions such as diabetes. Understanding of the etiology of these diseases and development of treatments is hampered by the lack of normal and healthy human PSNs for study, which are only available from abortuses or rare surgical procedures.

Human embryonic stem cells (hESCs) are an ideal source of cells for generating normal PSNs for study of disease and drug development, since they can be grown virtually indefinitely in tissue culture and have the potential to form any cell type in the body. Several years ago, we generated human neurons with the molecular characteristics of PSNs from hESCs at low (less than 1%) yields (Pomp et al., Stem Cells 23:923–930, 2005). The present chapter details our most recently improved method that uses 2 rounds of PA6-induction to rapidly generate PSNs at more than 25% purity (Pomp et al., Br. Res. 1230: 50–60, 2008).

The neural crest (NC) is a transient multipotent embryonic stem cell population that is the source of PSNs. NC cells give rise to diverse and important tissues in man, but human NC has not been studied because of the difficulty in obtaining 3–5 week human embryos. The methods described in this chapter can also be used to quickly generate large numbers of human NC for study.

Key words: Peripheral sensory neurons, neural crest, neurospheres, PA6 stroma, human embryonic stem cells.

1. Introduction

Generation of neurons from hESCs has focused on CNS phenotypes with clinical importance, such as spinal motoneurons and mesencephalic dopaminergic neurons (i.e., (1, 2)). However, many pathologies of PNS neurons exist, such as Familial Dysautonomia, infections of PSNs by varicella virus, and damage to PSNs and/or their processes resulting from diseases such as diabetes. We therefore set out several years ago to establish methods for generating PNS neurons for preclinical studies.

PSN differentiation from hESCs was first obtained (3) using the SDIA-PA6 cell-induction method developed for murine ESC by Sasai and colleagues (4, 5). hESCs cocultured with PA6 murine stromal cells for 4 weeks induced neurons that coexpressed (1) peripherin (peri) and Brn3a, and (2) peripherin and tyrosine hydroxylase, combinations characteristic of PSNs and sympathetic neurons, respectively (5). Subsequently, we evaluated the use of hESC-derived propagatable neural precursors ("neurospheres") for generation of PSNs. We initially found that use of noggin-induced neurospheres as starting material instead of naïve hESCs greatly enhanced PSNs generation (6). Propagatable neurospheres that can be differentiated to make CNS derivatives have been generated from murine ESCs using PA6 induction (7). This chapter describes a method to make propagatable PA6-induced human ESC-derived neurospheres using our own microsurgical selection method to enhance efficiency, and the use of these neurospheres to rapidly generate high yields of PSNs.

The progenitors of PSNs in amniote embryos are a transient population of migratory cells, the neural crest (NC) (8). In addition to PSNs, NC gives rise to a plethora of cell types, including peripheral sympathetic neurons, Schwann cells, melanocytes, mesenchymal cells of the skull, and others. Although the NC of several vertebrate species has been the intense topic of research for more than 50 years, very little is known about human NC since it is present only in embryos of 3–5 weeks gestation that are difficult to obtain from pregnancy terminations. We find that PA6-induced neurospheres contain NC-like cells, and that replating the neurosphere cells on PA6 generates more than 50% NC like cells in a matter of days.

Another method for generating NC-like cells from hESCs using FACS sorting has been reported (9). Expression of p75 and some other markers was used to isolate a population of stem-like cells that could be propagated and differentiated to cells expressing markers associated with a number of NC derivatives including PSNs, Schwann cells, mesenchymal cells, etc. However, there was little or no information provided on the purity of the differentiated derivatives, and in the case of PSNs, the time to obtain them far exceed the 1 week required in the methods described in the present chapter. Therefore, the use of PA6 induction, while initially less "pure" than use of FACS-isolated precursors of Lee et al., appears to be preferable at this time for the production of PSNs for modeling PNS disease and injury. Very recently, our results with PA6-induced neurospheres generating NC-like cells have been repeated and validated by another group that combined the induction with p75 sorting (10).

2. Materials

2.1. Culture

2.1.1. Cells

1. hESCs are available from many sources. We obtained line H9, (11) from the National Stem Cell Repository, Madison, WI, USA and HUES7, HUES9, HUES8 from Dr. Douglas Melton at Harvard University, Cambridge, MA, USA (12).

2. PA6 murine stromal cells, derived from murine skull bone marrow, are available from the Riken Cell Bank (Riken Bioresource Center Cell Bank, Koyadai, Japan, Cat. No. RCB1127).

3. Human foreskin fibroblasts can be obtained from many sources, including the ATCC (Manassas, VA, USA; Cat. No. SCRC-1041). The fibroblasts used in this study were a generous gift of Interpharm Corporation, Israel.

2.1.2. Materials

1. Tissue culture plastic: Tissue culture plate 24 well, Cat. No. 662160, tissue culture dish 100 mm, Cat. No. 664160, tissue culture dish 60 mm, Cat. No. 628160, petri dish 35 mm, Cat. No. 627102 (Greiner-Bio-One, Neuburg, Germany).

2. Coverslips: Menzel Glazer, Braunschweig, Germany; 13-mm cover glass thickness # 1, Cat. No. CB00130RA1 (*see* **Note 1**).

3. Microscalpels: Made by electrolysis of tungsten wire (A-M systems, Carlsborg, WA, USA; Cat. No. 7185) (*see* **Note 2**). One lead of a 12-V AC transformer is connected to a bath of saturated NaOH in a plastic petri dish by a paper clip and the other lead is attached to a pin holder (Fine Science Tools; Cat. No. 26016-12) with the tungsten wire inserted. The pin holder is held in any simple micromanipulator and the tip of the wire lowered just into the solution, while monitoring with an old binocular microscope (*see* **Note 3**). The current is then applied and the wire dipped and removed from the solution, and the tip examined with the binocular microscope, until the appropriate shape and sharpness are obtained. Varying the angle of the wire can produce many different shapes of scalpels (13), but we find that straight is fine for the application described here.

4. 3-ml plastic Pasteur pipette (Sigma-Aldrich, St. Louis, MO, USA; Cat. No. BSV1371).

5. Tuberculin syringe 1 cc, no needle, 30-g needles (Pic indolor, Artsana, Italy).

2.1.3. Solutions and Medium

1. Gelatin (Merck, Whitehouse Station, USA; Cat. No. 1.04078.1000). Prepare a 0.1% solution in dH_2O, autoclave at 121°C for 30 min. Sterile stock is stored in the refrigerator.

2. Dulbecco's phosphate-buffered saline with divalent cations (PBS+) (Biological Industries, Bet HaEmek, Israel; Cat. No. 020-020-1A).

3. Dulbecco's phosphate-buffered saline *without* divalent cations (PBS) (Biological Industries, Cat. No. 020-0231A).

4. Trypsin/EDTA 0.05% (Biological Industries, Cat. No. 030151B).

5. BHK-21 medium/Glasgow MEM (Gibco BRL, Paisley, Scotland; Cat. No. 21710025).

6. B27 supplement (Gibco BRL, Cat. No. 17504044).

7. DMEM (Biological Industries, Cat. No. 010551A).

8. DMEM knock-out (Gibco BRL, Cat. No. 10829018).

9. F12 (Biological Industries Bet Haemek, Israel, Cat. No. 010951A).

10. NGF (Alomone Labs, Jerusalem, Israel, Cat. No. N240).

11. N2 supplement (Gibco BRL, Cat. No. 17502048).

12. OptiMEM (Gibco BRL, Cat. No. 31985047).

13. Tissue culture medium additives from Biological industries: (i) penicillin-streptomycin solution, Cat. No. 03-031-1, (ii) L-Glutamine in saline solution, Cat. No. 03-020-1, (iii) sodium pyruvate solution, Cat. No. 03-042-1, MEM nonessential amino acids solution, Cat. No. 01-340-1.

14. 2-Mercaptoethanol (Merck Darmstadt, Germany, Cat. No. 1.15433).

15. Fibroblast growth factor -basic, human (Sigma, Cat. No. F0291).

16. Induction medium containing 90% BHK-21 medium/Glasgow-modified Eagle medium, 10% knockout serum replacement, 2 mM glutamine, 1 mM pyruvate, 0.1 mM nonessential amino acid solution, and 0.1 mM β-mercaptoethanol.

17. Neurosphere medium consisting of DMEM/F12 (1:1), B27 supplement (1:50), glutamine 2 mM, 50 units/ml penicillin, 50 μg/ml streptomycin, and 20 ng/ml bFGF.

18. DRG medium consisting of DMEM/F12 (1:1), B27 supplement (1:50), 2 mM glutamine, 50 units/ml penicillin, 50 μg/ml streptomycin, and 10 ng/ml NGF.

2.2. Immunocytochemistry

1. 4% Paraformaldehyde (see **Note 4**). For 200 ml, add 8 gm paraformaldehyde granules (Electron Microscopy Sciences; Hatfield, PA, USA; Cat. No. 19208) to 100 ml dH$_2$O. Heat to 60°C while stirring in a fume hood. Add 1 M NaOH one drop at a time until clear and cool to ambient temperature. Bring volume to 200 ml with 2x PBS.
2. Antibodies

Antibody	Dilution	Host	Supplier
AP2	2	Mouse	DSHB (Iowa City, Iowa, USA; Cat. No. 3B5)
Brn3a	110	Mouse	Chemicon (Temecula CA, USA; Cat. No. MAB1585)
E-cadherin	60	Mouse	Thermo Scientific (Fremont, CA, USA; Cat. No. MS1862)
GFAP	100	Mouse	Thermo Scientific, Cat. No. MS1407
GFP	150	Rabbit	Santa Cruz (Santa Cruz, CA, USA; Cat. No. SC8334)
HNK1	no dilution	Mouse	ATCC (Manassas, VA, USA; supernatant of hybridoma line, Cat. No TIB-200)
Islet-1	5	Mouse	DSHB, Cat. No. 40.2D6
NCAM	300	Rabbit	Chemicon, Cat. No. AB5032
Neurofilament 200 kd	600	Rabbit	Sigma, Cat. No. N4142
OCT4	80	Mouse	Santa Cruz, Cat. No. SC5279
Peripherin	500	Rabbit	Chemicon, Cat. No. AB1530
P75	2	Mouse	Santa Cruz, Cat. No. A1603
βIIItubulin	600	Mouse	Promega, Cat. No. G7121

Dilutions shown are for staining coverslips with cultured cells. For immunostaining of paraffin-embedded tissues, the antibody concentration was double of that used for immunostaining on coverslips (for example –E-cadherin was diluted 1–30 for immunostaining of paraffin-embedded tissue).

3. Secondary antibodies: (i) Alexa Fluor 488 goat anti-mouse IgG (H+L) conjugate (Invitrogen, Molecular Probes; Carlsbad, CA, USA; Cat. No. A-11001), (ii) Alexa Fluor 594 goat anti-rabbit IgG (H+L) conjugate (Invitrogen Molecular Probes; Cat. No. A-11012).

4. Humid chamber: Immuno Moist Chamber, Black (Ted Pella Inc, Redding CA, USA; Cat. No. 21049-B).

5. Glazed porcelain plate, 12 well, black EMS (Cat. No. 71575-12BK).

2.3. RT-PCR

2.3.1. Kits and Enzymes

1. Gene Elute Mammalian Total RNA Miniprep Kit (Sigma, Cat. No. NA0100).
2. Tri-Reagent (Sigma, Cat. No. T9424).
3. PCR GoTaq Master Mix (Promega Madison, WI, USA; Cat. No. M7121).
4. Blue Taq DNA Polymerase (EURx LTD Gdasnk, Poland; Cat. No. E2510).

2.3.2. Primers

βIII TUBULIN (ATGCGGGAGATCGTGCACAT, CCCCTGAGCGGACACTGT, 238 bp).

CGRP (AGA CAT CCA GCA AGCAAC AGA AC, CCA GCC AAG AAA ATA ATA CCACAT T, 444 bp).

E-CADHERIN (TTCCCTCGACACCCGATTCAAAGT, AGCTGTTGCTGTTGTGCTTAACCC 876 bp).

FOXD3 (CAAGCCCAAGAACAGCCTAGTGAA, TGACGAAGCAGTCGTTGAGTGAGA, 202 bp).

GALANIN (CAAGGAAAAACGAGGCTG, AGGCAAAGAGAACAGGAA, 400 bp).

GAPDH (CTTTTAACTCTGGTAAAGTGG, TTTTGGCTCCCCCCTGCAAAT, 287 bp).

HB9 CCTAAGATGCCCGACTTCAACTC, GCCTTTTTGCTGCGTTTCCATTTC, 236 bp.

KI67 (GCCCCAACCAAAAGAAAGTCT, AGCTTTGTGCCTTCACTTCCA, 133 bp).

MSX1 (CCTTCCCTTTAACCCTCACAC, CCGATTTCTCTGCGCTTTTCT, 284 bp).

NCX/HOX11L1 (GTGCCTGGGCCCTCGGGTTTG, AGCACCTGTGAGCGGGA, 208 bp).

OCT4 (GAGAACCGAGTGAGAGGCAACC, CATAGTCGCTGCTTGATCGCTTG, 165 bp).

OTX2 (CGCCTTACGCAGTCAATGGG, CGGGAAGCTGGTGATGCATAG, 641 bp).

PAX3 (CAGCACCAGGCATGGATTT, ATACTGTAGCCTGTGGTGC, 309 bp)

SNAI1 (GCTGCAGGAGGACTCTAATCCAGAGTT, GACAGAGTCCCAGATGAGCATTG, 107 bp).

SOX9: AGTGGGTAATGCGCTTGGATAGGT, CGAAGATGGCCGAGATGATCCTAA, 203 bp.

SUBSTANCE-P (CGACCAGATCAAGGAGGA, CAGCATCCCGTTTGCCCATT, 121 bp).

SOX1 (ATGCACCGCTACGACATGG, GCTCATGTAGCCCTGCGAGTTG, 65 bp).

TRKA (CAGCCGGCACCGTCTCT, TCCAGGAACTCAGTGAAGATGAAG, 80 bp).

RUNX-1 (CACGCACGAATTTTCAG, GTGGTCCTATTTAAGCCAGC, 90 bp).

2.4. Paraffin Histology

1. Superfrost Plus microscope slides (Menzel-Glaser, Cat. No. 041300).

2. Paraplast+ (Sigma-Aldrich, Cat. No. P-3683).

3. PBS without divalent cations. For 500 ml, dissolve 80 g NaCl, 2 g KCl, 11.5 g Na_2HPO_4 (anhydrous), 2 g KH_2PO_4, 1 g Thimerosal (Sigma-Aldrich, Cat. No. T-5125) in 450 ml dH_2O. Bring volume to 500 ml when dissolved. This makes a 20× stock that is diluted before use.

4. Microwave oven (*see* **Note 5**).

5. Staining dishes. (i) Plastic for deparaffinization, Hematoxylin, and Eosin staining and dehydration (Tissue-Tek, Torrance, CA, USA; Cat. No. 4451); (ii) Glass for immunocytochemistry (Electron microscopy sciences, Cat. No. 70312 20) x6; (iii) Polypropylene staining dish for microwave antigen retrieval (Electron microscopy sciences, Cat. No. 70321-10); Staining Rack, (Electron microscopy sciences, Cat. No. 70321-20).

6. Citrate buffer (diluted from 10x solution). For 500 cc 10X solution: add 14.5 g tri-sodium citrate dihydrate (Carlo Elba, Milan, Italy, Cat. No. 479487) to 400 cc ddH_2O. Bring to pH 6.0 with 1 N HCl.

7. Blocker: 1% bovine serum albumin (USB, Cleveland, OH, USA; Cat. No. 70195), 0.5% Triton-×100 (USB, Cat. No. 22686) in PBS+. Make up 100 ml and freeze in sterile 15-ml tubes.

9. Mounting medium: 75% glycerol (Sigma, Cat. No. G-8773) 25% PBS-, add 1% n-propyl gallate (Sigma, Cat. No. P-3130) for antibleaching.
10. 0.1 mg/ml Hoechst 33258 (Bisbenzimide) (dilute 1:1000 from stock solution). For 10 ml stock solution: add 10 mg Hoechst 33258 (Sigma; Cat. No. B-2883) to 100 ml ddH$_2$O. Working solution can be reused and kept in light-protected container at 4°C until nuclear staining weakens.
11. Nail polish (*see* **Note 6**).
12. Ethanols (70%, 95% denatured diluted with dH$_2$O, 100% anhydrous analytical reagent).
13. Xylene.
14. Cedarwood oil (Sigma, Cat. No. 318086).
15. Paraplast Plus (McCormick Scientific St. Louis, MO, USA; Cat. No. 502004).
16. Low melting temperature agarose (Sigma, Cat. No. A-4018).

2.5. Photography

1. Digital cameras (Scion, Frederick, MD, USA; models no. CFW1310M (monochrome) and CFW 1310C (color)).
2. Zeiss LSM confocal laser scanning microscope (Carl Zeiss, Gottingen, Germany).
3. Olympus BX60 microscope (Olympus, Tokyo, Japan).
4. ImageJ (http://rsb.info.nih.gov/ij).

3. Methods

3.1. Culture

3.1.1. Human ES Cell Culture

Human ES cells (HUES7, HUES9, HUES8, and H9) are cultured on mitomycin-inactivated human fibroblast feeder layers on 0.1% gelatin-coated tissue culture dishes (coat 30' at 37°C) in hESC medium, as described by many others (including several chapters in this volume).

1. 10^5 hESCs are seeded per 6-cm dish containing 1×10^6 mitomycin-treated mitotically inactivated fibroblasts.
2. Medium is changed every other day and hESCs split 1:3 using 0.05% trypsin/EDTA (4' at 37°C) once a week to dishes containing freshly defrosted inactivated fibroblasts.

3.1.2. Stromal Cell Culture

PA6 mouse stromal cells are maintained as a monolayer in 10% FCS, 1% glutamine, and 1% penicillin-streptomycin, in DMEM (fibroblast medium). Confluent cultures are washed twice with

PBS without Ca^{+2} and Mg^{+2}, treated with 0.05% trypsin/EDTA for 5', and passaged approximately weekly at a ratio of 1:5 on 10-cm tissue culture dishes.

3.1.3. Generation and Propagation of NSPs

1. PA6 cells are seeded on gelatin-coated 10-cm tissue culture dishes and grown for 5–7 days until they reach approximately 95% confluence.
2. hESC cultures are trypsinized for 4' at 37°C, after which the trypsin is neutralized with medium.
3. The hESC culture is withdrawn and expelled five times using a large-mouthed 3 cc plastic Pasteur pipette. This mild trituration also separates the sheet of fibroblast feeders from the cell suspension, which allows the simple removal of the fibroblasts.
4. hESCs are counted, suspended in induction medium, 60,000 cells/dish seeded on the PA6 monolayer.
5. Medium is replaced on the fourth day of coculture and every 2 days thereafter.
6. On day in-vitro 12, two types of colonies are observed, homogeneous opaque colonies and heterogeneous colonies (**Fig. 15.1**). The homogeneous opaque-white colonies are removed with a microscalpel and transferred to bacteriological Petri dishes containing neurosphere medium.
7. After about 1 week, mostly homogeneous neurospheres (NSPs) are observed floating in the dishes. The NSPs sometimes contain pigment, which could be from neural crest-derived melanocytes or retinal pigment epithelial cells (not shown).
8. NSPs are maintained in suspension at 5–20 NSPs/ml; half of the medium is changed every 3 days.
9. Every 2–3 days or when the NSPs reach about 1-mm diameter, they are passaged by cutting into halves/quarters using a microsurgical scalpel. This is done under a dissecting microscope in a laminar flow (non-biohazard) hood. (*See* **Note 7**).
10. NSPs to be used for generation of neural crest or PSNs can be analyzed by histology (**Section 3.2.2**) or by RT-PCR (**Section 3.4**), or further differentiated from spheres of 2–8 weeks after colony dissection from the PA6 cells (**Section 3.1.4**). *See* **Note 8**.

3.1.4. Induction of NC and PSNs from NSPs by PA6 Coculture

1. Incubate NSPs with 0.05% trypsin/EDTA for 5' at 37°C. Add equal amount of medium to neutralize enzyme.
2. Dissociation: Remove the needle from a 1 cc tuberculin syringe and draw NSPs into the syringe.

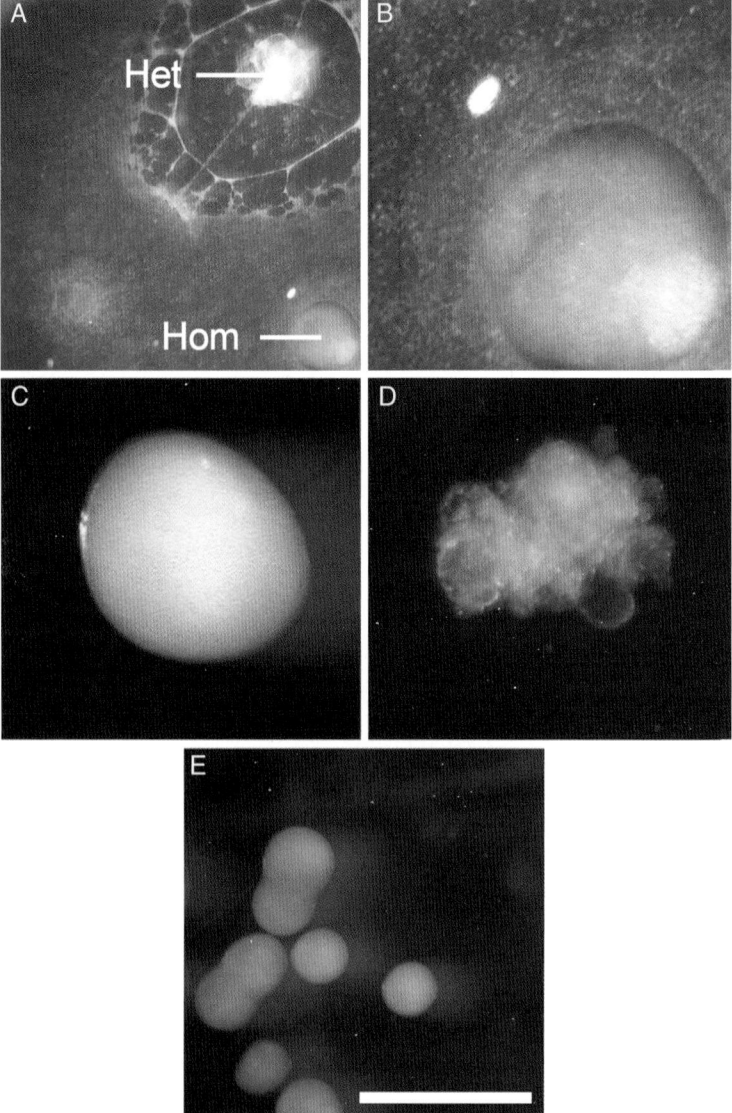

Fig. 15.1. Generation of neurospheres from hESCs using PA6 co-culture. (**A**) Two types of hESC colonies predominate after 12–14 days of coculture of hESCs and PA6 cells: heterogeneous (Het) and homogeneous (Hom) colonies. (**B**) Shows a homogenous colony from panel (**A**) at higher magnification. The Hom colonies gave rise to neurospheres (**C**), while the heterogeneous colonies developed into cystic structures similar to embryoid bodies (**D**) when grown in suspension in serum-free medium. A number of NSPs propagated for about 1 month, some of which have fused together, is shown in (**E**). Bar = 1 mm.

3. Re-attach 30 g needle and expel solution through the needle. This results in the formation of small aggregates and some single neural precursor cells.

4. NSP cells and aggregates (15 NSPs/24 wells) are then suspended in DRG medium and seeded onto PA6-coated coverslips in wells of a 24 well plate. (The coverslips are prepared by seeding 10–30,000 PA 6 cells onto gelatin-coated coverslips, and waiting 3–7 days until they reach 95% confluence).

5. Change medium every 3 days.

3.2. Immunostaining

3.2.1. Immunostaining of Cultures

1. Rinse coverslips in 24-well plate twice in PBS^{++} at RT temperature for 5'.

2. Fix with 4% buffered paraformaldehyde (PAF) for 30' at RT.

3. Rinse repeatedly with PBS with 0.5% Triton-x 100 (PBST).

4. Incubate coverslips for 1 h in blocking solution.

5. Remove the coverslips with a number 5 forceps by their edges and incubate with primary antibodies (see **Note 9**) for 1 h at room temperature or overnight at 4°C. (i) Cut a piece of parafilm, remove the backing, and place it "clean" side up on an inverted multiwell culture dish in a Tupperware or similar food storage container. (ii) Place 30 µl drops of diluted antibody on the parafilm and place the coverslip on top of the drop using a forceps. Add a small amount of water to the food storage container to maintain humidity.

6. Remove the coverslips from the antibody and rinse in PBST in a ceramic depression dish 3x 5' each. Multiple rinses are performed by changing the solution in a depression, in order to minimize the potential trauma to the adherent cells.

7. Incubate with secondary antibodies for 30' on parafilm in a humid chamber at RT.

8. Rinse 3x PBST 5' each in the ceramic dish.

9. Counterstain in the ceramic dish with Hoechst dye for 5' to visualize nuclei.

10. Rinse in PBST

11. Mount coverslips on microscope slides by placing them cells side down into a drop of PBS/glycerol with antifade.

12. Fix coverslips to slide with dots of nail polish. When stored at 4°C, fluorescence is stable for more than 1 year.

Many peripheral sensory-like neurons (Brn3a+/peripherin+) are observed using immunocytochemistry after 1 week of coculture (**Fig. 15.2A, B**). Long-term culture reveals that these cells fire action potentials and express a number of maturation markers of peripheral sensory neurons (14). NSPs reinduced by PA6 did not generate TH+/peripherin+ putative sympathetic neurons (3). In addition to PSNs, many cells also show a NC-like phenotype (AP2+/NCAM+) at this time (**Fig. 15.2C, D**).

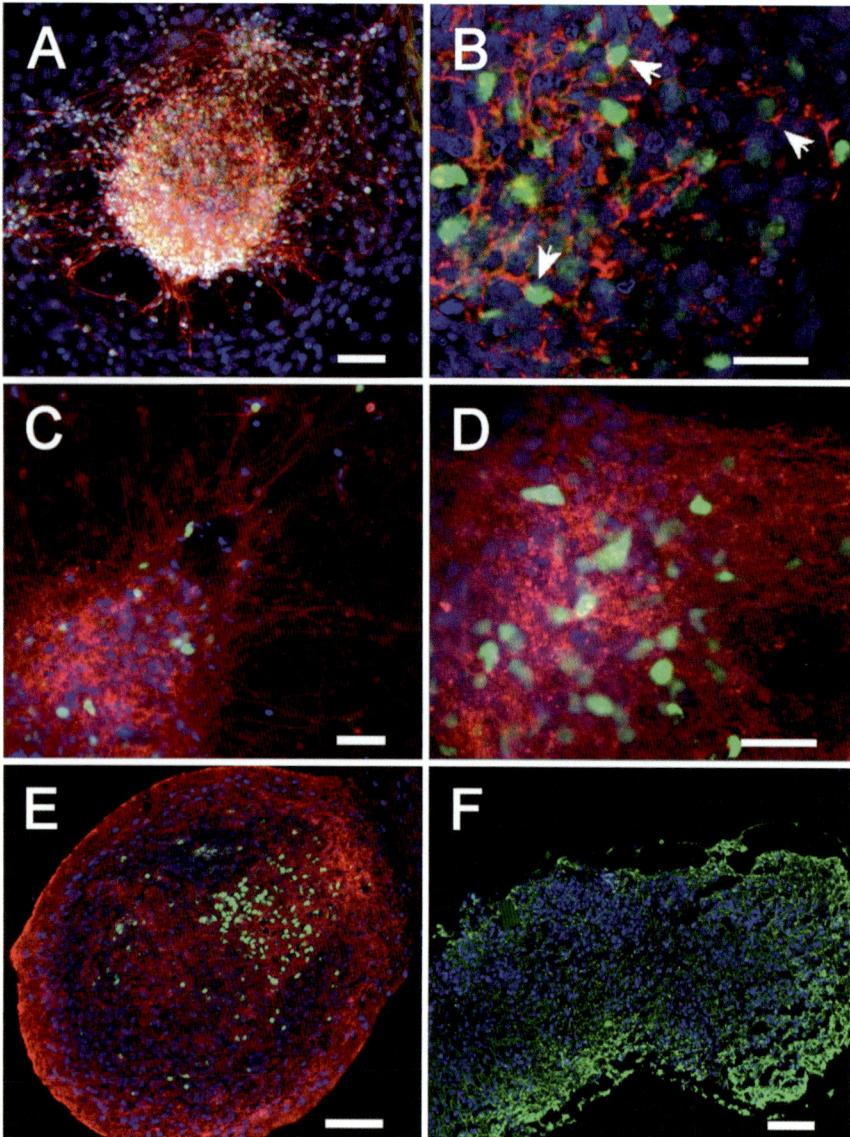

Fig. 15.2. Immunocytochemical identification of putative human peripheral sensory neurons and neural crest generated from hES-derived NSPs. (**A, B**) Expression of Brn3a (*green*) and peripherin (*red*) in colonies derived from PA6-NSP cocultured with PA6 cells. (**A**) Shows a low-magnification view of a whole colony and (**B**) a confocal image at higher magnification. The nuclei (*blue*) surrounding the colonies belong primarily to the PA6 cells. *Arrowheads* in **B** point to double-stained putative PSNs in the colony. (**C, D**) Putative human neural crest cells in 1-week PA6-NSP cocultures. A low magnification image of a colony of NCAM (*red*)-stained cells with nuclei expressing AP2α (*green*), a combination indicative of the neural crest phenotype, is shown in **C**. A confocal image of a similar colony with AP2+/NCAM+ stained cells is show in **D**, the AP2 staining is clearly nuclear. (**E, F**) Immunostainings of sections of PA6-NSPs passaged for 1 month. (**E**) Is a section through a NSP containing putative NC cells double-immunopositive for NCAM (*red*) and AP2 (*green*). In (**F**), a section through a fused pair of NSPs stained for the early neuronal marker βIII tubulin (*green*) reveals the presence of neurons and axons. Nuclei are stained blue with Hoechst nuclear dye. Scale bars=A, E, F=100 μm, B, C, D=50 μm.

3.2.2. Immunostaining of Paraffin-Embedded NSPs

3.2.2.1. Embedding and Sectioning of NSPs

1. Fix NSPs in 4% buffered PAF overnight at 4°C.
2. Prepare a cone of tinfoil and fill with 2% low melting temperature agarose dissolved in dH_2O with 5% sucrose at 60°C. Let cool to approximately 40°C.
3. Transfer fixed NSPs to agarose and let them sink to the bottom of the cone.
4. Gel agar at RT, remove from foil, and refix with 4% PAF for 1 h.
5. Dehydrate agar blocks in an ethanol series 70, 95, and 100% EtOH, 2×10 min each on a bench top shaker and clear in xylene.
6. To soften the agar, place in cedarwood oil (CWO). Blocks initially float, but sink after the oil replaces the xylene. The blocks can be left overnight or indefinitely in CWO.
7. After the CWO treatment, infiltrate in paraffin 0.5 h and 2×1 h and embed.
8. Prepare sections and adhere to Superfrost Plus slides.

3.2.2.2. Staining of Paraffin Sections of NSPs

1. Bake slides in paraffin oven (56°C) for 30 min (*see* **Note 10**).
2. Deparaffinize by $2 \times 15'$ incubations in Toluene, followed by $5'$ each in 100, 95, and 70% ethanol. Wash for 10 min in dH_2O.
3. Incubate in a microwave for antigen retrieval for $15'$ in sodium citrate buffer in a polypropylene container. Cover the slides with 200 ml buffer and start the microwave at full heat. Watch the container until the solution reaches boiling and then lower the power to 30–50%. If the level of liquid falls below the top of the slides, stop the microwave procedure midway and add more buffer. When finished, remove from oven and allow the container to cool to room temperature.
4. Wash for 15 min in PBS.
5. Incubate in blocker for 15 min.
6. Drain blocker (do not rinse) and then add the primary antiserum diluted in blocker (*see* **Note 11**).
   Incubate overnight or 48 h at 4°C in a humid chamber.
7. Warm for 1 h and then remove coverslips by placing the slides slowly into the slide carrier in a glass staining dish containing 200 ml PBS. After about $2'$, slowly raise the slide carrier with the metal handle and the coverslips will fall off by themselves. Rinse another $3'$ and then transfer to second PBS rinse bath for $5'$.
8. Drain slides and add secondary antibody (ies); incubate for 1–2 h in a humid chamber at RT.

9. Remove coverslips as in Step 7 and rinse in PBS 2×5′.
10. Counterstain with Hoechst for 5–10 min.
11. Coverslip in anti-fade medium and seal coverslips with nail polish.
12. Let nail polish dry in a covered box for 10–15 min and view with a fluorescence microscope (*see* **Note 13**).

Figure 15.2E, F shows immunostained sections of paraffin-embedded NSP using the techniques described above.

3.3. Microscopic Analysis

3.3.1. Microscopic Observation and Photography

Coverslips and sectioned NSPs are viewed with an Olympus fluorescence microscope or confocal microscopy and photographed with a digital camera and ImageJ software.

3.3.2. Quantification of Cells

For quantification of double-labeled cells (AP2/NCAM putative NCs and Brn3a/peripherin putative peripheral sensory neurons), nuclei stained for the transcription factors (AP2 or Brn3a) were counted within zones of stained cytoplasm (NCAM or peripherin).

1. PSNs are defined operationally as Brn3a+/peripherin+ cells, and NC-like cells as AP2×/NCAM×.
2. Quantitation can be performed by counting colonies containing double-stained cells per total colonies.
3. Percentages of stained cells can sometimes be obtained in a standard fluorescence microscope by determining the number of double-stained cells/total Hoechst-stained nuclei. However, when the cells are very densely packed, serial optical sections taken with a Zeiss Apotome or confocal microscope are necessary for counting.

Figure 15.3A shows a comparison of the purity of generation of PSNs between three methods used in our laboratory – direct induction from hESCs, induction from NSPs made with noggin, and NSPs made using PA6 as described here.

3.4. Nucleic Acid Analyses

3.4.1. Reverse Transcription

1. Extract and purify total RNA using the Tri-Reagent or Gene Elute Mammalian Total RNA Miniprep Kit according to the manufacturers' instructions.
2. Read RNA concentration in a spectrophotometer.
3. Prepare reverse transcription mixture of oligodT primers, dNTPs, and buffer + 100 ng RNA.

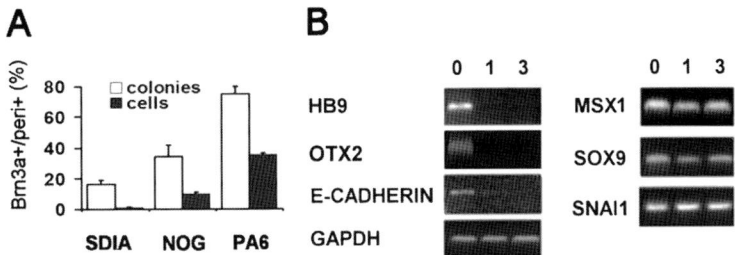

Fig. 15.3. Quantitation of peripheral neurons and gene expression in PA6-hES-NSP cultures. (**A**) Shows the percentage of colonies containing putative PSNs (*open bars*) and the percentage of putative PSNs of the total human cells (*filled bars*) in: hESCs co-cultured with PA6 stromal cells for 4 weeks (SDIA, (1)), noggin-induced neurospheres (NOG) co-cultured with PA6 for 3 weeks (7), and PA6-NSPs co-cultured for 1 week with PA6 stromal cells (PA6, (2). (**B**) RT-PCR analysis of PA6-NSPs in suspension (0) and after 1 and 3 weeks of coculture with PA6 stromal cells. *HB9*, a spinal motoneuron marker and *OTX2*, a marker for anterior CNS, are expressed in the neurospheres, as is *E-CADHERIN*, a marker of epithelial cells (column 0). The expression of these genes is shut down with reculture of PA6-NSP cells on PA6. In contrast, NC markers *MSX1, SOX9, and SNAI1* are expressed in naïve PA6-NSPs as well as after 1 and 3 weeks of coculture with PA6 (columns 1,3).

4. Reverse transcription is carried out for 35' at 42°C, using reverse transcriptase. The generated, cDNA was used for PCR analysis.

3.4.2. PCR Amplification

1. Add primers at a concentration of 500 nM to the cDNA in a volume of 25 µl with blue Taq DNA Polymerase and the appropriate buffer.

2. Carry out PCR reaction under the following conditions: denaturation, 95°C for 30 s; annealing, 47–62°C for 30 s; elongation, 72°C for 1 min. The number of cycles varies between 30 and 40 (*see* **Note 14**).

3. Perform negative controls (for RT-PCR, the absence of DNA contamination) without reverse transcriptase (RT) to confirm the absence of DNA contamination.

4. Separate amplified products on agarose gels and visualize by staining with ethidium bromide.

5. For semiquantitative comparison of expression of mRNAs, the amount of cDNA used in each reaction is normalized to GAPDH mRNA. Bands are analyzed densitometrically using ImageJ (*see* **Note 15**).

Figure 15.3B shows expression of several differentiation marker mRNAs in naïve neurospheres (0), and after 1 and 3 weeks of re-coculture of neurospheres with PA6 (1,3).

4. Notes

1. Sterilizing coverslips: Spread a number of coverslips out in a petri dish in a laminar flow hood. Cover with 100% ethanol and let evaporate. Cover dish. Place coverslips in wells of a 24-well culture plate and rinse once with PBS before use.

2. This is a Teflon-coated wire we had in the lab. The Teflon must be stripped off first by scraping with scissors. Noncoated wire can also be purchased.

3. The NaOH splatters when the current is applied, so eye protection and a lab coat should be worn, and a microscope that is near the end of its life used.

4. Any paraformaldehyde can be used. We buy the crystals in order to reduce exposure to this toxic substance; the powder is easily spread around the lab, even using a hood, while the granules, even if they spill cannot be carried by wind currents. Concentrated solutions of paraformaldehyde can be purchased and diluted with PBS as well.

5. After having one oven rust out, we bought one with an enamel interior.

6. Others have claimed that certain brands interfere with fluorescence; we buy bargain-counter samples and have not had a problem.

7. If cells have been infected with lentiviruses or other biohazardous material, we cut in a biohazard hood using an analogue video camera with close-up lenses, attached to a ring stand while looking at a TV screen located outside the hood.

8. Beyond 8 weeks, generation of neurons is reduced and that of glia increased (3), which could of course be taken advantage of for studying glial cell types.

9. We apply mixtures of antibodies if the primary antibodies are either from two different species (mouse and rabbit) or two different mouse Ig subtypes (IgG and IgM).

10. For routine histology (hematoxylin and eosin, Feulgen, etc.), conventional microscope slides coated with polylysine solution are fine. For immunostaining, and most particularly, for microwave antigen retrieval, Superfrost Plus slides are absolutely required. Otherwise, the sections will fall off the slides during the microwaving procedure.

11. Antibody dilutions vary from batch to batch; we use the manufacturer's suggestion as a lower guide and perform tests on each vial received.

12. We use 75 μl/slide and carefully coverslip with 24 × 50 coverglass to keep the antibody in place.
13. The slides are stored at 4°C in regular plastic slide boxes; Alexa, Cy2, and Texas-Red fluorescence are stable for more than 6 m.
14. The sequences of all the human primers used in this study were compared to the murine database and confirmed that there would not be significant adherence to potential murine templates (except for actin primers that recognize mouse transcripts as well as human- and mouse-specific m8f primers). Control experiments showed that the murine PA6 cells or human fibroblasts grown alone did not express any of the mRNAs for these markers.
15. Detailed instructions for densitometry http://rsb.info.nih.gov/ij/docs/menus/analyze.html#gels

Acknowledgments

Our thanks to Chaya Morgenstern for her devoted technical and logistic help that enabled us to perform these experiments. Supported by the Israel Academy of Sciences grant 158/07, Chief Scientist's Office of the Ministry of Health/Public Committee for Allocation of Estate Funds, Ministry of Justice, Israel, the Dr. Miriam and Sheldon Adelson Medical Research Foundation, and the Taubenblatt Foundation to RSG.

References

1. R.N. Singh, T. Nakano, L. Xuing, J. Kang, M. Nedergaard, and S.A. Goldman, (2005) Enhancer-specified GFP-based FACS purification of human spinal motor neurons from embryonic stem cells, Exp. Neurol. 196: 224–234.
2. N.S. Roy, C. Cleren, S.K. Singh, L. Yang, M.F. Beal, and S.A. Goldman, (2006) Functional engraftment of human ES cell-derived dopaminergic neurons enriched by coculture with telomerase-immortalized midbrain astrocytes, Nat. Med. 12: 1259–1268.
3. O. Pomp, I. Brokhman, I. Ben-Dor, B. Reubinoff, and R.S. Goldstein, (2005) Generation of peripheral sensory and sympathetic neurons and neural crest cells from human embryonic stem cells, Stem Cells 23: 923–930.
4. H. Kawasaki, K. Mizuseki, S. Nishikawa, S. Kaneko, Y. Kuwana, S. Nakanishi, S.-I. Nishikawa, and Y. Sasai, (2000) Induction of midbrain dopaminergic neurons from ES cells by stromal cell-derived activity, Neuron 28: 31–40.
5. K. Mizuseki, T. Sakamoto, K. Watanabe, K. Mugruma, M. Ikeya, A. Nishiyama, A. Arakawa, H. Suemori, N. Nakatsugi, H. Kawasaki, F. Murakami, and Y. Sasai, (2003) Generation of neural crest-derived peripheral neurons and floor plate cells from mouse and primate embryonic stem cells., Proc. Natl. Acad. Sci. USA 100: 5828–5833.
6. I. Brokhman, L. Gamarnik-Ziegler, O. Pomp, M. Aharonowiz, B.E. Reubinoff, and R.S. Goldstein, (2008) Peripheral sensory neurons differentiate from neural

precursors derived from human embryonic stem cells, Different 76.
7. H. Kitajima, S. Yoshimura, J. Kokuzawa, M. Kato, T. Iwama, T. Motohashi, T. Kunisada, and N. Sakai, (2005) Culture method for the induction of neurospheres from mouse embryonic stem cells by coculture with PA6 stromal cells, J. Neurosci. Res. 80: 467–474.
8. N.M. LeDouarin, and C. Kalcheim, (1999) The Neural Crest, Cambridge University Press, Cambridge.
9. G. Lee, H. Kim, Y. Elkabetz, S.G. Al, G. Panagiotakos, T. Barberi, V. Tabar, and L. Studer, (2007) Isolation and directed differentiation of neural crest stem cells derived from human embryonic stem cells, Nat. Biotechnol. 25: 1468–1475.
10. X. Jiang, Y. Gwye, S.J. McKeown, M. Bronner-Fraser, C. Lutzko, and E.R. Lawlor, (2008) Isolation and characterization of neural crest stem cells derived from in vitro differentiated human embryonic stem cells, Stem Cells Dev. (available online doi: 10.1089/Scd. 2008. 0362).
11. J.A. Thomson, J. Itskovitz-Eldor, S.S. Shapiro, M.A. Waknitz, J.J. Swiergiel, et al., (1998) Embryonic stem cell lines derived from human blastocysts, Science 282: 1145–1147.
12. C.A. Cowan, I. Klimanskaya, J. McMahon, J. Atienza, J. Witmyer, J.P. Zucker, S. Wang, C.C. Morton, A.P. McMahon, D. Powers, and D.A. Melton, (2004) Derivation of embryonic stem-cell lines from human blastocysts, N. Engl. J. Med. 350: 1353–1356.
13. G.W. Conrad, J.A. Bee, S.M. Roche, and M.A. Teillet, (1993) Fabrication of microscalpels by electrolysis of tungsten wire in a meniscus, J. Neurosci. Methods 50: 123–127.
14. O. Pomp, I. Brokhman, L. Ziegler, M. Almog, A. Korngreen, M. Tavian, and R.S. Goldstein, (2008) PA6-induced human embryonic stem cell-derived neurospheres: a new source of human peripheral sensory neurons and neural crest cells, Br. Res. 1230: 50–60.

Chapter 16

Embryonic Stem Cells as a Model for Studying Melanocyte Development

Susan E. Zabierowski and Meenhard Herlyn

Abstract

Melanocytes are neural crest-derived pigment-producing cells that reside in the inner ear, in the uveal tract, in hair follicles, and in the skin. The main function of melanocytes is to provide pigmentation through melanin production and secretion to the immediate surrounding area. Although much is known about mature melanocyte function and regulation, particularly in the skin, little is known with regard to the signals and gene expression patterns that ensue upon melanocyte development and differentiation from embryonic precursors. The ability to examine these patterns in an in vitro specified setting through the use of embryonic stem cells holds great potential for understanding melanocyte biology. In this chapter, we outline our procedures for the differentiation of human embryonic stem cells toward mature pigment-producing melanocytes that express the appropriate melanocytic markers and home to the epidermal basal layer in 3D skin reconstructs.

Key words: Pigment, melanocytes, neural crest, keratinocytes, migratory, Wnt3A, SCF, ET3, skin reconstructs.

1. Introduction

In vertebrate development, melanocytes originate from migratory neural crest cells that emerge from the neural plate during embryogenesis. These neural crest cells give rise to committed unpigmented melanoblasts that home to specified destinations where they ultimately give rise to mature pigment-producing melanocytes (reviewed in 1, 2). Although melanocytes reside in the eye, inner ear, hair matrix, mucous membranes, and central nervous system, the most well-studied human melanocytes are those in the skin. Skin-dwelling melanocytes reside on the basement membrane of the epidermal–dermal junction where they maintain close contact with surrounding keratinocytes (3). The primary function of these epidermal melanocytes is to produce and transport the pigment melanin to surrounding keratinocytes providing the skin pigmentation and protection from harmful ultraviolet rays (2, 4).

Defects in melanocytes can lead to numerous pigmentary disorders including, oculocutaneous, albinism, piebaldism, vitiligo, and Waardenburg syndrome (5). Quail-chick chimeric studies and mice bearing mutations specifically affecting neural crest derivates have been instrumental in the identification of genes involved in the development, migration, survival, and differentiation of melanocytes from neural crest cells. Many of these studies have indicated that Wnt (6, 7), endothelin 3 (8), and stem cell factor (SCF) (9, 10) are all critically involved in the differentiation of neural crest precursors toward pigmented cells. The essential role of Wnt signaling in melanocyte development has been demonstrated by studies on Wnt1 and Wnt3a-null mice in which almost no Dct-positive melanoblasts were observed (11). Further studies in zebrafish indicate that Wnt signaling, through activated beta-catenin, promotes a fate decision between either melanocytes or glia cells (12). The role of endothelins was recognized when knockout mice for either the endothelin-B receptor (EDNRB) or its ligand, endothelin-3, lacked melanocytes (13, 14). EDNRB is involved in the initial dispersal and migration of neural crest cells destined to become melanocytes, as its absence during this period leads to an almost complete loss of melanocytes in mutant offspring (15, 16). In quail primary neural crest cultures, addition of endothelin-3 strongly facilitates neural crest cell proliferation and differentiation into glial cells and melanocytes (10). Disruptions in the END3 system are associated with Waardenberg-Shah syndrome and Hirsch-sprung's disease type 2 in humans (17, 18). The mouse strains *white dominant spotting* (W) and *steel* (Sl), harboring mutations in the Kit tyrosine kinase receptor (c-Kit) and its ligand SCF, respectively, have revealed that this receptor and its ligand play a complex role in the survival, migration, and differentiation of embryonic melanoblasts during their homing to the skin, which appears to differ among species (19–21). Mouse mutants for the c-kit receptor or SCF exhibit various degrees of pigmentation defects and in humans c-kit mutations are associated with piebaldism (22).

Despite the vast number of mouse and chick/quail studies undertaken to monitor melanocyte development and survival, major architectural differences exist between human and mouse skin. Whereas mouse melanocytes are localized to the hair follicles deep in the dermis, human melanocytes reside at the basement membrane of the epidermal–dermal junction surrounded by keratinocytes. These environmental differences make it difficult to extend mouse or chick/quail melanocyte development to human melanocyte development. Thus, through the use of hESCs and their differentiation to melanocytes, the potential for learning about the factors required for proper human melanocyte development from embryonic precursors is vast.

We have recently defined a novel feeder-free approach for the differentiation of human melanocytes from human embryonic stem cells that can reproducibly generate a greater than 85% homogenous population of mature pigmented human melanocytes (23). We have identified that the combination of Wnt3a, SCF, and EDN3 are indispensable for proper melanocyte differentiation from human embryonic cells. These melanocytes exhibit the correct morphology, synthesize melanosomes, and express all the proper melanocyte markers.

2. Materials

2.1. Medium Components and Composition (see Note 1)

1. *MEF derivation medium (1000 ml):* 870 ml DMEM (Invitrogen #11965-092), 100 ml defined FBS (Invitrogen #16000-044; heat inactivate for 30 min at 57°C), 10 ml L-glutamine (Invitrogen #21051-024 – make 200 mM stock in water and store at –20°C), 10 ml nonessential amino acids 100X (Invitrogen #11140), 10 ml penicillin-streptomycin 100×.

2. *MEF medium (1000 ml):* 880 ml DMEM (Invitrogen # 11965-092), 100 ml defined FBS (Invitrogen # 16000-044; heat inactive for 30 min at 57°C), 10 ml L-glutamine (Invitrogen #21051-024 – make 200 mM stock in water and store at –20°C), 10 ml nonessential amino acids 100X (Invitrogen #11140).

3. *Cryopreservative medium for MEFs (100 ml):* 60 ml DMEM (Invitrogen # 11965-092), 20 ml defined FBS (Invitrogen # 16000-044; heat inactive for 30 min at 57°C), 20 ml DMSO (Fisher #F128-500).

4. *0.1% gelatin (500 ml):* 500 mg gelatin powder (Sigma #G1890), 500 ml endotoxin-free, reagent grade water. Autoclave gelatin solution for 45 min on a liquid cycle and store at room temperature (RT).

5. *Human embryonic stem cell medium (1000 ml):* 800 ml DMEM/F-12 (Invitrogen #11330-032), 200 ml Knockout-Serum Replacer (Invitrogen #10828-028), 10 ml 200 mM L-glutamine (Invitrogen #21051-024) + 7 µl β-mercaptoethanol/10 ml L-glutamine, 10 ml nonessential amino acids 100X (Invitrogen #11140), 1 ml basic

fibroblast growth factor (Fitzgerald Industries #RDI-118B-218 – make 4 μg/ml stock in 0.1% BSA in 1X DPBS and store at –70°C).

6. *Embryoid body (EB) medium (100 ml)*: 80 ml DMEM/F-12 (Invitrogen #11330-032), 20 ml Knockout-Serum Replacer (Invitrogen #10828-028), 1 ml 200 mM L-glutamine (Invitrogen #21051-024), + 0.7 μl β-mercaptoethanol/1 ml L-glutamine, 1 ml nonessential amino acids 100X (Invitrogen #11140).

7. *Mel-1 medium – for melanocyte differentiation (100 ml):* 20 μl Dexamethasone (Sigma #D-2915 – make up to 0.25 M in water and store at –20°C), 1 ml ITS Liquid Medium Supplement (Sigma #I-3146), 1 ml Linoleic Acid-BSA (Sigma #L-9530), 30 ml DMEM-Low Glucose (Invitrogen #11885), 20 ml MCDB201, 1 ml L-ascorbic acid (Sigma #A-4403), 50 ml Wnt3a conditioned medium (*see* **Section 3.7** and **Note 2**), 1 ml Stem Cell Factor (Fitzgerald Industries #RDI-118B-218 – make up to 10 μg/ml in 0.1% BSA in 1X DPBS and store stock at –70°C), 100 μl Basic Fibroblast Growth Factor (Fitzgerald Industries #RDI-118B-218 – make up to 4 μg/ml in 0.1% BSA in 1X DPBS and store stock at –70°C), 100 μl Endothelin-3 (American Peptide Co. #88-5-10 – make up 264 μg/ml in 0.1% BSA in 1X DPBS and store stock at –70°C), 150 μl Cholera toxin (Sigma #C-3012 – make up to 3.32 μg/ml in 0.1% BSA in 1X DPBS and store stock at 4°C), 12.5 μl TPA (Sigma #P-1583 – make up to 250 μg/ml in 0.1% BSA in 1X DPBS and store stock at –20°C).

8. *L-Wnt3a cell medium – 10% FBS (1000 ml)*: 900 ml DMEM (Cellgro #10-017-CM), 100 ml FBS, 8 μl of 50 mg/ml G418 (Sigma #G-8168).

9. *L-cell medium – 10% FBS (1000 ml)*: 900 ml DMEM (Cellgro #10-017-CM), 100 ml FBS.

10. *L-Wnt3A conditioning medium – 1% FBS (1000 ml)*: 900 ml DMEM (Cellgro #10-017-CM), 10 ml FBS.

11. 1X PBS Ca^{2+}/Mg^{2+} free (Cellgro #21-031-CM).

12. Hank's buffered saline solution (HBSS) without Ca^{2+}/Mg^{2+}.

13. 0.25% trypsin/EDTA.

14. 10 ng/ml Fibronectin.

2.2. Cells

1. CF-1 female mice (day 13–14 gestation) for MEF derivation (Jackson Laboratory)
2. L-cells (ATCC #CRL-2648)
3. L-Wnt-3A cells (ATCC #CRL-2647)
4. H9ES (WiCell Research Institute)

2.3. Additional Materials

1. 6-well plates
2. Cell lifters
3. Glass pickers
4. 70% ethanol
5. Forceps
6. Scissors
7. Iris scissors
8. 50-ml sterile conical tubes
9. 15-ml sterile conical tubes
10. 5-ml sterile glass pipets
11. T25 flasks
12. T75 flasks

3. Methods

3.1. Derivation of Mouse Embryonic Fibroblast

1. Sacrifice a female mouse (WiCell: CF-1) at day 13 or 14 of pregnancy.
2. Soak the abdomen with 70% ethanol.
3. Using a forceps pull up the skin separating the hide from the peritoneum and cut a nick in the skin with scissor.
4. Using a new set of forceps and scissors, cut the peritoneum to expose the abdominal cavity.
5. Grab hold of the uterine horns with a blunt-point forceps and using a scissors cut them from the abdominal cavity. Place the uterine horns in a petri dish that contains 10 ml of Ca^{2+}/Mg^{2+}-free PBS.
6. Wash the uterine horns 3 times with 10 ml of Ca^{2+}/Mg^{2+}-free PBS.
7. Using 2 fine-pointed forceps, tease open the uterine walls or cut using a scissor to release the embryos into the petri dish.
8. Separate the embryos from the placenta and fetal membranes.
9. Transfer the embryos to a new petri dish and wash them 3 times using 10 ml of Ca^{2+}/Mg^{2+}-free PBS.
10. Using a fine-tipped forceps, individually, dissect out and discard the viscera, liver, and heart, which appear as red spots, from each embryo.
11. Wash the embryos 3 times with 10 ml of Ca^{2+}/Mg^{2+}-free PBS.
12. Remove the Ca^{2+}/Mg^{2+}-free PBS and add 2 ml 0.25% Trypsin/EDTA solution to the washed embryos.

13. Using a curved Iris scissors, finely mince each embryo. Add 5 ml Trypsin/EDTA solution.
14. Incubate at 37°C for 20–30 min on shaker until individual cells are visible (using an inverted cell-culture microscope).
15. Add 20 ml of MEF derivation medium to the plate after the incubation. Transfer the individualized cells to a 50-ml conical tube.
16. Rinse remaining tissue in the plate with a 5 ml MEF derivation medium. Transfer to the 50-ml conical tube. Mix by pipetting a few times. Allow the debris in the suspension to settle to the bottom of the tube for 1 min.
17. Remove the top 12 ml of the suspension containing the individualized cells and spin down to remove all the trypsin, then resuspend the pellet and divide into T75 flask (3 embryos/flask).
18. Mix the remaining 8 ml with the debris, spin, resuspend, and add into one flask.
19. Add additional MEF derivation medium to each flask; bring the final volume to 20 ml/flask.
20. Incubate flasks in a 37°C, humidified incubator at 5% CO_2 for 2–3 days, until 80–90% confluent. At this time, the MEFs are ready to be harvested and frozen (*see* **Note 3**).

3.2. Freezing MEF

1. Aspirate medium from each flask and wash with 5 ml/T75 of Ca^{2+}/Mg^{2+}-free PBS.
2. Add 2 ml/T75 of 0.25% Trypsin/EDTA for 4–8 min at room temperature.
3. Detach the cells by tapping.
4. Add 4 ml/T75 of MEF derivation medium.
5. Pool into 50-ml centrifuge tubes.
6. Mix the pooled cells by pipetting a few times.
7. Allow the mixed cells to sit for about 2 min or until the large chunks settle on the bottom.
8. Pipette the cells suspended in the MEF derivation medium into 2 new 50-ml centrifuge tubes.
9. Centrifuge the tubes at $200 \times g$ for 5 min (1000 RPM, 5 min).
10. Remove supernatant.
11. Resuspend cells in a volume equal to one half of the total volume to freeze (freeze 2 vials/75 flask = $5-10 \times 10^6$ cells/vial).
12. Put the cells on ice.
13. Add an equal volume of 4°C MEF cryopreservation medium to an equal volume of MEF cells at 4°C and mix.

14. Label vials and aliquot 1 ml cells/vial.
15. Place the vials in –70°C freezer overnight (O/N).
16. The vials may be transferred into liquid phase of LN_2.

3.3. Thawing Mouse Embryonic Fibroblasts (MEF)

1. Coat one T75 flask with 7 ml 0.1% gelatin for at least 1 h at 37°C prior to thawing MEFs.
2. Remove a frozen stock vial of MEF from liquid nitrogen.
3. Immerse vial in a 37°C water bath. Swirl gently. When only a small sliver of an ice crystal remains, remove vial from water bath.
4. Immerse vial in 70% ethanol. While vial air-dries in hood, generously rub gloved hands with 70% ethanol.
5. Pipette 10 ml MEF medium into a 15-ml conical tube. Add cells drop-wise to medium to avoid osmotic shock. Gently pipette up and down to mix.
6. Centrifuge at $200 \times g$ for 5 min and aspirate supernatant.
7. Resuspend cell pellet in 10 ml MEF medium and place in gelatin-coated T75.
8. Incubate O/N at 37°C.
9. The next day aspirate medium containing any unattached dead cells and replace with 10 ml fresh MEF medium.
10. Monitor cells daily. Split when the cells have reached 95% confluence.

3.4. Splitting MEFs

MEF of ~95% confluence in T75 are usually split at 1:4.

1. Coat plates with 0.1% gelatin, 1 ml/well (6-well plates) or 7 ml/T75 for at least 1 h at 37°C prior to splitting MEFs.
2. Aspirate medium from MEFs and wash each T75 flask with 5 ml HBSS without Ca^{2+}/Mg^{2+}.
3. Add 4 ml 0.25% trypsin/EDTA solution per T75 and incubate at RT for ~5 min. Detach cells by tapping.
4. Add 8 ml/T75 MEF medium to neutralize.
5. Transfer to 50-ml conical tubes and spin at RT for 5 min at 2000 rpm.
6. Meanwhile, aspirate gelatin from coated flasks, add 9.5 ml MEF medium per new coated T75 flask.
7. Aspirate medium from pellet and resuspend in 2 ml of MEF medium for a 1:4 split.
8. Add 0.5 ml to each flask and incubate at 37°C.
9. Change medium every other day until MEFs are confluent and ready to split or irradiate (*see* **Notes 4** and **5**).

Attention: Keep one back up un-split flask. Just change the medium.

3.5. Irradiation of MEFs for hES Cell Seeding (Usually Irradiate on Wednesdays).

1. Coat 6-well plates with 1 ml/well of 0.1% gelatin and incubate at 37°C for at least 1 h prior to seeding irradiated MEFs.
2. Aspirate medium from 95% confluent T75 of MEFs and wash with 5 ml HBSS without Ca^{2+}/Mg^{2+}.
3. Add 4 ml 0.25% trypsin/EDTA solution per T75 and incubate at RT for approximately 5 min. Detach cells by tapping.
4. Add 8 ml/T75 MEF medium to neutralize.
5. Transfer to 50-ml conical tube and pellet cells for 5 min at 1300 rpm at RT.
6. Resuspend cells in 5 ml MEF medium and count cells.
7. Irradiate cells in 15-ml conical tube for 17 min at 1 Gy/min (*see* **Notes 6** and **7**).
8. Following irradiation, pellet cells for 5 min at 1300 rpm at RT.
9. Resuspend cells at 3.25×10^4 cells/ml MEF medium.
10. Aspirate gelatin from 6-well plates.
11. Seed 2 ml of 3.25×10^4 cell/well to give 7.5×10^4 cell/well.
12. Incubate overnight at 37°C (*see* **Note 8**).

3.6. Passaging hES Cells onto Irradiated MEFs (Usually on Thursdays)

3.6.1. To Irradiated MEF Cells on 6-Well Plates

1. Warm hES medium in 37°C water bath.
2. Aspirate MEF medium from irradiated MEFs.
3. Add 1 ml/well 37°C hES medium and incubate 2–3 min at 37°C to dissipate MEF medium.
4. Aspirate hES medium from irradiated MEFs and add 1 ml/well fresh hES medium.
5. Incubate at 37°C until ES cells are ready to be seeded.

3.6.2. ES Colony Picking and Plating

1. Using a microscope find undifferentiated hES colonies and from the bottom side of the plate, circle them with a Sharpie, foregoing the dark multilayer differentiated colonies (**Fig. 16.1**).
2. Once colonies have been chosen and circled, slowly remove old ES medium from each well using a pipette.
3. Slowly add 1 ml/well fresh 37°C hES medium.
4. Using an autoclaved pointed glass pipette scraper, scrape circled ES colonies from multiple wells into the hES medium.
5. Slowly seed approximately 10 scraped colonies (dropwise) onto one well of irradiated MEFs using a *glass* 5-ml pipette (*see* **Note 9**).
6. Obtain any residual scrapings by adding 1 ml/well fresh hES medium to just-scraped wells. Collect and evenly redistribute onto MEFs.

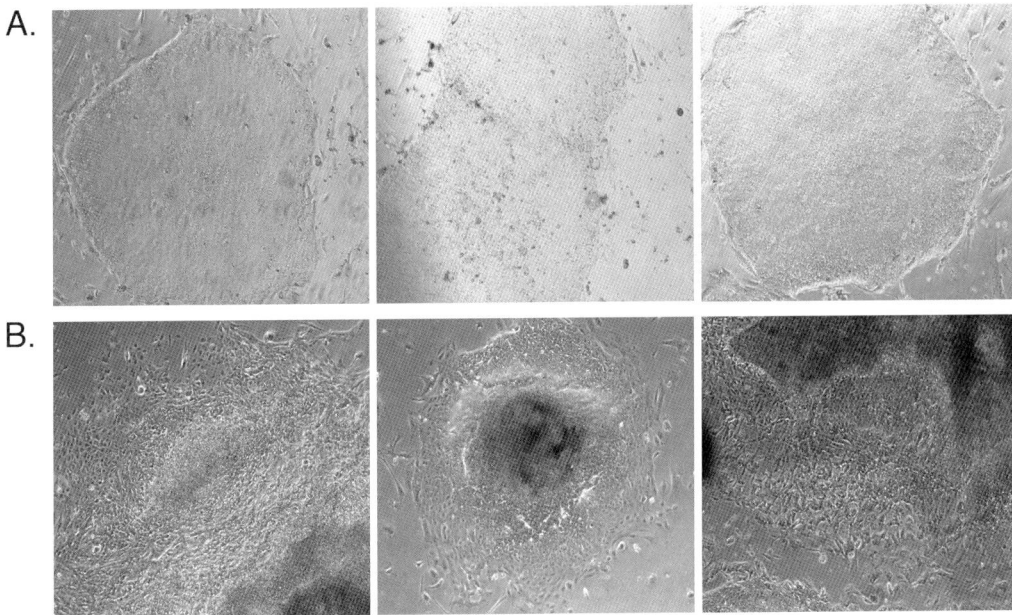

Fig. 16.1. Colonies of undifferentiated and differentiated cells of the human embryonic stem (hES) cell line H9. H9 hES cells were seeded onto irradiated mouse embryonic fibroblasts (MEFs) in 6-well plates and grown for 7 days under standard hES culture conditions. (**A**) Well-defined undifferentiated colonies. (**B**) Spontaneously differentiated colonies.

7. Add supplemental hES medium to newly seeded wells to get approximately 3 ml hES medium per well.

8. Add 2 ml/well hES medium to old wells that have just been scraped (*see* **Note 10**).

9. Incubate all plates at 37°C.

10. The next day, slowly add another 2 ml hES medium at 37°C to freshly passaged ES cells as well as the old ES cells.

11. The following day, slowly remove 3 ml of medium and replace with 2 ml fresh 37°C hES medium.

12. Change 2/3 hES medium (approximately 2 ml) everyday thereafter until next passage (the following Thursday when ES cells will be ready to split again).

3.7. Generating Embryoid Bodies (EBs) from ES Cells

1. Slowly pipette off all medium from 6-well plates of 7-day old ES cells.

2. Add 1 ml EB medium (does not need to be warmed to 37°C) per well of 6-well plates (*see* **Note 11**).

3. With plastic cell lifter, gently scrape the entire well of multiple wells, trying not to break up the ES colonies.

4. Using 5-ml glass pipette, collect scraped cells and place into a T25 flask. Use one T25 per one 6-well plate.

5. Add medium to 7.5 ml. Incubate overnight at 37°C.

6. The following day, change the medium and flask. To do this, transfer all floating material from a T25 to a 15-ml conical flask and discard the flask. Spin for 5 min at 600 rpm. Aspirate old medium. Gently resuspend in 7.5 ml fresh EB medium taking care not to break up cell clusters and place in new T25 flask (*see* **Note 12**).

7. Change 2/3 EB medium the following day. When changing medium, sit the flask upright and let larger clusters of cells settle to the bottom (approximately 3 min) and carefully pipette off 2/3 (approximately 5 ml) medium taking care not to disturb the settled clusters. Replace with 5 ml fresh EB medium.

8. Continue incubating at 37°C for another 3 days. By day 4, the cell clusters should have formed free-floating spheres of varying sizes that are ready to be differentiated (**Fig. 16.2A**).

3.8. Wnt3a-Conditioned Medium for Mel-1

1. The L-Wnt3a cell line (24) is normally cultured in L-Wnt3a cell medium (DMEM containing *10% FBS and 0.4 mg/ml G418*).

2. When cells reach confluence, split 1:10 into 10-cm² petri dishes with L-Wnt3a conditioning medium (1% FBS without G418). Each petri dish should contain 10 ml of L-Wnt3a conditioning medium.

3. Allow cells to grow for 4 days at 37°C (approximately until confluence).

4. Remove the medium and filter sterilize. This is Batch 1.

5. Add 10 ml fresh L-Wnt3a conditioning medium (with 1% FBS without G418) and culture for another 3 days.

6. Remove the medium and sterile filter. This is Batch 2.

7. Discard the cells.

8. Mix Batch 1 and 2 at a 1:1 ratio. This is the Wnt3a-conditioned medium. It is stable at 4°C for approximately 1 month. For long-term storage, the medium can be stored at −70°C.

3.9. L-Conditioned Medium (Control for Wnt3a CM)

1. The parental control cell line, L cells, are cultured in L-cell medium (DMEM containing 10% FBS without G418).

2. When cells reach confluence, split 1:10 into 10-cm² petri dishes, each containing 10 ml L-Wnt3a conditioning medium (1% FBS without G418).

3. Allow cells to grow for 4 days at 37°C (approximately until confluence).

4. Remove the medium and sterile filter. This is Batch 1.

5. Add 10 ml fresh L-Wnt3a conditioning medium (with 1% FBS without G418) and culture for another 3 days.

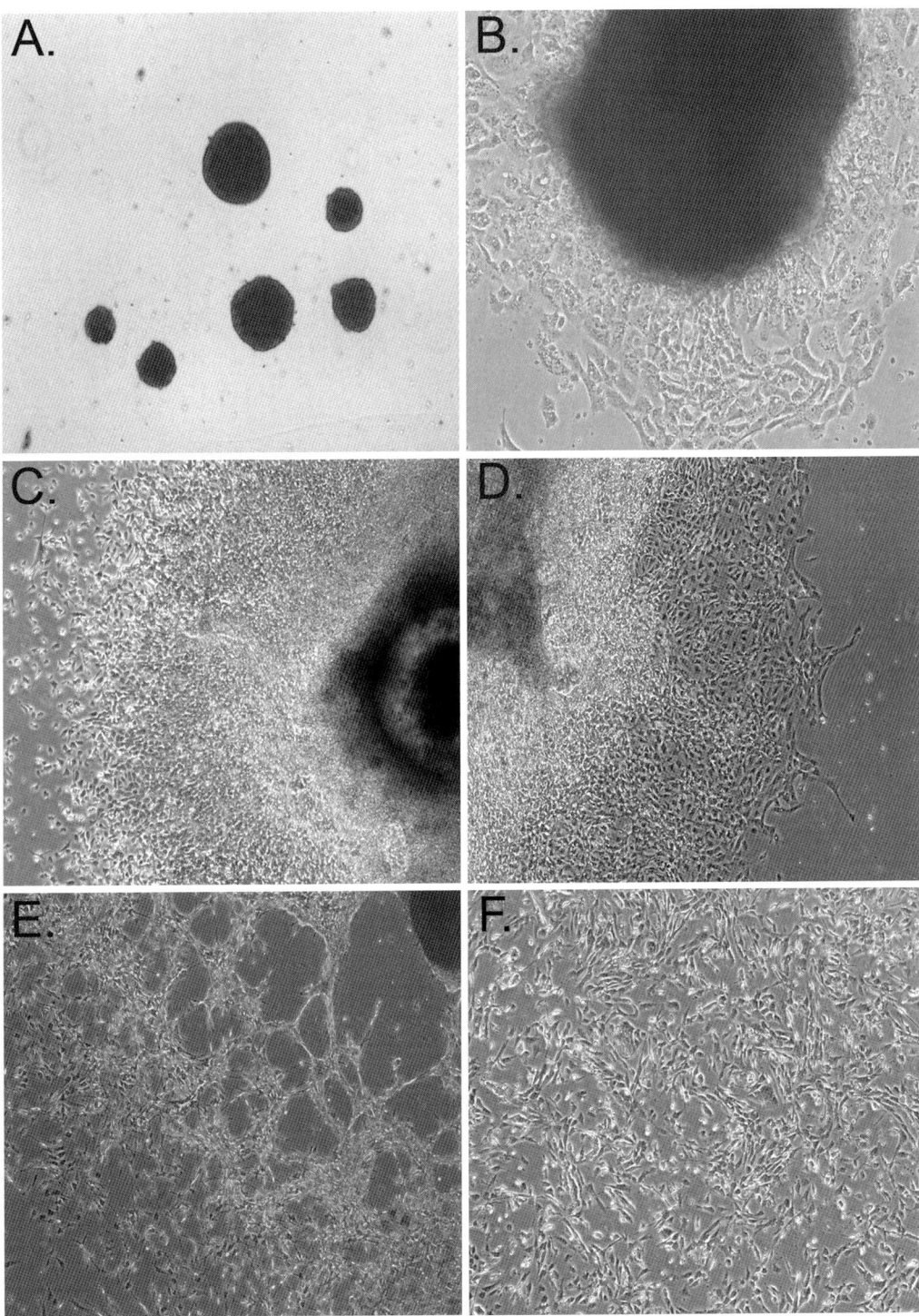

Fig. 16.2. Morphological changes of hES-induced melanocyte differentiation over a 50-day time course. (**A**) Day 0 in Mel-1 medium. (**B**) Day 2. (**C**) Day 10. (**D**) Day 20. (**E**) Day 30. (**F**) Day 50.

6. Remove the medium and sterile filter. This is Batch 2.
7. Discard the cells.
8. Mix Batch 1 and 2 at a 1:1 ratio. This is the L-conditioned medium. It is stable at 4°C for approximately 1 month. For long-term storage the medium can be stored at −70°C.

3.10. Confirmation of Wnt3a-Conditioned Medium

The culture supernatant containing the active form of mouse Wnt3a protein can be verified by the induction of β-catenin expression through Western blotting (25).

1. Seed normal L cells in 6-well culture dish at 1×10^5 cells per well in L-cell medium.
2. Incubate at 37°C overnight.
3. The following day, aspirate medium.
4. Prepare the following per well:

Well #	L-cell medium (ml)	L-Wnt3a-conditioned medium (mix of Batch 1 and 2) (ml)
1	2	0
2	1.5	0.5
3	1	1
4	0.75	1.25
5	0.5	1.5
6	0.25	1.75
7	0	2

5. Incubate for at least 12 h at 37°C.
6. Lyse the cells in 50 μl of RIPA lysis buffer.
7. Use 10 μl for immunoblotting to detect total β-catenin protein.

3.11. Melanocyte Differentiation

1. Coat 1 ml/well of 6-well plate or 3 ml/T25 with 10 ng/ml fibronectin O/N at 37°C.
2. Collect 4-day old EBs into 50-ml conical flask and let EBs settle to the bottom.
3. Gently remove as much old EB medium without disturbing settled EBs.
4. Add 7 ml Mel-1 to EBs. Gently pipette up and down twice and let EBs settle to the bottom. While EBs are settling, aspirate fibronectin from 6-well plates or T25 flasks.

5. Gently remove as much medium from EBs as possible without disturbing.
6. Wash EBs again with 7 ml Mel-1 medium.
7. Add 7 ml Mel-1 per T25 flask of EBs (i.e., If seeding EBs from three T25s, then add 21 ml Mel-1).
8. Gently evenly disperse EBs by pipette mixing. Immediately aliquot 7 ml of EBs per fibronectin-coated T25.
9. Incubate in a 37°C tissue culture incubator for 3 weeks, changing 2/3 of the medium every other day (i.e., from each T25 remove 5 ml of old medium and replace with 5 ml fresh Mel-1 every Monday, Wednesday, and Friday). *See* **Fig. 16.2** and **Note 13**.
10. Following 3 weeks of continuous culture (with cells reaching approximately 60% confluence), the cultures can be passaged at a ratio of 1:3. Cultures are dissociated into single-cell suspensions using 0.25% trypsin/EDTA and are replated onto fibronectin-coated flasks in Mel-1 medium *without TPA*.
11. Homogenous cultures of cells with a melanocyte-like dendritic morphology are established following an additional 4–6 weeks of maintenance in Mel-1 *without TPA* (*see* **Notes 14–17**).

4. Notes

1. We generally do not use antibiotics in our medium with the exception of MEF derivation medium. All medium are sterilized by using 0.22-μm CA (cellulose acetate membrane) filter. Medium can be stored for up to 1 month.
2. Wnt3a is an indispensable component of Mel-1 medium. If Wnt3a-CM is replaced by L-CM during differentiation, the cells that grow out exhibit a nonmelanocytic morphology, are unpigmented, and are negative for MITF (23). If the percentage of Wnt3a-CM in Mel-1 medium is reduced from 50 to 20%, melanocytic differentiation still occurs, but is delayed and the percentage of mature differentiated melanocytes decreases. Although we generally use Wnt3a-CM, soluble Wnt3a protein can be purified from Wnt3A cell-conditioned medium following the procedures as previously described (25) and can be used as an alternative to the conditioned medium.

3. We do see that there appear to be small colonies in our cultures, but we find that they are just small clumps of MEFs that did not get totally disaggregated with trypsin during derivation. We notice that they disappear over a few passages because they are being trypsinized and broken up again each time they are split. If we notice this happening, we will sometimes leave the mouse embryos in trypsin (after mincing) a little longer during the next derivation.

4. We sometimes notice that after a few passages a new cell type emerges that is distinct from the typical fibroblast morphology that begins to outgrow the MEFs (**Fig. 16.3**). If there are more than 20% of these cells in the MEF cultures, we discard the cultures and thaw a new vial of MEFs.

5. We use MEFs for up to 6 passages before discarding and thawing a new vial of MEFs. Sometimes the cells senesce earlier than passage 6. If this occurs we discard the cells and thaw a new vial.

6. Every machine is unique and requires irradiation curve (kill curve) data to determine the correct dosage. The rads of exposure needed to inactivate MEFs can also vary from lot to lot and must be tested.

7. Although mitomycin C inactivation of MEFs has been successfully used in other laboratories, we prefer to use radiation to reduce the possibility of altering the hES cells.

8. ES cells should be seeded onto irradiated MEFs the following day. Irradiated MEFs can be kept for 5 days.

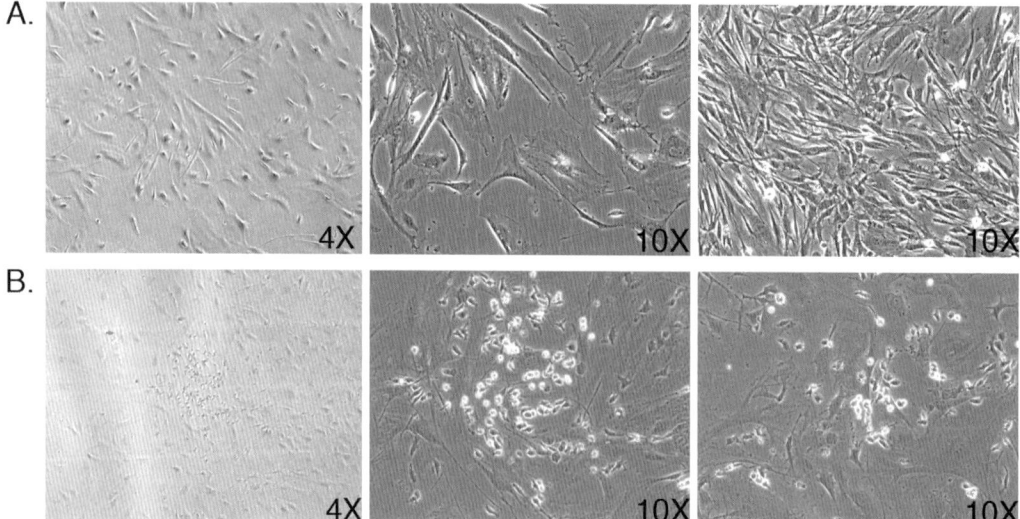

Fig. 16.3. CF-1 derived mouse embryonic fibroblasts. (**A**) Typical morphology of good quality MEF cultures. (**B**) A contaminating cell type clustered amongst MEFs.

9. Scraped ES colonies tend to stick to the sides of plastic pipettes. We have found that glass pipettes are a better alternative when passaging scraped ES colonies than plastic pipettes. Less material is lost during passaging.

10. We keep all old plates, scraped and unscraped, until we have determined that the newly passaged hES cells have attached and are growing.

11. EB medium is the same medium as hES medium without bFGF.

12. Any contaminating feeder layer cells will adhere to the flask and are discarded following serial passage of the cells into new flask.

13. After 24–48 h, some of the EBs form adhesive colonies, out of which bipolar cells migrate (**Fig. 16.2**). The differentiating hESCs continue to proliferate and reach 60% confluence after 3 weeks of continuous culture (**Fig. 16.2**).

14. Whereas EBs are unpigmented, hESC-derived melanocytes are highly pigmented by 9 weeks and retain their pigmented phenotype in long-term cultures over a 6-month period. These cells eventually undergo senescence at passage 40.

15. To confirm melanocytic phenotype, expression of melanocytic markers, including MITF, c-KIT, DCT, TYR, TYRP1, SILV, and S100 can be assessed using Western blotting or immunofluorescence.

16. To determine whether hESC-derived melanocytes are of neural crest origin or retinal-pigmented cell phenotype, RT-PCR using primers specific for the neural crest-derived MITF isoform can be used.

17. To determine whether hESC-derived melanocytes can home to the basal layer of the epidermis, hESC-derived melanocytes can be placed in 3D organotypic cultures that mimic human skin.

References

1. Dupin, E. and Le Douarin, N.M. (2003) Development of melanocyte precursors from the vertebrate neural crest. *Oncogene*, **22**, 3016–3023.

2. Yamaguchi, Y., Brenner, M. and Hearing, V.J. (2007) The regulation of skin pigmentation. *J Biol Chem*, **282**, 27557–27561.

3. Haass, N.K. and Herlyn, M. (2005) Normal human melanocyte homeostasis as a paradigm for understanding melanoma. *J Investig Dermatol Symp Proc*, **10**, 153–163.

4. Costin, G.E. and Hearing, V.J. (2007) Human skin pigmentation: melanocytes modulate skin color in response to stress. *Faseb J*, **21**, 976–994.

5. Tomita, Y. and Suzuki, T. (2004) Genetics of pigmentary disorders. *Am J Med Genet C Semin Med Genet*, **131C**, 75–81.

6. Dunn, K.J., Brady, M., Ochsenbauer-Jambor, C., Snyder, S., Incao, A. and Pavan, W.J. (2005) WNT1 and WNT3a promote expansion of melanocytes through distinct modes of action. *Pigment Cell Res*, **18**, 167–180.

7. Jin, E.J., Erickson, C.A., Takada, S. and Burrus, L.W. (2001) Wnt and BMP signaling govern lineage segregation of

melanocytes in the avian embryo. *Dev Biol*, **233**, 22–37.

8. Reid, K., Turnley, A.M., Maxwell, G.D., Kurihara, Y., Kurihara, H., Bartlett, P.F. and Murphy, M. (1996) Multiple roles for endothelin in melanocyte development: regulation of progenitor number and stimulation of differentiation. *Development*, **122**, 3911–3919.

9. Yoshida, H., Kunisada, T., Grimm, T., Nishimura, E.K., Nishioka, E. and Nishikawa, S.I. (2001) Review: melanocyte migration and survival controlled by SCF/c-kit expression. *J Investig Dermatol Symp Proc*, **6**, 1–5.

10. Lahav, R., Ziller, C., Dupin, E. and Le Douarin, N.M. (1996) Endothelin 3 promotes neural crest cell proliferation and mediates a vast increase in melanocyte number in culture. *Proc Natl Acad Sci USA*, **93**, 3892–3897.

11. Ikeya, M., Lee, S.M., Johnson, J.E., McMahon, A.P. and Takada, S. (1997) Wnt signalling required for expansion of neural crest and CNS progenitors. *Nature*, **389**, 966–970.

12. Dorsky, R.I., Moon, R.T. and Raible, D.W. (1998) Control of neural crest cell fate by the Wnt signalling pathway. *Nature*, **396**, 370–373.

13. Baynash, A.G., Hosoda, K., Giaid, A., Richardson, J.A., Emoto, N., Hammer, R.E. and Yanagisawa, M. (1994) Interaction of endothelin-3 with endothelin-B receptor is essential for development of epidermal melanocytes and enteric neurons. *Cell*, **79**, 1277–1285.

14. Hosoda, K., Hammer, R.E., Richardson, J.A., Baynash, A.G., Cheung, J.C., Giaid, A. and Yanagisawa, M. (1994) Targeted and natural (piebald-lethal) mutations of endothelin-B receptor gene produce megacolon associated with spotted coat color in mice. *Cell*, **79**, 1267–1276.

15. Lee, H.O., Levorse, J.M. and Shin, M.K. (2003) The endothelin receptor-B is required for the migration of neural crest-derived melanocyte and enteric neuron precursors. *Dev Biol*, **259**, 162–175.

16. Shin, M.K., Levorse, J.M., Ingram, R.S. and Tilghman, S.M. (1999) The temporal requirement for endothelin receptor-B signalling during neural crest development. *Nature*, **402**, 496–501.

17. Edery, P., Attie, T., Amiel, J., Pelet, A., Eng, C., Hofstra, R.M., Martelli, H., Bidaud, C., Munnich, A. and Lyonnet, S. (1996) Mutation of the endothelin-3 gene in the Waardenburg-Hirschsprung disease (Shah-Waardenburg syndrome). *Nat Genet*, **12**, 442–444.

18. Puffenberger, E.G., Hosoda, K., Washington, S.S., Nakao, K., deWit, D., Yanagisawa, M. and Chakravart, A. (1994) A missense mutation of the endothelin-B receptor gene in multigenic Hirschsprung's disease. *Cell*, **79**, 1257–1266.

19. Ito, M., Kawa, Y., Ono, H., Okura, M., Baba, T., Kubota, Y., Nishikawa, S.I. and Mizoguchi, M. (1999) Removal of stem cell factor or addition of monoclonal anti-c-KIT antibody induces apoptosis in murine melanocyte precursors. *J Invest Dermatol*, **112**, 796–801.

20. Morrison-Graham, K. and Weston, J.A. (1993) Transient steel factor dependence by neural crest-derived melanocyte precursors. *Dev Biol*, **159**, 346–352.

21. Murphy, M., Reid, K., Williams, D.E., Lyman, S.D. and Bartlett, P.F. (1992) Steel factor is required for maintenance, but not differentiation, of melanocyte precursors in the neural crest. *Dev Biol*, **153**, 396–401.

22. Spritz, R.A., Giebel, L.B. and Holmes, S.A. (1992) Dominant negative and loss of function mutations of the c-kit (mast/stem cell growth factor receptor) proto-oncogene in human piebaldism. *Am J Hum Genet*, **50**, 261–269.

23. Fang, D., Leishear, K., Nguyen, T.K., Finko, R., Cai, K., Fukunaga, M., Li, L., Brafford, P.A., Kulp, A.N., Xu, X. et al. (2006) Defining the conditions for the generation of melanocytes from human embryonic stem cells. *Stem Cells*, **24**, 1668–1677.

24. Shibamoto, S., Higano, K., Takada, R., Ito, F., Takeichi, M. and Takada, S. (1998) Cytoskeletal reorganization by soluble Wnt-3a protein signalling. *Genes Cells*, **3**, 659–670.

25. Willert, K., Brown, J.D., Danenberg, E., Duncan, A.W., Weissman, I.L., Reya, T., Yates, J.R., 3rd and Nusse, R. (2003) Wnt proteins are lipid-modified and can act as stem cell growth factors. *Nature*, **423**, 448–452.

Chapter 17

In Vitro Derivation of Chondrogenic Cells from Human Embryonic Stem Cells

Wei Seong Toh, Eng Hin Lee, Mark Richards, and Tong Cao

Abstract

Human embryonic stem cells (hESCs) have the ability to self-renew and differentiate into any cell lineage of the three germ layers, therefore holding great promise for regenerative medicine applications. However, directing lineage-restricted differentiation of hESCs and obtaining a homogenous differentiated cell population is still a challenge. We previously described a micromass culture system as a model system to study chondrogenic commitment of the hESCs. Using this system, various growth factors including BMP2 and TGFβ1 direct chondrogenic differentiation and modulate cartilage-specific matrix gene expression in a distinctive manner. Furthermore, a high percentage of differentiated cells exhibit typical morphological characteristics of chondrocytes and express cartilage matrix proteins such as type II collagen and proteoglycans. Chondrogenic cells can be further isolated and cultured to form functional cartilage tissue in vitro. Here, we describe in detail our established protocols to analyze chondrogenic differentiation of hESCs, and possible isolation of chondrogenic cells to form functional cartilaginous tissue.

Key words: Chondrogenic differentiation, embryonic stem cells, chondrocytes, cartilage, human.

1. Introduction

Human embryonic stem cells (hESCs), derived from the inner cell mass of the blastocyst stage embryos, represent a promising cell source for transplantation because of their unlimited self-renewal and ability to differentiate into various somatic cell lineages (1). However, one of the challenges in stem cell research is to understand, control, and develop an efficient and stable culture milieu for directing differentiation of hESCs into a defined chondrogenic lineage. There is still limited understanding of the factors, signals, and even the environment necessary to induce hESCs to specifically differentiate into the chondrogenic lineage (2).

Our group has previously established an effective high-density micromass system that allows chondrogenic commitment of embryoid body (EB)-derived cells (3, 4). This chapter presents the improved protocol for chondrogenic differentiation of hESCs and possible isolation of putative chondrogenic cells for functional cartilaginous tissue formation. Various analytical methods including RT-PCR, histology, and immunostaining will be presented.

2. Materials

2.1. Cell Culture

1. Dulbecco's Ca^{2+}- and Mg^{2+}-free phosphate-buffered saline (PBS).
2. Collagenase Type IV, Lyophilized (Cat. No. #17104-019; Gibco/BRL, Gaithersburg, MD, USA): Prepare the collagenase IV splitting medium (1 mg/mL) with DMEM/F12 medium. Filter sterilize before use.
3. Collagenase P. (Cat. No. 11 213 857 001, Roche): Prepare the collagenase P splitting medium (1.5 mg/ml) (0.15%) with DMEM/10% FBS medium. Filter sterilize before use.
4. 0.25% trypsin/1 mM ethylenediaminetetraacetic acid (EDTA) solution (Cat. No. #25200-072; Gibco/BRL).
5. 0.05% trypsin/0.53 mM ethylenediaminetetraacetic acid (EDTA) solution (Cat. No. #25300-054; Gibco/BRL).
6. Cell dissociation buffer (Cat. No. #13151-014; Gibco/BRL).
7. 6-well and 24-well Ultra Low Attachment Microplates (Cat. No. #3471 and #3473; Costar(r) Corning, Nagog Park Acton, MA, USA).
8. 6-well, 12-well, and 24-well microwell plates (Nunc, Wiesbaden, Germany).
9. Milli-Q-purified water (Millipore, Billerica, MA, USA).
10. Gelatin, Type A, from porcine skin, approximately 300 Bloom (Cat No. #G1890; Sigma) 0.1% (w/v) gelatin-coated tissue culture plates. Weigh out 0.5 g gelatin in an autoclavable bottle and add 500 mL of distilled water (dH_2O). Autoclave with cap loosely tightened, allow cooling to room temperature (RT), and pipette into the culture plates. Coat the plates overnight by incubating at 37°C until use.

11. 40-μm nylon cell strainer (Falcon Cat. No. #352340; BD Biosciences Inc., Franklin Lakes, NJ, USA).

12. 10-mL syringes (Becton Dickinson Inc.).

13. 22-G needles (Sterican B BRAUN).

14. 3D orbital shaker (Boeco, Hamburg, Germany).

2.1.1. Medium

1. Dulbecco's Modified Eagle's Medium (DMEM)/F-12 medium (Cat. No. #11330-032; Gibco/BRL).

2. DMEM high-glucose (Cat. No. #D1152; Sigma, St Louis, MO, USA).

3. Knockout™ Serum Replacement (KSR) (Cat. No. #10828-028; Gibco/BRL).

4. Fetal Bovine Serum (FBS) (Cat. No. #CH30160.03; Hyclone, Logan, UT).

5. ITS^{+1}; 100X (6.25 μg/mL insulin, 6.25 μg/mL transferrin, 6.25 ng/mL selenium, 1.25 mg/mL bovine serum albumin, 5.35 μg/mL linoleic acid) (Cat. No. #354352; BD Bioscience Inc, Franklin Lakes, NJ).

6. 2-Mercaptoethanol, 14.3 M (Cat. No. #M7522; Sigma).

7. GlutaMAX™-I Supplement, 200 mM; 100X (Cat No. #35050-061; Gibco/BRL).

8. MEM nonessential amino acids (NEAA) solution, 10 mM; 100X (Cat. No. #11140-050; Gibco/BRL).

9. Sodium pyruvate solution, 100 mM; 100X (Cat. No. #11360-070; Gibco/BRL).

10. Penicillin/Streptomycin (P/S), 10,000 U/10,000 μg; 100X (Cat. No. #15140-122; Gibco/BRL).

11. L-Proline (Cat. No. #P5607; Sigma).

12. Ascorbic acid 2-phosphate (AA2P, Cat. No. #A8960; Sigma) 100X stock: dissolve 50 mg in 10 mL PBS. Aliquot in working volumes and store at −20°C.

13. Dexamethasone (Cat. No. #D2915; Sigma) 1000X stock: Dissolve 0.0115 g in 20 mL ddH$_2$O. Aliquot and store at −20°C (*see* **Note 1**).

14. Medium 1: DMEM high-glucose supplemented with 10% FBS, 10% KSR, 1% sodium pyruvate, 1% NEAA, and 100 U/100 μg P/S.

15. Medium 2: DMEM high-glucose supplemented with 1% ITS^{+1} (6.25 μg/mL insulin, 6.25 μg/mL transferrin, 6.25 ng/mL selenium, 1.25 mg/mL bovine serum albumin, 5.35 μg/mL linoleic acid), 1% KSR (*see* **Note 2**), 2 mM GlutaMAX™, 40 μg/mL L-proline, 50 μg/mL AA2P, 1% sodium pyruvate, 1% NEAA, 10^{-7} M dexamethasone, and 100 U/100 μg P/S (*see* **Note 3**).

16. Medium 3: DMEM high-glucose supplemented with 10% FBS, 1% sodium pyruvate, 1% NEAA, and 100 U/100 μg P/S.

17. Medium 4: DMEM high-glucose supplemented with 1% ITS^{+1} (6.25 μg/mL insulin, 6.25 μg/mL transferrin, 6.25 ng/mL selenium, 1.25 mg/mL bovine serum albumin, 5.35 μg/mL linoleic acid), 2 mM GlutaMAX™, 40 μg/mL L-proline, 50 μg/mL AA2P, 1% sodium pyruvate, 1% NEAA, 10^{-7} M dexamethasone, and 100 U/100 μg P/S.

2.1.2. Growth Factors

The effects of growth factors, BMP2 and/or TGFβ1, on chondrogenic development of EBs were studied, with the source, preparation, and concentration used given in parentheses.

1. Recombinant human transforming growth factor-beta-1 (TGFβ1) (Cat. No. 240-B; R&D Systems): Prepare 1000X stock solution (10 μg/mL) by reconstituting in filter-sterilized 4 mM HCl containing 0.1% (w/v) BSA, follow by freezing in aliquots of working volume (see **Note 4**).

2. Recombinant human bone morphogenetic protein 2 (BMP2) (Cat. No. #355-BM; R&D Systems, Minneapolis, MN, USA): Prepare 1000X stock solution (100 μg/mL) by reconstituting in filter-sterilized 4 mM HCl containing 0.1% (w/v) BSA, follow by freezing in aliquots of working volume.

2.2. Analysis of mRNA

1. Total RNA extraction performed using RNeasy® Mini Kit (Cat. No. #74106; Qiagen, Chatsworth, CA, USA).

2. QIAshredder spin column (Cat. No. #79654; Qiagen).

3. RNase-free Dnase (Cat. No. #79254; Qiagen).

4. cDNA synthesis performed using iScript™ cDNA synthesis kit (Cat. No. #170-8891; Bio-Rad, Hercules, CA, USA).

5. Real-time RT-PCR reactions using Power SYBR Green PCR Master Mix System (Product No. #4367659; Applied Biosystems, Foster City, CA, USA).

6. Milli-Q-purified water (Millipore).

7. Diethyl pyrocarbonate-treated water (DEPC-H_2O): Add 500 μL DEPC (Cat. No. #D5758; Sigma) to 500 mL Milli-Q-purified water and stir overnight at RT. DEPC is inactivated by autoclaving for 20 min (see **Note 5**).

8. Sterile RNase-free reagents, polypropylene tubes, tips, and other materials.

9. Nanodrop ND-1000 spectrophotometer (NanoDrop Technologies, Wilmington, DE, USA).

10. PCR thermal cycler (Bio-Rad).

11. PCR thermocycler Applied Biosystems 7500 Real-Time PCR System (Applied Biosystems, Foster City, CA, USA).

12. Gel electrophoresis and standard apparatus (Bio-Rad).
13. Gel Doc EQ Imaging System (Bio-Rad).

2.3. Histochemical Staining

1. Fixative: 4% paraformaldehyde in PBS. Paraformaldehyde (PFA; Cat. No. P6148; Sigma).
2. Alcian Blue working solution (0.5% w/v): Dissolve 0.5 g Alcian Blue 8GX (Sigma, Cat. No. #A3157) in 100 mL 0.1 M hydrochloric acid solution (pH 1.0).
3. Mayer's hematoxylin.
4. Nuclear Fast Red solution.
5. Scott's tap water: Dissolve 3.5 g sodium bicarbonate and 20 g magnesium sulfate in 1 L dH_2O.
6. DePex mounting medium (Cat. No. #361254D; VWR International, Poole, Dorset, UK).
7. Tissue processor.
8. Paraffin.
9. Ethanol.
10. Xylene.

2.4. Immunostaining

1. UltraVision HRP Detection System (Cat. No. #TM-125-HL; Lab Vision Inc., Fremont, CA, USA).
2. Pepsin (Cat. No. #AP-9007-005; Lab Vision).
3. Hydrogen peroxide block (Cat. No. #TA-125-HP; Lab Vision).
4. Fixative: 4% PFA in PBS. Dissolve 4 g PFA in 100 mL of PBS, heat the mixture in 60°C waterbath, stir until the solution becomes clear. Cool to RT before use. Store at 4°C and use within 1 week. (*see* **Note 6**).
5. Triton-X-100 (Bio-Rad).
6. Goat serum (Cat. No. #S-1000; Vector Laboratories, Burlingame, CA, USA).
7. Bovine serum albumin (BSA) (Cat. No. #A9418; Sigma).
8. Blocking buffer – PBS/2% BSA/10% goat serum – Dissolve 0.2 g BSA in 9 mL PBS. Add 1 mL of goat serum to the mixture. Filter before use.
9. 24-well microwell plates (Nunc).
10. Vetashield mounting medium with DAPI for nuclear counterstaining (Cat. No. #H-1200; Vector Laboratories).
11. Monoclonal antitype II collagen antibody (Chemicon).
12. Monoclonal antitype I collagen antibody (Sigma).
13. Polyclonal antitype I collagen antibody (Chemicon).

14. Polyclonal anti-Oct4 antibody (Santa Cruz Biotechnology).
15. Alexa Fluor® 488 goat antimouse IgG (H+L) highly cross-adsorbed.
16. Alexa Fluor® 594 goat antirabbit IgG highly cross-adsorbed. The sources of antibodies and dilutions used are summarized in **Table 17.1**.

Table 17.1
List of antibodies used for immunostaining

Primary antibodies	Host	Dilution/Application	Source	Catalogue No.
Collagen II	Mouse	1:150 (IF) 1:500 (IHC)	Chemicon	MAB8887
Collagen I	Mouse	1:1000 (IHC)	Sigma	C2456
Collagen I	Rabbit	1:75 (IF)	Chemicon	AB745
Oct-4	Rabbit	1:100 (IF)	Santa Cruz Biotechnology	sc-9081
Secondary antibodies	Host	Dilution	Source	
Alexa Flour 488	Goat	1:200	Molecular Probes	A11029
Alexa Flour 594	Goat	1:200	Molecular Probes	A11037

Application = IF: Immunofluorescene; IHC: Immunohistochemistry

3. Methods

The methods pertaining to the (1) formation of embryoid bodies (EBs) from hESCs, (2) chondrogenic differentiation of EBs, (3) isolation of chondrogenic cells, analysis of the (3) mRNA expression, and (4) synthesis of cartilage-specific matrix proteins are described below:

3.1. Embryoid Body Formation

1. By day 6–7 of hESC culture (*see* **Note 7**), hESCs are allowed to form embryoid bodies and spontaneously differentiate into cells of ecto-, meso, and endodermal lineages (**Fig. 17.1A and B**).
2. Aspirate medium from hESCs and add 1 mL/well of collagenase IV splitting medium into each well of the 6-well plate.
3. After incubation at 37°C in incubator for 5 min, scrape the hESC colonies from the plate with a 5-mL pipette and transfer cells to a 15-mL sterile falcon tube.

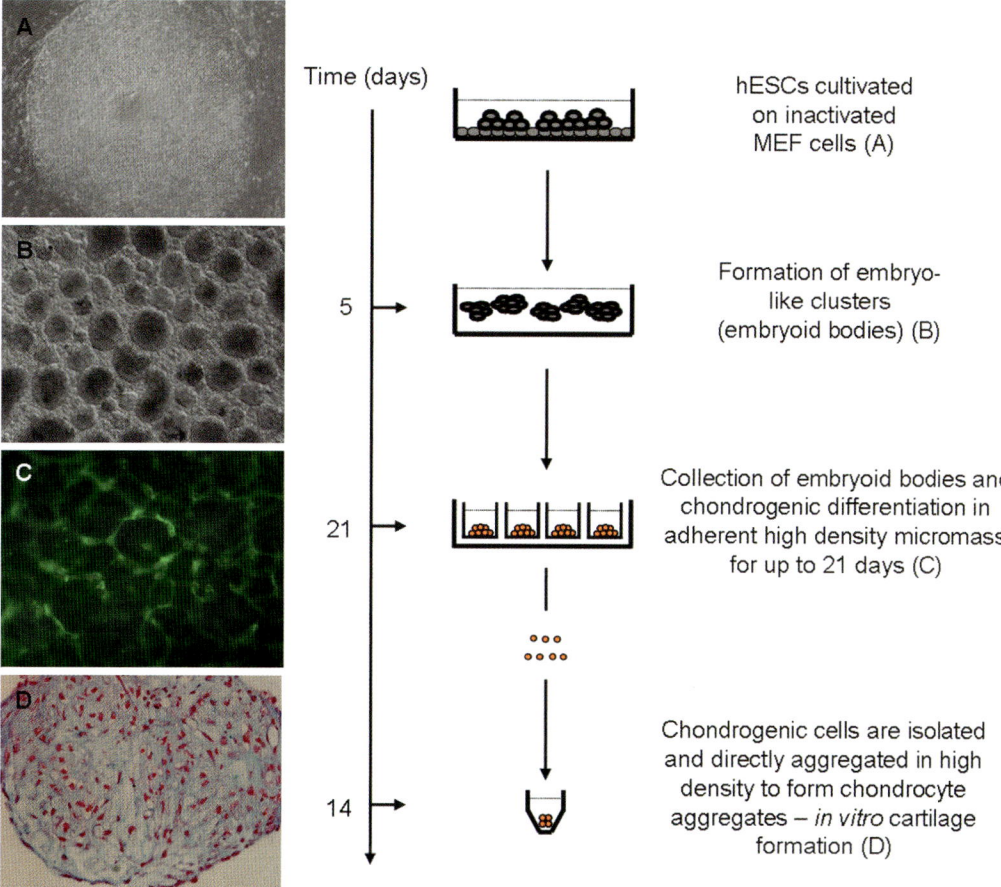

Fig. 17.1. Cultivation and differentiation of hESCs in vitro. **(A)** To keep hESCs undifferentiated and pluripotent, the cells were cocultivated with growth-inactivated MEF cells in the presence of bFGF. **(B)** After collagenase IV treatment, hESCs were cultured in suspension to form embryoid bodies (EBs) for a period of 5 days. 5-Day EBs were collected and dissociated into single cells and cultured at a high density of 3×10^5 cells per 15 μl spot in a 24-well plate precoated with 0.1% gelatin. **(C)** Collagen II producing chondrogenic cells that form a dense interconnecting filamentous matrix network could be observed by immunoflurorescene (*green*) staining. **(D)** Chondrogenic cells were isolated and allowed to aggregate at high density to form chondrocyte aggregates that are enriched in proteoglycans as detected by alcian blue staining.

4. Add 1 mL of EB differentiation medium – medium 1 (*see* **Section 2.1.1, Item 14**) into each well. Scrape off the remaining cells in the well with a cell scraper and pool the cells together in the 15-mL falcon tube. Gently pipette the cells up and down a few times in the tube, to further break up the colonies.

5. Allow the cell clumps to settle down for 5–10 min. Aspirate the supernatant and then resuspend the cells with medium 1 to give a final volume of 3 mL per well of 6-well ultra low-attachment microplates. The splitting ratio is set at 1:1 ratio, where one 6-well plate of hESCs is split to form one 6-well plate of EBs.

6. After overnight culture in suspension, hESCs form floating aggregates known as EBs. Culture medium is changed every second day. To change medium, transfer EBs into 50-mL falcon tube and let the aggregates settle for 5–10 min. Aspirate the supernatant, replace with fresh medium, and transfer back to the 6-well ultra low-attachment microplates for further culture.

7. At day 5, EBs are harvested for induction of chondrogenic differentiation.

3.2. Micromass Cultures of EB-Derived Cells

1. Transfer the 5-day EBs into a sterile 50-mL falcon tube and allow the EBs to settle for 3–5 min. Aspirate the supernatant and wash once with PBS (*see* **Note 8**).

2. 5-Day EBs are then dissociated into single cells by dissociation using a solution mixture containing 0.05% trypsin/EDTA and cell dissociation buffer at 7:3 ratio for 5 min at 37°C, followed by passing the cell suspension through a 22-G needle and then a 40-μm cell strainer to obtain single cells suspended in prewarmed medium 1. The serum-containing medium helps to inactivate the trypsin (*see* **Note 9**).

3. Wash twice with the same medium by spinning down the cells at 250 g for 5 min and resuspending in medium. At the second spin, resuspend the cells in a lower volume of medium for cell count.

4. Culture cells at a high density of 3×10^5 cells per 15 μL spot in a 24-well plate precoated with 0.1% gelatin (*see* **Note 10**). After incubation for approximately 2 h, 500 μL of the same medium is carefully added to each well. Micromass cultures are incubated overnight at 37°C to enable cell attachment, prior to induction of differentiation the next day.

5. Following incubation for 24 h to allow cell attachment, replace with serum-free chondrogenic medium – medium 2 (*see* **Section 2.1.1, Item 15**) with or without growth factors TGFβ1 10 ng/mL (*see* **Section 2.1.2, Item 1**) and/or BMP2 100 ng/mL (*see* **Section 2.1.2, Item 2**) for a period of 21 days with medium change every alternate day (*see* **Note 11**). Cultures in the basic serum-free chondrogenic medium without growth factor supplementation will serve as the control. By day 21 of differentiation, chondrogenic cells organized in dense filamentous matrix can be detected by specific anti-type II collagen antibody staining, as shown in **Fig. 17.1C**.

3.3. Isolation of Chondrogenic Cells and In Vitro Cartilage Formation

1. Micromass cultures are digested overnight with Collagenase P (1.5 mg/ml) digestion to release the chondrogenic cells.

2. Pass the cells through the 40-μm nylon cell strainer.

3. Only single cells derived after collagenase digestion and passing through the strainer are used for further differentiation.

4. The cells are spun down at high density, approximately 2×10^5 cells per 500 μL in 15-ml falcon tubes to form aggregates, which were placed onto a 3D orbital shaker set at 30 rpm. Medium 3 (*see* **Section 2.1.1**, **Item 16**) is used to allow cell recovery at Day 0.

5. Aggregation of the chondrogenic cells could be observed the next day (Day 1).

6. At day 1, perform medium change to differentiation medium – medium 4 (*see* **Section 2.1.1**, **Item 17**) with TGFβ1 (10 ng/mL), 500 μL per aggregate.

7. Chondrogenic aggregates can also be pooled together and cultured in an ultra low-attachment 24-well microplate on the orbital shaker set for continuous gentle shaking.

8. Allow the chondrogenic aggregates to further differentiate for another 14 days to form a cartilage-like pellet. Functional cartilaginous formation by committed chondrogenic cells can be analyzed by histological analysis. **Figure 17.1D** shows the image of a chondrogenic aggregate stained with Alcian blue.

3.4. Analysis of Cartilage-Related Genes

3.4.1. Total RNA Extraction and cDNA Synthesis

1. Remove the medium from the cultures and wash once with PBS. Extract total RNA using the RNeasy Mini Kit, following the manufacturer's instructions. Scrape the cells with 350 μL of RNeasy Lysis Buffer per well of a 24-well plate.

2. Homogenize the cell suspension by passing through the QIAshredder column and spin down to obtain the homogenate. The homogenate can be frozen and stored at –80°C for subsequent RNA extraction.

3. Continue the RNA extraction protocol and DNase treatment according to the manufacturer's instructions and elute the sample RNA with RNase-free water provided in the kit. DEPC-H$_2$O can also be used (*see* **Section 2.2**, **Item 7**).

4. Determine the RNA quantity and quality by measuring the absorbance at 260 and 280 nm for each sample using the Nanodrop ND-1000 spectrophotometer. Calculate the RNA concentration: 1 unit at 260 nm corresponds to 40 μg of RNA per mL. Store the RNA samples at –80°C if cDNA synthesis is not carried out on the same day.

5. Reverse transcriptase reaction is performed using a PCR thermal cycler following the manufacturer's instructions. cDNA synthesis is performed using 500 ng of total RNA per 20 μL reaction volume with the addition of 5X iScript reaction mix and iScript reverse transcriptase over a 30-min incubation time at 42°C, followed by enzyme inactivation at 85°C for 5 min.

3.4.2. Semiquantitative Reverse Transcriptase-Polymerase Chain Reaction (RT-PCR)

1. To assess lineage restriction of the chondrogenically differentiated hESCs, RT-PCR is used to screen for markers of other germ layers including endoderm and ectoderm derivatives. PCR primers are listed in **Table 17.2**.
2. Perform PCR amplifications using 1 μL of the resulting cDNA samples by first denaturing at 95°C for 5 min, followed by 35 cycles of 95°C for 30 s, annealing at 55–58°C for 45 s, and extending at 72°C for 60 s, and followed by a final extension at 72°C for 5 min.
3. Analyze the PCR-amplified products on 2% (w/v) agarose gel-electrophoresis. The results are evaluated by ultraviolet detection of the ethidium bromide-stained gel. Images are taken using the Biorad Light Imaging System.
4. For analysis, mean pixel intensities of each band are measured using the NIH public domain imaging software – ImageJ and normalized to mean pixel intensities of the *β-actin* band.

3.4.3. Quantitative Real-Time Reverse Transcriptase-Polymerase Chain Reaction (RT-PCR)

1. *Col 1* and *Col 2* gene expression is analyzed by real-time RT-PCR reactions using Power SYBR Green PCR Master Mix System (Applied Biosystems, Foster City, CA, USA) on PCR thermocycler Applied Biosystems 7500 Real-Time PCR System (Applied Biosystems, Foster City, CA, USA).
2. Equal amounts of input cDNA must be used per reaction and all reactions performed in triplicates (*see* **Note 12**). Gene expression of each target gene is usually normalized to the reference endogenous housekeeping gene glyceraldehydes-3-phosphate dehydrogenase (*GAPDH*).
3. Real-time RT-PCR was performed at 95°C for 10 min, followed by 40 cycles of amplifications, consisting of a denaturation step at 95°C for 15 s, and an extension step at 60°C for 1 min. PCR primers are listed in **Table 17.3** (*5*). The level of expression of each target gene is then calculated as $2^{-\Delta\Delta Ct}$, as previously described (*6*).
4. 5-Day EBs were dissociated into single cells and cultured as high-density micromass for 21 days in the absence or presence of either TGFβ1 (10 ng/ml), BMP2 (100 ng/ml), or combination of both growth factors. Quantitative analysis of *Col 2* and *Col 1* was done by quantitative real-time RT-PCR at the end of 21 days, as shown in **Fig. 17.2**

3.5. Immunostaining

3.5.1. Immunofluorescent Staining

1. Cells are cultured in 24-well plates. Remove the medium completely from the wells. Rinse once with 1 mL of PBS.
2. Fix the cells with 4% PFA for at least 20 min at RT.
3. Wash the cells three times with PBS.
4. Permeabilize the cells with 0.2% Triton-X-100/PBS for 15 min at RT.

Table 17.2
Sequence of primers and conditions for cartilage-related markers and other germ layer markers used in conventional reverse transcriptase-polymerase chain reaction (RT-PCR)

Gene	Primer sequence	Product size (bp)
Sox 9	Sox9F:5'-GAACGCACATCAAGACGGAG-3' Sox9R:5'-TCTCGTTGATTTCGCTGCTC-3'	631
Col 2a1	Col2F:5'-TTCAGCTATGGAGATGACAATC-3' Col2R:5'-AGAGTCCTAGAGTGACTGAG-3'	472
Aggrecan1	hAggF: 5'-TGAGGAGGGCTGGAACAAGTACC-3' hAggR: 5'-GAGGTGGTAATTGCAGGGAACA-3'	350
Decorin	hDecoF:5'-CCTTTGGTGAAGTTGGAACG-3' hDecoR:5'-AAGATGTAATTCCGTAAGGG-3'	300 (55°C)
Biglycan	hBiglyF:5'-TGCAGAACAACGACATCTCC-3' hBiglyR: 5'-AGCTTGGAGTAGCGAAGCAG-3'	475
COMP	hCOMP-F: 5'-CAACTGTCCCCAGAAGAGCAA-3' hCOMP-R: 5'-TGGTAGCCAAAGATGAAGCCC-3'	588
Col 9a1	hCol9-F: 5'-ACAGCAGGACTCCCTGGA-3' hCol9-R: 5'-TGATCACCAGGTGCACCAG-3'	410
Link protein	LPF:5'-CCTATGATGAAGCGGTGC-3' LPR:5'-TTGTGCTTGTGGAACCTG-3'	618
Oct4	Oct4F: 5'-CGRGAAGCTGGAGAAGGAGAAGCTG-3' Oct4R: 5'-AAGGGCCGCAGCTTACACATGTTC-3'	247 (55°C)
NFH	hNFH-F: 5'- TGAACACAGACGCTATGCGCTCAG-3' hNFH-R: 5'- CACCTTTATGTGAGTGGACACAGAG-3'	398 (56°C)
K14	hK14F: 5'-ATGATTGGCAGCGTGGAG-3' hK14R: 5'-GTCCAGCTGTGAAGTGCTTG-3'	390
AFP	AFPF: 5'-CCATGTACATGAGCACTGTTG-3' AFPR: 5' CTCCAATAACTCCTGGTATCC-3'	338 (55°C)
VEGFR2	VEGFR2F:5'-AAAACCTTTTGTTGCTTTTTGA-3' VEGFR2R:5'-GAAATGGGATTGGTAAGGATGA-3'	237 (55°C)
Osteocalcin	hOC-F: 5'-ATGAGAGCCCTCACACTCCTC-3' hOC-R: 5'-GCCGTAGAAGCGCCGATAGGC-3'	294
β-actin	β-ActinF: 5'-CCAAGGCCAACCGCGAGAAGATGAC-3' β-ActinR: 5'-AGGGTACATGGTGGTGCCGCCAGAC-3'	587

The PCR conditions were as follows: 95°C/5 min, 95°C/30 s, 58°C/45 s (except where specified otherwise), 72°C/1 min and 72°C/5 min.

Table 17.3
Sequence of primers used for real-time RT-PCR

Gene	Primer sequence	Product size (bp)
Col 1	Col1F: 5'-CAGCCGCTTCACCTACAGC-3' Col1R: 5'-TTTTGTATTCAATCACTGTCTTGC C-3'	83
Col 2	Col2F: 5'-GGCAATAGCAGGTTCACGTACA-3' Col2R: 5'-CGATAACAGTCTTGCCCCACTT-3'	79
GAPDH	GAPDHF: 5'-ATGGGGAAGGTGAAGGTCG-3' GAPDHR: 5'-TAAAAGCAGCCCTGGTGACC-3'	119

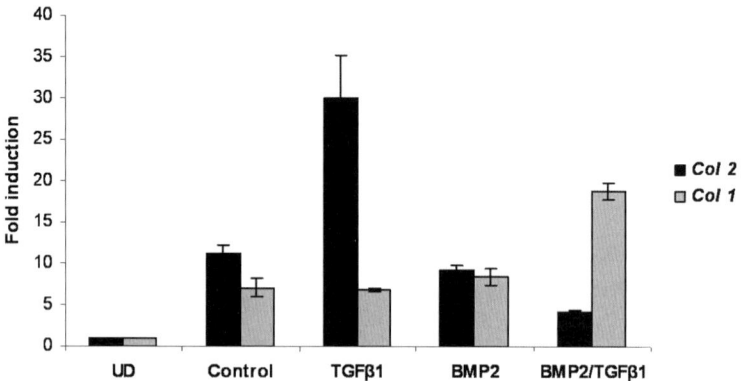

Fig. 17.2. Growth-factor modulation of chondrogenesis of EB-derived cells in high-density micromass cultures. 5-Day EBs were dissociated into single cells and cultured as high-density micromass for 21 days in the absence or presence of either TGFβ1 (10 ng/ml), BMP2 (100 ng/ml), or combination of both growth factors. Quantitative analysis of *Col 2* and *Col 1* was done by quantitative real-time RT-PCR at the end of 21 days. Results showed the mean magnitudes of mRNA levels normalized to *GAPDH*, and expressed relative to the expression level of undifferentiated (UD) hESCs.

5. Block with blocking buffer – PBS/2% BSA/10% goat serum for 45 min.
6. Incubate with primary antibody diluted in PBS/2% BSA for 2 h at RT. The list of antibodies and dilutions used in immunofluorescent staining is described in **Table 17.1**.
7. Wash the cells three times with PBS.
8. Incubate with the secondary antibody diluted in PBS/2% BSA for 1 h at RT in the dark.
9. Wash the cells three times with PBS and last wash with dH$_2$O.
10. Air dry for 5 min and apply Vectashield mounting medium with DAPI for counterstaining.
11. Antibody isotypes or no antibody sera served as the negative control.

3.6. Chondrogenic Aggregate Processing

1. Fix each chondrogenic cell aggregate in 4% PFA overnight at 4°C.
2. Dehydrate aggregates in successive ethanol washes of 70, 70, 80, 80, 95, and 95%, for 15 min each.
3. Briefly stain aggregates with eosin, followed by two brief rinses in 100% ethanol. This stain serves to visualize the aggregates and aid in their handling.
4. Incubate aggregates in two changes of 100% ethanol, 20 min each.
5. Transfer to two changes of xylene, 20 min each.
6. Transfer to three changes of paraffin, 20 min each.
7. Remove each aggregate and embed in paraffin block as per standard embedding procedure.
8. Cut 5-μm sections of each aggregate and transfer 2–3 sections onto each microscope glass slide.
9. Store the slides at RT for future staining.

3.7. Histological Analysis

3.7.1. Alcian Blue Staining

1. Deparaffinize the aggregate sections in two changes of xylene, 10 min each.
2. Rehydrate the sections in serial washes of 100% ethanol, 95% ethanol, and 70% ethanol, 2 min each.
3. Hydrate the sections in dH$_2$O for 5 min.
4. Stain the sections with alcian blue solution for 30 min (*see* **Section 2.3, Item 2**).
5. Rinse in tap water.
6. Counterstain with nuclear fast red for 5 min.
7. Rinse in tap water.
8. Dehydrate in successive ethanol washes of 70, 95, and 100% ethanol.
9. Clear with xylene.
10. Coverslip with xylene-based DePex mounting medium.
11. View by light microscopy.

3.7.2. Collagen I and II Immunohistochemistry

1. Deparaffinize the aggregate sections in two changes of xylene, 10 min each.
2. Rehydrate the sections in serial washes of 100% ethanol, 95% ethanol, and 70% ethanol, 2 min each.
3. Hydrate the sections in PBS for 5 min.
4. To facilitate antigen retrieval and antibody access, incubate with pepsin at 37°C for 20 min.
5. Rinse once with PBS, then block with hydrogen peroxide block for 15 min at RT to quench any endogenous peroxidase activity.

6. Wash 4 times with PBS, block with Ultra V Block prior to 1 h incubation at RT with either anticollagen II monoclonal antibody, Mab II-II6B3 diluted 1:500 in PBS or anticollagen I monoclonal antibody, COL-1 diluted 1:1000 in PBS.

7. At the end of primary antibody incubation, wash 4 times with PBS, before adding the prediluted biotin-conjugated goat derived antimouse secondary antibody and incubate at RT for 30 min (*see* **Section 2.4**, **Item 1**).

8. At the end of secondary antibody incubation, wash 4 times with PBS, before incubating with streptavidin-conjugated horseradish peroxidase at RT for 45 min.

9. Wash 4 times with PBS before adding the DAB chromogen to visualize the antibody-antigen reaction.

10. Counterstain with Mayer's hematoxylin for 2–3 min, then wash in tap water.

11. Blue in Scott's Tap water before rinsing in tap water.

12. Dehydrate in successive ethanol washes of 70%, 95%, and 100% ethanol.

13. Clear with xylene.

14. Coverslip with xylene-based DePex mounting medium.

15. View by light microscopy.

4. Notes

1. Stock preparation of dexamethasone is stable for 6 months at −20°C in a nondefrosting freezer.

2. 1% KSR instead of a strictly serum-free medium was used in this study, based on reports that mesodermal differentiation of ESCs may be inhibited under serum-free conditions (7).

3. Use chondrogenic differentiation medium for not more than 2 weeks after preparation, as components such as L-glutamine and AA2P degrade over time. Store the chondrogenic differentiation medium at 4°C and discard unused medium after 2 weeks.

4. Avoid repeated freeze-thaw cycles of the growth factors. Once thawed, keep the growth factor aliquot at 4°C and use within 1 week.

5. DEPC is slightly hazardous in case of skin or eye contact and of inhalation.

6. PFA is toxic. Work inside hood and use gloves when handling. Prepare fresh before use to avoid formic acid formation. Store at 4°C and use within 1 week.

7. HESC cultures (H1 and H9; NIH stem cell registry: http://stemcells.nih.gov/research/registry, also *see* reference (1)) follow exactly as recommended by the Wicell protocol. Refer to website http://www.wicell.org/ for protocols describing the expansion and propagation of hESCs on murine embryonic feeder cells.

8. Allow the EBs to settle down during the wash with PBS. Usually it takes less than 3 min for all the EBs to settle. Avoid pipetting too many times or foaming because the cells should be collected in clumps.

9. Dissociate the EBs into single cells for no more than 5 min and pass the cell suspension once through a 22-G needle and then 40-μm cell strainer. Prolonged or repeated dissociation is detrimental to cell viability.

10. The method for high-density micromass culture has been modified from previously published methods (8, 9). After removal of 0.1% gelatin from the wells, ensure the plates are completely dry before use. This is to prevent dispersion of cells when dispensing the cells in a single drop onto the well in a micromass culture.

11. Efficiency of chondrogenic differentiation and commitment of hESCs to chondrogenic lineage are enhanced with growth factor stimulation, especially TGFβ1.

12. Accurate pipetting is crucial to the success of the experiment and it is recommended that a dedicated set of pipettes be set aside for quantitative RT-PCR and be calibrated on a regular basis.

References

1. Thomson JA, Itskovitz-Eldor J, Shapiro SS, Waknitz MA, Swiergiel JJ, Marshall VS, Jones JM. (1998) Embryonic stem cell lines derived from human blastocysts. *Science* **282,** 1145–1147.
2. Heng BC, Cao T, Lee EH. (2004) Directing stem cell differentiation into the chondrogenic lineage in vitro. *Stem Cells* **22,** 1152–1167.
3. Toh WS, Yang Z, Liu H, Heng BC, Lee EH, Cao T. (2007) Effects of culture conditions and bone morphogenetic protein 2 on extent of chondrogenesis from human embryonic stem cells. *Stem Cells* **25,** 950–960.
4. Toh WS, Yang Z, Heng BC, Cao T. (2008) Differentiation of human embryonic stem cells towards the chondrogenic lineage. *Stem Cell Assays-Methods in Molecular Biology* **407,** 333–349.
5. Martin I, Jakob M, Schafer D, Dick W, Spagnoli G, Heberer M. (2001) Quantitative analysis of gene expression in human articular cartilage from normal and osteoarthritic joints. *Osteoarthritis Cartilage* **9,** 112–118.
6. Livak KJ, Schmittgen TD. (2001) Analysis of relative gene expression data using real-time quantitative PCR and the 2(-Delta Delta C(T)) Method. *Methods* **25,** 402–408.
7. Wiles MV, Johansson BM. (1999) Embryonic stem cell development in a chemically defined medium. *Experimental Cell Research* **247,** 241–248.
8. Ahrens PB, Solursh M, Reiter RS. (1977) Stage-related capacity for limb chondrogenesis in cell culture. *Developmental Biology* **60,** 69–82.
9. Mello MA, Tuan RS. (1999) High density micromass cultures of embryonic limb bud mesenchymal cells: An in vitro model of endochondral skeletal development. *In Vitro Cellular & Developmental Biology. Animal* **35,** 262–269.

Chapter 18

Vascular Differentiation of Human Embryonic Stem Cells in Bioactive Hydrogel-Based Scaffolds

Sharon Gerecht*, Lino S. Ferreira*, and Robert Langer

Abstract

The vascularization of tissue constructs remains a major challenge in regenerative medicine, as the diffusional supply of oxygen can support only 100–200 μm thick layers of viable tissue. The formation of a mature and functional vascular network requires communication between endothelial cells (ECs) and smooth muscle cells (SMCs). Potential sources of these cells that involve noninvasive methodologies are required for numerous applications including tissue-engineered vascular grafts, myocardial ischemia, wound healing, plastic surgery, and general tissue-engineering applications. Human embryonic stem cells (hESCs) can be an unlimited source of these cells. They can be expanded in vitro in an undifferentiated state without apparent limit, and hES-derived cells can be created in virtually unlimited amounts for potential clinical uses. Recently, vascular progenitor cells as well as endothelial and smooth muscle cells have been isolated from hESCs.

Key words: Vascular differentiation, endothelial cells, smooth muscle cells, scaffolds, tissue engineering, regenerative medicine.

1. Introduction

Despite the advances in the vascular differentiation of hESCs (1–5), there is a need for improved platforms for their vascular differentiation since in most cases the yield of cells is low (below 10% of the differentiated cells). In addition, there is a demand for scaffolds to allow the transplantation of these cells in vivo while improving their engraftment. The scaffold or cell carrier, although temporary, can be engineered to support migration, proliferation,

*Both authors contributed equally for this chapter.

and differentiation of vascular progenitor cells or differentiated cells and to aid in the organization of these cells in three dimensions (3D). In one application, progenitor or differentiated cells are expanded in vitro, placed onto biodegradable matrices in combination with factors that stimulate vascular differentiation, followed by implantation at the desired place. This chapter describes protocols for the vascular differentiation of hESCs using hydrogel-based scaffolds. In this case, undifferentiated hESCs are encapsulated into bioactive hydrogels to enhance their vascular differentiation and to act as a scaffold for their in vivo transplantation.

1.1. hESCs and vascular differentiation

Human ESCs are undifferentiated cells isolated from the inner cell mass of the developing blastocyte. To maintain their undifferentiated states in vitro, hESCs are cultured on mouse embryonic fibroblasts (MEFs) as a feeder layer (6) or indirectly as a source of conditioned medium in feeder-free culture conditions (7), or with a defined medium composition (8, 9). Vascular differentiation of hESCs can be induced via three methods (summarized in **Fig. 18.1**): (1) Two-dimensional (2D) culturing of the cells on ECM proteins or specific feeder layer, which can induce directed differentiation toward a specific lineage (10, 11); (2) growing hESCs in suspension in differentiation medium to form aggregates, leading to embryoid bodies (EBs). EBs induce spontaneous

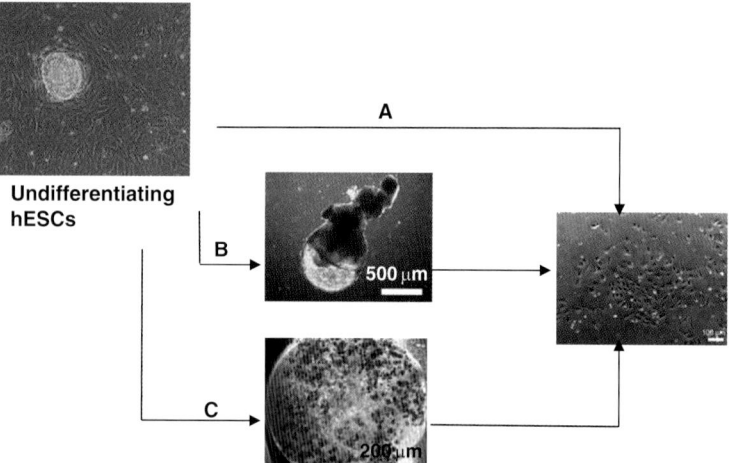

Fig. 18.1. Scheme showing three methods for stimulating the vascular differentiation of hESCs. (**A**) Undifferentiated hESCs are plated in a culture dish either in the presence or absence of a feeder layer or extracellular matrix that promotes vascular differentiation (2, 5). (**B**) Undifferentiated hESC aggregates are placed into suspension and differentiated as EBs in differentiation medium. Afterwards, cells can be plated onto a culture dish as single cells for further differentiation (13, 56). (**C**) Undifferentiated hESC aggregates are encapsulated in bioactive hydrogels for vascular differentiation. Afterwards, cells are removed from the gel by enzymatic digestion or other means and cultured onto a culture dish as single cells for further differentiation (17, 36, 57).

differentiation into the various cell types of the three germ layers in an organizational manner that parallels embryonic development (12). Vascular cells can be then isolated for further differentiation and maturation or for transplantation (3, 4, 13); and (3) encapsulating hESCs in 3D porous scaffolds, including bioactive hydrogels, followed by their release from the gel and subcultured for further differentiation (14–18).

Endothelial cells (ECs) can be characterized by their cobblestone cell morphology, formation of capillary-like structures when cultured on Matrigel, incorporation of DiI-labeled acetylated low-density lipoprotein (DiI-Ac-LDL), and expression of endothelial markers — such as vascular endothelial cadherin (VE-cad); platelet endothelial cell adhesion molecule-1 (PECAM1), also known as CD31; CD34; and vascular endothelial growth factor receptor 2 (VEGFR 2), also known as Flk1 or KDR. Functional ECs also express von Willebrand factor (vWF), endothelial nitric oxide synthase (eNOS), and E-selectin proteins (19). In contrast, smooth muscle cells (SMCs) are characterized by having a spindle-shaped morphology, being able to contract and relax in response to carbachol and atropine, rarely forming capillary-like structures when cultured in Matrigel, and expressing SM markers such as a-SM actin, SM myosin heavy chain, calponin, caldesmon, and SM22 (1).

ECs were isolated from 13-day-old human EBs using fluorescence-activated cell sorting (FACS) of $CD31^+$ cells. In addition to expressing CD34, Flk1, VE-cad, and taking up DiI-Ac-LDL, the isolated ECs also expressed mature endothelial protein vWF (20). Wang et al. later showed that $CD31^+Flk1^+CD45^-$ cells isolated from 10-day-old human EBs constituted a population of primitive precursors, which could be induced to mature into ECs (4). We have reported that progenitors isolated from 10-day-old EBs expressed CD34, suggesting a common origin for endothelial and hematopoietic cells (1). When cultured in endothelial growth medium (EGM)-2 supplemented with either $VEGF_{165}$ or PDGF-BB, those progenitors differentiated into ECs or SMCs, respectively. Human ESC-derived ECs and SMCs, when implanted into severe combined immunodeficient (SCID) mice, formed functional microvasculatures, which integrated with the host vasculatures (16).

The differentiation of hESCs into ECs has also been achieved using scalable 2D methods that avoid an EB intermediate. We have shown that hESCs can be directed to differentiate into ECs and SMCs by culturing them on collagen IV, with the supplementation of $VEGF_{165}$ or PDGF, respectively (2). Other researchers have further demonstrated different feeder layers or ECMs — including mouse bone marrow cells (S17, OP9, and MS-5) (21, 22), mouse yolk sac ECs (c166) (21), MEFs (5), and methylcellulose — which induce their differentiation toward early hematopoietic precursors with the potential to differentiate into ECs (23).

Recently, 3D scaffolding has emerged as another methodology to induce vascular differentiation via the 3D culturing (14, 18) or supplementation of bioactive molecules to induce vascular differentiation (15, 16).

1.2. 3D Scaffolding for hESCs

During normal embryogenesis, hESCs differentiate along different lineages in the context of complex 3D tissue structures, where the extracellular matrix (ECM) and different biomolecules (e.g., growth factors and other ligands) play an important role in this process (24). ECM provides structural support and biological signals that regulate cell differentiation. Some examples pointed out for this influence. hESCs cultured in alginate scaffolds express higher levels of vascular markers than in EBs (18). Moreover, the addition of neurotrophins/retinoic acid, transforming growth factor β, activin-A/insulin-like growth factor in hESCs cultured in 3D poly(α-hydroxy esters) scaffolds induced their differentiation into 3D structures with characteristics of developing neural tissues, cartilage, or liver, respectively, in contrast with cells cultured in a 2D culture system (25, 26).

Although these strategies contribute o some extent for the control of hESC proliferation and differentiation into specific cell lineages, they required extensive ex vivo cell culture and the presence of medium supplemented with growth factors. Alternatively, biodegradable bioactive scaffolds incorporating biomolecules including cell-adhesion epitopes, growth factors, or other ligands within the matrix could allow a better spatial control of cell differentiation and they could create chemical clues for stem cell differentiation in vitro and in vivo (27). In addition, the enzymatic degradation of such scaffolds would allow the isolation of enriched populations of differentiated cells, which is not possible using the previous synthetic scaffolds. Below we describe two protocols for the vascular differentiation of hESCs using bioactive hydrogel-based scaffolds.

1.3. Hydrogels for Cell Encapsulation

1.3.1. Type of Hydrogels for Cell Encapsulation

Hydrogels are 3D hydrophilic polymeric networks, able to swell with large quantities of water (water content $\geq$ 30% by weight) and biological fluids without the dissolution of the polymer (28, 29). In the last years, several hydrogel formulations have been developed from natural polymers including polysaccharides (ex: dextran, hyaluronic acid, alginate, chitosan, and agarose) and proteins (fibrinogen and collagen) as well as synthetic polymers [poly(acrylic acid), poly(ethylene glycol), and poly(vinyl alcohol)]. The structural integrity of hydrogels depends on crosslinks formed between polymer chains via chemical bonds or physical interactions. Hydrogels represent an important class of biomaterials in medicine because they exhibit excellent biocompatibility, cause minimal inflammatory responses and tissue damage, present mechanical properties similar to soft tissues, and have high permeability for oxygen and nutrients (28, 29). This makes hydrogels an interesting platform to culture and differentiate cells in 3D and to deliver them in vivo.

Hydrogels derived from natural polymers are frequently used in tissue engineering as scaffolds since the polymers are either components of or have macromolecular properties similar to the extracellular matrix. For example, hyaluronic acid (HA) is the simplest glycosaminoglycan and is found in nearly every mammalian tissue and fluid and is degraded enzymatically by hyaluronidases (30). It is a linear polysaccharide composed of a repeating disaccharide of (1-3) and (1-4)-linked β-D- glucuronic acid and N-acetyl-β-D-glucosamine units (30). This polymer is known to function as a microenvironmental cue that coregulates several cellular events including gene expression, signaling, proliferation, motility, metastasis, and morphogenesis (31, 32). Dextran is another natural polymer that is often used to prepare hydrogels for tissue-engineering applications. Dextran is a bacterial polysaccharide, consisting essentially of α-1,6 linked D-glucopyranose residues with a few percent of α-1,2, α-1,3, or α-1,4-linked side chains (33). Dextran-based hydrogels are biocompatible (34), biodegradable (33, 34), and cell-nonadhesive (34), which allows one to tailor its cell adhesiveness. Dextran-based hydrogels are an excellent platform to prepare bioactive scaffolds for stem cell differentiation, where the cell-adhesive properties of the scaffold can be determined exclusively by the immobilized bioactive factor, and not by the polymer.

1.3.2. Photopolymerizable Hydrogels as a Simple Platform to Encapsulate Cells

In some cases, hydrogel precursor solutions can be photopolymerized in the presence of photoinitiators and light (visible or ultraviolet), yielding a hydrogel (35). The light cleaves the photoinitiators, which will generate free radicals that subsequently will initiate the polymerization reaction to form cross-linked hydrogels. Photopolymerizable hydrogels are being evaluated for a number of tissue-engineering applications including cell encapsulation. In most cases, cells are not affected by the photopolymerization process, including hESCs (36). This allows one to seed cells throughout the scaffold material and form hydrogels in situ. In the case of stem cells, these hydrogels can be used as a 3D culture system to study in vitro the differentiation of stem cells, or they can be used as a vehicle for the in vivo transplantation and differentiation of stem cells. One major advantage of using photopolymerizable hydrogels for stem cell transplantation is that cell-containing hydrogels can be created in situ in a minimally invasive manner, for example using laparoscopic devices (35).

1.3.3. Hydrogels for Vascular Differentiation of hESCs

HA is highly expressed during embryonic development and participates in many morphogenetic steps during vertebrate development (31). The expression of HA in the very earliest stage of embryogenesis has been associated to the maintenance of an undifferentiated state. During the differentiation of the embryo, the expression of HA decreases and at this stage HA has been associated to the regulation of angiogenic responses (37). Specifically low molecular weight degradation products (3-10 disaccharide units) of HA have been reported to stimulate endothelial cell proliferation, migration, and sprout formation (37). Recently, we explored HA hydrogel culture systems for vascular differentiation of hESCs (36). hESCs were encapsulated in photopolymerizable HA hydrogels and cultured in MEF-conditioned medium for 1 week, after which the medium was replaced by angiogenic differentiation medium containing vascular endothelial growth factor (VEGF). After 1 week of differentiation, staining with specific vascular markers revealed that most sprouting cells were positive for smooth muscle actin whereas few were positive for CD34 (36).

Recently, we also reported the use of dextran hydrogels for the vascular differentiation of hESCs (36). For that purpose, we developed photopolymerizable dextran-based hydrogels comprising either insoluble (RGD sequences) or soluble (vascular endothelial growth factor ($VEGF_{165}$)) factors for the preferential differentiation of hESCs into the vascular lineage. The RGD peptide is the adhesive motif found in fibronectin, the earliest and most abundantly expressed extracellular matrix molecule during embryonic vascular development (38). $VEGF_{165}$ has been reported to induce the differentiation of ESCs into endothelial cells and involved as a survival factor of ESCs during hypoxia (13, 39). Our results showed that dextran-based hydrogels with different compositions can support the differentiation of hESCs and lead to over a 22-fold increase in the percentage of cells expressing the VEGF receptor KDR/Flk-1, a vascular marker, as compared to spontaneously differentiated EBs. When the cells were removed from these networks and cultured in media promoting further vascular differentiation, they contained higher fraction of vascular cells than the spontaneously differentiating EBs (36).

2. Materials and Equipment

2.1. Equipment

1. UV Lamp Blak-Ray 100AP (UVP Inc Cat# 95-0127-01)
2. Biosafety Cabinet
3. Fume Hood
4. Freeze-dryer
5. Magnetic stirrers
6. Sonicator Vibra Cell (Sonics & Materials Inc., Danbury)

7. Homogeneizer L4RT-A (Silverson)
8. Centrifuge
9. Spectrophotometer for 96-well plates
10. FACScan (BD Biosciences, NJ, USA)
11. Fluorescence microscopy (Nikon TE300 inverted microscope) or confocal laser scanning microscopy (Zeiss LSM510 Laser scanning confocal)

2.2. Differentiation Media (see Note 1)

1. Knockout DMEM (Invitrogen, Cat# 10829-018)
2. Fetal Bovine Serum (FBS) (Hyclone, Cat# SH30070.03)
3. L-glutamine (200 mM in 0.85% NaCl) (Invitrogen, Cat# 25030-081)
4. 2-mercaptoethanol (55 mM in PBSA) (Invitrogen, Cat# 21985-023)
5. Nonessential amino acid solution (Invitrogen, Cat# 11140-050)
6. Millipore* Stericup* Vacuum Filter 500 ml (Millipore, Cat# SCGPU05RE)

2.3. Growth and Differentiation of hESCs in Photocured HA-Based Hydrogels

1. Sterile Methacrylated HA (preparation has been reported elsewhere ((40); see **Note 2**)
2. 0.05% (w/v; in PBS) Irgacure 2959 (Cyba) solution
3. 1-ml syringes
4. Tissue culture-treated Petri dishes
5. MEF-conditioned media (7) or defined media for hESC growth (Stem Cell Technologies, mTeSR™1)
6. EGM-2 (Lonza, Cat# CC-4176)
7. $VEGF_{165}$ (50 ng/ml, R&D, Cat# 293-VE-010)
8. bFGF (4 ng/ml; Invitrogen, Cat# 13256-029)
9. Hyaluronidase (Sigma, Cat#H3884)
10. Medium: undifferentiating or differentiating medium (as described in **Section 6.1**)
11. For EB: nonadherent Petri dishes
12. For reculture: tissue culture-treated Petri dishes

2.4. Differentiation of hESCs in Photocured Dextran-Based Hydrogels

1. Poly(lactic acid-*co*-glycolic acid) (PLGA, Boehringer Ingelheim, Cat# Resomer 502H; see **Note 3**)
2. Methylene chloride (Sigma-Aldrich, Cat# 443484)
3. Poly(vinyl alcohol) 88 mol% hydrolyzed, Mw ~ 25,000 Da (PVA, Polysciences, Cat# 02975)
4. $VEGF_{165}$ (10 µg, R&D, Cat# 293-VE-010)

5. Magnesium hydroxide (Fluka, Cat# 63078)
6. Bovine serum albumin (Sigma-Aldrich, Cat# A9418)
7. Differentiation medium (as described in **Section 5.1**)
8. Dextran-acrylate with a degree of substitution of 20% (preparation has been reported elsewhere (41))
9. Acr-PEG-RGD (preparation has been reported elsewhere (36))
10. 0.5% (w/v; in PBS) Irgacure 2959 (Cyba) solution
11. 1-mL Syringes
12. Tissue culture-treated Petri dishes

2.5. Characterization of Encapsulated hESCs

1. MTT (Sigma-Aldrich, Cat# M5655-1G)
2. DMSO (Sigma-Aldrich, Cat# 472301)
3. Pellet Pestle (VWR, Cat# KT749521-1590)
   a. mL Microtube (VWR, Cat# KT749510-1590)
4. 96-well plate
5. FACS tubes (BD, Cat# 352235)
6. FBS (Hyclone, Cat# SH30070.03)
7. PBS without Ca^{2+} and Mg^{2+} (Invitrogen, Cat# 14190-250)
8. Collagenase B (Roche, Cat# 11088815001)
9. Nonenzymatic cell dissociation solution (Sigma, Cat# c5914-100 mL)
10. 40-μm cell strainer (BD, Cat# 352340)
11. Antibody monoclonal mouse antihuman SSEA4-PE (R&D Systems, Cat# FAB1435P; see **Note 4**)
12. PE-conjugated mouse IgG_3 isotype control (BD Pharmingen, Cat# 559926)
13. Antibody monoclonal mouse antihuman PECAM1-FITC (BD Pharmingen, Cat# 555445)
14. FITC-conjugated mouse $IgG_{1,k}$ isotype control (BD Pharmingen, Cat# 555748)
15. Antibody monoclonal mouse antihuman CD34-PE (Miltenyi Biotec, Cat# 130-081-002)
16. PE-conjugated mouse IgG_{2a} isotype control (Miltenyi Biotec, Cat# 130-091-835)
17. Antibody monoclonal mouse antihuman KDR/Flk-1-PE (R&D Systems, Cat# FAB357P)
18. PE-conjugated mouse IgG_1 isotype control (R&D Systems, Cat# IC002P)
19. Propidium iodide (Sigma-Aldrich, Cat# 70335)

20. Formalin-Free Fixative, Accustain® (Sigma)
21. Fetal Bovine Serum (FBS) (Hyclone, Cat# SH30070.03)
22. Anti-CD34 (1:20; Dako, Cat# M7165)
23. Anti-CD31 (1:20; Dako, Cat# M0823)
24. Anti-SMA (1:100; Dako, Cat# M0851)
25. Anti-VE-cad (1:200; Santa Cruz, Cat#sc-9989)
26. Anti-KDR (1:200; Santa Cruz, Cat# sc-6251)
27. Anti mouse Cy3 (1:50; Sigma, Cat# C2181)
28. Dapi (1:1000; Roche, Cat# 10236276)

3. Methods

3.1. Differentiation Media

Mix into a 500-ml Stericup: 400 mL Knockout DMEM, 100 mL of FBS, 2.5 mL of L-glutamine, 1 mL of 2-mercaptoethanol, and 3 mL of nonessential amino acid. Filter and store at 4°C (to be used up to a month after preparation).

3.2. Encapsulation and Culture of hESCs in Photocured HA-Based Hydrogels

1. Dissolve methacrylated HA at a concentration of 2 wt.% in PBS containing 0.05 wt.% Irgacure 2959 at 37°C for approximately 20 min. For this protocol, use total volume of 400 µl.
2. Removed hESCs from their feeder layer by incubation with collagen type IV (1 mg/mL in DMEM) for up to 1 h. Collect cell aggregates into 15-mL conical tube (BD Falcon).
3. Centrifuge at 800 rpm for 3 min.
4. Aspirate supernatant and suspend the cells in precursor solution (2 wt.% methacrylated HA as described above) to yield a final concentration of $0.5-1 \times 10^7$ cells per mL.
5. Pipette well and transfer 50 µL of the suspension into a sterile mold (end cut 1 mL syringe to obtain discs measuring 5 mm in diameter × 2 mm thick), and photopolymerize (~10 mW/cm^2 UV light, BlakRay) for 10 min.
6. Remove the gel from the syringe using the plunger and place it in a 24-well plate containing 2 mL of MEF-conditioned medium supplemented with 4 ng/ml bFGF (7) or defined media for hESC growth.
7. Culture the HA-hESCs constructs at 37°C, in a humidified atmosphere at 5% CO_2 (cell incubator), and replace media every day to maintain undifferentiating hESC colonies, or every second day to induce vascular differentiation.

Analysis: Throughout the cultivation in construct in undifferentiating medium, hESCs remained viable, undifferentiated, and proliferating (15). Exposure of encapsulated hESCs to differentiation medium containing VEGF resulted in cell sprouting and elongation after 48 h. After 1 week of differentiation, elongated cells were found to express specific endothelial and smooth muscle markers, indicating on their vascular lineage fate.

3.3. Release of hESCs from Photocured HA-Based Hydrogels (see Note 5)

1. Prepare release medium of hESC growth or differentiation medium containing 1000 or 2000 U/ml hyaluronidase.
2. Remove hESC-HA constructs from their culture media and into the release media.
3. Incubate for 24 h in a humidified atmosphere at 5% CO_2 (cell incubator).
   (i) For reculture, collect cells, centrifuge, wash three times with PBS to remove any hydrogel residues, resuspend in growth medium, and culture on MEF-coated dishes using standard methods (42).
   (ii) For EB formation, cultivate release hESCs in nonadherent Petri dishes in differentiation medium (as described in **Section 2.2**).

Analysis: At any time during culture, encapsulated hESCs could be released using hyaluronidase. Importantly, the release of hESCs from the HA hydrogels is associated with the full preservation of cell viability and undifferentiated state. Colonies released from the hydrogels readily adhered to MEFs and proliferated at rates normally seen in standard monolayer cultures. Furthermore, hESCs that were cultured in HA hydrogels for 30 days, released using hyaluronidase and subsequently cultured in suspension, were found to form EBs containing cell types representative of all three germ layers (15).

3.4. Encapsulation of hESCs in Photocured Dextran-Based Hydrogels

3.4.1. Preparation and Characterization of VEGF-Containing Micropartices

1. Prepare 2 L of 1% (w/v) PVA solution. From this solution, prepare 1 L of 0.5% (w/v) PVA solution. Keep these solutions at $-4°C$ until needed.
2. Dissolve 10 μg of $VEGF_{165}$ in 100 μL of BSA solution (50 mg/mL).
3. Prepare a solution of PLGA (100 mg in 2 mL of methylene chloride) in a 15-mL Falcon tube. Make this solution in the fume hood.
4. Add 4 mg of magnesium hydroxide (Note: magnesium hydroxide does not dissolve in methylene chloride).
5. Add 50 μL of $VEGF_{165}$ aqueous solution to the organic polymer solution and emulsify the aqueous and organic

phases by sonication (Vibra Cell, Sonics & Materials Inc., Danbury). Adjust the power output to 50 W and the time to 10 s.

6. Transfer the emulsion to a beaker containing 50 mL of 1% (w/v) PVA and homogenize at 7,500 rpm for 30 s (homogeneizer L4RT-A Silverson, equipped with a $^3/_4''$ tubular frame with a square hole high shear screen).

7. Add the previous emulsion to 100 mL 0.5% PVA and stir this suspension at 150 rpm for 3 h, at room temperature, to evaporate the organic solvent. This step should be done in the fume hood.

8. Transfer all the contents of the beaker to 50-mL Falcon tubes and centrifuge at 1,200 rpm for 7 min.

9. Suspend the particles in 40 mL distilled water and centrifuge at 1,200 rpm for 7 min. Repeat this step one more time.

10. Suspend the particles in 10 mL of distilled water and then freeze-dry the particle suspension for 2 days.

Analysis: Analysis of the microparticles by Scanning Electron Microscopy (SEM) allows one to have an estimate of their diameter (**Fig. 18.2A**). This protocol yields particles with a diameter around 6–7 μm. VEGF concentration and bioactivity within these particles can be assessed by using protocols reported elsewhere (36).

3.4.2. Encapsulation of hESCs in Dextran-Based Hydrogels (see Note 6)

1. UV sterilize the following gel components: 100 mg of dextran-acrylate macromonomer, 1.94 mg of Acr-PEG-RGD, and VEGF-containing microparticles for 30 min.

2. Dissolve the 100 mg of dextran-acrylate in 400 μL of PBS, at 37°C, for approximately 20 min.

3. Dissolve 1.94 mg of Acr-PEG-RGD in the dextran-acrylate solution (this will yield a solution of 0.5 mM of RGD).

4. hESC colonies were dissociated from the MEF feeder layer by collagen type IV (1 mg/mL in DMEM) during approximately 1 h (time for hESCs cultured in 10 cm Petri dishes; 30 min would be likely enough for hESCs in 6-well plates). The cell aggregates were collected in a 15-mL Falcon tube.

5. Wait 5 min until the cell aggregates form a pellet and remove the supernatant (most of the cells in the supernatant are MEFs and small hESC aggregates).

6. Wash the cell aggregates with 10 mL of PBS and centrifuge at 800 rpm for 3 min.

7. Remove the supernatant and suspend the cells in PBS to yield a final concentration of 30×10^6 cells per mL.

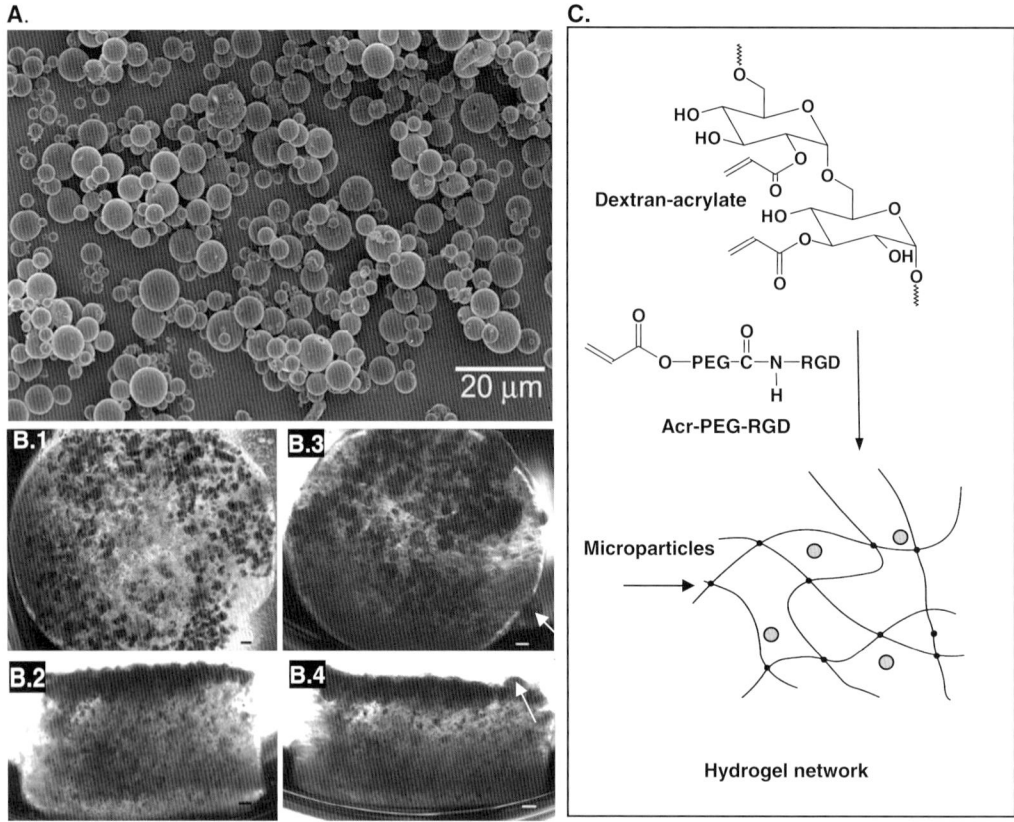

Fig. 18.2. hESC aggregates encapsulated within bioactive dextran-based hydrogels. (**A**) Scanning electron micrographs of 7-μm PLGA microparticles after preparation. (**B**) Distribution of hESC aggregates on dextran-based hydrogels with 0.5 mM Acr-PEG-RGD and 5 mg/mL VEGF-loaded microparticles, at day 0 (C.1, C.2) and day 10 (C.3,C.4). *Top* (C.1, C.3) and *side* (C.2, C.4) views. *Arrows* in C.3 and C.4 indicate clump of cells proliferating outside the hydrogel. (**C**) Schematic representation for the preparation of dextran-based hydrogels. *Scale bar* corresponds to 200 μm (Images reproduced from Ferreira et al. (36)).

8. Mix 400 μL of dextran-acrylate solution with the other components needed for differentiation including the sterilized microparticles by pipetting up and down.

9. Mix 400 μL of the cell suspension with the microparticle suspension prepared above and finally add 100 μL of photoinitiator.

10. Transfer 50 μL of the previous suspension to a 1-mL syringe (5 mm in diameter) and photopolymerize the suspension using a UV lamp Blak-Ray (4 mW/cm^2) for 10 min.

11. Remove the gel from the syringe using the plunger and place it in a 24-well plate containing 2 mL of differentiation medium (*see* **Section 2.2**).

12. Let the differentiation of cell constructs to proceed for 10 days at 37°C, in a humidified atmosphere at 5% CO_2 (cell incubator), and change the media every 3–4 days.

Analysis: Human ESC aggregates distribute across the network albeit preferentially located at one of the sides of the hydrogel, i.e., the farthest one from the UV light (**Fig. 12.2B**). This is due to the deposition of the hESC aggregates before the cross-linking reaction takes place. When these constructs are incubated in differentiation medium, cell aggregates at the surface proliferate and clumps of cells start to proliferate outside the network (**Fig. 18.2B.3 and B.4**).

3.5. Characterization of Viability and Proliferation of the Encapsulated Cells

1. Remove the differentiation media from the wells containing the cell-hydrogel constructs and add 1 mL of MTT solution (0.45 mg/mL in differentiation medium).

2. Incubate the samples at 37°C in a 5% humidified chamber for 4 h.

3. Transfer the cell-hydrogel construct to a 1.5-mL polypropylene tube using a spatula and add 0.2 mL of DMSO.

4. Break apart the gel with a pellet pestle to allow the solubilization of formazan crystals within the cells that presented mitochondrial metabolic activity.

5. Spin down the tubes at 1200 rpm for 3 min.

6. Transfer the supernatant to a 96-well plate and read at 540 nm.

Analysis: Human ESC aggregates encapsulated in bioactive dextran or HA hydrogels keep their viability overtime (15, 36). For bioactive dextran hydrogels, no statistical differences were observed for encapsulated cells between day 2 and day 10 (36). Human ESCs encapsulated in HA hydrogel maintain their viability for over 30 days in culture (15).

3.6. Characterization of Vascular Differentiation of Encapsulated hESCs

3.6.1. Analysis by Fluorescence-Activated Cell Sorting (FACS)

1. Remove the differentiation medium from the low-adhesion 24-well plate containing cell constructs (2–3 constructs per well) and place it in a 15-mL Falcon tube.

2. Separate dextran gel mechanically with a sterilized spatula into small pieces (as many as possible to increase surface area for digestion) or use hyaluronidase to release hESCs encapsulated in HA hydrogels (as described in **Section 6.2**).

3. Add 2 mL of a 0.4 U/mL collagenase B solution per well during 2 h, at 37°C, to dissociate the cells in the construct.

4. Remove the medium carefully from each well, wash the gel pieces with 5 mL PBS, and allow them to settle down for few minutes.

5. Remove the PBS and add 2 mL of cell dissociation solution for 10 min at 37°C, 5% CO_2, and finally break down the gel pieces with the pipette.

6. Filter the cell suspension through a 85-μm mesh strainer to remove cell clumps and gel pieces.

7. Centrifuge the cells at $1,200g$ for 3 min at room temperature, resuspend the cells in a solution of PBS containing 5% (v/v) FBS, and then count the cells in a hemocytometer.

8. Centrifuge the cells, suspend them in 5% (v/v) FBS in PBS at a concentration of 1.25×10^6 cells per mL, and aliquot 100 μL of the cell suspension (1.25×10^5 cells) in 15-mL Falcon tubes. For each antigen-specific staining, an isotype control staining should be done. (If your antibody is not already conjugated to the fluorescent marker and you will have to apply a secondary antibody containing the fluorescent marker, reserve also a fraction for secondary antibody only).

9. Add appropriate antibody solution (follow the antibody manufacturer guidelines) and place the tubes on ice for 30 min. Every 10 min, flick the tubes to make sure that mixing occurs.

10. Add 10 mL of 5% FBS in PBS and centrifuge the cells at $1,200g$ for 3 min, at room temperature.

11. Add 200–250 μL of 5% FBS in PBS to the centrifuged cells and transfer this labeled cell suspension to a FACS tube.

12. Add 2 μL of propidium iodide solution (0.01% v/v) in 5% FBS in PBS to the cell suspension.

13. Take the tubes to the FACS facility and run the FACS experiment.

Analysis: For differentiation experiments from hESCs, it is important to examine the percentage of hESCs that remain during differentiation. Typical markers for undifferentiated hESC include octamer 4 (Oct-4), stage specific embryonic antigens 3 or 4 (SSEA-3 or SSEA-4), alkaline phosphatase, and nanog. Oct-4 is a transcription factor of the POU family and is important for the pluripotency of hESCs (43, 44). Oct-4 is a nuclear transcriptional factor and it will be localized to the cell nucleus. Human ESCs which express Oct4 are undifferentiated. SSEA-3 and SSEA-4 are glycoproteins expressed early in embryonic development and in pluripotent stem cells. SSEA-3 and SSEA-4 are expressed in undifferentiated hESCs while SSEA-1 is expressed during differentiation of hESCs (45). Alkaline phosphatase is a hydrolase enzyme that removes phosphate groups from many molecules including nucleotides and proteins. Undifferentiated hESCs exhibit a strong signal for ALP (46). Nanog is a transcription factor that is involved with self-renewal of undifferentiated ESCs (47).

A combination of different markers is generally used to monitor the vascular differentiation of hESCs encapsulated in hydrogels. Specifically for endothelial cells, there are few, if any, protein/RNAm markers that are both specifically and uniformly expressed in the endothelium (48). For example, platelet endothelial cell adhesion molecule-1 (PECAM1; also called CD31), one of the leading endothelial markers, is expressed on endothelial cells and monocytes; vascular endothelial- cadherin (VE-CAD), a Ca^{2+}-dependent protein that functions to mediate cell–cell binding, is expressed in endothelial cells, trophoblasts, and some macrophages; CD34, a transmembrane protein is expressed on endothelial cells as well as hematopoietic stem cells (49). In addition, endothelial phenotype changes according to their differentiation/maturation stage and their final localization in the human body. For example, we recently reported that endothelial progenitor cells isolated from hESCs express different levels of CD34 and Flk-1/KDR overtime during their differentiation (13). Moreover, some markers are only expressed after activation by inflammatory cytokines or growth factors (49). Taken together, to identify and monitor endothelial progenitor cells, we should use a combination of markers including, PECAM1, CD34, Flk-1/KDR, VE-CAD, vWF, Tie-2, Ang2, acetylated-LDL uptake, among others (49, 50).

3.6.2. Analysis by Immunofluorescence (see Note 7)

1. Remove hESC constructs (in HA or Dextran hydrogels) from culture and place in 1 mL fixative (Accustain) for 20 min at 37°C.
2. Rinse twice with PBS and incubate with 2% (w/v) BSA in PBS for 1 h.
3. Incubate with desirable primary antibodies diluted in PBS (see above reagent and materials section for antibodies and dilutions) for 1 h at 37°C.
4. Rinse three times with PBS and incubate for 30 min with antimouse Cy3 secondary antibody at 37°C in dark.
5. Rinse three times with PBS and add DAPI to the last rinse.
6. Visualize the immuno-labeled hESCs using either fluorescence microscopy or confocal laser scanning microscopy.

Analysis: $VEGF_{165}$ contributing to the vascular differentiation of hESCs was examined using both dextran and HA hydrogels. Differentiation of hESCs encapsulated in VEGF-releasing dextran hydrogels was compared to EB differentiation in medium supplemented with VEGF. Either in EBs or encapsulated hESC aggregates, KDR/Flk-1+ cells did not form vascular networks, however, encapsulated hESCs express higher levels of KDR/Flk-1 marker than EBs (**Fig. 18.3A, B**). For hESCs encapsulated in HA hydrogels, sprouting and cell elongation were observed when cultured for one week in EGM-2 differentiation medium supplemented with $VEGF_{165}$. Staining with specific vascular markers revealed that few were positive for CD34 while most sprouting cells were positive for smooth muscle actin (**Fig. 18.3C, D**) (15).

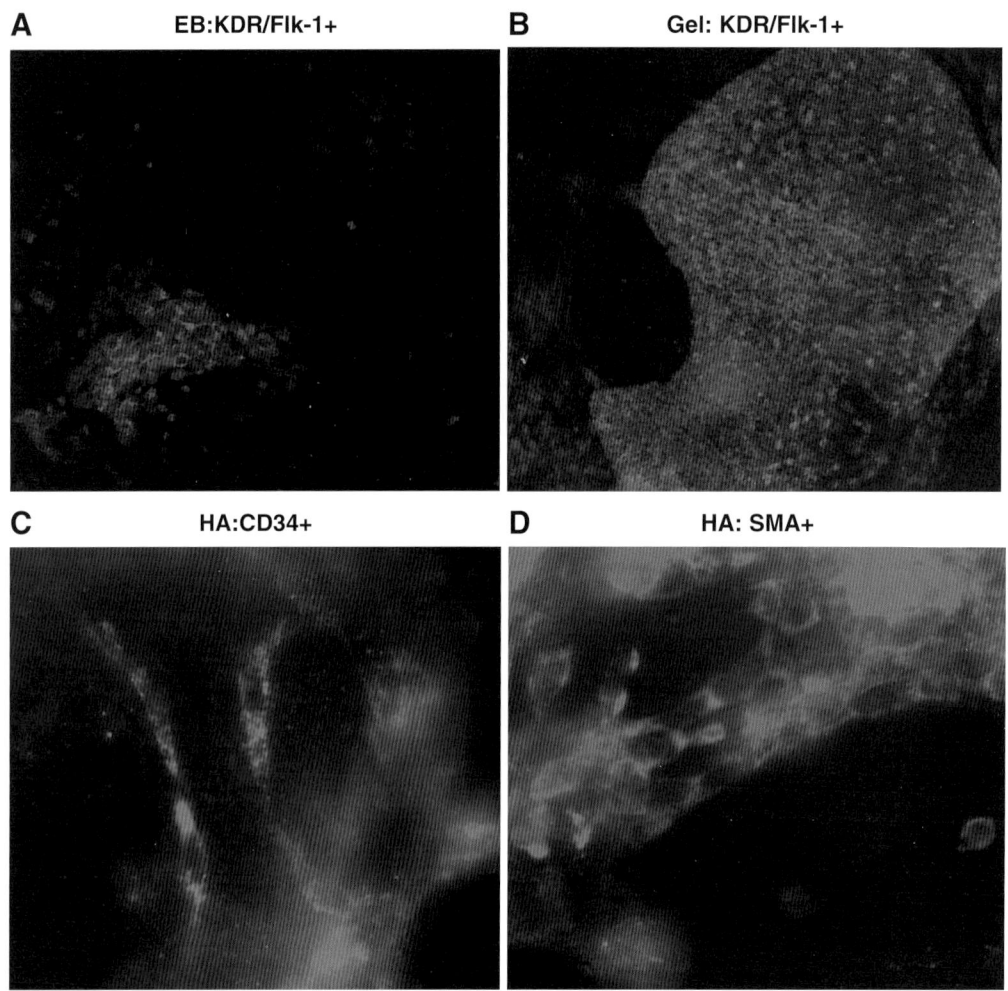

Fig. 18.3. Localization and organization of vascular markers on encapsulated hESCs. Confocal images of: (**A**) KDR/Flk-1$^+$ cells from EBs and from (**B**) hESC aggregates encapsulated in dextran-based hydrogels with 0.5 mM Acr-PEG-RGD containing 5 mgmL^{-1} of 7-μm microparticles loaded with VEGF$_{165}$ (Images reproduced from Ferreira et al. (36); (**C**) CD34+ and (**D**) SMA+ cells sprout from hESCs in HA cultured in medium supplemented with VEGF$_{165}$.

3.7. Conclusions and Future Perspectives

In recent years, tremendous advances have been made in the field of vascular differentiation of hESCs, positioning them as a reliable source for vascular therapeutics and vascular engineering. Various methods have been developed and have demonstrated the efficient isolation of vascular progenitors and their functional derivatives, ECs and SMCs, from hESCs. Development of biomaterials capable of presenting specific environmental cues of choice offers a unique opportunity to generate instructive 3D environments for vascular assembly. Therefore, hESC-based vascular engineering is in an advantageous position to develop new methods for designing and controlling in vitro vascular microenvironments, for the fundamental insight into effects of factors that guide differentiation and cellular organization, as well as for the practical outcome of constructs engineered in vitro for vascular therapeutic implications. Given the complex microenvironments involved in vasculogenesis during early embryogenesis, challenges still remain including to better monitor, control, and quantify the kinetics of vascular differentiation, maturation, tissue organization, and functionality within the 3D scaffolds.

4. Notes

1. For simplicity, we focus herein on the vascular differentiation conditions. In all our studies, undifferentiated hESCs were propagated on inactivated MEFs in growth medium consisting of 80% KnockOut DMEM, supplemented with 20% Knock-Out Serum Replacement, 4 ng/ml basic fibroblast growth factor, 1 mM L-glutamine, 0.1 mM β–mercaptoethanol, and 1% nonessential amino acid stock (Invitrogen Corporation, Carlsbad, CA, USA). hESCs were passaged every 4–6 days using 1 mg/ml type IV collagenase (Invitrogen Corporation, Carlsbad, CA, USA).

2. In this protocol, we describe HA hydrogels that were prepared from methylacrylate HA synthesized as previously described (40). In vitro and in vivo studies have shown that these gels are biocompatible (40, 51, 52). During in vitro culture, undifferentiated hESCs expressed high levels of HA receptors, CD44 and CD168 (**Fig. 18.4A**). The addition of human fluorescein-HA to the culture of hESCs on MEFs resulted in the localization of HA receptors to the cell membranes, first at the edges of cell colonies and then at their centers (**Fig. 18.4B**). Fluorescein-HA was observed to be internalized and localized within the cells. No internalization of fluorescein-HA could be observed once anti CD44 was added to the cultures of hESCs, indicating receptor-mediated internalization of HA by hESCs. Immunofluorescence of hESC colonies cultured on MEFs revealed that densely packed colonies expressed human hyaluronidase Hyal 1&2. RT-PCR analysis corroborated that hESCs express high levels of Hyal 2, one of the isoforms of human hyaluronidase (**Fig. 18.4C**). Using HA hydrogel, encapsulated hESCs can be maintained as undifferentiated when cultured in MEF-conditioned media supplemented with 4 ng/ml bFGF (7) or defined media as previously demonstrated (8). To induce differentiation, encapsulated hESCs can be transferred or directly cultured in differentiation medium of EGM-2 supplemented with $VEGF_{165}$ (15).

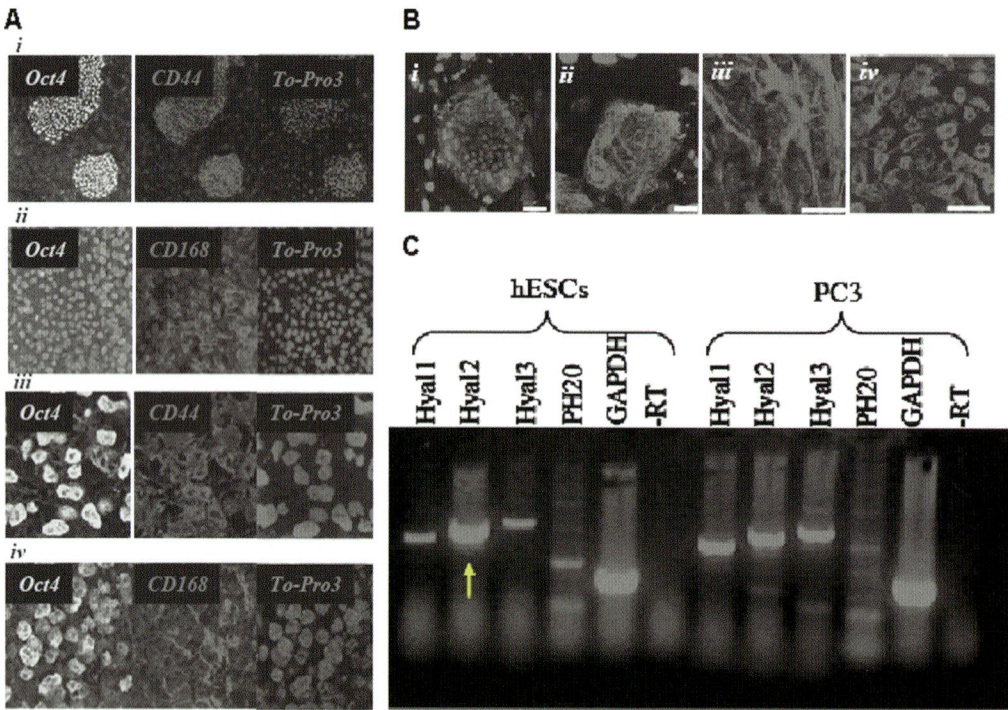

Fig. 18.4. HA putative role in cultures of undifferentiated hESCs. (**A**) (**i**)–(**ii**). Undifferentiated hESC colonies showed to express undifferentiated cell markers Oct4 and CD44 or CD168, respectively. (**iii**)–(**iv**). Higher magnification revealed intracellular expression of CD44 and either membrane or intracellular expression of CD168. (**B**) Confocal analysis revealed relocalization of HA receptors in hESC membranes of both (**i**) CD44 and (**ii**) CD168 (nuclei are shown as well) in response to addition of human fluorescein-HA to the growth medium. Higher magnification of CD168 localization is shown with addition of human HA (**iii**) and without (**iv**). (**C**) RT-PCR analysis revealed high expression levels of a hyaluronidase isomer, Hyal 2, in undifferentiated hESCs. PC3 line served as positive control. *Scale bars* = 100 μm (Images reproduced from Gerecht et al. (15)).

3. PLGA polymer has been approved by FDA for use in medical devices and controlled release systems (53). PLGA microparticles present several advantages for the controlled release of growth factors, including their inherent biocompatibility, degradation at physiologic pH, and release rates up to days, weeks, or months (53). However, PLGA microparticles might have a negative effect on growth factor stability during the preparation and storage, primarily due to the acids (lactic and glycolic acids) released during their degradation (54). The accumulation of acids in the interior of microparticles and the consequent reduction of pH might denature the encapsulated growth factor. In our protocol, magnesium hydroxide is incorporated in the microparticle formulation to buffer the acidic environment during their degradation (54). PLGA polymers are commercially available from different vendors. There are four major suppliers of good manufacturing practice-grade PLGA polymers including Boehringer

Ingelheim (trade name: Resomer), Purac (trade name: Purasorb), Alkermes (trade name: Medisorb), and Absorbable Polymers International (trade name: Lactel). Depending on the ratio of lactide to glycolide in PLGA, the degradation and growth factor release kinetics is affected (36). The degradation is related to the monomer ratio in the copolymer, i.e., the higher the content of glycolide units, the lower the time required for degradation. In this protocol, we use PLGA 50:50 a copolymer (Resomer 502H; polymer inherent viscosity of 0.16-0.24) (36) whose composition is 50% lactic acid and 50% glycolic acid. Several methods are available to prepare PLGA microparticles (55). The protocol below describes a double-emulsion approach to prepare microparticles of 7 μm in diameter.

4. The antibody that is utilized in this protocol has already been conjugated to a fluorescent marker. However, the unavailability of fluorescence-conjugated monoclonal antibodies for certain antigens requires that the cell surface antigen would be bound by an antibody that would then be attached to a fluorescence-conjugated secondary antibody. Protocol variations for this have been noted below.

5. To enable the use of HA hydrogels for research and cellular therapy, hESCs need to be released from the hydrogel with preservation of high viability. This can be achieved by treatment with hyaluronidase. First, hESCs were examined for their viability during a long-term treatment (i.e., 24 h). Human ESC colonies incubated with growth medium containing 100–2000 U/ml hyaluronidase preserved their normal morphology with no apparent loss of viability (15). Incubation of HA-hESC constructs in growth medium containing 2000 U/ml hyaluronidase resulted in complete degradation of the hydrogel. We also found that the viability of hESCs incubated with 2000 U/ml HAase for 24 h was comparable with that measured for incubation with 1 mg/ml collagenase IV for 30 min (76.5 ± 8% vs. 70 ± 4.5%, respectively). In contrast, hyaluronidase concentrations of <1000 U/mL resulted in only partial degradation of HA hydrogels over a 24 h period and were associated with low efficiency of hESC retrieval. Importantly, the release of hESCs from the HA hydrogels was associated with the full preservation of cell viability and undifferentiated state.

6. In this protocol, we describe dextran hydrogels that were prepared from dextran-bearing acrylate groups (**Fig. 18.2C**) that were synthesized by a novel procedure reported by us (41). In vitro and in vivo biocompatibility studies have shown that these gels are biocompatible, however, cell-non-adhesive (34). To create cell adhesive sites in dextran gels, we cross-linked

dextran-bearing acrylate groups with RGD peptide attached to an acrylated poly(ethylene glycol) moiety (acryloyl poly(ethylene glycol)-Arg-Gly-Asp (Acr-PEG-RGD)) (36) (**Fig. 18.2C**). The protocol described in **Section 3.4** does not include the preparation of dextran macromonomers as well as the preparation of RGD-PEG-Acr. To control the differentiation of hESCs, VEGF-microparticles were incorporated in the gel network (**Fig. 18.2C**).

7. This protocol is for the in situ staining of encapsulated hESCs in either HA or dextran hydrogels. Several markers for vascular lineages are proposed, though other markers could be examined as well. For analysis, we have been using either fluorescence or confocal microscopy depending on the resolution and the imaging complexity required.

References

1. Ferreira LS, Gerecht S, Shieh HF, Watson N, Rupnick MA, Dallabrida SM, et al. Vascular progenitor cells isolated from human embryonic stem cells give rise to endothelial and smooth muscle-like cells and form vascular networks in vivo. *Circulation Research* 2007; 101(3):286–294.

2. Gerecht-Nir S, Ziskind A, Cohen S, Itskovitz-Eldor J. Human embryonic stem cells as an in vitro model for human vascular development and the induction of vascular differentiation. *Lab Invest* 2003; 83(12):1811–1820.

3. Levenberg S, Golub JS, Amit M, Itskovitz-Eldor J, Langer R. Endothelial cells derived from human embryonic stem cells. *Proc Natl Acad Sci USA* 2002; 99(7):4391–4396.

4. Wang L, Li L, Shojaei F, Levac K, Cerdan C, Menendez P, et al. Endothelial and hematopoietic cell fate of human embryonic stem cells originates from primitive endothelium with hemangioblastic properties. *Immunity* 2004; 21(1):31–41.

5. Wang ZZ, Au P, Chen T, Shao Y, Daheron LM, Bai H, et al. Endothelial cells derived from human embryonic stem cells form durable blood vessels in vivo. *Nat Biotechnol* 2007; 25(3):317-318.

6. Thomson JA, Itskovitz-Eldor J, Shapiro SS, Waknitz MA, Swiergiel JJ, Marshall VS, et al. Embryonic stem cell lines derived from human blastocysts. *Science* 1998; 282 (5391):1145–1147.

7. Xu C, Inokuma MS, Denham J, Golds K, Kundu P, Gold JD, et al. Feeder-free growth of undifferentiated human embryonic stem cells. *Nat Biotechnol* 2001; 19(10):971–974.

8. Ludwig TE, Levenstein ME, Jones JM, Berggren WT, Mitchen ER, Frane JL, et al. Derivation of human embryonic stem cells in defined conditions. *Nat Biotechnol* 2006; 24(2):185–187.

9. Ludwig TE, Bergendahl V, Levenstein ME, Yu J, Probasco MD, Thomson JA. Feeder-independent culture of human embryonic stem cells. *Nat Methods* 2006; 3(8):637–646.

10. Reubinoff BE, Pera MF, Fong CY, Trounson A, Bongso A. Embryonic stem cell lines from human blastocysts: somatic differentiation in vitro. *Nat Biotechnol* 2000; 18(4):399–404.

11. Gerecht-Nir S, Ziskind A, Cohen S, Itskovitz-Eldor J. Human embryonic stem cells as an in vitro model for human vascular development and the induction of vascular differentiation. *Lab Invest* 2003; 83(12): 1811–1820.

12. Itskovitz-Eldor J, Schuldiner M, Karsenti D, Eden A, Yanuka O, Amit M, et al. Differentiation of human embryonic stem cells into embryoid bodies compromising the three embryonic germ layers. *Mol Med* 2000; 6(2):88–95.

13. Ferreira LS, Gerecht S, Shieh HF, Watson N, Rupnick MA, Dallabrida SM, et al. Vascular progenitor cells isolated from human embryonic stem cells give rise to endothelial and smooth muscle like cells and form vascular networks in vivo. *Circ Res* 2007; 101(3):286–294.

14. Levenberg S, Huang NF, Lavik E, Rogers AB, Itskovitz-Eldor J, Langer R. Differentiation of human embryonic stem cells on

three-dimensional polymer scaffolds. *Proc Natl Acad Sci USA* 2003; 100(22): 12741–12746.

15. Gerecht S, Burdick JA, Ferreira LS, Townsend SA, Langer R, Vunjak-Novakovic G. Hyaluronic acid hydrogel for controlled self-renewal and differentiation of human embryonic stem cells. *Proc Natl Acad Sci USA* 2007; 104(27):11298–11303.

16. Ferreira LS, Gerecht S, Fuller J, Shieh HF, Vunjak-Novakovic G, Langer R. Bioactive hydrogel scaffolds for controllable vascular differentiation of human embryonic stem cells. *Biomaterials* 2007; 28(17):2706–2717.

17. Gerecht S, Townsend SA, Pressler H, Zhu H, Nijst CLE, Bruggeman JP, et al. A porous photocurable elastomer for cell encapsulation and culture. *Biomaterials* 2007; 28(32):4826–4835.

18. Gerecht-Nir S, Cohen S, Ziskind A, Itskovitz-Eldor J. Three-dimensional porous alginate scaffolds provide a conducive environment for generation of well-vascularized embryoid bodies from human embryonic stem cells. *Biotechnol Bioeng* 2004; 88(3): 313–320.

19. Levenberg S, Zoldan J, Basevitch Y, Langer R. Endothelial potential of human embryonic stem cells. *Blood* 2007; 110(3):806–814.

20. Levenberg S, Golub JS, Amit M, Itskovitz-Eldor J, Langer R. Endothelial cells derived from human embryonic stem cells. *Proc Natl Acad Sci USA* 2002; 99(7):4391–4396.

21. Kaufman DS, Hanson ET, Lewis RL, Auerbach R, Thomson JA. Hematopoietic colony-forming cells derived from human embryonic stem cells. *Proc Natl Acad Sci USA* 2001; 98(19):10716–10721.

22. Vodyanik MA, Bork JA, Thomson JA, Slukvin II. Human embryonic stem cell-derived CD34+ cells: Efficient production in the coculture with OP9 stromal cells and analysis of lymphohematopoietic potential. *Blood* 2005; 105(2):617–626.

23. Zambidis ET, Peault B, Park TS, Bunz F, Civin CI. Hematopoietic differentiation of human embryonic stem cells progresses through sequential hematoendothelial, primitive, and definitive stages resembling human yolk sac development. *Blood* 2005; 106(3):860–870.

24. Philp D, Chen SS, Fitzgerald W, Orenstein J, Margolis L, Kleinman HK. Complex extracellular matrices promote tissue-specific stem cell differentiation. *Stem Cells* 2005; 23(2):288–296.

25. Levenberg S, Burdick JA, Kraehenbuehl T, Langer R. Neurotrophin-induced differentiation of human embryonic stem cells on three-dimensional polymeric scaffolds. *Tissue Eng* 2005; 11(3–4):506–512.

26. Levenberg S, Huang NF, Lavik E, Rogers AB, Itskovitz-Eldor J, Langer R. Differentiation of human embryonic stem cells on three-dimensional polymer scaffolds. *Proc Natl Acad Sci USA* 2003; 100(22):12741–12746.

27. Lutolf MP, Hubbell JA. Synthetic biomaterials as instructive extracellular microenvironments for morphogenesis in tissue engineering. *Nat Biotechnol* 2005; 23(1):47–55.

28. Peppas NA, Bures P, Leobandung W, Ichikawa H. Hydrogels in pharmaceutical formulations. *Eur J Pharm Biopharm* 2000; 50(1):27–46.

29. Drury JL, Mooney DJ. Hydrogels for tissue engineering: scaffold design variables and applications. *Biomaterials* 2003; 24(24): 4337–4351.

30. Laurent TC, Fraser JR. Hyaluronan. *Faseb J* 1992; 6(7):2397–2404.

31. Toole BP. Hyaluronan in morphogenesis. *Semin Cell Dev Biol* 2001; 12(2):79–87.

32. Toole BP. Hyaluronan: from extracellular glue to pericellular cue. *Nat Rev Cancer* 2004; 4(7):528–539.

33. Mehvar R. Dextrans for targeted and sustained delivery of therapeutic and imaging agents. *J Control Release* 2000; 69(1):1–25.

34. Ferreira L, Rafael A, Lamghari M, Barbosa MA, Gil MH, Cabrita AM, et al. Biocompatibility of chemoenzymatically derived dextran-acrylate hydrogels. *J Biomed Mater Res A* 2004; 68(3):584–596.

35. Nguyen KT, West JL. Photopolymerizable hydrogels for tissue engineering applications. *Biomaterials* 2002; 23(22):4307–4314.

36. Ferreira LS, Gerecht S, Fuller J, Shieh HF, Vunjak-Novakovic G, Langer R. Bioactive hydrogel scaffolds for controllable vascular differentiation of human embryonic stem cells. *Biomaterials* 2007; 28(17):2706–2717.

37. Slevin M, Kumar S, Gaffney J. Angiogenic oligosaccharides of hyaluronan induce multiple signaling pathways affecting vascular endothelial cell mitogenic and wound healing responses. *J Biol Chem* 2002; 277(43): 41046–41059.

38. Francis SE, Goh KL, Hodivala-Dilke K, Bader BL, Stark M, Davidson D, et al. Central roles of alpha5beta1 integrin and fibronectin in vascular development in mouse embryos and embryoid bodies. *Arterioscler Thromb Vasc Biol* 2002; 22(6):927–933.

39. Brusselmans K, Bono F, Collen D, Herbert JM, Carmeliet P, Dewerchin M. A novel role for vascular endothelial growth factor as an autocrine survival factor for embryonic stem cells during hypoxia. *J Biol Chem* 2005; 280(5):3493–3499.

40. Burdick JA, Chung C, Jia X, Randolph MA, Langer R. Controlled degradation and mechanical behavior of photopolymerized hyaluronic acid networks. *Biomacromolecules* 2005; 6(1):386–391.

41. Ferreira L, Gil MH, Dordick JS. Enzymatic synthesis of dextran-containing hydrogels. *Biomaterials* 2002; 23(19):3957–3967.

42. Amit M, Itskovitz-Eldor J. Derivation and spontaneous differentiation of human embryonic stem cells. *J Anat* 2002; 200(Pt 3):225–232.

43. Donovan PJ. High Oct-ane fuel powers the stem cell. *Nat Genet* 2001; 29(3):246–247.

44. Niwa H, Miyazaki J, Smith AG. Quantitative expression of Oct-3/4 defines differentiation, dedifferentiation or self-renewal of ES cells. *Nat Genet* 2000; 24(4):372–376.

45. Ginis I, Luo Y, Miura T, Thies S, Brandenberger R, Gerecht-Nir S, et al. Differences between human and mouse embryonic stem cells. *Dev Biol* 2004; 269(2):360–380.

46. Draper JS, Pigott C, Thomson JA, Andrews PW. Surface antigens of human embryonic stem cells: changes upon differentiation in culture. *J Anat* 2002; 200(Pt 3):249–258.

47. Chambers I, Colby D, Robertson M, Nichols J, Lee S, Tweedie S, et al. Functional expression cloning of Nanog, a pluripotency sustaining factor in embryonic stem cells. *Cell* 2003; 113(5):643–655.

48. Aird WC. Phenotypic heterogeneity of the endothelium: I. Structure, function, and mechanisms. *Circ Res* 2007; 100(2):158–173.

49. Garlanda C, Dejana E. Heterogeneity of endothelial cells. Specific markers. *Arterioscler Thromb Vasc Biol* 1997; 17(7):1193–1202.

50. Rafii S, Lyden D. Therapeutic stem and progenitor cell transplantation for organ vascularization and regeneration. *Nat Med* 2003; 9(6):702–712.

51. Chung C, Mesa, J., Randolph, M.A., Yaremchuk, M., Burdick, J.A. Influence of gel properties on neocartilage formation by auricular chondrocytes photoencapsulated in hyaluronic acid networks. *J Biomed Mater Res A* 2006; 77A:518–525.

52. Ifkovits JL, Burdick JA. Review: photopolymerizable and degradable biomaterials for tissue engineering applications. *Tissue Engineering* 2007; 13(10):2369–2385.

53. Mundargi RC, Babu VR, Rangaswamy V, Patel P, Aminabhavi TM. Nano/micro technologies for delivering macromolecular therapeutics using poly(D,L-lactide-co-glycolide) and its derivatives. *J Control Release* 2008; 125(3):193–209.

54. Zhu G, Mallery SR, Schwendeman SP. Stabilization of proteins encapsulated in injectable poly (lactide- co-glycolide). *Nat Biotechnol* 2000; 18(1):52–57.

55. Astete CE, Sabliov CM. Synthesis and characterization of PLGA nanoparticles. *J Biomater Sci Polym Ed* 2006; 17(3): 247–289.

56. Vodyanik MA, Thomson JA, Slukvin, II. Leukosialin (CD43) defines hematopoietic progenitors in human embryonic stem cell differentiation cultures. *Blood* 2006; 108(6):2095–2105.

57. Gerecht S, Burdick JA, Ferreira LS, Townsend SA, Langer R, Vunjak-Novakovic G. Hyaluronic acid hydrogel for controlled self-renewal and differentiation of human embryonic stem cells. *Proc Natl Acad Sci USA* 2007; 104(27): 11298–11303.

Chapter 19

Differentiation of Neural Precursors and Dopaminergic Neurons from Human Embryonic Stem Cells

Xiao-Qing Zhang and Su-Chun Zhang

Abstract

Directed differentiation of human embryonic stem cells (hESCs) to a functional cell type, including neurons, is the foundation for application of hESCs. We describe here a reproducible, chemically defined protocol that allows directed differentiation of hESCs to nearly pure neuroectodermal cells and neurons. First, hESC colonies are detached from mouse fibroblast feeder layers and form aggregates to initiate the differentiation procedure. Second, after 4 days of suspension culture, the ESC growth medium is replaced with neural induction medium to guide neuroectodermal specification. Third, the differentiating hESC aggregates are attached onto the culture surface at day 6–7, where columnar neural epithelial cells appear and organize into rosettes. Fourth, the neural rosettes are enriched by detaching rosettes and leaving the peripheral flat cells attached and expanded as neuroepithelial aggregates in the same medium. Finally, the neuroepithelial aggregates are dissociated and differentiated to nearly pure neurons. This stepwise differentiation protocol results in the generation of primitive neuroepithelia at day 8–10, neural progenitors at the second and third week, and postmitotic neurons at the fourth week, which mirrors the early phase of neural development in a human embryo. Identification of the primitive neuroepithelial cells permits efficient patterning of region-specific progenitors and neuronal subtypes such as midbrain dopaminergic neurons.

Key words: Neural induction, embryonic stem cells, differentiation, cell replacement, dopaminergic neurons.

1. Introduction

Directed differentiation of specific lineages has been a focal point in the field of human embryonic stem cell (hESC) research (1, 2). The differentiation paradigms are key to revealing the cellular and molecular mechanisms underlying early human development, exploring tissue engineering, and devising cell replacement therapy and drug screening. Ideally, the differentiation protocol should be based on the developmental principles and truly directed (controllable). It should also be simple, defined, and reproducible.

We have developed a reproducible, chemically defined protocol that allows directed differentiation of hESCs to a synchronized population of neuroectodermal cells and then a nearly pure population of neurons (3, 4). The design of the protocol rests essentially on the fundamental principle of neuroectodermal induction (5). As such, the specification of neuroepithelial cells and subsequent neuronal and glial differentiation mirror the time course of human embryo development predicted from animal embryological studies and limited human specimens. We avoid any unknown components, such as sera or stroma cells, in the system so that the culture is chemically defined, allowing flexible modification for mechanistic analyses. Because of the simplicity, the protocol has been proven highly reproducible for the past decade. The neural differentiation is also robust with over 90% efficiency (3, 4). This endows the system particularly amenable for biochemical analysis that often requires materials of large amount. Technically, the adherent colony culture also permits continual observation of morphological transformation along the differentiation processes.

Under this protocol, differentiating cells manifest five morphologically identifiable stages, the ESCs, ESC aggregates, neuroepithelia in the form of neural tube-like rosettes, neural progenitors in neuroepithelial aggregates, and postmitotic neurons. The undifferentiated state of the starting cell population is essential for the successful replication of the protocol. The quality of hESCs is often overlooked when it comes to differentiation. In reality, partially differentiated hESCs will unavoidably result in unsynchronized differentiation, such as the generation of postmitotic neurons within the first week along with neuroepithelial generation. Human ESCs, cultured on mouse embryonic fibroblast (MEF) feeder layers, are usually passaged every 5–6 days before differentiation. At this stage, the ESC colonies are generally uniform in size and they form aggregates of similar size after detaching from the MEF layers. These ESC aggregates are initially suspended in the ESC growth medium for 4 days to promote cell survival and then the medium is replaced with the serum-free neural induction medium to guide the cells toward the ectoderm fate. This is evidenced by transient expression of FGF5 and uniform expression of Otx2 and Sox2, but lack of brachyury and α-fetoprotein (AFP), meso-endodermal markers (4). Prolonged culture in the ESC growth medium will lead to the formation of "embryoid bodies," which is not ideal for neural differentiation.

Unlike most of the neural differentiation protocols that involve long-term aggregated culture or coculture with stroma cells, in our protocol, the ESC aggregates are reseeded onto a culture dish (without feeder) and they reform individual monolayer colonies.

This permits individual cells to be exposed to the culture environment evenly for synchronized differentiation. It also allows morphological observation. After 8–10 days of differentiation, neuroepithelia, characterized by columnar epithelial morphology, appear. Their identity is confirmed by expression of a host of neuroectodermal transcription factors including Pax6, Lhx2, Six3, and Sox2, and the lack of other germ layer markers (4). By two weeks of differentiation, these epithelial cells organize into rosettes, reminiscent of the cross section of the neural tube. These cells express Sox1 in addition to the above neural markers and exhibit polarity by concentrated expression of N-cadherin in the lumen (3, 4). This stage of differentiation can be subdivided into two morphologically distinct stages. The early columnar neuroepithelia, which we term primitive anterior neuroepithelia, uniformly express anterior transcription factors, but can be readily specified to other regional identities (6–9). Therefore, the primitive neuroepithelia behave like true neural stem cells. We will use midbrain dopaminergic neuron differentiation to illustrate the point. The late columnar neuroepithelia, usually with a lumen inside and expression of definitive neural marker Sox1, cannot be patterned to region-specific progenitors and are termed as definitive neuroepithelia (7, 9).

The neuroepithelia in the form of neural tube-like rosettes can be readily detached from the culture surface and expanded in suspension like "neurospheres" although these cells are relatively primitive comparing to the progenitor aggregates derived from the embryonic or adult brain. These cells, upon dissociation and plating onto substrate, will differentiate to postmitotic neurons.

2. Materials

2.1. Supplies

1. Polystyrene multidishes, 6-well and 24-well (Nunc; Cat. Nos 140675 and 142475).
2. 60 × 15-mm Petri dish (Fisher Scientific; Cat. No. 08-757-13A).
3. T25 flasks, the polystyrene flasks with polyethylene filter cap (TPP, Trasadingen, Switzerland; Cat. No. 90026).
4. T75 flasks, the polystyrene flasks with polyethylene filter cap (Nunc; Cat. No. 156499).
5. Polystyrene conical tube, 15- and 50-ml (Research Products International Corp; Cat. Nos 163224 and 163228).
6. Serological pipettes 5-, 10-, and 25-ml (Fisher Scientific; Cat. Nos 13-678-11D, 13-678-11E, and 13-678-11).
7. 9" Pasteur pipettes (Fisher Scientific; Cat. No. 13-678-20D).

8. Steriflip sterile disposable vacuum filtration system (Millipore, Cat. No. SCGP00525).

9. Stericup vacuum-driven disposable filtration system (Millipore, Cat. No. SCGPU05RE).

2.2. Stock Solutions

1. L-Glutamine solution (200 mM) (Gibco-BRL, Rockville, MD; Cat. No. 25030). Make aliquots of 5 mL and store at –20°C.

2. MEM nonessential amino acids solution (Gibco-BRL; Cat. No. 11140).

3. Knockout serum replacer (Gibco-BRL; Cat. No. 10828). Store stock in –80°C. Make aliquots of 50 mL and store at –20°C if it cannot be used up in a week after thawed.

4. Dulbecco's modified eagle medium: Nutrient mixture F-12 1:1 (DMEM/F12) (Gibco-BRL; Cat. No. 11330).

5. β-Mercaptoethanol (14.3 M) (Sigma; Cat. No. M7522).

6. Neurobasal medium (Gibco-BRL; Cat. No. 21103).

7. Recombinant human FGF basic (Invitrogen; Cat. No. 13256-029) is dissolved in human ESC growth medium at a final concentration of 10 μg/ml. Aliquot 100 μl into sterilized tubes and store at –80°C. A final concentration of 4 ng/ml (1:2500) is used to culture human ESCs.

8. Dispase solution (1.5 U/ml); dissolve 75 U of Dispase (Gibco-BRL; Cat. No. 17105-041) in 50 ml DMEM/F12 in a water bath for 15 min and filter sterilize the dispase solution with a 50-ml Steriflip.

9. Accutase in DPBS/0.5 mM EDTA (Innovative Cell Technology, San Diego, CA; Cat. No. AT104).

10. 0.05% trypsin-EDTA solution (Gibco-BRL; Cat. No 25300).

11. Trypsin inhibitor (1 mg/ml); dissolve 50 mg trypsin inhibitor (Gibco-BRL; Cat. No. 17075-029) in 50 ml DMEM/F12 and filter through 50-ml Steriflip.

12. Heparin (1 mg/ml); dissolve 10 mg heparin (Sigma; Cat. No. H3149) in 10 ml DMEM/F12 medium. Aliquot 0.5 ml into sterilized tubes and store at –80°C.

13. 100X N2 supplement (Gibco-BRL; Cat. No. 17502-048).

14. 50X B27 supplement without vitamin A (Gibco-BRL; Cat. No. 12587-010).

15. Fetal bovine serum (FBS) (Gibco-BRL; Cat. No. 16000-044).

16. Laminin from human placenta (Sigma; Cat. No. L6274).

17. Ascorbic acid (200 μg/ml); dissolve 2 mg ascorbic acid (Sigma; Cat. No. A-4403) in 10 ml PBS. Aliquot 0.5 ml into sterilized tubes and store at –80°C.

18. Cyclic AMP (1 mM); dissolve 4.914 mg cyclic AMP (Sigma; Cat. No. D-0260) in 10 ml sterilized water. Aliquot 0.5 ml into sterilized tubes and store at −80°C.

19. Sonic hedgehog (SHH, 100 μg/ml); dissolve 0.1 mg SHH (R&D; Cat. No. 1845-SH) in 1 ml sterilized PBS with 0.1% BSA. Aliquot 30 μl into sterilized tubes and store at −80°C.

20. Recombinant BDNF, GDNF, or IGF-1 (all are 100 μg/ml); dissolve 200 μg BDNF, GDNF, or IGF-1 (Pepro Tech Inc; Cat. Nos 450-02, 450-10, 100-11) in 2 ml sterilized water. Aliquot 30 μl into sterilized tubes and store at −80°C.

21. Recombinant FGF8 (100 μg/ml); dissolve 200 μg FGF8 (Pepro Tech Inc; Cat. No. 100-25) in 2 ml sterilized PBS with 0.1% BSA and 2 μg/ml heparin. Aliquot 30 μl into sterilized tubes and store at −80°C.

22. TGFβ3 (10 μg/ml); dissolve 10 μg TGFβ3 (R&D; Cat. No. 243-B3) in 1 ml PBS. Aliquot 30 μl into sterilized tubes and store at −80°C.

2.3. Media

1. Human ESC growth medium. Sterilely combine 392.5 ml DMEM/F12, 100 ml Knockout serum replacer, 5 ml MEM nonessential amino acids solution, 2.5 ml of 200 mM L-glutamine solution (final concentration of 1 mM), and 3.5 μl 14.3 M β-Mercaptoethanol (final concentration of 0.1 mM). Medium can be stored at 4°C for up to 7–10 days (*see* **Note 1**).

2. Neural induction medium (DMEM/F12/N2). Sterilely combine 489 ml of DMEM/F12, 5 ml N2 supplement, 5 ml MEM nonessential amino acids solution, and 1 ml of 1 mg/ml Heparin. Medium can be stored at 4°C for up to 2 weeks.

3. Neural differentiation medium. Sterilely combine 485 ml of Neurobasal, 5 ml N2 supplement, and 10 ml B27 supplement. Medium can be stored at 4°C for up to 4 weeks. Add laminin (1 μg/ml), cAMP (1:10000), ascorbic acid (1:1000), BDNF (1:10000), GDNF (1:10000), and IGF-1 (1:10000) before use.

3. Methods

3.1. Make ESC Aggregates

1. Ten minutes before start, warm up DMEM/F12, dispase, and ES cell growth medium in a 37°C water bath.

2. Aspirate medium from each well of the 6-well plate containing the ESCs (*see* **Fig. 19.1A** and **Note 2**).

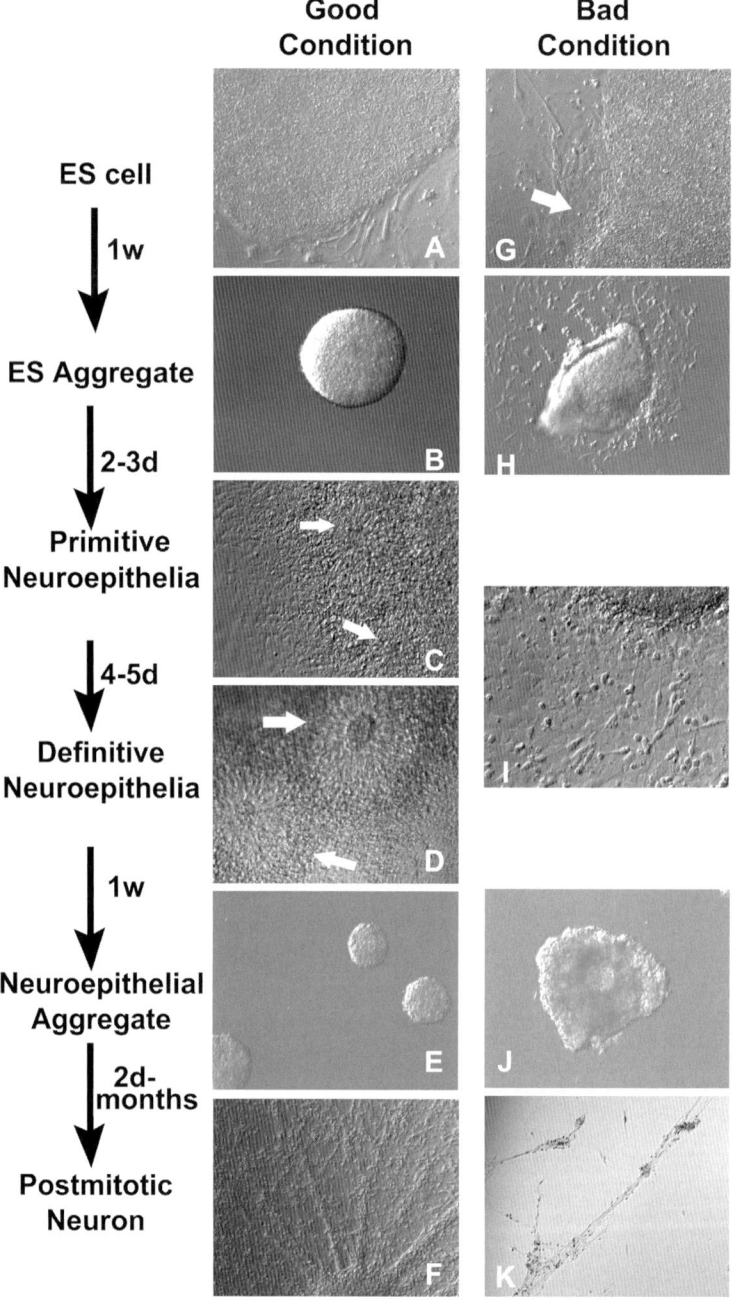

Fig. 19.1. Five morphologically distinct steps of neuronal differentiation from human ESCs. (**A–F**) Phase-contrast image showing ESC colony, ESC aggregate, primitive neuroepithelia, definitive neuroepithelia, neuroepithelial aggregate, and postmitotic neuron stages. *Arrows* show rosette formation in primitive or definitive neuroepithelial stages. (**G–K**) Phase-contrast image showing a partially differentiated ESC colony (*arrow*), differentiated ESC aggregates, asynchronized neural differentiation with neurons appearing at the neuroepithelial stage, neuroepithelial aggregates, and postmitotic neurons.

3. Wash the cells with DMEM/F12 once to remove serum replacement.
4. Aspirate DMEM/F12 and add to each well 1 ml dispase.
5. Incubate in a CO_2 incubator for 3 min; observe the cells under a phase contrast microscope.
6. When the edges of ESC colonies begin to curl, aspirate the dispase-containing medium from the 6-well plate and wash the cells gently with 2 ml DMEM/F12.
7. Collect the ESC colonies using a 10-ml serological pipette with ESC growth medium. If some ESC colonies remain attached, blow the colonies off gently and pool all the cells to a 50-ml conical tube.
8. Pipette up and down against the bottom of the tube to break up the colonies into ~200-μm pieces (*see* **Note 3**).
9. Centrifuge at 200*g* for 1 min at room temperature.
10. Aspirate the medium off the cell pellet.
11. Resuspend the ESCs in the ESC growth medium and transfer them to flasks (for one entire 6-well plate, resuspend the cells in 50 ml and culture them in a T75 flask; for 1–3 wells of cells use 15 ml ESC growth medium and transfer to a T25 flask).
12. Within 12–18 hours, replace most of the medium to remove dead cells and switch cells to a new flask to remove attached MEF if necessary.
13. Continue feeding the ESC aggregates every day for a total of 4 days. Set the flask down at a tilted angle so that the ESC aggregates settle in one corner of the flask. Aspirate off about half of the medium and add fresh ESC growth medium to replace the amount aspirated.

3.2. Neuroepithelial Differentiation (Formation of Neural Tube-Like Rosettes)

1. After 4 days, aspirate the ESC growth medium completely after the ESC aggregates are settled down to the corner of the flask. Rinse the aggregates with the neural induction medium once and resuspend the aggregates in the neural induction medium.
2. Change half of the medium 2 days later.
3. The next day, precoat the 6-well plate with 300 μl neural induction medium supplemented with 20 μg/ml laminin each well (*see* **Note 4**). For polyornithine-treated coverslips in 24-well plates, coat with 50 μl neural induction medium/ laminin (*see* **Note 5**).
4. Transfer the ESC aggregates into a 50-ml conical tube and let the aggregates settle to the bottom of the tube. Aspirate off most of the medium and transfer all the cell aggregates to a 60-mm Petri dish. Get rid of abnormal aggregates, such as pigmented and strange-shaped ones (*see* **Fig. 19.1H**).

5. Plate 30–40 aggregates to each well of a precoated 6-well plate or 3–4 aggregates on each coverslip in a 24-well plate and let the aggregates attach overnight.

6. Aspirate the medium containing laminin and add 2 ml neural induction medium to each well of the 6-well plate or 0.5 ml to each well of a 24-well plate on the next day.

7. Change most medium every other day.

8. Observe the culture daily. Cells in the colony will become elongated in 2–3 days after attachment and will arrange radically like rosettes in the center with a few flat cells in the periphery (see **Fig. 19.1C**). At this period, we call them primitive anterior neuroepithelia. With further culturing in the same medium for another 4–5 days, the area and thickness of rosettes increase and almost every colony contains multiple neural tube-like rosettes with lumens; we call them definitive neuroepithelia (see **Fig. 19.1D**).

3.3. Enrichment and Expansion of Neuroepithelial Cells

1. After formation of neural tube-like rosettes at day 14–15, the multicell layered structures need to be subcultured in order to avoid cell death and to promote cell growth.

2. Aspirate off old medium and add 1 ml new neural induction medium to each well.

3. Gently blow off the clusters with a 5-ml serological pipette and collect them to a 15-ml conical tube.

4. Triturate the clumps with a 5- or 10-ml serological pipette up and down twice, but not to break up the clumps.

5. Transfer the floating cells (mostly aggregates of neural rosette cells) to a new tube. Collect the large clumps that are settled on the bottom of the tube to a 1.5-eppendorf tube and use a P200 pipette to break them into small clusters.

6. Pool all the cells together and centrifuge at $50g$ for 2 min at room temperature.

7. Resuspend the cell pellets with 15 ml neural induction medium supplemented with B27 (1X) and transfer them to a T25 flask.

8. The rosette aggregates will roll up to form round spheres after about 1 day (see **Fig. 19.1E**).

9. Change half of the medium and feed the neuroepithelial aggregates every other day with the neural induction medium with B27.

3.4. Differentiation of Neurons from Neural Progenitors

1. The neuroepithelial aggregates can be kept in suspension for several weeks or months (see **Note 6**). How long the neuroepithelial cells should be expanded in suspension depends on

research objectives. They usually generate large projection neurons in the first month, followed by smaller interneurons, astrocytes, and oligodendrocytes.

2. Precoat culture plates or polyornithine-treated coverslips with laminin (20 μg/ml, diluted in neural differentiation medium) at 37°C for at least 3 h. For coverslips in 24-well plates, coat one coverslip with 50 μl laminin medium. For 1 well of 6-well plates, coat with 300 μl laminin medium in the center.

3. Warm up accutase, trypsin, trypsin inhibitor, DMEM/F12, and neural differentiation medium in a 37°C water bath.

4. Collect the neuroepithelial aggregates into a 15-ml conical tube and let the aggregates settle to the bottom of the tube. Aspirate off most of the medium and transfer all the aggregates to a 1.5-ml eppendorf tube. Gently remove the remaining medium.

5. Add 500 μl accutase/trypsin (1:1) to the cells and incubate at 37°C for 2–3 min.

6. Add 500 μl trypsin inhibitor to the cells and centrifuge at 50g for 2 min at room temperature.

7. Remove supernatant and wash with DMEM/F12 once.

8. Add 100–200 μl neural differentiation medium to the tube and pipette up and down against the bottom of the tube to break up the cells into small clusters (50–100 μm) or single cells with a P200 pipette (see **Note 7**).

9. Plate the cells in the precoated plates or coverslips. Check under the microscope about the density of the cells. Similar to seeding ESC aggregates, plate 30–40 aggregates in each well of a 6-well plate or 3–4 aggregates on each coverslip. For single-cell plating, resuspend the cells in neural differentiation medium at the density of 400,000 cells/ml. Then put 25 μl cells onto each coated coverslip or 150 μl onto each well of 6-well plate.

10. After the cells attach to the bottom (2–3 h later after plating), add 1 ml or 3 ml of neural differentiation medium to each well of the 24-well plates or 6-well plates, respectively.

11. Feed the cells every other day with the same medium and change half of the medium each time.

3.5. Differentiate Human ESCs to Midbrain Dopaminergic Neurons

1. Differentiate primitive neuroepithelial cells from hESCs as described in **Sections 3.1**, **3.2**, and **3.3**. This usually happens at around 8–10 days after differentiation from hESCs, with the appearance of columnar cells in the culture.

2. Aspirate off old medium and add 3 ml neural induction medium supplemented with FGF8b (50 ng/ml) and SHH (100 ng/ml).

3. Continue feeding the cells every other day with the same medium with FGF8b and SHH for 7 days until formation of neural tube-like rosettes.

4. Enrich and expand the neuroepithelia as described above (**Section 3.3**) but in neural induction medium containing FGF8b (50 ng/ml), SHH (100 ng/ml), B27 (1X), and ascorbic acid (200 μM).

5. Feed the neural progenitor aggregates every other day with the neural induction medium with FGF8b, SHH, B27, and ascorbic acid.

6. After 7 days of culture in suspension, the neural progenitor aggregates are dissociated to single cells after digestion with accutase/trypsin (1:1) and plated onto laminin precoated surface as mentioned above (**Section 3.4**).

7. Feed the cells every other day with the neural differentiation medium containing FGF8b (50 ng/ml), SHH (100 ng/ml), ascorbic acid (200 μM), cAMP (1.0 μM), TGFβ3 (1 ng/ml), BDNF (10 ng/ml), GDNF (10 ng/ml), and Wnt3a conditioned medium (1X) for 3 weeks (*see* **Note 8**).

8. Withdraw FGF8b, SHH and Wnt3a conditioned medium at day 44 after differentiation and maintain the cells in neural differentiation medium with ascorbic acid, cAMP, TGFβ3, BDNF, and GDNF.

9. At this point, numerous multipolar dopaminergic neurons can be detected through immunostaining with TH antibody.

4. Notes

1. If filtration is needed, we recommend using disposable receipt bottle (Stericup, Millipore) other than reused glass bottle.

2. The undifferentiated state of ESCs is very important for efficient neural induction. If there are differentiated cells in the ESC culture, floating ESC aggregates are easy to attach to the flask and neurons will be easily found in the culture dish when these aggregates are plated down (*see* **Fig. 19.1H** and **I**)

3. Usually the size of the clusters is around twice the size for ESCs for passaging/splitting. Cells will die and the yield of ESC aggregates will be quite low if the size of the clusters is too small. The differentiation will be unsynchronized if the size is too large.

4. As an alternative to the attachment of ESC aggregates using laminin, FBS can also be used. Add neural induction medium containing 10% FBS to the ESC aggregates as mentioned in **Section 3.2**. Plate 1–1.5 ml of medium containing 30–40 aggregates to each well of a 6-well plate and let the cells attach in the incubator overnight. Right after the cells attach, wash the cells with warm DMEM/F12 once and feed with neural induction medium. Please note that it is better to leave the cells in the presence of serum for as short time as possible.

5. Coverslips need be pretreated with nitric acid for 1 h and then with 95% ethanol for 30 min after washing with distilled water. These coverslips then need be coated with 0.1 mg/ml poly-ornithine overnight before use (*for details see* www.wicell.org).

6. Flame polish a cotton-plugged Pasteur pipette to narrow the opening end slightly and make a 20–30° bend at the narrow part of the shaft. Pipette up and down 1–3 times to let the aggregates go through the bended pipette and shear them into small pieces. In this way, neuroepithelial aggregates can be expanded from 1 flask into 2.

7. Cells can be dissociated to either single cells or small clusters. Small clusters tend to survive better than single cells. For dopaminergic differentiation, we dissociate the cells to single cells and plate at a high density.

8. L cells that express Wnt3a (ATCC, Manassas, VA, USA) are conditioned in neural differentiation medium for 24 h and the conditioned medium is used at a final concentration of 50% from day 24 to day 44 to promote the proliferation of specified DA progenitors.

Acknowledgments

This study was supported by the National Institute of Neurological Disorders and Stroke (NS045926, NS046587), the Michael J. Fox Foundation, and partly by a core grant to the Waisman Center from the National Institute of Child Health and Human Development (P30 HD03352).

References

1. Zhang SC (2003) Embryonic stem cells for neural replacement therapy: prospects and challenges. *J. Hematother. Stem Cell Res.* **12**, 625–634.
2. Du ZW, and Zhang SC (2004) Neural differentiation from embryonic stem cells: which way? *Stem Cell Development.* **13**, 372–381.
3. Zhang SC, Wernig M, Duncan ID, Brüstle O, and Thomson JA (2001) *In vitro* differentiation of transplantable neural precursors from human embryonic stem cells. *Nat. Biotechnol.* **19**, 1129–1133.
4. Pankratz MT, Li XJ, Lavaute TM, Lyons EA, Chen X, Zhang SC. (2007) Directed

neural differentiation of human embryonic stem cells via an obligated primitive anterior stage. *Stem Cells.* **25**, 1511–1520.

5. Hemmati-Brivanlou A, Melton D. (1997) Vertebrate embryonic cells will become nerve cells unless told otherwise. *Cell,* **88**, 13–17.

6. Li XJ, Du ZW, Zarnowska ED, Pankratz M, Hansen LO, Pearce RA, Zhang SC. (2005) Specification of motoneurons from human embryonic stem cells. *Nat Biotechnol.* **23**, 215–221.

7. Li XJ, Hu BY, Jones SA, Zhang YS, Lavaute T, Du ZW, Zhang SC. (2008) Directed differentiation of ventral spinal progenitors and motor neurons from human embryonic stem cells by small molecules. *Stem Cells.* **26**, 886–893.

8. Yang D, Zhang ZJ, Oldenburg M, Ayala M, Zhang SC. (2008) Human embryonic stem cell-derived dopaminergic neurons reverse functional deficit in parkinsonian rats. *Stem Cells.* **26**, 55–63.

9. Yan Y, Yang D, Zarnowska ED, Du Z, Werbel B, Valliere C, Pearce RA, Thomson JA, Zhang SC. (2005) Directed differentiation of dopaminergic neuronal subtypes from human embryonic stem cells. *Stem Cells.* **23**, 781–790.

Chapter 20

Transplantation of Human Embryonic Stem Cells and Derivatives to the Chick Embryo

Ronald S. Goldstein

Abstract

Traditional methods of studying the differentiation of human embryonic stem cells (hESCs) include generation of embryoid bodies, induced differentiation in vitro, and transplantation to immune-deficient mice. The chick embryo is a well-studied and accessible experimental system that has been used for many years as a xenograft host for mammalian cells. Several years ago, we performed experiments transplanting colonies of hESC into organogenesis-stage chick embryos to establish a novel system for studying the developmental programs and decisions of pluripotent human cells. Fluorescent hESC were used, in order to permit identification of the hESC in living embryos. We transplanted hESC into the trunk of chick embryos, both into and instead of developing somites. Our results showed that hESC survive, migrate, and integrate into the tissues of the chick embryo. Some of the hESC differentiated and the type of embryonic microenvironment that the implanted cells were exposed to modified their differentiation. Several other laboratories have subsequently xenografted hESC-derived cells to chick embryos for evaluating their differentiation in vivo. Therefore, the hESC-chick embryo system is a useful xenograft system complementing studies in rodents and in vitro, as well as uniquely shedding light on early processes in the development of human cells in the embryonic context.

Key words: Human embryonic stem cells, xenograft, human embryogenesis, chick embryo, somites, in ovo microsurgery.

1. Introduction

The most clinically important, and imagination-stimulating application of human embryonic stem cells (hESCs), is the generation of normal differentiated cells for replacement of those lost in disease or injury. However, there are several additional important ways in which hESC can be used productively. Use of these cells has the potential to reveal important details about early human embryonic development, since early human embryos, for both practical and ethical reasons, are not accessible to researchers.

The NIH guidelines and the National Stem Cell Institute/Wicell MTA forbid the grafting of hESC to vertebrate embryos. Prevention of the generation of chimeras, with a large human contribution, is apparently the motivation for this rule. Only one report exists in the literature of attempting to make blastocyst chimeras using hESC, where the host blastocyst was murine (1), non-NIH lines were used, and IRB approval by the institution was obtained. Surprisingly, this attempt was quite unsuccessful and yielded very little information. Apparently because of the ethical arguments against performing such experiments, they have not been repeated. By contrast, transplantation of undifferentiated hESC into adult animals for testing pluripotency by formation of teratomas is routine today and does not appear to raise any ethical problems. However, the uninjured adult tissue environment may lack many of the factors required to direct differentiation of the hESC. In addition, teratomas are isolated by a capsule from the host, which impedes migration and integration of the cells into the surrounding tissue.

The chick embryo has been studied intensively at the anatomical, cell biological, and molecular levels, and the precise timing and placement of many of the inductions and morphogenetic events underlying its development are known. In addition, experimental manipulations of the embryo are quite simple, due to the easy accessibility of the embryo by simply opening a "window" in the shell. A number of studies have shown that mammalian cells survive and differentiate in response to environmental signals of avian embryos. For example, when chick neural tube is replaced with that of mouse embryos, mouse neural structures develop, including a spinal cord, and neural crest and their derivatives, the peripheral ganglia (2). Mouse embryonic stem cells (mESCs) and their derivatives have also been transplanted successfully to the chick embryo. In one elegant study, motoneuron precursors derived from mESC were transplanted to the neural tube of the chick, and mouse motoneurons not only differentiated, but sent their axons out chick nerves, and innervated peripheral muscles (3).

Several years ago, we transplanted hESC colonies into the developing tissues of organogenesis-stage chick embryos, in an attempt to provide hESC with an environment of developing tissues, while avoiding the ethical problem of generating embryos with extensive chimerism (4). We found that placement of the hESC in direct contact with the host neural tube consistently resulted in the generation of neural tube-like structures ("neural rosettes") that grew with the same orientation as the host CNS, with their axis parallel to that of the chick. In contrast, interspersing the hESC within somites led to more integration of individual human cells into chick structures such as peripheral ganglia and vertebrae.

There have been two additional studies of grafting undifferentiated hESC to chick embryos (5, 6). In this work, hESC were injected into the amnion of embryos whose neural tube was opened as a model for neural tube closure defects. The cells adhered to the outside of the neural tube lesions, and somewhat enhanced its closure. However, they did not integrate into the neural tube. In their second study, the authors performed immunocytochemistry for neuronal and glial markers and found little neural differentiation by the injected hESC.

Several recent studies have also included grafts of differentiated hESC into chick embryos and onto the chorioallatoic membrane of chick eggs. Studer's group transplanted hESC-derived p75 sorted putative "neural crest stem cells" (NCSC) to the intersomitic space of St 10-12 chick embryos (7). The authors observed extensive migration of the grafted cells three days after grafting, and integration and differentiation appropriate to the host in the case of the sympathetic ganglion. In another study by this group, a population of hESC-derived neurons, including putative spinal motoneurons, was grafted into the dorsal neural tube of St 16-16 chick embryos (8). Some of the cells expressed their in vitro-induced motoneuron identity (i.e., transcription factor HB9 expression) in spite of dorsal signals from their ectopic location. Some of the cells migrated ventrally, and long axons were observed to extend into the periphery, although it was not clear whether these belonged to motoneurons or other transplanted neurons (NCAM/GFP staining).

Another study transplanted hESC-derived CD34+ hematopoeitic stem/progenitor cells (HSPC) into early (gastrulation/neurulation stage) chick embryos. Substantial engraftment into several tissues including the bursa of Fabricius and thymus was observed. Very surprisingly, the study did not report whether the HSPC engrafted in the bone marrow, although the embryos were sacrificed at a time when virtually all hematopoeisis was already taking place there (9). In yet another study, neural precursors generated from hESC were cografted with preinnervation gut to the choriallantoic membrane of chick eggs. The neural precursors migrated through the gut wall, and differentiated neuronally (10).

The present chapter details the techniques that were involved in our experiments grafting hESC into 1.5–2 day chick embryos, with the primary focus on the surgical techniques themselves (*see* **Note 1**). In addition to summarizing our published results with naïve hESC, we present in this updated chapter previously unpublished results of transplanting embryoid bodies (EBs) into 2-day chick embryo somites (**Fig. 20. 3**).

2. Materials

2.1. Culture

1. 3.5-cm culture tissue culture plates (Nunc, Roskilde, Denmark; Cat. No. 153066).
2. Gelatin (Merck, Whitehouse Station, USA; Cat. No. 1.04078.1000). Prepare a 0.1% solution in dH_2O, autoclave at 121°C for 30 min. The sterile stock is stored in the refrigerator.
3. Vybrant CFDA SE Cell Tracer Kit (Molecular Probes, Eugene, OR, USA; Cat. No. V12883). Make stock solution by dissolving the contents of vial of powder in 90 µl supplied DMSO.

2.2. Surgery

1. Fertile chick eggs (*see* **Note 2**).
2. Surgical instruments (*see* **Note 3**): Fine scissors (Fine Science Tools, North Vancouver, Canada; Cat. No. 91460-11); Dumont #5 forceps (Fine Science Tools; Cat. No. 91150-20); Pin holder (Fine Science Tools; Cat. No. 26016-12); x2; Sterilization case with rubber insert (Fine Science Tools; Cat. No. 20311-21); Minuten pins (Fine Science Tools# 26002-10).
3. Microscalpels are made in one of two ways: (**i**) For very fine dissection, including opening of the ectoderm etc., they are made by electrolysis of tungsten wire (A-M systems, Carlsborg, WA, USA; Cat. No. 7185) (*see* **Note 4**). One lead of a 12 V AC transformer is connected to a bath of saturated NaOH in a plastic petri dish by a paper clip, and the other lead is attached to a pin holder (Fine Science Tools; Cat. No. 26016-12) with the tungsten wire inserted. The pin holder is held in any simple micromanipulator and the tip of the wire lowered just into the solution, while monitoring with an old binocular microscope (*see* **Note 5**). The current is then applied, and the wire dipped and removed from the solution, and the tip examined with the binocular microscope, until the appropriate shape and sharpness are obtained. Varying the angle of the wire can produce many different shapes of scalpels (11), but we find that straight is fine for the operation described here. (**ii**) For removal of somites and manipulation of tissue during the implantation, pins (Fine Science Tools; Cat. No. 26000-40,-45) are sharpened on a stone (Fine Science Tools; Cat. No. 29008-22) using a drop of paraffin oil. The scalpel blade should look like a butter knife when finished. That is, it should be thin and not very sharp at the end. Obtaining a thin blade requires holding the pin almost parallel to the surface of the stone and rubbing it back and

forth laterally, switching sides periodically. For sharpening, the pin should be inserted into the holder most of the way, only leaving 2–3 mm outside for sharpening. Otherwise, it just bends when it is rubbed against the stone. When the sharpening is finished, the pin is pulled out with forceps, leaving about 5 mm inside the holder. The longer the pin, the more visibility there is during the operation, since the pin holder can block the field of view if it is too close.

4. Cellotape (*see* **Note 6**).
5. Egg incubator (GQF Manufacturing, Savanah, GA, USA; Cat. No. 1550).
6. Stereo microscope with bifurcated fiber-optic light guides for surgery and epifluorescence for viewing graft in vivo.
7. Cold plate that can maintain 12°C (Home-made from a Pelletier device, digital thermostat, and black-anodized aluminum plate. Plexiglas cover retains cooling).
8. Dulbecco's phosphate-buffered saline with divalent cations (PBS+) (Biological Industries, Bet HaEmek, Israel; Cat. No. 020-020-1).
9. Pancreatin (Invitrogen, Carlsbad, CA, USA; Cat. No. 610-8728-AE).
10. Penicillin-streptomycin-amphotericin (Biological Industries; Cat. No. 03-033-1).
11. India ink (Pelikan Tushce A, Hanover, Germany; Cat. No. 201 665).
12. Gel-loading tips (USP, San Leandro, CA, USA; Cat. No. TGL-1000).
13. Embryo dish (Electron Microscopy Sciences, Hatfield, PA, USA; Cat. No. 70543-30).

2.3. Fixation and Embedding

1. 4% Paraformaldehyde (*see* **Note 7**). For 200 ml, add 8 g paraformaldehyde granules (Electron Microscopy Sciences; Cat. No. 19208) to 100 ml dH$_2$O. Heat to 60°C while stirring in a fume hood. Add 1 M NaOH one drop at a time until clear and cool to ambient temperature. Bring volume to 200 ml with 2x PBS without divalent cations (double the quantity of salts in **Section 2.4.1** below).
2. Paraplast+ (Sigma-Aldrich, St. Louis, MO, USA; Cat. No. P-3683).
3. Sylgard elastomer (World Precision Instruments, Sarasota, FA, USA; Cat. No. SYLG184).

2.4. Immunocytochemistry

1. PBS without divalent cations (PBS). For 500 ml, dissolve 80 g NaCl, 2 g KCl, 11.5 g Na_2HPO_4 (anhydrous), 2 g KH_2PO_4, 1 g Thimerosal (Sigma-Aldrich; Cat. No. T-5125) in 450 ml dH_2O. Bring volume to 500 ml when dissolved. This makes a 20×stock that is diluted before use.

2. Microwave oven (*see* **Note 8**).

3. Staining dishes. (i) Plastic for deparaffinization, hematoxylin and eosin staining, and dehydration (Tissue-Tek, Torrance, CA, USA; Cat. No. 4451); (ii) glass for immunocytochemistry (Electron Microscopy Sciences; Cat. No. 70312-20) x6; (iii) polypropylene staining dish for microwave antigen retrieval (Electron Microscopy Sciences; Cat. No. 70321-10); staining rack, (Electron Microscopy Sciences; Cat. No. 70321-20).

4. 1X Citrate buffer (diluted from 10x solution). For 500 cc 10X solution, add 14.5 g tri-sodium citrate dihydrate (Carlo Elba, Milan, Italy; Cat. No. 479487) to 400 cc ddH_2O. Bring to pH 6.0 with 1 N HCl.

5. Blocker 1% bovine serum albumin (USB, Cleveland, OH, USA; Cat. No. 70195), 0.5% Triton-X100 (USB; Cat. No. 22686) in PBS+. Make up 100 ml and freeze in sterile 15-ml tubes.

6. Antibodies: Antimammalian neurofilament-M (Developmental Studies Hybridoma Bank, Iowa City, IA, USA; Cat. No. 2H3,); Anti-βIII Tubulin mAb (Promega, Madison, USA; Cat. No. G712A); Anti-human Desmin, (DAKO, Glostrup, Denmark; Cat. No. M 0760); Anti-Islet-1 (Developmental Studies Hybridoma Bank; Cat. No. 40.2D6); Anti-GFP (Santa Cruz Biotechnology, Inc., Santa Cruz, CA, USA; Cat. No. sc-8334); Anti-HNK (supernatant from growing ATTC, Manassas, VA, USA; hybridoma # TIB-200, or Sigma; Cat. No. # C0678); Alexa Fluor 488 goat anti-mouse IgG (H+L) conjugate (Molecular Probes; Cat. No. A-11001); and Alexa Fluor 594 goat anti-rabbit IgG (H+L) conjugate (Molecular Probes; Cat. No. # A-11012).

7. Slides (Superfrost Plus) (Menzel-Glaser, Braunschweig, Germany; Cat. No. 041300).

8. Glass coverslips 24 × 50-mm, 24 × 60-mm.

9. Mounting medium: 75% glycerol (Sigma; Cat. No. G-8773) 25% PBS, add 1% n-propyl gallate (Sigma; Cat. No. P-3130) for antibleaching.

10. 0.1 mg/ml Hoechst 33258 (Bisbenzimide) (dilute 1:1000 from stock solution). For 10 ml stock solution: add 10 mg Hoechst 33258 (Sigma; Cat. No. B-2883) to 100 ml ddH_2O. Working solution can be reused and kept in light-protected container at 4°C until staining is no longer strong.

11. Nail polish (*see* **Note 9**).

12. Polylysine for coating slides (Sigma; Cat. No. P8920).

2.5. Histology

1. Harris Hematoxylin (Sigma; Cat. No. HHS-16).

2. Eosin Yellowish (Gurr, London, England; Cat. No. 22232).

3. Cold Schiff: dissolve 10 g Basic Fuchsin (MCB, Norwood, OH. USA; Cat. No. B300/BX135) in 300 ml 1 N HCl (24 ml HCl conc. + 276 ml dH$_2$O). Dissolve 10 g sodium metabisulfite (Sigma; Cat. No. S9000) in 1700 ml dH$_2$O. Mix the solutions. Add activated charcoal (BDH DARCO G60, Poole, England; Cat. No. 33187) and mix. Let stand for 24 h at room temperature. Filter with #1 Whatman filter paper, solution should be colorless. Store at 4°C in a well-closed bottle.

4. Sulfurous rinse solution: prepare a stock of 10% wt/vol. sodium metabisulfite. Working solution: 10 ml stock + 10 ml 1 N HCl + 180 ml dH$_2$O.

5. Fast Green FCF (Sigma; Cat. No. F7258).

6. Entellan (Merck; Cat. No. 1.07961.0500).

7. Ethanols (70%, 95% denatured diluted with dH$_2$O, 100% anhydrous analytical reagent).

8. Xylene.

9. Toluene.

2.6. Digital Photography

1. Digital cameras (Scion, Frederick, MD, USA; models no. CFW1310M (monchrome) and CFW 1310C (color).

2. ImageJ (http://rsb.info.nih.gov/ij).

3. Methods

3.1. Culture of ES Cells

3.1.1. Growth of Human ES Cells and Generation of Embryoid Bodies

The growth of hESC used was as described (12) and has been expanded upon in several chapters in this volume. In some experiments, hESC that constitutively expressed EGFP (13) were used for grafting. Embryoid bodies generation was described in (12), and techniques for their generation are also described in this volume.

3.1.2. Partial Removal of Fibroblasts and Initiation of Differentiation

Before grafting, hESC were cultured for 1–3 days on gelatin-coated plates (cover bottom of plate with gelatin solution in sterile hood at least 1 h at room temperature, then rinse with dH$_2$O and medium) without fibroblast feeder lines in order to obtain a purer population of hESC for grafting and to begin the process of hESC differentiation.

3.1.3. Vital Staining of hESC

In some experiments, cells were vitally stained with CFDA.

1. Warm 1.5 ml PBS+ to 37°C.
2. Add 3 µl CFDA stock solution (*see* **Note 10**) and mix.
3. Replace the medium in the 35-mm culture plate containing the hESC colonies with the dye and incubate for 15 min in a CO_2 incubator.
4. Rinse with warm PBS+.
5. Change to 1.5 cc Dulbecco's minimum essential medium (DMEM) and incubate at least 30′ in the CO_2 incubator.
6. When ready for use, change to PBS+, and collect colonies as in **Section 3.2.6** below.

3.2. Microsurgery for Transplants

3.2.1. Incubation and Preparation of Embryos

1. Incubate fertile chicken eggs WITHOUT TURNING in a horizontal position from 40–45 h to obtain embryos of 10–20 somite pairs.
2. Mark the approximate position of the embryo by drawing a small pencil mark on the topmost position on the shell. Care is exercised from this point on that the eggs should not rotate in any direction.
3. At least 20 min before beginning surgery, lower the embryo away from the shell by removing 1–1.5 cc of albumin. This is accomplished by wiping the point of the egg with a tissue moistened with 70% ethanol and making a small hole in the point by tapping sharply with pointed fine surgical scissors. Insert a 21-G needle attached to a 2-cc syringe into the hole and withdraw the albumin very slowly, since rapid removal of the albumin makes large bubbles above the embryo, making subsequent surgery very difficult.
4. Open the shell by cutting a 1 cm circular hole with fine dissection scissors.

3.2.2. Visualizing the Embryo

Although some use vital dyes to stain chick embryos before surgery, in our experience the best visualization is obtained when black ink is injected in the subblastodermal space. A small amount of India ink diluted 1:3 in PBS+ is injected subblastodermally using a hand-drawn pipette (*see* **Notes 11 and 12**).

3.2.3. Exposure of the Embryo for Surgery

1. In the experiments described here, hESC are transplanted to the area of the somites (**Fig. 20.1A** and **B**, *see* **Note 13**).
2. Make a slit in the shell membrane above the 3–4 most recently formed somites with a microscalpel. Intrinsic tension makes the membrane separate and reveal the underlying ectoderm of the embryo.
3. *Immediately* moisten the shell membrane with PBS+ after opening the egg in order to prevent damage to the embryo.

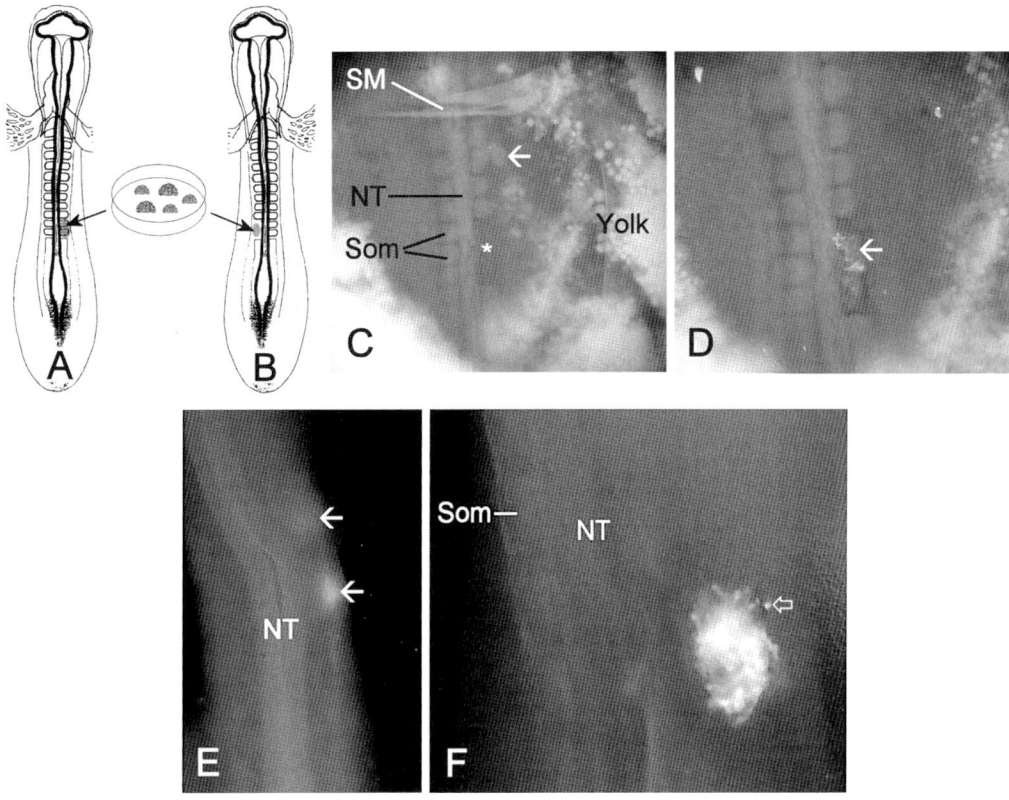

Fig. 20.1. Diagrams of graft of hESC into (**A**) and instead of (**B**) epithelial somites of chick embryos of about 36 h incubation. (**C** and **D**) show a portion of the trunk of a live embryo being grafted. An *asterisk* is in the middle of a gap made by the removal of 3 intact epithelial somites in **panel C**. One of the removed somites is indicated by the *white arrow*. **Panel D** shows the same embryo, with the gap filled in with hESC (*arrow*). The bright white material below the *arrow* in **D** is a small piece of plastic removed from the petri dish along with the hESC colony. (**E** and **F**) are images of live embryos that had been implanted with GFP-expressing hESC 24 h previously. In E, 2 clumps of fluorescent cells are visible (*arrows*), the upper clump is less bright because the cells are less superficial. In (**F**), individual cells, such as the one indicated by the *open arrow*, can be seen migrating away from the graft. SM – shell membrane, NT – neural tube, Som – somites.

4. When there is a thin layer of PBS+ over the membrane, pierce it gently with the microscalpel held at about 45° and then lift up the scalpel with a sharp movement – this results in a small tear of the membrane.

3.2.4. Preparation of Embryos for Grafts of hESC into Somites

1. Three slits are cut in the ectoderm in order to access the somites, one between the somites to be damaged and the neural tube, one between the last-formed somite and the segmental plate, and one between the most rostral somite to be damaged and the somite rostrad to it.

2. Cut the ectoderm by piercing with a sharp movement at one end of the cut and lifting the scalpel upwards.

3. After making the 3 cuts, insert the scalpel between the somites to be damaged and the flap of ectoderm formed by the cuts, lifting up the flap and exposing the somites.

4. Gently, make several slices in the somites using the scalpel, taking care not to cut so deep that the endoderm is damaged. If the endoderm is cut (it is only 10-μm thick at this stage), the ink that was injected for contrast (**Section 3.2.2**) comes seeping up through the cut, and the embryo is likely not to survive the surgery (*see* **Note 14**).

3.2.5. Preparation of Embryos for Grafts of hESC in Place of Somites

1. The shell membrane is removed as in **Section 3.2.3**, and a longitudinal slit is made in the ectoderm between the somites to be removed and the neural tube as in the previous section.

2. A 10 μl drop of pancreatin is then dripped on this region in order to loosen the somites from their surrounding tissues. The tip of the pipette should not touch the embryo, the drop should "fall" from a short distance above. When the drop is placed correctly, a small pool of liquid forms.

3. The enzyme is then allowed to work for 5–10 min, during which time the aperture in the egg shell is covered (but not sealed) with a large glass coverslip to prevent desiccation of the embryo (*see* **Note 15**). The somites are ready for removal when a gentle nudge from the microscalpel moves them out of place (**Fig. 20.1C** and *see* **Note 16**). The enzyme is neutralized with a drop (from a Pasteur pipette) of 5% heat-inactivated serum in PBS+ (*see* **Note 17**).

3.2.6. Transfer and Implantation of hESC to the Chick Embryo

1. The petri dish containing the hESC is removed from the CO_2 incubator, and the culture medium replaced with PBS+ at room temperature (*see* **Note 18**).

2. The colonies are cut into small pieces under the dissecting microscope and scraped off the bottom of the culture dish with a microscalpel.

3. The pieces of colony are collected and transferred, several at a time, into an embryo dish containing PBS+ that is kept on a cooling plate maintained at 12–15°C (*see* **Note 19**). One can prepare the colonies before beginning to operate on the eggs, or while waiting for the pancreatin to act. In our experience, when maintained at 12–15°C, the hESC remain viable for at least 6 h.

4. Immediately before transfer of the hESC to the egg, place a drop of PBS+ in the graft area, again making a "pool" of liquid between the embryo and the shell membrane.

5. Now, as quickly as is practical, select a few pieces of hESC colony from the embryo dish and aspirate them into a gel-loading tip in a volume of 5 μl.

6. Transfer the tissue by placing the tip into the "pool" of PBS in the egg, and gently expel liquid until the hESC leaves the tip (*see* **Note 20**).

7. Set down the pipettor and using the microscalpel (preferably a dull one) coax the tissue into the damaged somite or space left by the removal of the somites (**Fig. 20.1D**). In the case of somite replacement, it is advisable to try to fill the gap as well as possible with hESC, because of the tremendous regenerative ability of somites (14). This regeneration results in the chick cells out-competing the foreign (hESC) cells and preventing their expansion.

8. Once the hESC tissue is in place, the ectodermal flap is smoothed over the area of the graft to act as a "blanket" holding the transplanted cells in place.

9. Seal the egg with cellotape, label it with the exact details of the experiment using a graphite pencil, and return the egg carefully to the incubator.

3.2.7. Follow-Up of Grafts

1. The morning after the operation, the eggs are opened aseptically and viewed with a binocular microscope equipped with epifluorescence for visualizing GFP (**Fig. 20.1E** and **F**) or CFDA. The implanted hESC are still relatively superficial at this point, and the embryos can be photographed under both fluorescence and brightfield illumination.

2. 4 or 5 drops of PBS+ containing 10 concentrated antibiotic solution is then gently dripped onto the embryo to prevent dehydration and combat infection (*see* **Note 21**).

3.3. Fixation and Embedding

3.3.1. Fixation

1. After the second incubation period, embryos are removed from the egg.

2. Transfer the embryos to a petri dish with a 5-mm layer of Sylgard elastomer and pin them out with Minuten pins in PBS+ at room temperature (*see* **Note 22**). The GFP fluorescence, although visible after paraformaldehyde fixation, is stronger before fixation, so this is a good time to photograph the embryo for future localization of transplanted cells (see **Section 3.3.3**). One can pin out and fix several embryos at once, by drawing a "map" of their positions in the dish before fixation.

3. After photography/observation, the PBS+ is aspirated carefully with a Pasteur pipette, and 4% paraformaldehyde added.

4. The embryos become fairly stiff after about 20-min fixation at room temperature, and are carefully unpinned and placed into fresh fixative in scintillation or similar glass vials. If the surgery is not near the head, it is worthwhile to decapitate the

embryo, to improve the subsequent dehydration and embedding. Fixation can be performed for 2–3 h at room temperature, or overnight at 4°C.

3.3.2. Embedding and Sectioning

1. Fixed embryos are rinsed 2 × 5 min in dH$_2$O and dehydrated via ethanol, 2 changes each 70%, 95%, and 100% AR. The times are dependent on the size of the embryo, each step for 4-day embryos is 15 min, for 9-day embryos, the times should be doubled.
2. The embryos are then cleared in 2 changes of Xylene of 15–30 min according to the size of the embryos.
3. 3 changes of paraffin follow, and finally the embryo is embedded longitudinally, with the neck positioned in such a way that it will be cut in cross-section.

3.3.3. Serial Sectioning and Searching for Graft

One of the major challenges in this entire experimental procedure is finding the relatively few-grafted ES and their progeny within the "background" of an entire embryo. The cells can be positively identified without ambiguity in sections in one of three ways: (1) immunostaining for human specific markers such as human mitochondrial marker; (2) using cells marked either genetically (i.e., expressing a marker such as GFP) or with a vital dye such as CFDA; and (3) using mammalian or human specific antibodies to differentiation markers (such as mammalian neurofilament protein).

1. Long ribbons of serial sections are prepared, and laid out in long flat boxes (*see* **Note 23**).
2. The ribbons are cut into strips about 2/3 the length of a microscope slide, using an old, blunt scalpel blade; a paper or cardboard "ruler" of the right length is a help for the "guestimating."
3. The ribbons are then mounted onto slides in a staggered manner. That is, if the width of the ribbon is such that it is possible to mount 4 ribbons/slide, every 4th ribbon is mounted on slide 1, the next of the series on slide 2, etc. This allows sampling of large extents of the embryo in order to find the graft. Initial staining with anti-GFP or antihuman markers is performed on one slide from each set, and the stained sections scanned until the graft is found (*see* **Note 24**).

3.3.4. Feulgen Staining

The nuclei of mammalian cells, like those of quail cells, are much larger than those of the chick embryo (i.e., (4)). The graft can therefore often be easily identified by Feulgen staining (**Fig. 20.2A** and **B**), which gives a sharper and more intense nuclear staining than Hematoxylin and Eosin. In addition, Feulgen combined with Fast-Green counterstaining often gives better general histology than hematoxylin and eosin.

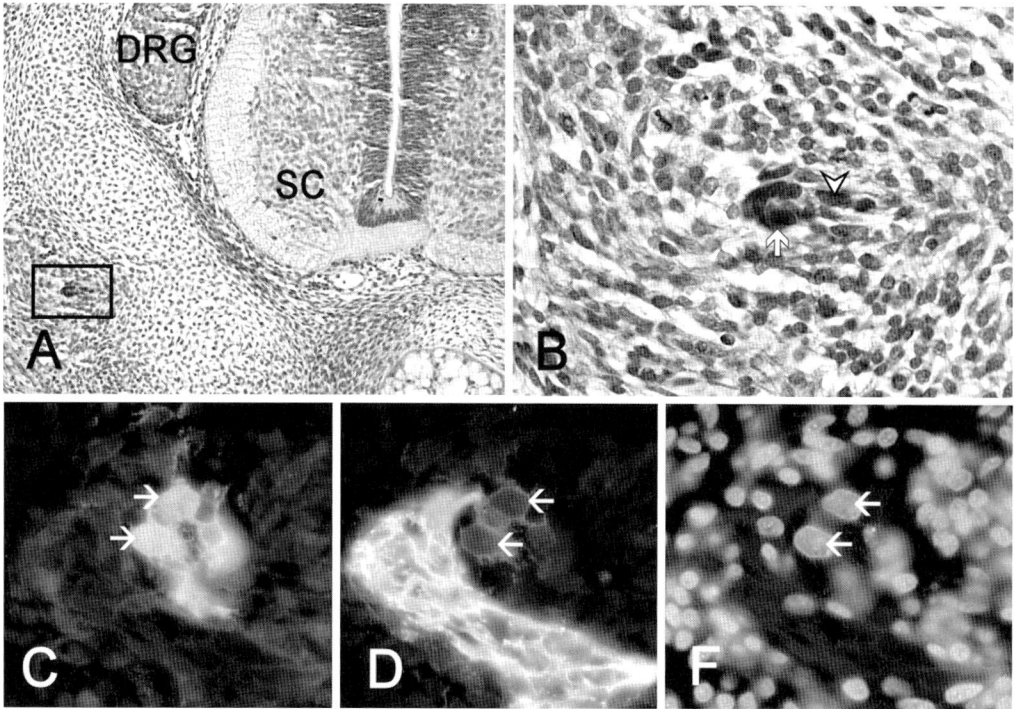

Fig. 20.2. Feulgen-staining of a section through a grafted embryo is shown in **A** and **B**. A small cluster of human cells (in the *black rectangle*) is easily observed even at low magnification, using Feulgen nuclear staining. At higher magnification in **B**, a cluster of a few human cells, with their larger, more intensely staining nuclei are indicated by the *open arrow*. The *arrowhead* points to a single human cell that can be identified in the background of the smaller chick nuclei. Panels **C, D,** and **E**, are fluorescent micrographs of a field stained with antibodies to GFP and neuron-specific tubulin, and Hoechst nuclear stain, respectively. The arrows in each panel point to a pair of human neurons. In **F**, their large nuclei can easily be distinguished from those of the surrounding chick cells. The large fluorescent structure in D is a peripheral nerve of the chick. DRG – dorsal root ganglion, SC – spinal cord.

Feulgen staining procedure:
1. Rehydrate to dH_2O.
2. Incubate in 5 N HCl for 30 min (40 ml HCl/60 ml dH_2O).
3. Incubate in tap water for 15 min.
4. Rinse in dH_2O for 5 min.
5. Incubate for 90 min in Cold Schiff reagent.
6. Incubate in Sulfurous wash for 1 min then 4 min.
7. Wash in gently running tap water for 15 min.
8. Wash in dH_2O for 5 min.
9. Counterstain lightly with Fast green (1% aqueous solution) for 5 min.
10. Dip in 70% ethanol.
11. Dehydrate and mount with Entellan.

3.3.5. Immunocytochemistry

For absolute identification of the ES cells, it is necessary to perform antibody staining.

1. Bake slides in paraffin oven (56°C) for 30 min (*see* **Note 25**).
2. Deparaffinize by 2 × 15′ incubations in toluene, followed by 5′ each in 100%, 95%, and 70% ethanol. Wash for 10′ in dH_2O.
3. Incubate in a microwave for antigen retrieval for 15′ in 10 mM sodium citrate buffer in a polypropylene container. Cover the slides with 200 ml buffer and start the microwave at full heat. Watch the container until the solution reaches boiling, then lower the power to 30–50%. If the level of liquid falls below the top of the slides, stop the microwave procedure midway and add more buffer. When finished, remove from oven and allow the container to cool to room temperature.
4. Wash for 15′ in PBS.
5. Incubate in blocker for 15′.
6. Drain blocker (do not rinse) and then add the primary antiserum diluted in blocker (*see* **Note 26**).
   Incubate overnight or 48 h at 4°C in a humid chamber (*see* **Note 27**).
7. Warm for 1 h, then remove coverslips by placing the slides slowly into the glass staining dish carrier containing 200 ml PBS. After about 2′, slowly raise the slide carrier with the metal handle, and the coverslips will fall off by themselves. Rinse for another 3″, and then transfer to second PBS rinse bath for 5″.
8. Drain slides, and add secondary antibody(ies); incubate for 1–2 h in a humid chamber.
9. Remove coverslips as in Step 7 and rinse in PBS 2 × 5″.
10. Counterstain with Hoechst 1 µg/ml 5–10′.
11. Coverslip in antifade medium and seal coverslips with nail polish.
12. Let nail polish dry in a covered box for 10–15 min and view with a fluorescence microscope (*see* **Note 28**).

Figure 20.2C–E shows 2 human neurons adjacent to a nerve of the chick. The human cells are double-stained for GFP and neuron-specific tubulin, while the chick nerve is only stained with the neural marker. **Figure 20.3** shows the results of transplanting a 1-week embryoid body in place of somites.

3.3.6. Photomicrography

Bright field images are collected using a Scion CFW1310 color Firewire camera and fluorescence images with a Scion monochrome fivewive camera.

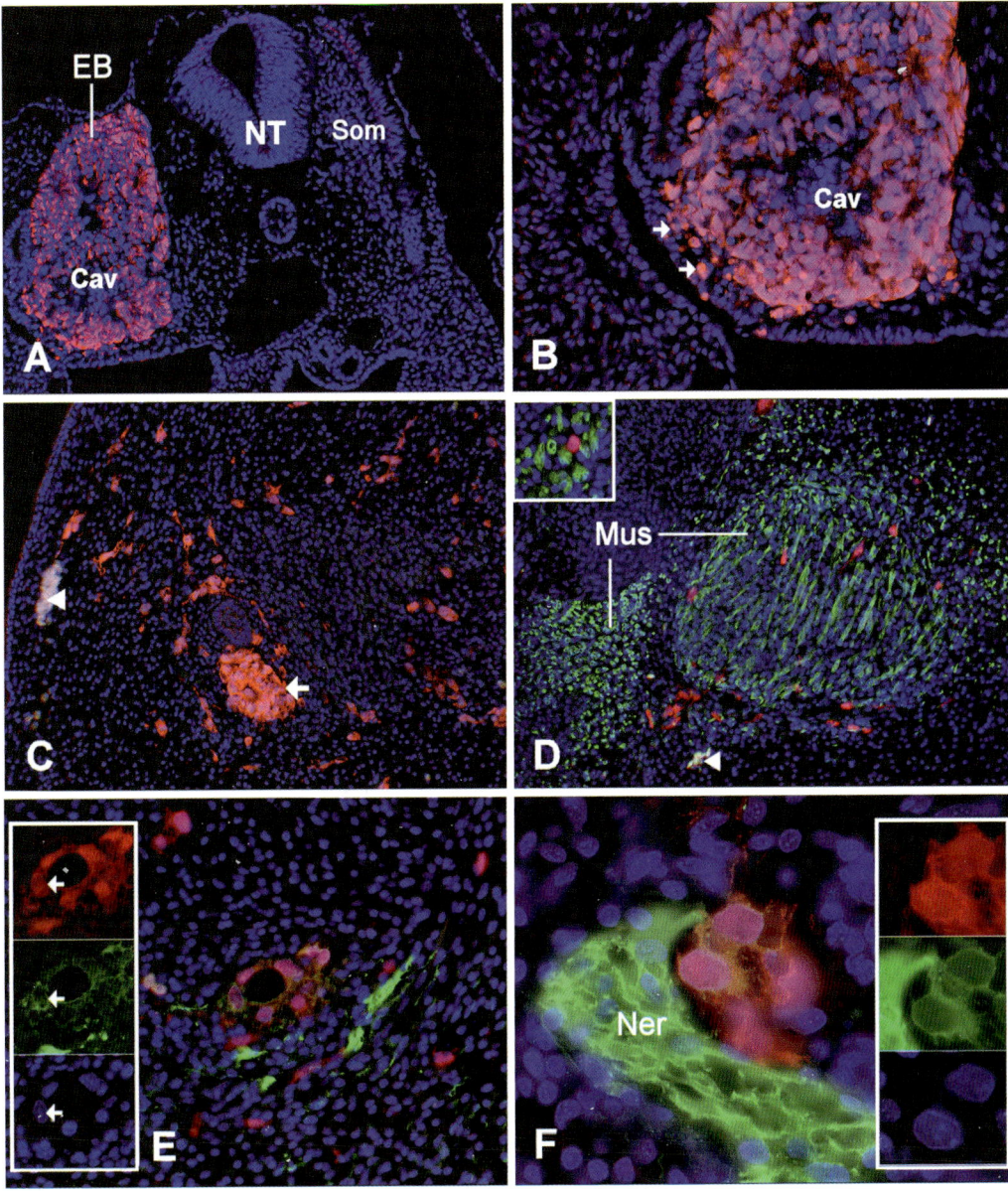

Fig. 20.3. Human embryoid bodies (EBs) dissociate and migrate into host tissues when transplanted into chick somites. (**A**) A GFP-expressing (*red staining*) EB one day after transplantation. A cavity (Cav) is starting to form in the EB. On the contralateral, unoperated side, the somite (Som) has dissociated into sclerotome and dermomyotome. NT= neural tube. (**B**) When the EB is examined at higher magnification, several cells (*arrows*) are observed to already have begun to migrate away. (**C**) EB-derived cells have widely dispersed within the chick tissues five days after the operation. Only a small proportion of the grafted cells remain as a clump (*arrow*). (**D**) Muscle is not formed by EB-derived cells 5 days after surgery. Double-immunostaining for GFP (*red*) and vertebrate desmin (*green*) shows that grafted cells that have integrated into muscles (Mus) have not differentiated into muscle. The *inset* shows a single human cell surrounded by chick muscle cells. *Arrowheads* in (**C**) and (**D**) point to red blood cells which autofluoresce. (**E**) Neurons form from human EBs implanted into somites. Double-immunostaining for GFP and neuron-specific tubulin shows that a small EB-derived tubule contains several immature neurons. The inset shows the tubule in each of the individual fluorescence channels: *red* – GFP, *green* – β-3-tubulin, and *blue* – Hoescht nuclear stain. The *arrows* in the *inset* point to the neuron with the highest level of tubulin expression. (**F**) A pair of EB-derived neurons adjacent to an intensely stained chick nerve (Ner). The *inset* shows the cells in each of the fluorescent channels as described in (**E**).

4. Notes

1. An additional method of introducing hESC to embryos is injection of a suspension of cells.
2. We obtained freshly laid fertile eggs from a local farm (Moshav Sitria). The eggs were kept at 12°C for up to 10 days before incubation, without noticeable ill effects.
3. Instruments should be cleaned promptly after each use, and rinsed with dH_2O and 70% ethanol to remove water. Dry sterilization is performed at 180°C for 90 min. With careful cleaning and dry sterilization, instruments last for many years.
4. This is Teflon-coated wire we had in the lab. The Teflon must be stripped off first by scraping with scissors. Noncoated wire can also be purchased.
5. The NaOH splatters when the current is applied, so eye protection and a lab coat should be worn, and a microscope that is near the end of its life used.
6. There are many brands of cellotape on the market; they vary greatly in their ability to remain adherent to egg shell in high-humidity and 37°C. It is worthwhile buying a few types and testing to see which sticks best after 3–4 days in the incubator.
7. Any paraformaldehyde can be used. We buy the crystals in order to reduce exposure to this toxic substance; the powder is easily spread around the lab, even using a hood, while the granules, even if they spill cannot be carried by wind currents. Many labs use formalin and dilute 1:10 with PBS; we have not tried this.
8. After having one microwave oven rust out, we bought one with an enamel interior.
9a. Others have claimed that certain brands interfere with fluorescence, we buy bargain-counter samples, and have not had a problem.
9b. Another dye that could be used for this purpose is CM-DiI (i.e., C7001, Life Technologies Corporation, Carlsbad, CA, USA), which has the advantage of being retained after paraffin processing (15, 16). Xenografted hESC cells/derivatives can also be labeled by infection with adenoviral vectors that confer constitutive GFP expression (10).
10. Although the manufacturer recommends using the stock dye only immediately after reconstitution with DMSO, we obtained substantially identical results with a 1-week-old stock solution.

11. Although the ES cells have been tested for hepatitis and AIDS, for extra precaution the ink was injected using a graduated microsyringe (GS-1200 (VX-07844-00) Gilmont® Micrometer Syringes, Barnant, Barrington, Il, USA) connected to the pipette via a 19-G steel needle (whose tip had been cut off with electrician's pliers) and mineral oil-filled polyethylene tube.

12. A tuberculin syringe can also be used for injecting ink. However, in our experience, making too large a hole with the ink injection pipette is one of the primary causes of loss of operated embryos. The surface tension of the external membranes is very strong and cause a continual tearing, if the hole is not small enough to repair itself within a day or so.

13. Of course, transplantation of the HES can be made virtually anywhere in the chick embryo. Many years of study have elucidated the precise positions and timing of development of its various tissues and organs.

14. If embryos older than about 20 somite pairs are used, excessive damage of the somites with the scalpel can cause bleeding, an occurrence that reduces the chance of survival after surgery.

15. If the enzyme is permitted to act too long, the neural tube begins to look like an "accordion," and survival chances are small.

16. It is advisable to have one microscalpel with a dull tip for removing the somites in order to minimize the possibility of damage to the endoderm (and embryo).

17. Whatever type is cheapest and most easily available in the lab.

18. All of the preceding steps can be performed in a laminar flow biological hood with positive air-flow. However, it is not advised to use such a hood for work with human biological material. Therefore, the blower should be shut down before the HES is brought to the hood. Alternatively, a biohazard hood can be used, with holes cut for the oculars of the binocular dissecting microscope.

19. In order to prevent the sticking of the tissue to the pipette tip, aspirate/collect some medium without tissue into the tip, keep the plunger of the pipettor fixed, and then up the tissue in the remaining volume by allowing the plunger to spring up until it stops. It is advisable to cut many more pieces than will actually be transplanted, since there is invariably loss of tissue both in transit to the embryos, and within the embryos themselves.

20. This is a critical step, if the expelling is done too rapidly, the tissue "swims away," and is lost within the embryo!

21. Survival of the embryos is between 50 and 100% 1 day after surgery, at 5 days 20–50%.

22. A slotted spoon is a valuable aid in removing the delicate younger embryos without damage.
23. Such as those that 8 × 10 inch photographic printing paper is sold in
24. For routine histology, (Hematoxylin and Eosin, Feulgen, etc.), conventional microscope slides coated with polylysine solution are fine. For immunostaining, and most particularly, for microwave antigen retrieval, Superfrost Plus slides are absolutely required. Otherwise, the sections will fall off the slides during the microwaving procedure.
25. The use of Superfrost Plus slides is absolutely critical, otherwise sections fall off in the microwave step.
26. Antibody dilutions vary from batch to batch; we use the manufacturer's suggestion as a lower guide and perform tests on each vial received.
27. We use 75 µl/slide, and carefully coverslip with 24 × 50 coverglass to keep the antibody in place. Two antibodies from different species (mouse/rabbit) or 2 monoclonal subtypes (IgM, IgG, etc.) can be mixed and applied simultaneously.
28. The slides are stored at 4°C in regular plastic slide boxes; Alexa, Cy2, and Texas-Red fluorescence are stable for more than 6 months.

Acknowledgements

My deep appreciation to Nissim Benvenisty for introducing me to the marvels of hESC, the opportunity to work on this project, and many stimulating discussions. I would like to thank Camila Avivi, Victoria Zismanov, and Ariel Chipman for technical assistance with the histological analysis. hESC used in most experiments were grown by Micha Drukker in the Benvenisty lab at the Hebrew University who also prepared EBs. In a few experiments GFP-expressing hESC were grown by Orna Singer in the Reubinoff lab of the Hadassah Embryonic Stem Cell Research Center.

References

1. D. James, S.A. Noggle, T. Swigut, and A.H. Brivanlou, Contribution of human embryonic stem cells to mouse blastocysts, Dev. Biol. 295 (2006) 90–102.
2. J.C. Fontaine-Perus, P.Halgand, Y.Cheraud, T. Rouaud, M.E. Velasco, C.C. Diaz, and F. Rieger, Mouse chick chimeras: a developmental model of murine neurogenic cells., Dev 124 (1997) 3025–3036.
3. H. Wichterle, I. Lieberam, J.A. Porter, and T.M. Jessell, Directed differentiation of embryonic stem cells into motor neurons., Cell 110 (2002) 385–397.

4. R.S. Goldstein, M. Drukker, B.E. Reubinoff, and N. Benvenisty, Integration and differentiation of human embryonic stem cells transplanted to the chick embryo, Develop. Dynam. 225 (2002) 80–86.

5. D.H. Lee, S. Park, E.Y. Kim, S.K. Kim, Y.N. Chung, B.K. Cho, Y.J. Lee, J. Lim, and K.C. Wang, Enhancement of re-closure capacity by the intra-amniotic injection of human embryonic stem cells in surgically induced spinal open neural tube defects in chick embryos, Neurosci. Lett. 364 (2004) 98–100.

6. D.H. Lee, E.Y. Kim, S. Park, J.H. Phi, S.K. Kim, B.K. Cho, J. Lim, and K.C. Wang, Reclosure of surgically induced spinal open neural tube defects by the intraamniotic injection of human embryonic stem cells in chick embryos 24 hours after lesion induction, J. Neurosurg. 105 (2006) 127–133.

7. G. Lee, H. Kim, Y. Elkabetz, S.G. Al, G. Panagiotakos, T. Barberi, V. Tabar, and L. Studer, Isolation and directed differentiation of neural crest stem cells derived from human embryonic stem cells, Nat. Biotechnol. 25 (2007) 1468–1475.

8. H. Lee, G.A. Shamy, Y. Elkabetz, C.M. Schofield, N.L. Harrsion, G. Panagiotakos, N.D. Socci, V. Tabar, and L. Studer, Directed differentiation and transplantation of human embryonic stem cell-derived motoneurons, Stem Cells 25 (2007) 1931–1939.

9. T.S. Park, E.T. Zambidis, J.L. Lucitti, A. Logar, B.B. Keller, and B. Peault, Human embryonic stem cell-derived hematoendothelial progenitors engraft chicken embryos, Exp. Hematol. 37 (2008) 31–41.

10. I. Brokhman, O. Pomp, L. Shaham, T. Tennenbaum, M. Amit, J. Itskovitz-Eldor, and R.S. Goldstein, Genetic modification of human embryonic stem cells with adenoviral vectors: differences of infectability between lines and correlation of infectability with expression of the coxsackie and adenovirus receptor, Stem Cells Dev. 18 (2008) 447–456.

11. G.W. Conrad, J.A. Bee, S.M. Roche, and M.A. Teillet, Fabrication of microscalpels by electrolysis of tungsten wire in a meniscus, J. Neurosci. Methods 50 (1993) 123–127.

12. J. Itskovitz-Eldor, M. Schuldiner, D. Karsenti, A. Eden, O. Yanuka, M. Amit, H. Soreq, and N. Benvenisty, Differentiation of human embryonic stem cells into embryoid bodies comprising the three embryonic germ layers., Mol. Med. 6 (2000) 88–95.

13. R. Eiges, M. Schuldiner, M. Drukker, O. Yanuka, J. Itskovitz-Eldor, and N. Benvenisty, Establishment of human embryonic stem cell-transfected clones carrying a marker for undifferentiated cells., Curr. Biol. 11 (2001) 514–518.

14. A.L. Liu and K.M. Bagnall, Regeneration following somite removal in chick embryos, Anat. Embryol. (Berl) 192 (1995) 459–469.

15. K. Hemmrich, M.Meersch, H.D. von, and N.Pallua, Applicability of the dyes CFSE, CM-DiI and PKH26 for tracking of human preadipocytes to evaluate adipose tissue engineering, Cells Tissues. Organs 184 (2006) 117–127.

16. W. Andrade, T.J. Seabrook, M.G. Johnston, and J.B. Hay, The use of the lipophilic fluorochrome CM-DiI for tracking the migration of lymphocytes, J. Immunol. Methods 194 (1996) 181–189.

Chapter 21

Genetic Manipulation of Human Embryonic Stem Cells

Silvina Epsztejn-Litman and Rachel Eiges

Abstract

One of the great advantages of embryonic stem (ES) cells over other cell types is their accessibility to genetic manipulation. They can easily undergo genetic modifications while remaining pluripotent and can be selectively propagated, allowing the clonal expansion of genetically altered cells in culture. Since the first isolation of ES cells in mice, many effective techniques have been developed for gene delivery and manipulation of ES cells. These include transfection, electroporation, and infection protocols, as well as different approaches for inserting, deleting, or changing the expression of genes. These methods proved to be extremely useful in mouse ES cells, for monitoring and directing differentiation, discovering unknown genes and studying their function, and are now being initiated in human ES (HESC) cells. This chapter describes the different approaches and methodologies that have been applied for the genetic manipulation of HESCs and their applications. Detailed protocols for generating clones of genetically modified HESCs by transfection, electroporation, and infection will be described, with special emphasis on the important technical details that are required for this purpose.

Key words: Human ES cells, genetic manipulation, transfection, electroporation, infection, overexpression, targeted mutagenesis, homologous recombination, knock-down.

1. Introduction

1.1. Genetic Modification Approaches and Their Potential Applications

There are basically four types of strategies that can be applied for genetic engineering of HESCs: overexpression, knockout, knock-in, and knock-down experiments.

1.1.1. Overexpression

Overexpression of genes is usually based on random integration of an exogenous DNA sequence into the genome. It can be applied for constitutive or facultative expression of either cellular or foreign genes. It may also be used for the introduction of reporter or selection genes, under the regulation of tissue-specific promoters. These procedures allow to label and track specific cell lineages following induced differentiation of human embryonic stem cells (HESCs) in culture. Moreover, they can be employed for the isolation of pure populations of specific cell types, by the use of selectable markers. The marker gene may either be a selectable reporter, such as green fluorescent protein (GFP), which can be selected for by fluorescent-activated cell sorter (FACS), or a drug-resistance gene (1, 2). Indeed, transgenic fluorescent reporters driven by tissue-specific promoters have been successfully employed to identify various HESCs-derived cell types including neurons (3), cardiomyocytes (4) as well as hepatic and pancreatic committed cells (5, 6). Likewise, the introduction of selectable reporters under the regulation of an inner cell mass-specific promoter may allow the selection for or against undifferentiated cells in culture. This has been previously demonstrated by introduction of EGFP (enhanced GFP) into HESCs under the regulation of both *OCT4* (7) and *Rex1* (8), resulting in the production of green glowing cells only when they are in an undifferentiated state (8). The ability to isolate pure populations of specific cell types and eliminate undifferentiated cells prior to transplantation has great importance in cell-based therapy; this is because transplantation of undifferentiated cells may lead to teratoma formation.

Overexpression experiments may also be employed for directing the cell fate of differentiating ES cells in culture. This can be achieved by introducing master genes that play a dominant role in cell commitment, forcing the cells to differentiate into specific lineages that otherwise are rarely obtained among many other cell types in culture (9–11). In fact, it was possible to show that ectopic expression of HOXB4, a master regulator in blood cell differentiation, can drive HESCs to commit into hematopoietic lineages as they differentiate in vitro (*12*).

Random integration of promoter-driven transgenes may be employed for the generation of cell-based delivery systems by producing therapeutic agents at the site of damaged tissue. The use of ES-derived cells as therapeutic vectors has been previously shown to be feasible in mice, where grafting of ES-derived insulin-secreting cells normalized glycemia in streptozotocin-induced diabetic mice (13). Apart from tagging, selecting, and directing the differentiation of specific cell types, it is possible to inactivate endogenous genes to study their function. This can be achieved either by disrupting both copies of the gene or by downregulating its activity *in trans*.

1.1.2. Knockout

The most widely used technique for inactivating genes in ES cells is site-directed mutagenesis. This procedure involves the replacement of a specific sequence in the genome by a mutated copy through homologous recombination with a targeting vector.

The targeting vector that contains the desired mutation and a selectable marker, flanked by sequences that are interchangeable with the genomic target, pairs with the wild-type chromosomal sequence and replaces it through homologous recombination. By targeting both alleles, using distinct selection markers, it is possible to create "loss-of-function" or so-called knockout phenotypes in ES cells that can be used for functional studies of specific genes. This technology has been well practiced in mice for gene function studies, in which genetically altered cells are introduced into wild-type embryos, resulting in the creation of germ-line transmitting chimeras (14). The genetically manipulated animals can be further mutated to generate animals that are homozygous for the desired mutation. The creation of HESCs with a null genotype for specific genes may have great importance for modeling human diseases, and for the study of crucial developmental genes that in their absence are embryonic lethal. One example for generating a HESC-based disease model by homologous recombination is the targeting of the X-linked gene hypoxanthine phosphoribosyl transferase 1 *(HPRT1)*, which when mutated is responsible for the development of Lesch-Nyhan syndrome. This was performed by introducing a large deletion at the *HPRT1* locus in HESCs of an XY karyotype. The resulting cell lines recapitulate the major biochemical defect that characterizes Lesch-Nyhan affected individuals, which involves the accumulation of uric acid (*15*). Thus, these cells should be valuable for basic research, but more importantly for exploration of new gene therapy-based treatments and drug discovery.

1.1.3. Knock-In

Similar to the knockout strategy, it is possible to generate clones of HESCs in which the gene of interest is deleted by inserting a promoterless reporter gene through homologous recombination. The method, termed knock-in, allows the positioning of a reporter gene under the regulation of a native gene. Therefore, it can be applied to monitor the expression of a target gene in situ during ES cell differentiation. Accordingly, Zwaka and Thomson have created human knock-in ES cell lines that express either GFP or a neomycin-resistance gene under the regulation of the endogenous OCT4 promoter (16). The *OCT4* gene encodes for a transcription factor that is specifically expressed by pluripotent stem cells. Thus, by replacing *OCT4* with such reporters, the authors were able to monitor and select for undifferentiated HESCs in culture. Likewise, GFP cDNA was inserted into the locus of *MIXL1*, which is a developmentally regulated gene that is transiently expressed in the primitive streak during embryogenesis (*17*). In this case, the reporter knock-in reflected the expression of the endogenous *MIXL1*, enabling to identify primitive streak-like cells and isolation of primitive hematopoietic precursors from differentiating HESCs.

The relative ease by which ES cells can be genetically manipulated has made them particularly useful for the search of unknown genes whose pattern of expression suggests that they might have developmental importance. The identification of such genes is performed by the gene trap method, which is based on the random disruption of endogenous genes (reviewed by (18)). As opposed to targeted mutagenesis, it involves the random insertion of a reporter gene that lacks essential regulatory elements into the genome. Because the expression of the reporter gene is conditioned by the presence of an active endogenous regulatory element, it may serve to identify only transcribed sequences. Using this method, a large-scale gene disruption assay is possible, allowing the discovery of new genes and the creation of a wide variety of mutations.

Finally, the recent identification of the human homolog of the mouse *Rosa26* locus has facilitated in the generation of *Cre*-mediated HESC lines that can be utilized for inserting transgenes into a broadly expressed locus. The importance of this system is that it overcomes problems of gene silencing as well as gene disruption through insertional mutagenesis. Moreover, it does not require drug selection and thus may be useful for therapeutic applications in cases where drug-resistance cassettes are not desirable.

1.1.4. Knock-Down

Downregulation of particular genes can also be achieved by overexpressing specific RNA molecules that inhibit the activity of a given gene through the generation of small interfering RNA molecules (siRNAs). Because siRNAs operate *in trans* and are not involved in the modification of the targeted gene, it is relatively simple to achieve transient or conditional gene silencing using this method. The use of RNA interference (RNAi) was demonstrated to be feasible in mouse ES cells to inactivate genes and shown to be equally effective as the knockout models in the generation of null mutant embryos (*19*). Downregulation by RNAi in HESCs was demonstrated for the *HPRT*, *β2-microglobulin*, *OCT4*, *SOX2*, and other genes (20, 21). Applications of this loss-of-function approach will have widespread use, not only to study developmental roles of specific genes in human, but also for their utility in modulating HESC differentiation in vitro.

1.2. Methods for Genetic Manipulation

There are many factors that may influence transfection efficiency: phase of cell growth, number of passages, size and source of the transgene, vector type and size, and the selection system. However, the most important factor is the transfection method. Several gene-transfer techniques are now available for manipulating gene expression in HESCs. The latter include chemical-based (transfection), physical (electroporation), and viral-mediated (infection) techniques.

1.2.1. Transfection

Transfection is probably the most commonly used method for introducing transgenes into HESCs (**Fig. 21.1**). It is straightforward, relatively easy to calibrate, provides a sufficient number of cells for clonal expansion, can be performed on adherent cell cultures, and allows the insertions of constructs of virtually unlimited size. This system is based on the use of carrier molecules that bind to foreign nucleic acids and introduce them into the cells through the plasma membrane. In general, the uptake of exogenous nucleic acids by the cell is thought to occur through endocytosis, or in the case of lipid-based reagents, through fusion of lipid vesicles to the plasma membrane. The first study to describe stable

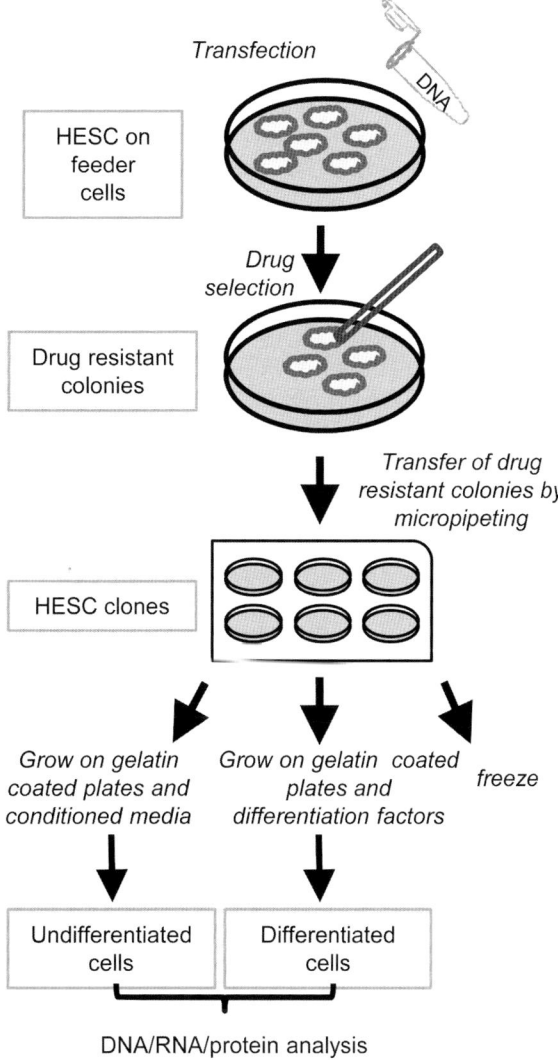

Fig. 21.1. Schematic illustration describing the methods for generating genetically modified HESCs by transfection.

transfection in HESCs was based on the use of a commercially available reagent, ExGen 500, which is a linear polyethylenimine (PEI) molecule that has a high cationic charge density (8). The unique property of this molecule is due to its ability to act as a "proton sponge," which buffers the endosomal pH, leading to endosome rupture and DNA release. This method routinely produces transient transfection rates of approx 10–20% and stable transfection efficiencies of $1:10^{-5}$–10^{-6} (8). Since then, other chemical-based transfection methods have been found to be equally effective. The calcium phosphate precipitation method is a widely used method for transfecting many different cell types. It is also based on negatively charged molecules that interact with DNA to form precipitates that are incorporated by the cells. The calcium phosphate transfection system seems to be slightly more efficient in gene delivery in comparison to ExGen 500. Lipofectamine 2000 reagent is a positively charged cationic lipid compound that forms small unilamellar liposomes and was recently shown to be useful in obtaining transient and stable transfections in HESCs as well (20, 21).

1.2.2. Electroporation

Electroporation is a method that employs the administration of short electrical impulses that create transient pores in the cell membrane, allowing foreign DNA to enter into the cells. Although efficient and most popular in mouse ES cells, this procedure gave poor results in HESCs, both in transient and stable transfection experiments. This is most probably due to the low survival rates of HESCs after the voltage shock. Zwaka and Thomson reported a protocol to increase the yield of electroporation 100-fold, thereby achieving an integration rate of approximately $1:10^{-5}$ (16). This was performed by carrying out the procedure on cell clumps rather than on single-cell suspension. In addition, electroporation was performed in standard cell culture media, which is a protein-rich solution, instead of PBS and altering the parameters of the protocol used in mouse ES cells. Using this method, 3–40% homologous recombination events among resistant clones were reported, subject to vector properties (22). A substantial number of HESC clones obtained by homologous recombination have been created thus far using different constructs, demonstrating the feasibility of this technique for site-directed mutagenesis in HESCs.

1.2.3. Infection

Unlike in all nonviral-mediated methods (transfection and electroporation), gene manipulation by viral infection can produce a very high percentage of modified cells (**Fig. 21.2**). To date, genetic manipulation of HESCs by viral infection has been reported by several groups using adeno- as well as Baculovirus and lenti-viral vectors (23–26). Infection studies with RNA and DNA viruses have demonstrated that these viral vectors have two distinct advantages over other systems: high efficiency of DNA transfer and single-copy

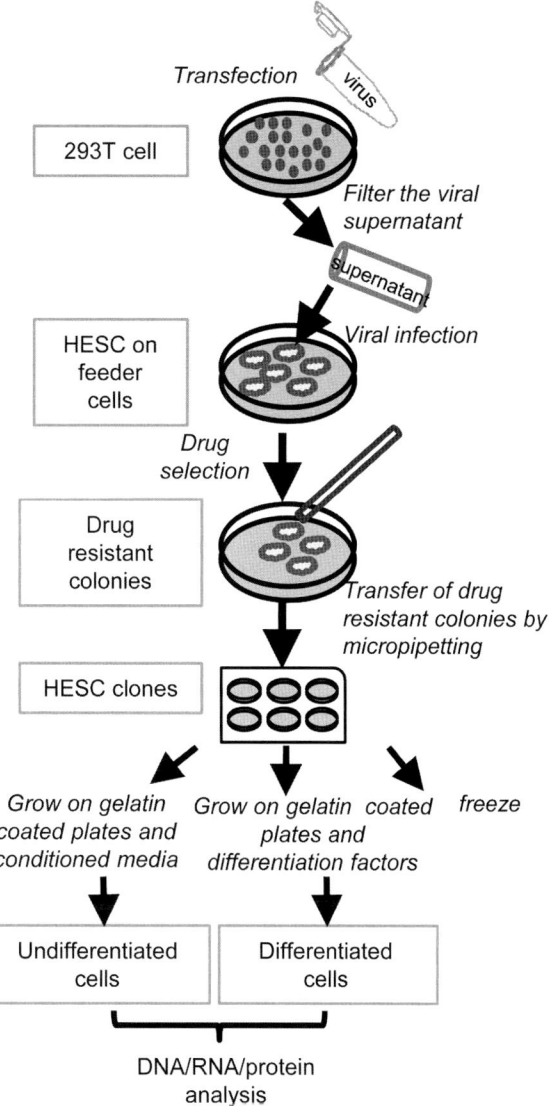

Fig. 21.2. Schematic illustration describing the methods for generating genetically modified HESCs by infection.

integrations. However, integration occurs randomly and cannot be targeted to a specific site in the genome. Yet, because of its high efficiency, this method could prove useful for bypassing the need for selection and time-consuming clonal expansion, as well as for experiments that aim for random insertion mutagenesis or gene trap.

Lentiviral-based vectors offer an attractive system for efficient gene delivery into HESCs. These vectors are derived from lentiviruses, a group of complex retroviruses that cause slow chronic immunodeficiency diseases in humans and animals. Lentiviral vectors (LVVs) can transduce both dividing and nondividing cells and were shown to drive gene expression efficiently in various types of "stem" cells. Gene delivery into HESCs by vectors derived from lentiviruses has the following advantages: (1) lentiviral vectors efficiently transduce HESCs; (2) they integrate into the host-cell genome, thus promoting stable transgene expression; (3) transgene expression is not significantly silenced in undifferentiated HESCs as well as following differentiation; and (4) transduced HESCs retain their self-renewal and pluripotent potential.

In order to improve vector biosafety and performance, first all pathogenic coding sequences were deleted from the vector, resulting in a replication-defective vector containing only the transgene and several essential regulatory viral sequences, such as the encapsidation signal and the viral LTR. Second, the proteins necessary for the early steps of viral infection (entering into the host cell, reverse transcription, and integration) were provided *in trans* by two additional plasmids: a packaging plasmid expressing the *gag*, *pol*, and *rev* genes and an envelope plasmid expressing a heterologous envelope glycoprotein of the vesicular stomatitis virus (VSV-G). Third, a large deletion was introduced into the U3 region of the viral LTR abolishing the viral promoter/enhancer activity. The self-inactivating (SIN) vector that was generated contained a heterologous internal promoter driving the expression of the transgene (27, 28). These steps resulted in a vector that could only undergo one round of infection and integration, a process termed transduction. Moreover, they minimized the risk of generation of wild-type HIV-1 by recombination.

Human immunodeficiency virus (HIV)-1-based lentiviral vectors were the first viral vectors used to genetically engineer HESCs. However, because of the severe pathogenic effects of HIV-1 replication in humans, the potential emergence of replication-competent retrovirus from HIV-1-based vectors raises concerns over their use in clinical settings. In addition, random chromosome integration of lentiviral vectors poses the risk of insertional mutagenesis, oncogene activation, and cellular transformation. In addition, lentiviral vectors may not be suitable for transient transgene expression.

Viral vectors derived from adenovirus and adenoassociated virus (AAV) have a much lower risk of insertional mutagenesis and have been tested in HESCs, but their transduction efficiencies were less satisfactory (25).

The insect baculovirus *Autographa californica multiple nucleopolyhedrovirus* (AcMNPV)-based vectors have recently been introduced as a new type of delivery vehicle for transgene expression in mammalian cells (29). The virus can enter mammalian cells but does not replicate, and it is unable to recombine with pre-existing viral genetic materials in mammalian cells. One significant advantage of using baculovirus AcMNPV as a gene delivery vector is the large cloning capacity to accommodate up to 30 kilobases (kb) of DNA insert, which can be used to deliver a large functional gene or multiple genes from a single vector.

1.2.4. Short- vs Long-Term Expression

Gene transfer experiments can be subdivided into short-term (transient) and long-term (stable) expression systems. In transient expression, the foreign DNA is introduced into the cells and its expression is examined within 1–2 days. The advantage of this assay is its simplicity and rapidity. Furthermore, because the foreign DNA remains episomal, there are no problems associated with site of integration and the copy number of the transgene. Yet, it does not allow conducting experiments over long periods. Moreover, transfection efficiency usually does not exceed 20%. For short-term induction, efficient transient expression can be achieved through the insertion of supercoiled plasmid DNA rather than the linear form. Transient expression in HESCs usually peaks roughly 48 h after transfection, and frequently results in high expression levels attributed to the high copy number of plasmid DNA molecules that occupy the cell. During long-term assays, one isolates a clone of HESCs that has stably integrated the foreign DNA into its chromosomal genome. The major advantage of this method is the ability to isolate stable ES cell lines that have been genetically modified and can be grown indefinitely in culture. In this type of experiment, it is important to linearize the vector, leading to greater integration and targeting efficiencies. When the target gene is nonselectable, one must introduce also a positive selection marker under the regulation of a strong constitutive promoter. This can be performed either by cotransfecting the selectable marker on a separate vector, or as is frequently done, by fusing the selectable marker to the targeting vector. Selection should not be carried out immediately after transfection but at least 24 h later, giving the cells time to recover, integrate the foreign DNA, and express the resistance-conferring gene.

2. Materials

2.1. Tissue Culture (see Notes 1 and 2)

1. Knockout DMEM-optimized Dulbecco's modified Eagle's medium for ES cells (Gibco BRL, Carlsbad, CA, USA; Cat. No. 10829-018).
2. DMEM 4.5 g/L glucose (Sigma, Dorset, UK; Cat. No. D5796).
3. 1 M β-mercaptoethanol (Sigma; Cat. No. M7522).
4. Nonessential amino acids 100X stock (Biological Industries, Kibutz Beit-Haemek, Israel; Cat. No. 01-340-1B).
5. Insulin-transferrin-selenium 100X (Gibco BRL; Cat. No. 41400-045).
6. Bovine serum albumin (Sigma; Cat. No. A-4919).
7. Mitomycin C (Sigma; Cat. No. M-0503).
8. 0.1% gelatin (Sigma; Cat. No. G-1890).

9. Hygromycin B (Sigma; Cat. No. H-3274).
10. 6-Thioguanine (Sigma; Cat. No. A-4660).
11. Knockout SR – serum-free formulation (Gibco BRL; Cat. No. 10828-028).
12. Fetal calf serum (Biological Industries).
13. l-Glutamine 100X stock (200 mM/L, Biological Industries; Cat. No. 03-020-1).
14. Penicillin (10,000 U/mL) and streptomycin (10 mg/mL) 100X stock (Biological Industries; Cat. No. 03 031-1B).
15. Human basic fibroblast growth factor (bFGF) stock solution (2 ng/µL) (human recombinant; Gibco BRL; Cat. No. 13256029).
16. Trypsin-EDTA: 0.25% trypsin and 0.05% EDTA (Biological Industries; Cat. No. 03-052-1).
17. G418 (Geneticin; Sigma; Cat. No. G-9516).
18. Puromycin (Sigma; Cat. No. P8833).
19. Dimethylsulfoxide (DMSO; Sigma; Cat. No. D-2650).
20. 1X phosphate-buffered saline (PBS) without Ca^{2+}/Mg^{2+}. For 1 L, mix 3.58 g sodium phosphate ($Na_2HPO_4 \bullet 12H_2O$), 0.24 g potassium phosphate monobasic (KH_2PO_4), 8 g sodium chloride (NaCl), 0.2 g potassium chloride (KCl), in a final volume of 1 L of double-distilled water (ddH_2O). Aliquot 200 mL solution per bottle and autoclave; store at room temperature.
21. 10 mM β-mercaptoethanol: dilute 1:100 in PBS, filter, sterilize, and store at 4°C.
22. 50X Mitomycin-C: dissolve 2 mg in 4 mL MEF medium and store at 4°C.
23. bFGF solution: add 10 µg of bFGF solution to 5 mL of filter-sterilized 0.1% bovine serum albumin dissolved in 1X PBS (with Ca^{2+}/Mg^{2+}), to give a final concentration of 2 µg/mL; store 1-mL aliquots at −20°C.
24. 0.1% gelatin solution: add 0.1 g of gelatin into a bottle containing 100 mL distilled water and autoclave immediately. The gelatin is dissolved while boiling in the autoclave; store at 4°C.
25. MEF media: add to a 500-mL bottle of DMEM (high glucose and l-glutamine) 50 mL fetal calf serum, 2.5 mL penicillin/streptomycin, 5 ml glutamine.
26. HESC medium: add to a 500-mL bottle of knockout DMEM, 75 mL knockout SR, 6 mL nonessential amino acids, 6 mL glutamine (2 mM), 3 mL insulin-transferrin-selenium, 60 µL

β-mercaptoethanol (0.1 mM), 3 mL penicillin/streptomycin, and 1.2 mL bFGF. ES media should be protected from light (*see* **Note 3**) and stored at 4°C up to 1 month.

27. Freezing medium: add 1 mL of DMSO to 9 mL of appropriate media (either hES or MEF media). Media should be prepared fresh.
28. Leishman's stain (BDH, Poole, England) in 100% methanol.
29. 293T cells medium: add to a 500-mL bottle of DMEM (high glucose and L-glutamine) 50 mL fetal calf serum, 2.5 mL penicillin/streptomycin, 5 mL glutamine.
30. FuGENE 6 (Roche 11-988-387).
31. Hexadimethrine bromide (polybrene) (Sigma H9268-5G).
32. Laminar flow hood.
33. Humidified incubator set at 37°C and 5% CO_2.
34. Phase-contrast microscope (objective range from 10 to 40).
35. Liquid nitrogen storage tank.
36. Refrigerator (4°C) and freezers (–20°C, –70°C).
37. 37°C water bath.
38. Electroporator (Biorad, Gene Pulser II System).
39. Swing-out centrifuge for conical tubes (15- and 50-mL).
40. Cell counter.
41. Gene pulser cuvette 0.4-cm electrode gap (Bio-rad Cat. No. 165-2088).
42. Pipetmen (2, 10, 20, 200, and 1000 µL) designated for tissue culture use only.
43. Sterile forceps and scissors for dissecting mouse embryos.
44. Falcon tissue-culture plates (100 ./'20 mm) and 6-, 12-, and 24-multiwell trays (Falcon, Bedford, MA; Cat. Nos 353047, 353047, 353043, 353046).
45. Falcon 15-mL and 50-mL (Falcon; Cat. Nos 352097, 352098) polypropylene conical tubes.
46. Cryo vials (1.8-mL CryTube; Nunc, Roskilde, Denmark; Cat. No. 363401).
47. Plastic pipettes (1, 2, 5, and 10 mL).
48. Tips for 2-, 10-, 20-, 200- and 1000-µL pipetmen.
49. Eppendorf tubes (1.5 mL).
50. Disposable filter unit FP 30/0.45 CA-S, white rim 0.45 µm, cellulose acetate sterile (whatman Ref No. 1046200).
51. Syringes sterile 20 ml.

2.2. Transfection

1. 2X HBS: 50 mM HEPES and 280 mM NaCl; dissolve 1.57 g NaCl and 1.19 g HEPES in approx 80 mL sterile ddH$_2$O. Adjust pH to 6.8 and bring to a final volume of 100 mL with ddH$_2$O. Filter-sterilize and store in 15-mL aliquots at 20°C.
2. 70 mM Na$_2$HPO$_4$: dissolve 2.5 g of Na$_2$HPO$_4$•12H$_2$O in 100 mL of ddH$_2$O. Filter-sterilize and store in 15-mL aliquots at –20°C.
3. Transfection buffer: mix 485 μL of 2X HBS with 15 μL of 70 mM Na$_2$HPO$_4$.
4. 2 M CaCl$_2$: dissolve 27.75 g CaCl$_2$ in ddH$_2$O to a final volume of 100 mL. Filter-sterilize and store 15-mL aliquots at −20°C.
5. Humidified incubator set at 34°C, 3% CO$_2$.
6. Tips for 2-, 10-, 20-, 200- and 1000-μL pipetmen.
7. 10-mL tubes.
8. Eppendorf tubes (1.5-mL).
9. ExGen 500 (Fermentas, Hanover, MD, USA; Cat. No. R0511).
10. Vortex.
11. Swing out centrifuge for microplates.

2.3. Infection

1. DMEM growth medium with 10% FCS and glutamine (1 mg/mL), without penicillin/streptomycin.
2. FuGENE 6 27μl (Roche).
3. Hexadimethrine bromide (polybrene), 5 μl (8 mg/ml).
4. Humidified incubator set at 34°C, 3% CO$_2$.
5. Tips for 2-, 10-, 20-, 200- and 1000-μL pipettes.
6. 10-mL tubes.
7. Eppendorf tubes (1.5 mL).
8. Tissue culture plates.

2.4. Colony Picking

1. HESC medium (*see* **Section 2.1**, **Item 26**).
2. G418 (200 μg/mL).
3. Puromycin (0.5–1 μg/mL).
4. Hygromycin (100 μg/mL).
5. 6-Thioguanine (1 μg/mL).
6. 6-, 12-, and 24-well Falcon tissue culture plates (*see* **Section 2.1.1, Item 11**).
7. Mouth apparatus consisting of an aspirator mouthpiece, tubing, and Pasteur pipette pulled on flame for collecting single colonies (*see* **Note 4**).

3. Methods

3.1. Tissue Culture (see Notes 5 and 6)

The special growth conditions that are required for supporting undifferentiated growth of HESCs in culture rely mostly on the presence of inactivated fibroblasts, serving as a feeder layer. The feeder layer sustains undifferentiated growth by secreting unknown growth factors and by serving as a growth matrix that allows the cells to adhere and grow as a monolayer culture. So far, primary mouse embryonic fibroblasts (MEFs) were the most commonly used in the propagation and derivation of HESCs. However, STO cells (30), fetal muscle (31), foreskin fibroblasts (32, 33), and marrow cells (34) were also reported to be equally effective in supporting undifferentiated growth. The feeders are prepared only from early passage MEFs (up to passage 5). Their mitotic inactivation is carried out by the treatment with mitomycin-C (35), but can also be achieved through irradiation (36). Normally, we prepare MEFs from 13.5-day-old ion cyclotron resonance-derived embryos. However, inactivated primary fibroblasts are required not only for routine maintenance of ES cells in culture, but also for stable transfection experiments, where drug selection is applied. Therefore, it is a prerequisite that feeder cells be resistant to the drug employed. For this purpose, one must separately prepare MEFs from different strains of mice that bear resistance to the desired drug or alternatively, use feeders that carry multidrug-resistant genes by intercrossing between different strains. For instance, the transgenic strain of mice DR-4 expresses four different drug-selected genes and can be used for the production of MEFs, which confer resistance to G418, puromycin, hygromycin, and 6-thioguanine drugs (37). The DR-4 strain, therefore, represents a suitable and an economical donor for the production of drug-resistant MEFs and is especially advantageous for gene-targeting experiments, which normally involve sequential selection for multidrug-resistant markers. There may be a significant variability between various batches of MEFs, with respect to their capacity for supporting undifferentiated proliferation of HESCs. To overcome this problem, the competence of different batches of MEFs to support undifferentiated growth can be assessed by testing their ability to maintain undifferentiated proliferation of mouse or primate ES cell lines before their use.

3.1.1. Isolation of MEFS

1. Coat plates with 0.1% gelatin by incubation for 1 h at room temperature.
2. Collect 13.5-day-old fetuses from pregnant mice using sterile equipment: sacrifice pregnant mice and dissect the embryos by removing the uterus and transferring it into a sterile PBS-containing Petri dish.

3. Rinse twice in PBS and relocate all work to laminar flow hood.

4. Using sterile tweezers and scissors, remove the fetuses from the uterus, separate them from extraembryonic tissues (amniotic and yolk sacs), and transfer them to a clean petri dish with PBS.

5. Count the number of collected fetuses and prepare, for later use, 1X 10-cm gelatin-coated tissue culture dish for every three fetuses.

6. Remove head and internal parts (liver, heart, kidney, lung, and intestine) with sterile tweezers under a stereomicroscope.

7. Cut the remaining tissues into small pieces in a minimal volume of PBS (1–2 mL) and transfer into a sterile 50-mL Falcon tube.

8. Disaggregate the cell clumps obtained by passing them through a 5-mL syringe with an 18-gauge needle, no more than 10 times.

9. Add MEF media to reach 10 mL per three embryos, distribute cell suspension evenly into 10-cm tissue culture dishes, and incubate.

10. Change media the following day. When plates are confluent (2–3 days after dissection), split 1:3 by trypsinization.

11. Change media (10 mL) every 2 days. When cell density reaches confluence, trypsinize the cells and freeze each 10-cm plate in one cryovial, store in liquid nitrogen.

3.1.2. Mitomycin-C Inactivation of MEFS

1. Thaw contents of one cryotube into 3X 10-cm culture dishes.

2. Grow the cells to confluence by changing the media every other day.

3. Further propagate the cells by splitting them twice at a 1:3 dilution (sums to 27 plates).

4. To inactivate the cells, add 40 µL of mitomycin-C stock solution (1 mg/mL) to 5-mL culture media (final concentration of 8 µg/mL) and incubate at 37°C, 5% CO_2, for 3 h.

5. Aspirate the mitomycin-containing medium and wash the plates twice with 6 mL PBS.

6. Tripsinize cells by adding 1 mL of trypsin-EDTA and incubate at 37°C, 5% CO_2, for 5 min.

7. Add 5 mL medium and suspend the cells by vigorous pipetting.

8. Collect cell suspension into a 50-mL Falcon tube.

9. Centrifuge mitomycin-treated cell pool at $1000g$ for 5 min.

10. Aspirate supernatant and add fresh medium to reach a final cell concentration of 4×10^6 cells/10-cm dish. Feeder plates can be stored in the incubator for 3–4 days, but should be examined under the microscope before use.

11. It is possible to freeze mitomycin-C treated MEFs and keep them for later use. For this purpose, freeze $1.5–7 \times 10^6$ cells in each cryotube and later thaw and plate to give 1–5X 10-cm dishes, respectively.

3.1.3. Maintenance of HESCs and Genetically Modified Clones

The maintenance of HESCs in culture relies on the continuous and selective propagation of undifferentiated cells. Controlling culture conditions and minimizing the effect of spontaneous differentiation, which constantly occurs, can achieve this. When passing the cells, care must be taken so that the cell number will not drop below a certain density, because this increases their tendency to differentiate, possibly from a lack of autocrine signaling. The differentiation status of the cultures should be followed daily by observation through a phase-contrast microscope. Undifferentiated colonies are easily recognized by their typical appearance, which includes small and equal-sized cells that are defined by a discrete border, pronounced nucleus, and clear cellular boundaries. As differentiation begins, the cells at the periphery of the colonies lose their typical morphology. At that stage, splitting must be performed (*see* **Note 7**).

3.1.4. Subculture of HESCs

1. Remove culture media and rinse with 6 mL PBS.
2. Add 1 mL of trypsin-EDTA and incubate for 5 min.
3. Add 5 mL growth medium and suspend the cells by vigorous pipetting.
4. Collect suspension into a conical tube and pellet by centrifugation $1000g$ for 5 min.
5. Resuspend with fresh media and plate on mitotically inactivated feeders prepared the previous day.

3.1.5. Freezing HESCs

1. Trypsinize HESCs and pellet them, as described in **Section 3.1.2.1, Steps 1–4**.
2. Resuspend cells in an appropriate amount of growth media supplemented with 10% DMSO.
3. Mix the cells gently by pipetting up and down and place in a properly marked cryotube.
4. Store at –70°C in a low-temperature vial container filled with isopropanol for at least 1 day.
5. For long-term storage, vials must be kept in liquid nitrogen.

3.1.6. Thawing HESCs
(see Note 8)

1. Incubate the frozen cryovial in a 37°C water bath until it is completely thawed.
2. Transfer and resuspend the cells with 5 mL growth media in a conical tube.
3. Pellet the cells by centrifugation at $1000g$ for 5 min.
4. Resuspend again in an appropriate amount of fresh media.
5. Plate cells and incubate overnight.

3.1.7. Mouse ES Cells Clonal Assay to Test Competence and Quality of KO-Serum Batch

Batch-to-batch variability in the competence of the KO-serum replacer to support undifferentiated proliferation may be remarkable. Clonal assays with mouse ES cells may be used to test the quality of the serum substitute batch before its use. An established culture of mouse ES cells is used as previously described (38) and all medium components should be those that will be used to culture the HESCs (*see* **Note 9**).

1. Trypsinize mouse ES cells (38) and plate individual cells in pregelatinized 6-cm petri culture dishes at a low density (1000 cells per plate).
2. Culture either with the medium that was in current use or the new tested medium at 37°C in a 5% CO_2 atmosphere.
3. Change medium once on the fifth day after plating.
4. On the seventh day, rinse the cultures with PBS and stain for 5 min with 0.15% Leishman's fix and stain.
5. Wash the stained cultures thoroughly with water and let them air dry.
6. Compare the number of colonies per plate as well as the size and degree of differentiation and select the batch of serum with the best performance compared with the batch in use.

3.2. Transfection

3.2.1. DNA Preparation for Transfection

1. Prepare DNA vector by any commonly used technique to obtain OD280/OD260 absorption ratio value of 1.8 or greater (*see* **Note 10**) (*see* **Fig. 21.1** and **Table 21.1**).
2. Linearize the vector by digesting it with the appropriate restriction enzyme.
3. Assess the completion of the restriction digest by electrophoresis of a small aliquot on a 1% gel agarose.
4. Ethanol precipitates the DNA and resuspend in a small volume (20–50 µL) of TE or sterile water. Adjust concentration to 1 µg/µL.

3.2.2. Growing HESCs for Transfection

1. Split (1:2 or 1:3) a morphologically undifferentiated and confluent HESC cell culture 2 days before transfection (*see* **Note 11**).

Table 21.1
Transfection protocol timetable

Days	
1	Plate MEF resistant cells
2	Split/thaw a vial of HESCs to high density
4	Transfect HESCs (high density cultures of 8–32 cells/colony)
5	Begin selection
6–10	Change selection media every day
11–15	Change selection media every other day
16–18	Screen for resistant colonies Pick up selected colonies and plate them on MEF-resistant feeder in 1X 24-well tissue culture trays
20–30	Split 1:2 and plate on MEF-resistant feeder in 1X 12-well twice Freeze and/or screen/further propagate in 1X 6-well trays

MEF, mouse embryonic fibroblasts; HESC, human embryonic stem cell.

3.2.3. Transfection by Calcium Phosphate

1. Harvest HESCs and split 1:4 into 10-cm culture dishes containing MEFs that were plated the previous day (*see* **Note 12**).

2. Prepare for each 10-cm plate transfection buffer and DNA in separate tubes. Dilute 10–20 μg of DNA in 240 mM $CaCl_2$ by bringing the DNA to a final volume of 0.5 mL with DDW and then slowly adding 60 μL of 2 M $CaCl_2$ (and not the reverse order).

3. Add very slowly the DNA solution (one to two drops/s) to the transfection buffer, while gently mixing by generating small air bubbles with a sterile disposable tip.

4. Incubate for 10 min at room temperature (*see* **Note 13**).

5. Add the 1-mL solution dropwise onto the cells without swirling or rotating the dish.

6. Incubate at 34°C, 3% CO_2, for 4 h and then change the growth media by aspirating it and washing twice with PBS. Add fresh media and return to the incubator.

7. Apply selection the following day by adding the appropriate drug to the growth media.

8. Refeed the cells with selection media when the medium starts to turn yellow, usually every day during the first 5 days and then every other day. By days 10–12 of selection, colonies should visible and large enough to be picked for further expansion and analysis.

3.2.4. Transfection by Exgen 500 (see Note 14)

1. Two days before transfection by Exgene 500, harvest and split HESCs into 6-well trays containing inactivated and drug-resistant MEFs.
2. About 1 h before transfection, change the growth media by rinsing the cells with PBS and adding 1 mL of fresh media to each well.
3. For each well of a 6-well tissue culture tray, prepare a tube containing 2 µg of DNA to a final volume of 50 µL of 150 mM NaCl and vortex.
4. In a separate tube, mix 10 µL ExGen 500 to 40 µL of 150 mM NaCl and vortex.
5. Mix DNA and transfecting agent by rapidly adding diluted ExGen 500 to DNA (not the reverse order). Vortex-mix the solution immediately for 10 s and then incubate for 10 min at room temperature.
6. Add 100 µL of ExGen/DNA mixture to each well.
7. Gently rock the plate back and forth to equally distribute the complexes on the cells.
8. Centrifuge culture trays immediately for 5 min at $280g$.
9. Incubate at 37°C, 5% CO_2, for 30 min.
10. Wash twice with PBS and return to incubator (*see* **Note 15**).

3.2.5. Electroporation (Essentially according to Zwaka and Thomson)

1. Grow healthy and undifferentiated cells in a 10-cm culture dish until they reach cell density greater than 70% confluence.
2. Trypsinize cells to collect clumps of undifferentiated HESCs by adding 1 ml of trypsin-EDTA for 3–5 min.
3. Add 10 ml HESC growth medium.
4. Collect cell suspension into a 15-ml Falcon tube.
5. Centrifuge cells at $600g$ for 5 min.
6. Aspirate supernatant and gently resuspend in 0.8 ml of hES fresh media, containing 20–30 µg linearized DNA vector, to reach a final cell concentration of $1-3 \times 10^7/0.8$ ml.
7. Transfer cell/DNA mix into pre-cooled 0.4-cm cuvettes.
8. Electroporate cells using the following parameters: 320 V, 250 µF. The time constant should be between 9.0 and 13.0.
9. Immediately after electroporation, allow cells to recover by standing in the cuvette on ice for 10 min.
10. Transfer contents, using 1-ml glass pipette, into a 15-ml tube containing 2 ml of prewarmed HESC media.
11. Pellet cells by centrifugation of $600g$ for 5 min.
12. Aspirate supernatant and gently resuspend pellet in 10 ml HESC media.

13. Plate cells onto two 10-cm culture dishes preseeded with 2.5×10^6 inactivated MEF feeders and return to incubator.
14. The following day, remove cell debris by washing twice with PBS and then add fresh HESC media.
15. Apply selection the following day (day 2 postelectroporation).
16. Change drug-containing HESC media once a day (5 days) and then every other day.

3.3. Infection

3.3.1. Retrovirus/Lentivirus Production

1. Plate 293T cells in 10-mm tissue culture dish (Dulbecco's modified Eagle's medium (DMEM) supplemented with 10% FBS, glutamine, PenStrep) 24 h before transfection so that they are 80% confluent for transfection (see **Fig. 21.2** and **Table 21.2**).

Table 21.2
Infection protocol timetable

Days	
1	Plate 293T cells 2×10^6 cells per plate
2	Transfect the 293T cells with the viral vectors (FUGENE 6) Split/thaw a vial of HESCs to high density
3	Change the medium of the 293T cells
4	Filter the viral supernatant (48 h) and infect the HESCs Add new medium to the 293T cells
5	Filter the viral supernatant (72 h) and infect the HESCs
6–10	Change selection media every day
11–15	Change selection media every other day
16–18	Screen for resistant colonies Pick up selected colonies and plate them on MEF-resistant feeder in 1X 24-well tissue culture trays
20–30	Split 1:2 and plate on MEF-resistant feeder in 1X 12-well twice Freeze and/or screen/further propagate in 1X 6-well trays

MEF, mouse embryonic fibroblasts; HESC, human embryonic stem cell.

2. Cotransfect 293T cells with 3 μg retroviral/lentiviral vector, 2 μg packaging plasmid, 1 μg VSV-G expression vector, and 18 μl FuGENE 6 (Roche) per plate according to the suppliers' conditions. Transfection of the cells has to be done in medium without antibiotics.
3. After 24 h, change medium to full medium (with antibiotics).
4. Collect virus supernatant from all plates 48 and 72 h after transfection with plastic pipettes and filter supernatant through a 0.45-μm filter.

3.3.2. Retroviral and Lentiviral Gene Transfer into Human ES Cells

1. Cultivate HESCs cultures on mouse embryo fibroblast feeder cells (MEF) or on matrigel in basic fibroblast growth factor (bFGF) supplemented MEF conditioned medium.

2. Plate 1×10^5 HESCs on a tissue culture plate pretreated with Matrigel or Gelatin and MEF-attached cells. In the case of Matrigel, add MEF-conditioned medium supplemented with bFGF (4 ng/ml) to keep the HESCs undifferentiated.

3. Collected and filtered the viral supernatant, after 48 h of cells transfection, together with 6 μg/ml hexadimethrine bromide (polybrene).

4. Culture the cells with the virus for 24 h, wash three times with PBS, and then add fresh media or the 72 h viral supernatant, for another 24 h in order to increase the infection efficiency.

5. On day 3 after infection, measure for transgene activity and continue the culture on MEFs or matrigel.

3.4. Colony Picking and Expansion

After 10–12 days in selection media, individual HESC-resistant clones become visible and are big enough to be isolated for expansion.

1. Screen transfected culture plates using an inverted microscope for the presence of resistant clones and mark their location at the bottom of the dish.

2. Manually pick selected HESC colonies (*see* **Note 16**).

3. Disconnect the cell colony from the feeders by dissociating it into small cell pieces using the sharp edge of the glass micropipette while collecting them by aspiration into the tip of the pipette.

4. Plate the small cell clumps on fresh drug-resistant feeder layer, in a single well of a 24-well culture tray and return to incubator for further growth. The replated cell clumps, which have originated from a single-cell clone, give rise to round flat colonies with well-defined borders in 3–5 days, while changing the selection media as necessary (*see* **Note 17**).

5. Scale up the clone population by splitting 1:2 with trypsin, twice.

6. When the wells (2×12-well) are approaching confluence, freeze each well in individual cryovials. The remaining cells can either be further expanded (**Fig. 21.3C**), by splitting 1:4, or directly used for DNA, RNA, or protein extraction (*see* **Note 18**) (**Table 21.1**).

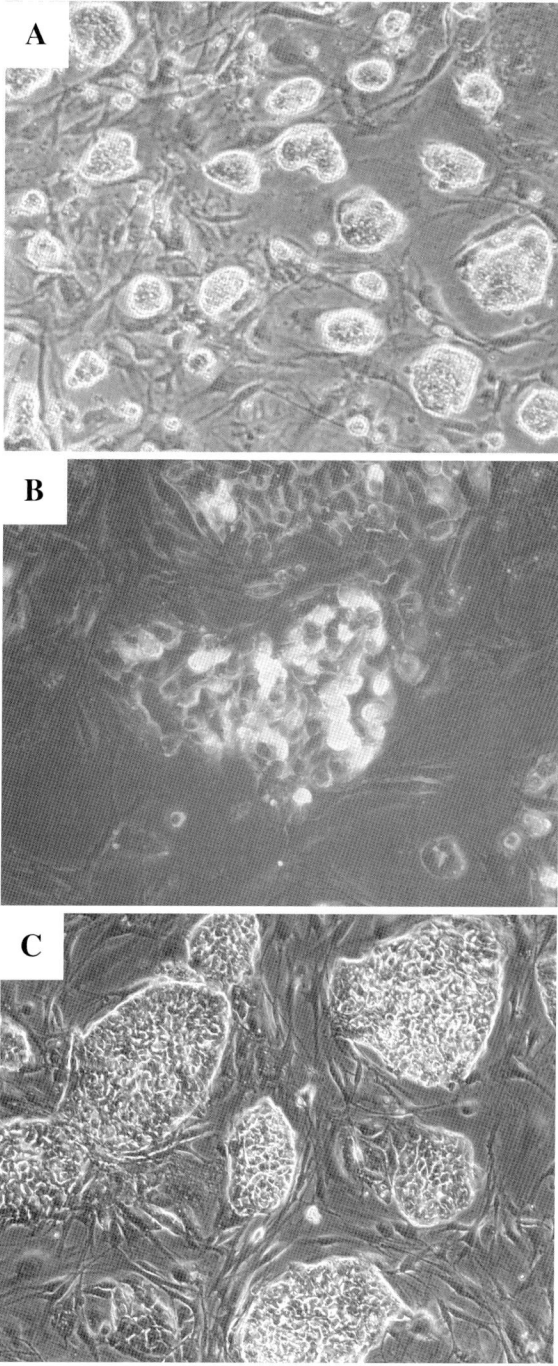

Fig. 21.3. **(A)** Human embryonic stem (HESC) cell culture on day of transfection. The culture should be composed of many small (8–32 cells) colonies. **(B)** Transient expression of CMV-EGFP in HESCs after 48 h to transfection. **(C)** Established cell line of HESCs after transfection, selection, and clonal expansion of genetically modified cells.

4. Notes

1. **Section 2.1, Items 1–10** are stored at 4°C, **Items 11–18** at −20°C, and **Item 19** at room temperature. As a rule, all tissue-culture protocols must be performed under sterile conditions, in a laminar flow hood, using sterile disposable plastics and clean, detergent-free glassware.
2. Media should be stored at 4°C and can be used for up to 1 month.
3. Serum replacement is sensitive to light. Protect supplemented HESC media by covering it with aluminum foil.
4. The mouth-controlled device is the same as the one that is commonly used for handling oocytes and preimplantation embryos in mice. The mouthpiece is available as a part of an aspiration tube assembly from Drummond (model no. 2-000-0001). Sterile glass Pasteur pipettes are pulled on a flame to create long tubing with a narrow opening. Soften the glass tubing by rotating it in a fine flame until the glass becomes soft. Then, withdraw the glass quickly from the heat and pull both ends smoothly to produce a tube with an internal diameter of about 200 µm. Neatly break the tube and fire polish its tip by quickly touching the flame.
5. All tissue-culture procedures are performed under sterile conditions, using prewarmed media and gelatin-precoated plates.
6. Protocols for cell freezing, thawing, and splitting are basically the same for all cell types (feeders and HESCs).
7. As in other cell lines growing in vitro, chromosomal aberrations may occur. Working with cells of low passage number can minimize this. Thus, it is advisable to monitor the karyotype of the cells following prolonged growth in culture and subsequent to stable transfection.
8. Cell thawing must be performed as quickly as possible.
9. The culture medium is supplemented with 10% of the tested batch of knockout serum substitute (instead of 15%) and mouse recombinant LIF at 1000 U/mL.
10. The purity of the DNA is very critical for successful transfection.
11. The cells should be transfected during the lag phase of cell division. The transfection rate is most efficient when the cell density reaches 50–70% and the colonies are small (8–32 cells per colony) (**Fig. 21.3A**). The colonies should have discrete borders and be composed of similar sized cells, with a pronounced nucleus.

12. The calcium phosphate (CaPO$_4$) transfection protocol is basically similar to the protocols used for other cell types.

13. At this time, fine DNA-calcium phosphate precipitates should be formed without agitation.

14. ExGen 500 (polyethylenimine, PEI) is a cationic polymer, which is capable of transfecting a wide range of cell types at relatively high efficiency. It interacts with the negatively charged DNA molecules by forming small, stable, and highly diffusible particles, which settle on the cell surface by gravity and absorb into the cell by endocytosis.

15. In parallel to the experiment, one may consider to carrying out transient transfection on a small number of cells with a construct carrying a constitutive expressed reporter gene, such as *CMV-EGFP*, to assess transfection efficiency before applying selection (**Fig. 21.3B**).

16. The colonies are picked up by the aid of a mouth apparatus connected to a sterile pulled and fire polished Pasteur pipette, as is commonly used for handling oocytes and preimplantation embryos (*see* **Note 4**).

17. We find this pickup method more suitable and efficient for isolating single HESC colonies than the method applied in mouse, where individual ES colonies are collected with a disposable tip, trypsinized, and then plated.

18. In some cases, it is crucial that no feeders will be present during the screen. For this purpose, cells must be propagated in feeder-free gelatinized plates, for at least one passage. Under such conditions, the cells must be grown in conditioned media (CM, HESC cell media conditioned by MEFs for 24 h), preventing from differentiation and consequently culture loss.

References

1. Li M, Pevny L, Lovell-Badge R, Smith A. Generation of purified neural precursors from embryonic stem cells by lineage selection. Curr Biol 1998;8:971–4.

2. Klug MG, Soonpaa MH, Koh GY, Field LJ. Genetically selected cardiomyocytes from differentiating embryonic stem cells form stable intracardiac grafts. J Clin Invest 1996;98:216–24.

3. Singh Roy N, Nakano T, Xuing L, Kang J, Nedergaard M, Goldman SA. Enhancer-specified GFP-based FACS purification of human spinal motor neurons from embryonic stem cells. Exp Neurol 2005;196:224–34.

4. Huber I, Itzhaki I, Caspi O, et al. Identification and selection of cardiomyocytes during human embryonic stem cell differentiation. FASEB J 2007;21:2551–63.

5. Lavon N, Yanuka O, Benvenisty N. Differentiation and isolation of hepatic-like cells from human embryonic stem cells. Differentiation 2004;72:230–8.

6. Lavon N, Yanuka O, Benvenisty N. The effect of overexpression of Pdx1 and Foxa2 on the differentiation of human embryonic stem cells into pancreatic cells. Stem Cells 2006;24:1923–30.

7. Gerrard L, Zhao D, Clark AJ, Cui W. Stably transfected human embryonic stem cell

clones express OCT4-specific green fluorescent protein and maintain self-renewal and pluripotency. Stem Cells 2005;23:124–33.

8. Eiges R, Schuldiner M, Drukker M, Yanuka O, Itskovitz-Eldor J, Benvenisty N. Establishment of human embryonic stem cell-transfected clones carrying a marker for undifferentiated cells. Curr Biol 2001;11:514–8.

9. Levinson-Dushnik M, Benvenisty N. Involvement of hepatocyte nuclear factor 3 in endoderm differentiation of embryonic stem cells. Mol Cell Biol 1997;17:3817–22.

10. Dekel I, Magal Y, Pearson-White S, Emerson CP, Shani M. Conditional conversion of ES cells to skeletal muscle by an exogenous MyoD1 gene. New Biol 1992;4:217–24.

11. Kim JH, Auerbach JM, Rodriguez-Gomez JA, et al. Dopamine neurons derived from embryonic stem cells function in an animal model of Parkinson's disease. Nature 2002;418:50–6.

12. Bowles KM, Vallier L, Smith JR, Alexander MR, Pedersen RA. HOXB4 overexpression promotes hematopoietic development by human embryonic stem cells. Stem Cells 2006;24:1359–69.

13. Soria B, Roche E, Berna G, Leon-Quinto T, Reig JA, Martin F. Insulin-secreting cells derived from embryonic stem cells normalize glycemia in streptozotocin-induced diabetic mice. Diabetes 2000;49:157–62.

14. Capecchi MR. Altering the genome by homologous recombination. Science 1989;244:1288–92.

15. Urbach A, Schuldiner M, Benvenisty N. Modeling for Lesch-Nyhan disease by gene targeting in human embryonic stem cells. Stem Cells 2004;22:635–41.

16. Zwaka TP, Thomson JA. Homologous recombination in human embryonic stem cells. Nat Biotechnol 2003;21:319–21.

17. Davis RP, Ng ES, Costa M, et al. Targeting a GFP reporter gene to the MIXL1 locus of human embryonic stem cells identifies human primitive streak-like cells and enables isolation of primitive hematopoietic precursors. Blood 2008;111:1876–84.

18. Stanford WL, Cohn JB, Cordes SP. Gene-trap mutagenesis: past, present and beyond. Nat Rev Genet 2001;2:756–68.

19. Kunath T, Gish G, Lickert H, Jones N, Pawson T, Rossant J. Transgenic RNA interference in ES cell-derived embryos recapitulates a genetic null phenotype. Nat Biotechnol 2003;21:559–61.

20. Vallier L, Rugg-Gunn PJ, Bouhon IA, Andersson FK, Sadler AJ, Pedersen RA. Enhancing and diminishing gene function in human embryonic stem cells. Stem Cells 2004;22:2–11.

21. Hay DC, Sutherland L, Clark J, Burdon T. Oct-4 knockdown induces similar patterns of endoderm and trophoblast differentiation markers in human and mouse embryonic stem cells. Stem Cells 2004;22:225–35.

22. Costa M, Dottori M, Sourris K, et al. A method for genetic modification of human embryonic stem cells using electroporation. Nat Protoc 2007;2:792–6.

23. Ma Y, Ramezani A, Lewis R, Hawley RG, Thomson JA. High-level sustained transgene expression in human embryonic stem cells using lentiviral vectors. Stem Cells 2003;21:111–7.

24. Pfeifer A, Ikawa M, Dayn Y, Verma IM. Transgenesis by lentiviral vectors: lack of gene silencing in mammalian embryonic stem cells and preimplantation embryos. Proc Natl Acad Sci USA 2002;99:2140–5.

25. Smith-Arica JR, Thomson AJ, Ansell R, Chiorini J, Davidson B, McWhir J. Infection efficiency of human and mouse embryonic stem cells using adenoviral and adeno-associated viral vectors. Cloning Stem Cells 2003;5:51–62.

26. Gropp M, Itsykson P, Singer O, et al. Stable genetic modification of human embryonic stem cells by lentiviral vectors. Mol Ther 2003;7:281–7.

27. Miyoshi H, Blomer U, Takahashi M, Gage FH, Verma IM. Development of a self-inactivating lentivirus vector. J Virol 1998;72:8150–7.

28. Zufferey R, Dull T, Mandel RJ, et al. Self-inactivating lentivirus vector for safe and efficient in vivo gene delivery. J Virol 1998;72:9873–80.

29. Zeng J, Du J, Zhao Y, Palanisamy N, Wang S. Baculoviral vector-mediated transient and stable transgene expression in human embryonic stem cells. Stem Cells 2007;25:1055–61.

30. Park JH, Kim SJ, Oh EJ, et al. Establishment and maintenance of human embryonic stem cells on STO, a permanently growing cell line. Biol Reprod 2003;69:2007–14.

31. Richards M, Fong CY, Chan WK, Wong PC, Bongso A. Human feeders support prolonged undifferentiated growth of human inner cell masses and embryonic stem cells. Nat Biotechnol 2002;20:933–6.

32. Amit M, Margulets V, Segev H, et al. Human feeder layers for human embryonic stem cells. Biol Reprod 2003;68:2150–6.
33. Hovatta O, Mikkola M, Gertow K, et al. A culture system using human foreskin fibroblasts as feeder cells allows production of human embryonic stem cells. Hum Reprod 2003;18:1404–9.
34. Cheng L, Hammond H, Ye Z, Zhan X, Dravid G. Human adult marrow cells support prolonged expansion of human embryonic stem cells in culture. Stem Cells 2003;21:131–42.
35. Reubinoff BE, Pera MF, Fong CY, Trounson A, Bongso A. Embryonic stem cell lines from human blastocysts: somatic differentiation *in vitro*. Nat Biotechnol 2000;18:399–404.
36. Thomson JA, Itskovitz-Eldor J, Shapiro SS, et al. Embryonic stem cell lines derived from human blastocysts. Science 1998;282:1145–7.
37. Tucker KL, Wang Y, Dausman J, Jaenisch R. A transgenic mouse strain expressing four drug-selectable marker genes. Nucleic Acids Res 1997;25:3745–6.
38. Robertson, E.J. (1987). *Teratocarcinomas and Embryonic Stem Cells:* A Practical Approach. IRL Press, Oxford.

Chapter 22

Genetic Manipulation of Human Embryonic Stem Cells in Serum and Feeder-Free Media

Stefan R. Braam, Chris Denning, and Christine L. Mummery

Abstract

Generic methods for genetic manipulation of human embryonic stem cells (hESCs) are important for both present research and future commercial applications. To date, differences in cell derivation and culture have required independent optimization of transfection and transduction protocols and some lines have remained refractile to all methods. Here we describe a culture protocol that has been extensively tested in 12 different hESC lines (1, 2) and shown to support efficient gene transfer independent of the method of gene delivery or history of the cell line. The system is based on Matrigel monolayer culture and conditioned medium from mouse embryonic feeder cells (MEFs) and entails transient high-density culture followed by rapid adaptation to low density for gene transfer. Under these conditions, plasmid transfection, virus infection, and siRNA transfection are highly effective. Stable genetically modified hESC lines can be generated with plasmid transfection, viral infection, or electroporation without loss of pluripotency or differentiation potential. The majority of lines generated in this system display a normal karyotype.

Key words: Human embryonic stem cell (hESC), genetic manipulation, homologous recombination, lentivirus, adenovirus, plasmid transfection.

1. Introduction

Realizing the full promise of hESCs requires efficient methods to manipulate their genomes during basic research. Progress in certain areas has lagged behind expectations because initial culture conditions resulted in poor transfection and low single-cell cloning efficiencies. Basic technology available to research on mouse ESCs, such as gene targeting and methods of gene knockdown, has been challenging in hESCs. To date, lentiviral infection has been the most efficient method for gene transfer, but it has major limitations, including incompatibility with homologous recombination, silencing of the transgene (3), and costly, time-consuming, large-scale virus production. Several other gene-transfer techniques including adenovirus (4) and baculovirus (5) have been tried and tested, but none have been successfully used in multiple hESC lines. In addition, straightforward plasmid transfection resulted in highly variable transfection efficiencies reported between different lines (3–35% transfection efficiency).

More importantly, genetic manipulation by electroporation, a crucial intermediate step for successful homologous recombination, has only been described for a limited number of hESC lines. Efficient homologous recombination will facilitate the generation of hESC lines expressing selectable cassettes from endogenous promoters to select for specific (differentiated) cell types. Fluorescent proteins can easily be visualized on standard fluorescent microscopes and detected using flow cytometry. This allows sorting of living cells for further characterization or tracking their fate. The use of drug-metabolizing enzymes (e.g., puromycin N-acetyltransferase) holds great promise as an efficient method for enriching the cell type of interest. Targeted gene disruption to introduce clinically relevant mutations or deletions will facilitate modeling of human diseases. The practical feasibility of homologous recombination in hESCs was first proven by Zwaka and colleagues in 2003 (6). They targeted *HPRT1* and generated an OCT4-GFP reporter line using a gene trap strategy. Since then surprisingly few targeted lines have been reported. Recently, Davis et al. (7) targeted the primitive streak gene *MIXL1* with GFP and Irion et al. (8) identified and targeted the human Rosa26 locus. The mouse Rosa26 locus has become a preferred integration site for transgenes and various reporter constructs, as it can be targeted by homologous recombination with relative ease, supports strong, ubiquitous expression of inserted sequences, and it is not subject to gene-silencing effects. Further research will establish whether the human Rosa26 locus has similar characteristics.

Recently, we showed using mass spectroscopy and immunofluorescence microscopy that hESCs express proteins belonging to epithelium-related cell–cell adhesion complexes, including adherens junctions, tight junctions, desmosomes, and gap junctions (9, 10). We found that hESCs passaged using trypsin and cultured on Matrigel retained these junctions, but changed morphology and upregulated vimentin, a protein associated with EMT and mesoderm differentiation but also with increased cell stress. Nevertheless, these cells are in fact still highly undifferentiated, as shown by their homogeneous expression of stem cell markers and, in this state, are highly receptive to multiple methods of genetic manipulation. Using this approach, we have optimized protocols for plasmid transfection, siRNA transfection, electroporation, lentiviral and adenoviral transduction (1). After genetic manipulation, the cells can be easily cultured in any system that supports undifferentiated growth including those in which they originally propagated. These protocols were tested on 12 independent lines, grown under conditions ranging from mechanical passage on MEFs to enzymatic passage and human fibroblast feeder cells. Following rapid adaptation to low-density, feeder-free culture, all lines showed reproducible high-transfection efficiencies and retained their stem cell characteristics.

2. Materials

2.1. Cell Culture Medium + Reagents

1) **hESC medium:** D-MEM/F-12 (1:1) basis with GlutaMAX™, supplemented with 15% (v/v) KnockOut™ Serum Replacement, 1% (v/v) non essential amino acids, 0.1% (v/v) penicillin/streptomycin, 100 μM β-mercaptoethanol, and 4 ng ml^{-1} basic fibroblast growth factor.

2) **Medium for MEFs:** D-MEM basis supplemented with 10% (v/v) fetal calf serum, 0.5% (v/v) penicillin/streptomycin, 1% (v/v) glutamine, 1% (v/v) non essential amino acids.

3) **Freezing medium:** 20% Hybrimax dimethylsulphoxide, 80% (v/v) FCS.

4) **Antibiotics;** Geneticin and Puromycin from Invitrogen.

5) **Transfection reagents:** Genejammer from Stratagene, Lipofectamine from Invitrogen, and Opti-MEM from Invitrogen.

6) 0.05% Trypsin-EDTA from Invitrogen.

2.2. Immunofluorescent Staining

1) Coverslips or chamberslides from Nunc.
2) **Fixative:** 4 gPFA/100 ml PBS. Heat for 20 min at 65°C. The solution can be frozen at –20°C for long-term storage.
3) **Permeabilization buffer:** 0.1% SDS in PBS.
4) **Blocking buffer:** 4% Normal Goat Serum in PBS.

2.3. FACS

1) **Fixation buffer:** 2 gPFA/100 ml PBS. Heat for 20 min at 65°C. The solution can be frozen at –20°C for long-term storage.
2) **FACS wash buffer:** 2% serum in PBS.
3) TrypLE from Invitrogen.

3. Methods

For efficient genetic manipulation, it is extremely important to culture hESCs in a monolayer on Matrigel-coated substrates. The adaptation of the cells to these culture conditions is a stepwise protocol; first, adaptation to Matrigel and second, adaptation to trypsin (if necessary). Success, particularly for cell lines cultured using "cut and paste," is critically dependent on very high-density culture during the first passages. After initial up-scale, the cells can be plated down at low density for efficient gene transfer without loss of pluripotency or massive cell death.

The quality of cells can be monitored using immunofluorescence or FACS for stem cell-specific cell surface markers such as Tra-1-60, GCTM2, SSEA4, and the transcription factors OCT3/4a and SOX2.

3.1. Preparation of MEFS for the Production of Conditioned Medium

1. Inactivate confluent mouse embryonic fibroblasts (MEFs) for 2.5 h with mitomycin C (10 μg/ml^{-1} in MEF medium).
2. Wash cells with medium and then twice with PBS to remove any traces of mitomycin C.
3. Trypsinize the cells, count them, and resuspend the cells at 1×10^6 cells per ml in MEF medium.
4. Seed the mitomycin C-inactivated MEFs on tissue-culture flask at a density of 6.4×10^4 cells/cm^2 and allow the cells to attach for a minimum of 4-h, preferably 24-h. Wash with PBS once and replace medium for hESC medium. Use 25 ml for a T75 flask.
5. Harvest MEF-conditioned medium (CM) after 24 h and replace with fresh unconditioned hESC medium. Resupplement the fresh CM with 4 ng ml^{-1} bFGF. The CM may be harvested for up to seven consecutive days. Filtration may be considered but is not essential; it helps to remove any dead fibroblast cells. CM can be used fresh or stored frozen at –20°C or –80°C for up to 6 months.

3.2. Preparation of Matrigel-Coated Tissue-Culture Plastics

a) Thaw and dilute Matrigel 1/100 (v/v) by repetitive pipetting in ice cold DMEM/F12 (1:1) (directly from the refrigerator).

b) Pipette the diluted Matrigel immediately into tissue-culture plastics and allow to polymerize for at least 45 min at room temperature. Note that the layer of polymerized Matrigel is only very thin and should not be visible. Appearance of lumpy areas indicates premature polymerization. Use 0.5 ml per well Matrigel + medium for 24-well plates, 1 ml for IVF organ culture and 12-well plate, 2 ml for 6-well plates, 5 ml for T25, and 12 ml for T75.

c) Plates can be used immediately or stored at 4°C. Before use, aspirate excess medium and un-polymerized Matrigel, and then rinse once with PBS (*see* **Note 1**).

3.3. Adaptation of hESCs to Feeder-Free Culture

1. This step is exclusively for cells maintained by mechanical "cut and paste" or any other culture system depending on colony formation. For cells adapted to trypsin passage, go to Step 5. Successful adaptation is critically dependent on a high-quality undifferentiated hESC culture. Using a dissecting tool, slice at least 10 colonies in small pieces, release the cells by vigorously pipetting with a P1000 Gilson pipette, and transfer them to two Matrigel coated dishes containing CM (the splitting ratio is ½) (**Fig. 22.1a, and b**). At this stage, dispase may be beneficial to release the cells from the feeders. Dispase is not crucial to release cells, but may enhance the process.

2. Keep refreshing CM daily for 4–5 days while colonies are growing and expanding (*see* **Note 2**).

3. Start removing 3D differentiated areas with a dissection tool at the stage that colonies start to touch each other.

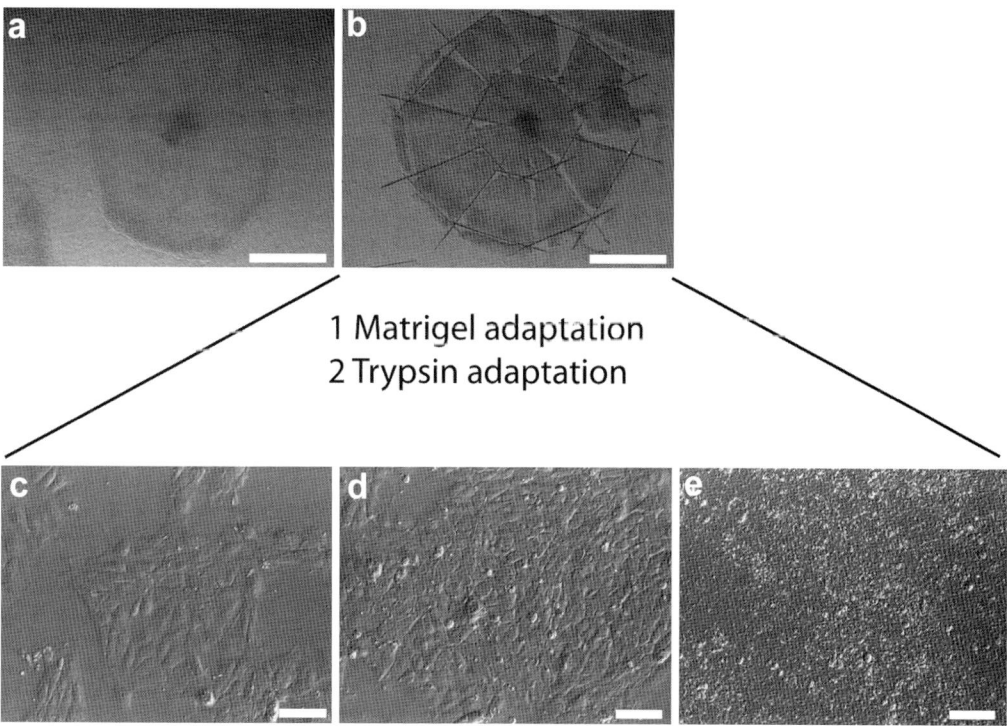

Fig. 22.1. Pictures of hESCs at different stages of adaptation. (**a, b**) Undifferentiated hESCs on feeders. (**b**) Colony fragments ready for plating on Matrigel. (**c,d,e**) hESCs growing in monolayer on Matrigel after adaptation (**c**) low density, ready for transfection, (**d**) example of semiconfluent cells, and (**e**) fully compacted hESC cells.

4. Wash the undifferentiated cells with PBS and add trypsin for 2–3 min at 37°C. Watch the cells closely at this stage; short incubation with trypsin will not release the cells while too long treatment will decrease the survival.

5. Dilute the trypsin + cells in at least 5 volumes of hESC medium and resuspend the cell suspension vigorously to release the cells. Collect the cells into a centrifuge tube and pellet the cell suspension ($180 \times g$, room temperature, 4 min). This procedure should yield a heterogeneous cell suspension consisting of smaller and larger hESC clumps.

6. Following centrifugation, resuspend the cells in CM and replate them at an approximate density of ~ 1–2.10^5 per cm^2. As a guideline, we usually take cells from two dishes and replate them on one tissue-culture coated dish. A small proportion of cells die during this first enzymatic passage.

7. Refresh the medium the next day. At this stage, it is crucial to watch the morphology of the cells closely. Use tissue-culture microscopy and compare cells in colonies with your monolayer culture at 100–400× magnification. Keep refreshing the medium daily until the cells form a confluent monolayer with morphology similar to colonies. Generally this takes 1–2 days (*see* **Note 2**) (**Fig. 22.1c–e**).

8. At the stage when cells become confluent, wash cells with PBS and add enough trypsin to cover the tissue-culture area for exactly 1.5 min Dilute cells/trypsin in at least 5 volumes of hESC medium and centrifuge immediately to remove the trypsin ($180 \times g$, room temperature, 4 min). Resuspend the cell pellet in CM and seed the cells at Matrigel-coated plates. Depending on cell line, split ratios are usually 1 in 3. Cells should reach confluence within 48–72 h. Some fast growing lines can be split occasionally in a 1 in 5 ratio (*see* **Note 3**).

9. Scale culture up to larger tissue-culture surface areas without exceeding a 1 in 3–5 split ratio and with daily replacement of CM. The following options can be used to transfect/transduce the cultured hESCs with desired sequences to obtain stably transfected transgenic cells.

3.4. Cryopreserving Cells

1. Trypsinize a confluent culture flask for 1.5 min and resuspend the cells in at least 5 volumes of hESC medium.

2. Centrifuge cells ($180g$, 4 min) and resuspend the pellet in 100% FCS on ice (the amount of FCS is dependent on the number of ampoules to be generated; in general, we resuspend cells from one T25 in 750 μl, which is enough for 3 ampoules).

3. Add 250 μl freezing medium per ampoule; mix the contents gently (final concentration of DMSO is 10%).

4. Transfer the ampoules immediately to a Nalgene "Mr. Frosty" cryopreservation container containing propan-2-ol and transfer the container to −80°C.

5. Transfer the ampoules after 24 h to liquid nitrogen for long-term storage.

3.5. si-RNA Transfection

1. The day before transfection, trypsinize cells as described **Section 3.3**, resuspend the cells in CM and seed at a density of $1-2 \times 10^5$ cells per well onto a Matrigel-coated 12-well plate.

2. Prepare for each sample as follows: (*see* **Note 4**).
   - Dilute 3 μl siRNA in 75 μl of Opti-MEM I medium and mix by flicking the tube. Note: this assumes an siRNA stock concentration of 20 mM.
   - Mix Lipofectamine 2000 by inversion before use and then dilute 1.5 μl in 75 μl of Opti-MEM I Medium. Mix gently and incubate for 5 min at room temperature.

3. After the 5-min incubation, combine the diluted siRNA with the diluted Lipofectamine 2000, which brings the total volume to 154.5 μl. Mix the contents of the tube gently and incubate for 20 min at room temperature to allow the siRNA:lipid complexes to form.

4. Aspirate medium from cells prepared at Step 1 and replace with 450 μl of CM.

5. Add the 154.5 μl of siRNA:Lipofectamine complexes dropwise to each well. This gives a final concentration of 100 nM siRNA (lower concentrations may be considered). Mix gently by rocking the plate back and forth.

6. Incubate the plate for 4 h in a tissue-culture incubator.

7. Add an additional 1.5 ml CM; this increases the total volume to ~2 ml per well.

8. Incubate the cells at 37°C for 48 h changing CM daily and analyze cells.

3.6. Genejammer Plasmid Transfection

1. The day before transfection, trypsinize cells as described in **Section 3.3**, resuspend the cells in CM, and seed at a density of $1-2 \times 10^5$ cells per well onto a Matrigel-coated 12-well plate.

2. Prepare each sample as follows: (*see* **Note 4**)
   - Dilute 5.25 μl Genejammer in 75 μl Opti-MEM and mix by flicking the tube.
   - Incubate at room temperature for 10 min.
   - Add 1.75 μg target vector to the solution and mix gently; incubate at room temp for a further 10 min.

3. Aspirate medium from cells prepared at Step 1 and replace with 450 μl of CM.

4. Add the 82 µl of DNA:Genejammer complexes dropwise to each well. This gives a final concentration of 3.6 ng/ml plasmid. Mix gently by rocking the plate back and forth.
5. Incubate the plate for 4 h in a tissue-culture incubator.
6. Add an additional 1.5 ml CM; this increases the total volume to ~2 ml per well.
7. Incubate the cells at 37°C for 48 h changing CM daily and analyze cells.

3.7. Electroporation

1. Trypsinize a confluent T25 culture flask as described in **Section 3.3**.
2. Following centrifugation ($180 \times g$, room temperature, 4 min), resuspend the cells in 800 µl CM containing 15–50 µg linearized DNA.
3. Transfer the cell/DNA mix to an 800 µl/4 mm gap electroporation cuvette and incubate for 5 min at room temperature.
4. Flick the cuvette to ensure a homogeneous cell suspension and electroporate at 320 V / 240 µF.
5. Incubate for 5 min at room temperature and resuspend the cells in at least 10–15 ml of CM (see **Note 5**).
6. Plate cells on 2–3 Matrigel-coated 60-mm dishes. This is largely dependent on growth characteristics of a particular hESC line.
7. Incubate the cells at 37°C for 48 h changing CM daily and analyze cells.

3.8. hESC Viral Transduction

1. The day before infection, trypsinize cells as described in **Section 3.3**, resuspend the cells in CM, and seed at a density of $1-2 \times 10^5$ cells per well onto a Matrigel-coated 12-well plate.
2. Aspirate medium from the hESC and replace with CM. Add increasing amounts of the concentrated virus (i.e., MOI 1, 5, 10, 25, and 100).
3. Incubate the cells at 37°C for 48 h changing CM daily and analyze cells.

3.9. Selection of Stably Transfected Cells

1. Initiate antibiotic selection 48 h after electroporation, Genejammer transfection, or viral transduction while changing CM daily. The onset of drug selection may be delayed when the dish is <90% confluent. However, if cells are too confluent, the action of the antibiotics will be delayed or ineffectual and death of untransfected cells will be severely reduced. Recommended final concentration for the antibiotics is 50 µg ml^{-1} for G418 and 300 ng/ml^{-1} for Puromycin (see **Note 6**).

2. It may be considered for slow growing cell lines to suspend antibiotic selection when approximately 50–80% of the cells are killed. This slows the kill rate so that residual untransfected cells act as feeder cells to the stably transfected cells to reduce the stress of clonal growth.

3. Restart antibiotic selection after an additional 2 days, while changing CM daily. It is highly recommended to maintain selection thereafter to eliminate cells that may silence the transgene (*see* **Note 7**).

4. After approximately 14 days from the start of the experiment, individual colonies should be visible.

3.10. Colony Transfer and Screening

1. Let the colonies grow until they are approximately 3 mm in diameter or until they start to fuse. Using a P200 pipette or glass needle, slice colonies into a grid motif and transfer (choose option a or b):
   a) 1/3 and 2/3 to 2 48-well plates, respectively
   b) 1/2 to a 48-well plate and 1/2 to DNA lyses buffer

2. Analyze DNA by PCR.

3. Pick a number of colonies (based on DNA analyses) for upscaling. Characterize cells for stem cell markers, differentiation potential, and karyotype.

3.11. FACS Analysis for Stem Cell Markers

1. Trypsinize hESCs with TrypLE for 4 min.
2. Add to single-cell suspension 5 ml cold FACS wash buffer.
3. Only for intracellular FACS! Add to single-cell suspension, 5 ml fixation buffer for 20 min, spin the cells, and resuspend in 5 ml cold FACS buffer.
4. Only for intracellular FACS! Spin again and add 100 µl 0.1% triton in PBSO, 5 min, RT, add 5 ml FACS wash buffer.
5. Spin for 4 min 400g at 4°C.
6. Add 5 ml FACS wash buffer, split the cells in several samples (depending on experiment), and spin for 4 min 200g at 4°C. Remember to save some cells to stain single-color controls for the set-up of the machine. You must have enough unstained cells and cells stained with each color in order to set up the machine.
7. Aspirate buffer and add to pellet: 50 µl 1st Ab in FACS wash buffer, incubate for 45 min 1st Ab, on ice.
8. Add 5 ml cold FACS wash buffer for 4 min 200g at 4°C.
9. Aspirate buffer and add to pellet: 50 µl 2nd Ab in FACS wash buffer, 45 min 2nd Ab, on ice in the dark.
10. Add 5 ml cold FACS wash buffer for 4 min 200g at 4°C.
11. Aspirate buffer and resuspend the pellet in 200 µl FACS wash buffer.

12. Optional (propidium iodide staining) viability marker, only for living cells!
    - add 50 µl 1 mg/ml RNAse A, 5 min RT
    - add 40 µl 50 µg/ml PI
13. Set FACS gates with control stained cells and count at least 50,000 cells.

3.12. Immunofluorescence

1. Culture cells on coverslips or chamberslides.
2. Aspirate medium and rinse cells with PBS.
3. Fix the cells for 20 min at RT.
4. Permeabilize cells for 5 min at room temperature with permeabilization buffer.
5. Wash 3x with PBS.
6. Incubate cells with blocking buffer for 1 h at RT.
7. Wash with PBS.
8. Incubate cells with primary antibody in PBS.
9. Wash 3x 5 min in PBS.
10. Incubate cells with secondary antibody in PBS in the dark.
11. Wash 3x 5 min in PBS.
12. Wash briefly with distilled water to remove salt traces.
13. Add Vectashield Hardset with DAPI and mount coverslip on a microscope slide.

4. Notes

1. Never let Matrigel dry out as this causes irreversible loss of extracellular matrix properties.
2. hESC medium may turn very acidic (yellow). This is normal and does not influence the results.
3. Cells split at low density, which are cultured for at least 72 h might require slightly longer trypsin incubation. In general, using maximally a 1–3 split, 1.5 min trypsin should be enough to yield a cell suspension with some single cells and clumps of 3-10 cells.
4. It is highly recommended that a fluorescent control is transfected; this can be either a fluorescent siRNA or a plasmid (e.g., pPGK-GFP-IRES-Neo or pCAG-GFP-IRES-PAC) to monitor transfection efficiency.
5. If smaller numbers of cells are electroporated, resuspend cells in at least 5 ml CM to prevent cell death arising from debris and DNA toxicity. Spinning the cells after electroporation may be considered.

6. The antibiotic concentration may need to be titrated and optimized for different hESC lines.

7. Some hESC clones will progressively silence stably transfected constructs even when an appropriate promoter (PGK or CAG) is used. It is highly recommended that a reporter construct (e.g., fluorescent protein or drug resistance marker) in the vector be included to allow visualization or continuous drug selection of transgene expression.

Acknowledgements

SRB and CLM are funded by the Bsik Dutch Platform for Tissue Engineering, EU Heart Development and Heart Repair, and the Bsik programme Stem Cells in Development and Disease. CD is funded by Medical Research Council, Biotechnology and Biological Sciences Research, and the British Heart Foundation.

References

1. Braam, S.R., Denning, C., van den Brink, S., Kats, P., Hochstenbach, R., Passier, R., and Mummery, C.L. (2008b). Improved genetic manipulation of human embryonic stem cells. Nat Meth 5, 389–392.

2. Braam, S.R., Denning, C., Matsa, E., Young, L.E., Passier, R., and Mummery, C. (2008a). Feeder-free culture of human embryonic stem cells in conditioned medium for efficient genetic modification. Nature Protocols 3, 1435–1443.

3. He, J., Yang, Q., and Chang, L.J. (2005). Dynamic DNA methylation and histone modifications contribute to lentiviral transgene silencing in murine embryonic carcinoma cells. J Virol 79, 13497–13508.

4. Smith-Arica, J.R., Thomson, A.J., Ansell, R., Chiorini, J., Davidson, B., and McWhir, J. (2003). Infection efficiency of human and mouse embryonic stem cells using adenoviral and adeno-associated viral vectors. Cloning and Stem Cells 5, 51–62.

5. Zeng, J., Du, J., Zhao, Y., Palanisamy, N., and Wang, S. (2007). Baculoviral vector-mediated transient and stable transgene expression in human embryonic stem cells. Stem Cells 25, 1055–1061.

6. Zwaka, T.P., and Thomson, J.A. (2003). Homologous recombination in human embryonic stem cells. Nat Biotech 21, 319–321.

7. Davis, R.P., Ng, E.S., Costa, M., Mossman, A.K., Sourris, K., Elefanty, A.G., and Stanley, E.G. (2008). Targeting a GFP reporter gene to the MIXL1 locus of human embryonic stem cells identifies human primitive streak-like cells and enables isolation of primitive hematopoietic precursors. Blood 111, 1876–1884.

8. Irion, S., Luche, H., Gadue, P., Fehling, H.J., Kennedy, M., and Keller, G. (2007). Identification and targeting of the ROSA26 locus in human embryonic stem cells. Nat Biotech 25, 1477–1482.

9. Dormeyer, W., van Hoof, D., Braam, S.R., Heck, A.J.R., Mummery, C.L., and Krijgsveld, J. (2008). Plasma membrane proteomics of human embryonic stem cells and human embryonal carcinoma cells. J Proteome Res 7, 2936–2951.

10. Van Hoof, D., Braam, S.R., Dormeyers, W., Ward-van Oostwaard, D., Heck, A.J.R., Krijgsveld, J., and Mummery, C.L. (2008). Feeder-free monolayer cultures of human embryonic stem cells express an epithelial plasma membrane protein profile. Stem Cells, 11, 2777–2781.

Chapter 23

Human-Induced Pluripotent Stem Cells: Derivation, Propagation, and Freezing in Serum- and Feeder Layer-Free Culture Conditions

Hossein Baharvand, Mehdi Totonchi, Adeleh Taei, Ali Seifinejad, Nasser Aghdami, and Ghasem Hosseini Salekdeh

Abstract

The recent discovery of genomic reprogramming of human somatic cells to an embryonic stem (ES) cell-like pluripotent state provides a unique opportunity for stem cell research. The reprogrammed cells, named as induced pluripotent stem (iPS) cells, possess many of the properties of ES cells and represent one of the most promising sources of patient-specific cells for use in disease model, development of pharmacology and toxicology, screening teratogens, and regenerative medicine. Here we describe the detailed methods for the generation of undifferentiated human iPS (hiPS) cells in feeder layer- and serum-free conditions. This system eliminates direct contact of stem cells with MEFs and reduces use of unknown serum factors that may have undesired activities and enables consistency in large-scale and long-term expansion of undifferentiated hiPS cells. Our findings greatly simplify the method for induction of pluripotency and bring it one step closer to clinical applications. Moreover, the established hiPS cells showed chromosomal stability during long-term culture.

Key words: Human-induced pluripotent stem cells, serum- and feeder-free culture conditions, derivation, maintenance, freezing.

1. Introduction

In 2006, Yamanaka and coworkers surprised the scientific community when they reported that both mouse embryonic fibroblasts and tail tip fibroblasts could be reprogrammed to a pluripotent state similar to that observed in embryonic stem (ES) cells, by retroviral transduction of just four genes, POU class 5 homeobox 1 (*Pou5f1/Oct4*), SRY (sex determining region Y)-box 2 (*Sox2*), Kruppel-like factor 4 (*Klf4*), and *c-Myc* (1). The reprogrammed cells were named induced pluripotent stem (iPS) cells. Approximately 1 year after the initial report of mouse iPS cells, groups from the Shinya Yamanaka and James Thomson labs reported that iPS cells derived from human dermal fibroblasts were capable of differentiating into all three germ layers during embryoid body formation and in vivo teratoma analysis in immuno-compromised mice (2, 3). These human iPS (hiPS) cells can give rise to most tissue types in the human embryo and possess many of the properties of human ES cells. Later, generation of hiPS cells was reported by several other groups (4–13). Recently, the reprogramming protocols of mouse (14) and human (10) fibroblasts to iPS cells were reported.

Reprogramming of somatic cells to iPS cells may allow (i) the development of patient-specific cell therapy, (ii) analysis of the pathways that lead to disease pathogenesis based on a particular genetic trait at the cellular level, (iii) examination of the effects of known genotypes on the cellular phenotype and then facilitation of studies such as those to monitor drug safety once drugs are on the market, (iv) providing a model for normal development in humans, which could help to understand disease pathogenesis and to develop suitable treatments in some cases, (v) providing a powerful tool for analyzing the effect of various teratogens or of specific genetic traits on cellular processes during embryogenesis, (vi) providing a useful tool for screening potential teratogens by their in vitro differentiation, (vii) investigation of the significance of the genetic background of the patient in the development of cancer by iPS cell establishment from cancerous somatic cells, and (viii) finally, development of alternative methods that avoid genetic manipulation (e.g., by using peptides or chemical compounds) for epigenetic reprogramming (for review *see* (15)).

In all reports of hiPS cell generation, the transduced human and mouse somatic cells were conventionally reprogrammed and maintained in culture with feeder cells (i.e., mouse embryonic fibroblasts or MEFs) and/or mixtures of exogenous factors such as fetal bovine serum. However, future applications of hiPS-derived cells or tissues will require that the cells and tissues are produced without contact with any animal sources. Furthermore, the use of hiPS cells also relies on the availability of routine large-scale culturing protocols for undifferentiated hiPS cells. The derivation and growth of the hESC line on either MEFs or on extra-cellular matrices supplemented with or without conditioned medium from MEFs were shown previously (for review *see* (16, 17)). This system eliminates direct contact of stem cells with MEFs and reduces use of unknown serum factors that may have undesired activities (e.g., inducing differentiation or cell death) and enables consistency in large-scale and long-term expansion of undifferentiated iPS cells. We showed that serum- and feeder-free conditions facilitate reprogramming and that the resulting human iPS cells expressed pluripotency markers and showed in vitro differentiation into three germ layer derivatives. Our findings greatly simplify the method for induction of pluripotency and bring it one step closer to clinical applications. Moreover, the established hiPS cells showed chromosomal stability during long-term culture (18–20). Here, we describe the detailed methods for the generation of hiPS cells in serum- and feeder-free conditions.

2. Materials

2.1. Chemicals

1. Basic fibroblast growth factor (bFGF, Sigma, Cat. No. F0291)
2. β-Mercaptoethanol (Sigma, Cat. No. M7522)
3. Blasticidin S hydrochloride (Invitrogen, Cat. No. R21001)
4. Collagenase type I (Sigma, Cat. No. C0130)
5. Collagenase type IV (Gibco, Cat. No. 17104-019)
6. Dispase (Gibco, Cat. No. 17105-041)
7. DMEM (Invitrogen, Cat. No. 11965-092)
8. DMEM/F12 (Gibco, Cat. No. 21331-020)
9. Dimethyl sulfoxide (DMSO, Sigma, Cat. No. D2650)
10. Fetal bovine serum (FBS, Hyclone, Cat. No. SH30071.03)
11. FP medium (for fibroblasts and Plat-A cells)
12. Fugene 6 transfection reagent (Roche, Cat. No. 1815075)
13. Gelatin (Sigma, Cat. No. G2500)
14. Hank's balanced salt solution (HBSS, Gibco, Cat. No. 14185-045)
15. hES cell medium
16. Hexadimethrine bromide (Polybrene, Sigma, Cat. No. H9268)
17. Freezing medium
18. Insulin-transferrin-selenite (ITS, 5 mg/ml insulin, 5 mg/ml transferrin, 5 μg/ml selenium, Gibco, 41400-045)
19. Knockout serum replacement (KOSR, Gibco, Cat. No. 10828-028)
20. L-Glutamine (L-Gln, Gibco, Cat. No. 25030-024)
21. Matrigel (Sigma, Cat. No. E1270)
22. Nonessential amino acid solution (100x, Gibco, Cat. No. 11140-035)
23. PBS without Ca^{2+}/Mg^{2+} (Gibco, Cat. No. 21600-051)

24. PBS with Ca^{2+}/Mg^{2+} (Gibco, Cat. No. 14287-072)
25. Penicillin/streptomycin (Gibco, Cat. No. 15070-063)
26. Plat-A cells (Cellbiolab, Cat. No. RV-102)
27. pMXs retroviral vectors containing cDNAs of *hOct3/4*, *hSox2*, *hKlf4*, or *hc-Myc* (Addgene, Cat. No. Plasmid 17217, 17218, 17219, 17220, respectively)
28. Puromycin (Sigma, Cat. No. P8833)
29. ROCK inhibitor Y-27632 (Calbiochem, Cat. No. 688000)
30. Trypsin/EDTA (0.05%/0.53 mM, Invitrogen, Cat. No. 25300-054)
31. HiSpeed Plasmid Maxi Kit (Qiagen, Cat. No. 12663)

2.2. Disposables

1. 15-ml conical tubes (TPP, Cat. No. 91015)
2. 50-ml conical tubes (TPP, Cat. No. 91050)
3. 6-well tissue-culture plates (TPP, Cat. No. 92006)
4. 4-well tissue-culture plates (Nunc, Cat. No. 176740)
5. 60-mm tissue-culture dish (Falcon, Cat. No. 353004)
6. 100-mm tissue-culture dish (Falcon, Cat No. 351005)
7. 1-ml plastic disposable pipette (TPP, Cat. No. 94001)
8. 5-ml plastic disposable pipette (TPP, Cat. No. 94005)
9. 10-ml plastic disposable pipette (TPP, Cat. No. 94010)
10. 25-ml plastic disposable pipette (TPP, Cat. No. 94024)
11. 25-cm^2 tissue-culture flask (TPP, Cat. No. 90025)
12. 75-cm^2 tissue-culture flask (TPP, Cat. No. 90075)
13. 150-cm^2 tissue-culture flask (TPP, Cat. No. 90150)
14. 0.22-μm pore size filter (Millex GP, Millipore, Cat. No. SLGP033RS)
15. 0.45-μm filter unit Millex-HV Filter (Millipore, Cat No: SLHV033RB)
16. 0.22-μm vacuum filteration (500 ml, TTP, Cat. No. 99500)
17. Cell scraper (TPP, Cat. Nos 99002 and 99003)
18. Disposable syringes, 10 and 20 ml
19. Glass Pasteur pipettes, 9 inches. Sterilize by autoclave
20. Cryovial (Greiner bio-one)

2.3. Equipment

1. Centrifuge 3K30 Centrifuge (Sigma)
2. Centrifuge ROTOFIX 32 (Hettich)
3. Inverted tissue-culture microscope with phase-contrast microscope (4, 10, 20, and 40x objectives) (Olympus, CKX41)

4. Stereomicroscope (Olympus, SZX12)
5. Class II biosafety cabinet with aspirator for tissue culture (JAL, JLVH120RS, Iran)
6. Class I biosafety cabinet with aspirator for tissue culture, fitted for stereomicroscope (JAL, JLCH120, Iran)
7. Micropipette (1–10, 10–100, 100–1000 μl, Eppendorf)
8. Freezing container (Nalgene Labware, Cat. No. 5100)
9. Dissecting forceps. Sterilize by autoclave
10. Dissecting scissors. Sterilize by autoclave
11. Autoclave (TOMY, SX-700)
12. Mastercycler gradiant (Eppendorf)
13. SmartSpec Plus (BioRad)
14. Water bath (Lab-companion, BW-05G)
15. Nitrogen tank (MVE)

2.4. Reagent setup

1. *Plat-A cells.* Retroviruses are efficient tools for delivering heritable genes into the genome of dividing cells. However, conventional NIH-3T3-based retroviral packaging cell lines have limited stability and produce low viral yields, mainly due to poor expression level of the retroviral structure proteins (gag, pol, env) in the packaging cells. The Platinum-A (Plat-A) cell line, a potent retrovirus packaging cell line based on the 293T cell line, was generated using novel packaging constructs, which utilizes EF 1α promoter to ensure longer stability and high-yield expression of retroviral structure proteins (gag, pol, amphotropic env). Plat-A cells can be kept in good condition for at least 4 months in the presence of drug selection and can produce retroviruses with an average titer of 1×10^6 infectious units/ml by transient transfection. In addition, replication competent retroviruses (RCR) are virtually nonexistent because only coding sequences of viral structural genes are used, avoiding any unnecessary retroviral sequences.

2. *Puromycin.* Dissolve in distilled water at 50 mg/ml and sterilize through a 0.22-μm filter. Aliquot and store at –20°C. Final concentration is 1 μg/ml.

3. *Blasticidin S hydrochloride.* Dissolve in distilled water at 10 mg/ml and sterilize through a 0.22-μm filter. Aliquot and store at –20°C. Final concentration is 10 μg/ml.

4. *Hexadimethrine bromide or polybrene.* Dissolve in distilled water at 6 mg/ml and sterilize through a 0.22-μm filter. Store at 4°C. Final concentration is 6 μg/ml.

5. *β-mercaptoethanol*. Dissolve 70 μl in 10 ml PBS without Ca^{2+}/Mg^{2+} for a 1000x stock (100 mM). Final concentration is 0.1 mM. Caution: When used, avoid inhalation and skin contact.

6. *FP medium (for fibroblasts and Plat-A cells)*. DMEM containing 10% FBS (vol/vol), 2 mM L-Gln, 100 units/ml penicillin and 100 μg/ml streptomycin, 0.1 mM β-mercaptoethanol, and 1 mM (1%) nonessential amino acid. To prepare 500 ml of the medium, mix 50 ml FBS, 5 ml L-Gln, 5 ml penicillin/streptomycin, and 5 ml nonessential amino acid, then fill up to 500 ml with DMEM medium. Filter the medium with a bottle-top 0.22-μm filter and store at 4°C for 2 weeks. For Plat-A cells, add 1 μl of 10 mg/ml puromycin stock and 10 μl of 10 mg/ml blastcidin S into 10 ml FP medium.

7. *bFGF*. Reconstitute 25 μg bFGF in 1 ml sterile Tris-base. Store at −20°C in 250–500 μl aliquots. Final concentration is 100 ng/ml.

8. *hES cell medium*. DMEM/F12 containing 20% KOSR, 2 mM L-Gln, 0.1 mM β-mercaptoethanol, 1 mM (1%) nonessential amino acid, 100 units/ml penicillin and 100 μg/ml streptomycin, ITS, and 100 ng/ml basic-fibroblast growth factor. To prepare 500 ml of the medium, mix 100 ml KOSR, 5 ml L-Gln, 5 ml nonessential amino acids, 500 μl 2-mercaptoethanol, 5 ml penicillin/streptomycin, 5 ml ITS, and then fill up to 500 ml with DMEM/F12 medium. Aliquot 50 ml and store at 4°C (*see* **Note 1**). Add 200 μl bFGF to each 50 ml and use in 1 week (*see* **Note 2**).

9. *Collagenase type I*. Dissolve in DMEM/F12 medium at 1 mg/ml (final concentration) and sterilize through a 0.22-μm filter. Always make fresh.

10. *ROCK inhibitor Y-27632*. Dissolve in cooled distilled water at 10 mg/30 ml (100x) and sterilize through a 0.22-μm filter. Store at −20°C in 100–500 μl aliquots. Final concentration is 10 μM.

11. *Collagenase IV and dispase*. To prepare final concentration of collagenase IV (0.5 mg/ml)/dispase (1 mg/ml), add 5 mg collagenase IV and 10 mg dispase in 10 ml DMEM/F12 medium (*see* **Note 3**). Sterilize through a 0.22-μm filter. Always make fresh.

12. *Freezing medium*. 10% DMSO (vol/vol) and 90% FBS (vol/vol) or 90% KOSR (vol/vol). Sterilize through a 0.22-μm filter. Chill on ice before use. Always make fresh.

13. *Gelatin solution and coating the culture dishes*. Autoclave 0.1% (wt/vol) gelatin in PBS with Ca^{2+} and Mg^{2+} to make homogenous and sterile solution. Add enough 0.1% gelatin

solution to cover the dish surface. Incubate the dish for at least 30 min at 37°C. Before using, aspirate excess gelatin solution and wash with PBS.

2.5. Preparation of Matrigel-Coated Plates

1. Slowly thaw the Matrigel stock solution on ice at 4°C for at least 2 h to avoid gel formation.
2. Place sterile 1.5-ml tubes and tips into a freezer for at least 1 h before aliquoting the Matrigel.
3. Maintain the Matrigel on ice at all times and use chilled tips, aliquot 0.5 mg of Matrigel into frozen 1.5-ml tubes on ice. Immediately freeze tubes at −80°C. The concentration of Matrigel varies from lot to lot, so the volume of Matrigel needed to obtain 0.5 mg will vary accordingly. The stock vials can be stored at −80°C up to several months.
4. To prepare the working solution, dilute the Matrigel stock solution 1:30 in cold DMEM/F12 to make 0.3 mg protein per ml. Coat each 4- and 6-well plate and 60-mm plate with 200 μl, 700 μl, and 1.5–2 ml of Matrigel working solution, respectively.
5. Incubate the Matrigel-coated plates for 2 h at 37°C. Aspirate excess Matrigel from the plate immediately before plating cells. Rinse the plates before use by PBS with Ca^{2+} and Mg^{2+} or DMEM/F12. Plates should be discarded if the medium dries.

3. Methods

3.1. Preparation of Human Dermal Fibroblasts

1. Obtain a 4–10 mm^2 skin punch biopsy by a dermatologist or other appropriately trained physician.
2. Place the biopsy specimen immediately in sterile HBSS supplemented with penicillin/streptomycin on ice for transport to the laboratory.
3. Soak the specimen in 70% ethanol for 30 s to reduce contamination.
4. Perform all subsequent steps in a tissue-culture hood.
5. Wash the specimen with sterile HBSS supplemented with penicillin/streptomycin twice.
6. Use sterile forceps and dissecting scissors, cut the skin biopsy into 1–2 mm^3 size pieces.
7. Digest skin with 0.1% dispase at 4°C overnight to remove epidermal layers.

8. Further digest remaining dermal parts with 0.1% collagenase I at 37°C for 4 h.
9. Centrifuge the digested cells, and then resuspend in 2 ml of FP medium.
10. Seed the cells in 6-well plates at 1×10^3 cells/cm^2, and maintain at 37°C with 5% CO_2.
11. Renew FP medium to remove residual nonadherent cells, after 1 day.
12. The resulting adherent cells will be grown to confluency within 7 days.
13. Aspirate the medium, wash twice with 2 ml of PBS without Ca^{2+} and Mg^{2+}, and add 0.3 ml of trypsin/EDTA and incubate at 37°C.
14. Add 2 ml of FP medium, suspend the cells, and transfer and pool to a 15-ml tube. Centrifuge the cells at 200g (~1200 rpm) for 5 min.
15. Discard the supernatant; resuspend the cells with 6 ml of FP medium and plate to a T-25 flask (passage 2). For ease of use, we use T-25 flask for expansion.
16. Subculture the cells every 4–6 days using trypsin/EDTA for 3 min at 37°C. If desired, freeze cells as described below.

3.2. Freezing of Human Fibroblast Cells

1. When human fibroblast cells have reached 80% confluence, aspirate medium, wash once with PBS, cover cells with trypsin/EDTA (1 ml), and incubate for 3 min at 37°C and then inactivate enzyme with FP medium (2 ml).
2. Transfer the cell suspension to a 15-ml tube, count the cells using a hemocytometer, and spin the cells at 200 g for 5 min.
3. Discard the supernatant and break the cell pellet by finger tapping, resuspend the cells with freezing medium per T-25 flask, and aliquot it at 1 ml per vial. Each vial has $1–2 \times 10^6$ cells.
4. Put the vials in a cell-freezing container and keep it at –80°C overnight.
5. Transfer the vials to liquid nitrogen tank the next day for long-term storage.

3.2.1. Establishing Plat-A Cultures from Frozen Cells

1. Prepare 9 ml of FP medium in a 15-ml tube.
2. Remove a vial of frozen Plat-A stocks from the liquid nitrogen tank and put the vial in a 37°C water bath until most (but not all) cells are thawed.
3. Spray vial with 70% ethanol and wipe dry before placing in tissue-culture hood.

4. Open the cap and transfer gently the cell suspension to a tube prepared in Step 1.
5. Centrifuge at 200g at room temperature for 5 min, and then discard the supernatant.
6. Resuspend the cells with finger tapping the tube a few times and add 5 ml of FP medium into a 50-ml conical tube and gently pipette the cell suspension up and down.
7. Determine the number of Plat-A cells using a hemocytometer and adjust the concentration to 2×10^5 cells/ml.
8. Transfer 15 ml of cell suspension (3×10^6 cells) into a gelatin-coated 150-T flask.
9. Swirl the flask well to mix the cells, and then incubate the cells for 2–3 days before expansion.
10. Incubate the cells in a 37°C, 5% CO_2 incubator.
11. Next day, replace the medium with new FP medium supplemented with puromycin (final concentration; 1 µg/ml) and blastcidin S (final concentration: 10 µg/ml) (*see* **Note 4**).

3.2.2. Splitting of Plat-A Cells

1. Wash cells once with PBS.
2. Add 5 ml of trypsin/EDTA solution to a 150-T flask and incubate at 37°C for 1–3 min.
3. Remove the cells from the dish surface by tapping the rim of the culture dish.
4. Transfer 10 ml of the culture medium to a 50-ml tube.
5. Wash the dish with 15 ml culture medium and transfer the medium to the 50-ml tube.
6. Centrifuge the cells for 5 min at 200g.
7. Discard the supernatant and break the cell pellet by finger tapping.
8. Add appropriate volume of FP medium and break up the cells into a single cell suspension by pipetting up and down several times. Seed them to new 150-T flasks at 1:4–1:6 dilution. Typically 2×10^7 cells can be harvested from one 150-T flask. Cells should become confluent within 2–3 days.

3.2.3. Freezing of Plat-A Cells

Refer section "Freezing of human fibroblast cells."

3.3. Establishment of hiPS Cells

An overview of the main stages of the protocol and how these stages need to be coordinated over time has been presented in **Fig. 23.1**.

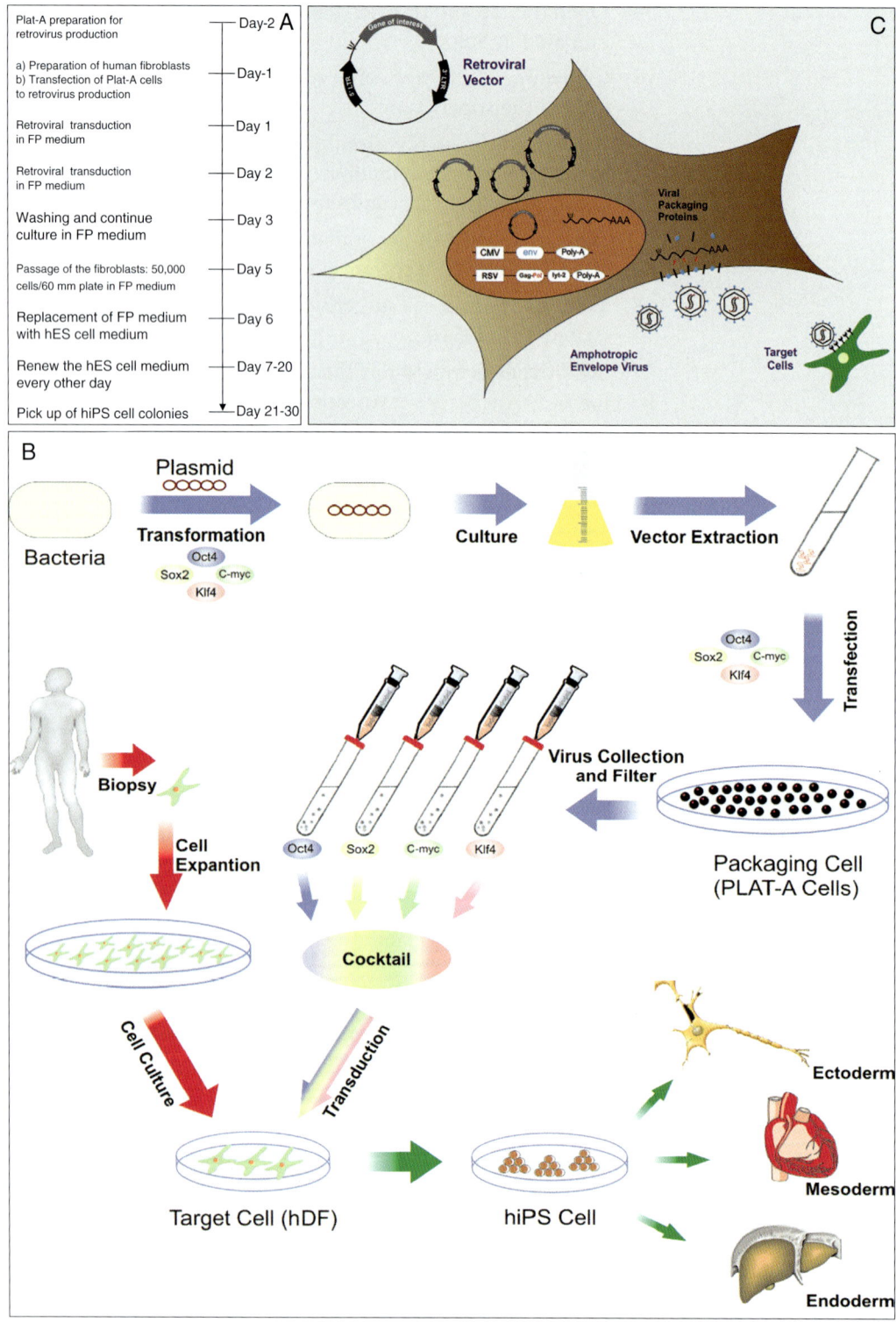

Fig. 23.1. A protocol summary of feeder- and serum-free production of human-induced pluripotent stem (hiPS) cells. (**A**) The timetable diagram of protocol. (**B**) Schematic diagram of protocol. (**C**) Packaging cells (Plat-A) and retroviral

3.3.1. Day 2: Plat-A Preparation for Retrovirus Production

1. Wash the cells with PBS, add 5 ml of trypsin/EDTA, and incubate for 1–3 min at room temperature.
2. After incubation, add 15 ml FP medium into the Plat-A flask, suspend the cells by gently pipetting, and transfer the cell suspension to a 50-ml tube. FP culture medium used in this period contains neither puromycin nor blasticidin S.
3. Centrifuge the cells at $200g$ for 5 min.
4. Discard the supernatant, break the pellet by finger tapping, and resuspend the cells in an appropriate amount of FP medium.
5. Count the number of cells and adjust the concentration to 6×10^5 cells per ml with FP medium.
6. Seed cells at 6×10^6 cells (10 ml) per 100-mm culture dish and incubate overnight at 37°C, 5% CO_2 (**Fig. 23.2A**) (*see* **Note 5**).
7. Continue to incubate the cells in a 37°C, 5% CO_2 incubator until they are 60–80% confluent in 1 day.

3.3.2. Day 1: Preparation of Human Fibroblast Cells

1. Culture human fibroblasts to ∼90% confluency in 25-T flask (**Fig. 23.2B**) (*see* **Note 6**).
2. Aspirate the culture medium and wash with 5 ml of PBS.
3. Discard PBS, add 1 ml per dish of trypsin/EDTA, and incubate at 37°C for 3 min.
4. Inactivate trypsin with 4 ml FP medium, suspend the cells by gently pipetting, and transfer to a 15-ml tube. Centrifuge the cells at $200g$ at room temperature for 5 min and discard the supernatant.
5. Resuspend the cells in 5 ml FP medium and determine cell number using hemocytometer and adjust the concentration to 4×10^4 cells per ml.
6. Transfer 2 ml of cell suspension (8×10^4 cells) to a well of 6-well plate. Incubate the dish overnight at 37°C, 5% CO_2.

3.3.3. Day 1: Transfection of Plat-A Cells for Retrovirus Production (B)

1. Transfer 0.3 ml of DMEM into a 1.5-ml tube.
2. Deliver 27 μl of Fugene 6 transfection reagent into the prepared tube in Step 1, mix gently by finger tapping, and incubate at room temperature for 5 min (*see* **Notes 7 and 8**).

Fig. 23.1. (continued) production. The viral *gag, pol,* and ampho *env* genes – necessary for particle formation and replication – are stably integrated into the genome of the PLAT-A packaging cell line. The retroviral expression vector provides the viral packaging signal (ψ) and a gene of interest. Transfection of the pMXs retroviral vector by Fugene 6 into the Plat-A packaging cell line produces replication-incompetent virus that can efficiently transfer genes into a variety of mammalian cell types. FP medium: medium for fibroblasts and Plat-A cells, hES cell: human embryonic stem cell.

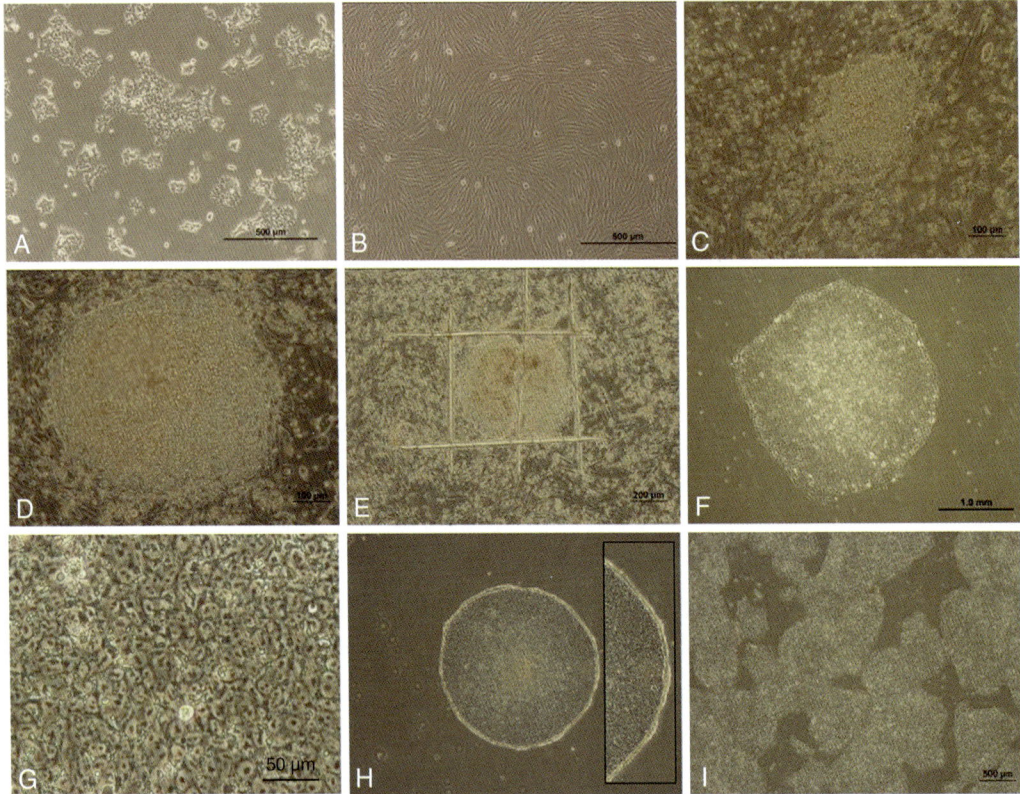

Fig. 23.2. Phase contrast micrograph of packaging cells, human dermal fibroblasts, an established hiPS cell colony, and passaging of hiPS cells. (**A**) Morphology of Plat-A packaging cells. (**B**) Morphology of human dermal fibroblasts used for hiPS cell induction. Morphology of an established hiPS cell colony on transduced cells at day 15 (**C**) and day 22 (**D**) in absence of feeder cells and serum. (**E**) Mechanical passaging by a flame-pulled Pasteur pipette. (**F**) Morphology of a hiPS cell colony on Matrigel at passage 2. Colony demonstrates a typical undifferentiated morphology with a clear border. (**G**) Higher magnification of hiPS cells. The cells display a compact morphology and a high nucleus-cytoplasmic ratio and containing prominent nucleoli typical of undifferentiated hES cells. (**H**) hiPS cell colony under phase contrast inverted microscope after collagenase/dispase treatment; colonies at the edge started dissociating from the bottom. Insert figure shows higher magnification. (**I**) The freezed-thawed hiPS cells after 5 passages.

3. Perform all subsequent steps under the containment of BL2+ biosafety.
4. Add 9 µg of retroviral vector (encoding *Oct4*, *Sox2*, *Klf4*, and *c-Myc*) drop-by-drop into the Fugene 6/DMEM-containing tube (*see* **Notes 9** and **10**).
5. Mix gently by finger tapping and incubate for 15–30 min.
6. Add the retroviral vector/Fugene 6 complex dropwise into the Plat-A dish and swirl the dish to ensure distribution over the entire plate surface.
7. Incubate 24 h at 37°C, 5% CO_2.

3.3.4. Days 1 and 2: Retroviral Transduction

1. Day 1: Collect the medium (10 ml) from the Plat-A dish by using a 10-ml sterile disposable syringe, filtering it through a 0.45-μm pore size cellulose acetate filter, and transferring into a 15-ml tube. Moreover, add 10 ml new FP medium in Plat-A dish.
2. Add 10 μl of 6 mg/ml polybrene solution into the each 10-ml filtrated virus-containing medium, and mix gently by pipetting up and down. The final concentration of polybrene will be 6 μg/ml (see **Note 11**).
3. Aspirate the medium from a human fibroblasts dish, and add 0.5 ml of each polybrene/virus-containing medium to each well of a 6-well plate. Totally 2 ml medium (0.5 ml of *Oct4*, 0.5 ml of *Sox2*, 0.5 ml of *Klf4*, 0.5 ml of c-Myc) per well was added.
4. Incubate the cells for 24 h at 37°C, 5% CO_2.
5. Day 2: Repeat steps 1-4. Plat-A cells were thrown out at this step.

3.3.5. Day 3: Medium Renewal

Aspirate medium, wash the cells three times with 3 ml PBS, add 2 ml fresh FP medium and incubate at 37°C, 5% CO_2, for 2 days.

3.3.6. Day 5: Replating of Transduced Cells

1. Aspirate FP medium from the cells, wash with PBS, and add 700 μl of trypsin/EDTA. Incubate at 37°C, 5% CO_2, for 3 min.
2. Inactivate medium with 3 ml FP medium and collect cells in a 15-ml conical tube.
3. Centrifuge at 200*g* at room temperature for 5 min.
4. Resuspend the cells in 3 ml FP medium and determine cell number using hemocytometer and adjust the concentration to 25×10^3 cells per ml.
5. Split all infected cells 5×10^4 cells per 60-mm Matrigel-coated plate.
6. Incubate the cells at 37°C, 5% CO_2, for 24 h.

3.3.7. Day 6: Medium Replacement

Aspirate medium and replace with hES cell medium, which conditioned on feeder cells. Also, it is possible to use only hES cell medium. We used both conditions and here the feeder-independent hES cell medium will explained. The conditioned hES cell medium was prepared as standard protocols on mouse embryonic fibroblasts (21).

3.3.8. Days 7–20: Culturing Infected Human Fibroblast Cells

Renew the hES cell medium every day (see **Note 12**). Over the next few days, some cells will grow into colonies that resemble hES cell colonies in morphology (i.e., with a clear border, composed of small, tightly packed cells). By day 15 colonies should appear (**Fig. 23.2C**).

3.3.9. Days 21–30: Picking and Expanding iPS Cell Colonies

1. When the colony reaches 1–3 mm in diameter (**Fig. 23.2D**), it should be passaged mechanically. Using a flame-pulled Pasteur pipette, cut the colony into 2–8 pieces and dislodge the pieces from the plate bottom (**Fig. 23.2E**). Perform this step using a stereomicroscope fitted in a tissue-culture biosafety cabinet. Use 10-µl pipette tip to transfer individual hiPS cell colony clumps on freshly plated Matrigel and deposit one colony per well of a 4-well plate with 0.5 ml hES cell medium.

2. Incubate the cells in 37°C, 5% CO_2, overnight.

3. Leave undisturbed for 2 days after passage and then feed every other day with hES cell medium. If a hiPS cell line is successfully derived, the clumps of cells will attach by 2 days, and over the next few days, the clumps will grow into colonies that resemble hES cells in morphology (i.e., flat, refractile, and composed of tightly packed small cells with a high nuclear to cytoplasmic ratio) (**Fig. 23.2F** and **G**)

4. After 7 days, the colony will become large enough to be passaged. We recommend using mechanical passage until line is well established in culture in 6-well plates.

5. Once derived, hiPS cells should be treated like hES cells. Instruction on how to passage, freeze, and characterize these cells can be found in the following.

3.4. Propagation of hiPS Cells

Ideally, hiPS cells passed either mechanically or with collagenase IV/dispase retain the same growth characteristics. A typical, well-established culture should be passaged every 7 days. We routinely are using the enzymatic passaging.

1. Prepare Matrigel-coated dishes 1 day before passage (*see* **Note 13**).

2. Aspirate the extra Matrigel and wash the plate with PBS and add 1.5 ml hES cell medium per well.

3. Aspirate medium from hiPS cells, and wash culture plate with 3 ml of PBS with Ca^{2+}/Mg^{2+} per dish.

4. Replace the PBS with 1 ml of collagenase IV/dispase mixture solution in each 60-mm plate.

5. Incubate the plate at 37°C for 5–10 min. Incubation time will vary depending on the batch of enzyme solution and the lines.

6. Observe the cells under phase contrast inverted microscope or stereomicroscope (darkfield) and continue the enzyme, when colonies at the edge start dissociating from the bottom (**Fig. 23.2 H**).

7. Remove the enzyme solution and wash the plates with 3 ml PBS.

8. Add 2 ml hES cell medium. Gently scrape with a 100-µl pipette tip the hiPS cells from the plate.

9. Dissociate the cells into small clusters (50–500 cells) by gentle pipetting. Do not reduce cells to a single-cell suspension. We recommend a split ratio 1:2 to 1:6, depending on hiPS cell colony concentration. We usually transfer 0.5 ml of the cell suspension to each well of 6-well plate containing 1.5 ml hES cell medium (see **Note 14**).

10. Return the plate to the incubator. Tilt the plate several times both in the horizontal and vertical directions to spread out cell clumps evenly.

11. Renew the medium every other day with 2 ml.

12. Passage when cells are 80% confluent (**Fig. 23.1**).

3.5. Freezing of hiPS Cells

Freezing of hiPS cells is performing based on our published paper (20).

1. When hiPS cells have reached 80% confluence, add 10 μM of the ROCK inhibitor Y-27632 to the culture medium 1 h before detaching the cells from the dish (10 μl/ml).

2. Aspirate the medium and wash the cells with 5 ml of PBS.

3. Remove PBS completely, add 0.8 ml of trypsin/EDTA, and incubate at 37°C for 5 min.

4. Add 2 ml of the hES cell medium and follow by flushing with a pipette several times in order to detach single cells.

5. Transfer the cell suspension to a 15-ml tube, count the number of cells, and spin the cells at 200 g for 5 min.

6. Discard the supernatant, resuspend the cells with freezing medium, and aliquot it at 250 μl per vial. Each vial has $10–20 \times 10^4$ cells.

7. Put the vials in a cell-freezing container and keep it at –80°C overnight.

8. Transfer the vials to liquid nitrogen tank next day for long-term storage.

3.6. Thawing of hiPS Cells

1. Prepare 4 ml of hES cell medium in a 15-ml tube.

2. Remove a vial of frozen hiPS cell stocks from the liquid nitrogen tank and put the vial in a 37°C water bath until most (but not all) cells are thawed.

3. Spray vial with 70% ethanol and wipe dry before placing in tissue-culture hood.

4. Open the cap and transfer gently the cell suspension to a tube prepared in Step 1.

5. Centrifuge at *200 g* at room temperature for 5 min, and then discard the supernatant.

6. Resuspend the cells with 3 ml of hES cell medium containing 10 μM of the ROCK inhibitor Y-27632 (10 μl/ml).

7. Transfer the cell suspension into a 60-mm dish in hES cell medium which is coated with Matrigel containing 10 μM (10 μl/ml) of the ROCK inhibitor Y-27632 (*see* **Note 15**).

8. Incubate the cells in a 37°C, 5% CO_2 incubator.

9. Next day, replace the medium with new medium without ROCK inhibitor Y-27632. **Figure 23.2I** demonstrates freezed-thawed hiPS cells at passage 5.

3.7. Characterization of hiPS Cells by Immunofluorescence Staining (SSEA3, SSEA4, Oct4, Nanog, Tra-1-60, Tra-1-80), Alkaline Phosphatase Staining, and Karyotyping

The isolated hiPS cells are morphologically very similar to hES cells and expressed similar gene and surface-antigen expression and developmental potential for differentiation (**Fig. 23.3**). Start characterization when the line has been expanded sufficiently.

1. Typically we perform these assays on colonies in a 4-well multidish. When the colonies are grown large enough to be passaged, wash three times with PBS and fix in fresh 4% (wt/vol) paraformaldehyde in PBS. For immunofluorescence staining, fix for 30 min; for alkaline phosphatase staining, fix for 5 min.

2. Wash three times with PBS, and then permeabilize with 0.2% (vol/vol) Triton X-100 in PBS for 30 min.

3. Block with 3% (wt/vol) BSA in PBS/0.2% Triton X-100 for 2 h.

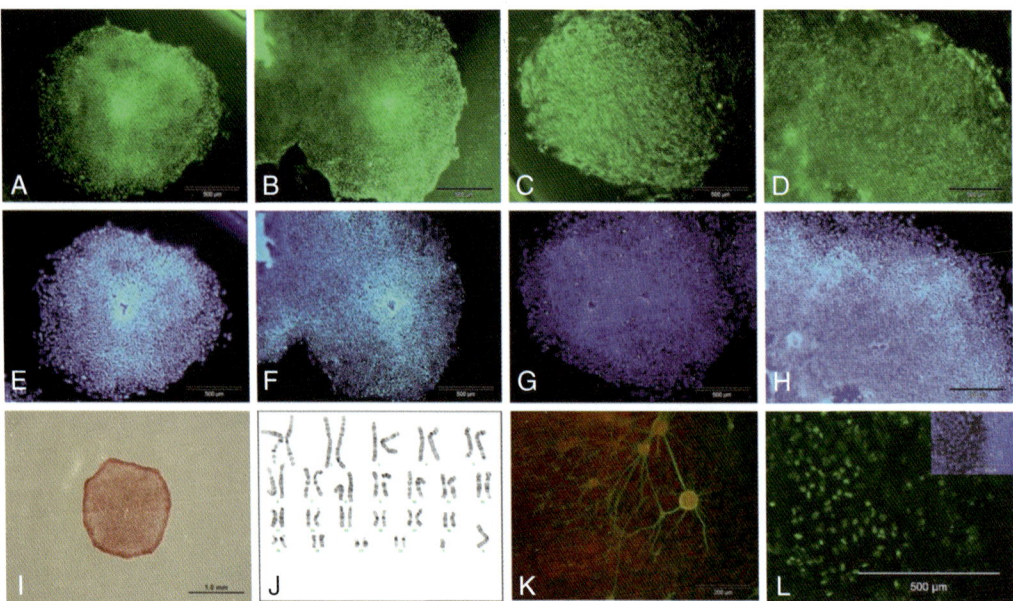

Fig. 23.3. Characteristics of established hiPS cells under feeder- and serum-free conditions. Representative immunofluorescence staining for Oct-4 (**A**), SSEA-4 (**B**), Tra-1-60 (**C**), and Tra-1-81 (**D**). Nuclei are stained blue with 4′,6-diaminidine-2-phenylidole dihydrochloride (DAPI) (**E-H**). The cells also expressed alkaline phosphatase (**I**) and had a normal karyotype (46 XY) (**J**). The hiPS cells were induced for differentiation into neural/ectodermal cells and endodermal cells as showed by microtubule associate protein 2 (K) and Sox17 (I). Nuclei are stained red by propidium iodide in (**K**) and DAPI in (**I**).

4. Incubate with primary antibodies (dilute in blocking solution appropriately) overnight at 4°C.

5. Wash with PBS three times and then incubate with appropriate secondary FITC antibody, diluted in blocking solution, for 2 h.

6. Wash four times with PBS.

7. If performing alkaline phosphatase staining, perform as per the manufacturer's (Sigma) recommendations.

For karyotype analysis, the cells were treated with thymidin (0.66 μM, Sigma) for 16 h at 37°C in 5% CO_2. After washing, the cells were left for 5 h and then treated with Colcemid (Gibco, 0.15 μg/ml, 30 min). Isolated hiPS cells were exposed to 0.075 M KCl at 37°C for 16 min and then were fixed with ice-cold 3:1 methanol:glacial acetic acid (repeated three times) and dropped onto precleaned chilled slides. Chromosome spreads were banded and Giemsa stained analyzed for chromosomal status. At least 20 metaphase spreads were screened and 10 banded karyotypes were evaluated for chromosomal rearrangements.

4. Notes

1. KSOR should be stored at −20°C and ideally used immediately after thawing. However, if stored at +4°C, they can be used 7 days after they have been made/thawed.

2. It is always best to thaw and add growth factor to the medium just prior to use.

3. You can keep collagenase/dispase solution at −20°C for a long time and thaw it before using. Enzymes should be dispensed in small aliquots to avoid repeated freezing and thawing. To passage hES cells, it is also possible to use only collagenase (1 mg/ml) or dispase (2 mg/ml) separately.

4. Do not culture cells to complete confluency. Split cells 1:4 to 1:6 every 2–3 days when the culture reaches 70–90% confluency.

5. For Plat-A cells transfection, they should be cultured on 0.1% gelatin-coated culture dishes.

6. To avoid replicative senescence, it is always best to use fibroblast cells within early passages.

7. To avoid adversely affecting transfection efficiency, do not allow undiluted Fugene 6 transfection reagent to come into contact with plastic surfaces other than pipette.

8. Fugene 6 should not be exposed to light for long periods of time.
9. The lentiviral vectors (Oct4, Sox2, Lin28, and Nanog) with related packaging cells (293FT cell line) have also been applied successfully for hiPS cell generation (3).
10. Retroviruses should be used freshly. The titer of retrovirus is absolutely important for hiPS cell generation. The freeze/thaw step decreases the titer of retrovirus. Best results are achieved when retroviruses are used freshly. Viruses may be kept at +4°C for 24 h prior to use.
11. Protamin sulfate (6 μg/ml) can be used in place of polybrene.
12. hES cell medium may be renewed every other day, also.
13. Coating of plates with Matrigel including ROCK inhibitor Y-27632 increases the plating efficiency (number of colonies per seeded hiPS cell clusters) during the passage of hiPS cells (19).
14. To avoid attachment of hES cell clusters, they should be washed carefully and passaged as quickly as possible.
15. Addition of ROCK inhibitor Y-27632 into the Matrigel culture medium increases the plating and cloning efficiency (Number of colonies per seeded single hiPS cells) after thawing (19).

Acknowledgments

This work was supported by Royan institute.

References

1. Takahashi K, Yamanaka S (2006) Induction of pluripotent stem cells from mouse embryonic and adult fibroblast cultures by defined factors. *Cell* 126:663–676.
2. Takahashi K, Tanabe K, Ohnuki M, Narita M, Ichisaka T, Tomoda K, Yamanaka S (2007) Induction of pluripotent stem cells from adult human fibroblasts by defined factors. *Cell* 131:861–872.
3. Yu J, Vodyanik MA, Smuga-Otto K, Antosiewicz-Bourget J, Frane JL, Tian S, Nie J, Jonsdottir GA, Ruotti V, Stewart R, Slukvin, II, Thomson JA (2007) Induced pluripotent stem cell lines derived from human somatic cells. *Science* 318:1917–1920.
4. Aasen T, Raya A, Barrero MJ, Garreta E, Consiglio A, Gonzalez F, Vassena R, Bilic J, Pekarik V, Tiscornia G, Edel M, Boue S, Belmonte JC (2008) Efficient and rapid generation of induced pluripotent stem cells from human keratinocytes. *Nat Biotechnol* 26:1276–1284.
5. Huangfu D, Osafune K, Maehr R, Guo W, Eijkelenboom A, Chen S, Muhlestein W, Melton DA (2008) Induction of pluripotent stem cells from primary human fibroblasts with only Oct4 and Sox2. *Nat Biotechnol* 26:1269–1275.
6. Tateishi K, He J, Taranova O, Liang G, D'Alessio AC, Zhang Y (2008) Generation of insulin-secreting islet-like clusters from human skin fibroblasts. *J Biol Chem* 283:31601–31607.
7. Park IH, Arora N, Huo H, Maherali N, Ahfeldt T, Shimamura A, Lensch MW, Cowan C, Hochedlinger K, Daley GQ (2008) Disease-specific induced pluripotent stem cells. *Cell* 134:877–886.

8. Do JT, Han DW, Scholer HR (2006) Reprogramming somatic gene activity by fusion with pluripotent cells. *Stem Cell Rev* 2:257–264.
9. Dimos JT, Rodolfa KT, Niakan KK, Weisenthal LM, Mitsumoto H, Chung W, Croft GF, Saphier G, Leibel R, Goland R, Wichterle H, Henderson CE, Eggan K (2008) Induced pluripotent stem cells generated from patients with ALS can be differentiated into motor neurons. *Science* 321:1218–1221.
10. Park IH, Lerou PH, Zhao R, Huo H, Daley GQ (2008) Generation of human-induced pluripotent stem cells. *Nat Protoc* 3:1180–1186.
11. Mali P, Ye Z, Hommond HH, Yu X, Lin J, Chen G, Zou J, Cheng L (2008) Improved efficiency and pace of generating induced pluripotent stem cells from human adult and fetal fibroblasts. *Stem Cells* 26:1998–2005.
12. Durcova-Hills G (2008) Induced reprogramming of human somatic cells into pluripotency: a new way how to generate pluripotent stem cells. *Differentiation* 76:323–325.
13. Nakagawa M, Koyanagi M, Tanabe K, Takahashi K, Ichisaka T, Aoi T, Okita K, Mochiduki Y, Takizawa N, Yamanaka S (2008) Generation of induced pluripotent stem cells without Myc from mouse and human fibroblasts. *Nat Biotechnol* 26:101–106.
14. Takahashi K, Okita K, Nakagawa M, Yamanaka S (2007) Induction of pluripotent stem cells from fibroblast cultures. *Nat Protoc* 2:3081–3089.
15. Nishikawa S, Goldstein RA, Nierras CR (2008) The promise of human induced pluripotent stem cells for research and therapy. *Nat Rev Mol Cell Biol* 9:725–729.
16. McDevitt TC, Palecek SP (2008) Innovation in the culture and derivation of pluripotent human stem cells. *Curr Opin Biotechnol* 19:527–533.
17. Baharvand H (2006) Embryonic stem cells: establishment, maintenance, and differentiation. In *Embryonic Stem Cell Research* Grier EV, Ed., Nova Science Publishers, Inc., Hauppauge, NY, pp. 1–63
18. Totonchi M, Taei A, Seifinejad A, Tabebordbar MSH, Rassouli H, Aghdami N, Gourabi H, Hosseini Salekdeh GH, Baharvand H (2009) Feeder- and serum-free establishment and expansion of human induced pluripotent stem cells. *Int J Dev Biol*, In press.
19. Pakzad M, Mollamohammadi S, Taei A, Totonchi M, Seifinejad A, Baharvand H (2009) Presence of a ROCK inhibitor in extracellular matrix supports more undifferentiated growth of feeder-free human empryonic and induced pluripotent stem cells upon passaging.
20. Mollamohammadi S, Taei A, Pakzad M, Totonchi M, Seifinejad A, Masoudi N, Baharvand H (2009) A simple and efficient cryopreservation method for feeder-free dissociated human induced pluripotent stem cells and human embryonic stem cells. *Hum Reprod.* [Epub ahead of print].
21. Schatten G, Smith J, Navara C, Park JH, Pederson R (2005) Culture of human embryonic stem cells. *Nature Methods* 2:455–463.

INDEX

A

Adeno-associated virus (AAV) 394
Agarose electrophoresis 145
Albinism 302
Alcian blue staining 329
Alkaline phosphatase 346, 440, 441
Annexin A4, identification of 174
Antibody(ies)
　dilutions 298, 384
　for immunostaining 322
　selection 420
　staining 380
ART, see Assisted reproductive technology
Assisted reproductive technology 4
Autographa californica multiple nucleopolyhedrovirus
　　(AcMNPV)-based vectors 394

B

Baculovirus 394
Basic fibroblast growth factor 110, 406, 430
Batch ion-exchange chromatography
　2D gels 174
　fractionation for 163–164
　protein desalting/concentration 154–155
　protein lysis buffer 154
bFGF, see Basic fibroblast growth factor
BG01v/R4 hESC platform line 237, 238, 261
　characterization 248–250
　using PCR 248–249
BIEX, see Batch ion exchange chromatography
Bioactive hydrogel-based scaffolds
　for cell encapsulation 336
　　photopolymerizable hydrogels 337
　　type of hydrogels for 336–337
　hESCs encapsulation and culture in 341–345
　　characterization 345–347
　release of hESCs from 342
　for vascular differentiation of hESC 337–338
Bio-Rad Tetra Cell mini gel system 166
Bipronuclear ova 5
Blasticidin S hydrochloride 429
Blastocysts
　classification 16

culture media for 25
endoderm cell 13
ICMs, embryonic stem cells 13
normal and abnormal 12
Blastomeres 2
Blot FastStain kit 157

C

Ca^{2+} dependent protein 347
Calibrator, definition 137
Cartilage-related genes, analysis of 325–326
β-Catenin 312
CD34+ hematopoeitic stem 369
cDNA synthesis
　reverse transcription buffer for 217
Cell adhesion
　extracellular matrix compounds 60
　quality of 105
Cell Apoptosis Assay
　gap junctions in hESC 215, 222–223
Cell encapsulation, bioactive hydrogel-based scaffolds
　for 336
　photopolymerizable hydrogels 337
　type of hydrogels for 336–337
Cell fragments 47
Cell lysis/basic fractionation 162–165
　batch ion-exchange chromatography,
　　fractionation 163–164
　2D electrophoresis, whole cell lysis 164
　free-flow isoelectric focusing, whole
　　cell lysis 164–165
　SDS-PAGE, fractionation 163
Cell markers 180
Cell Pluripotency Assay
　gap junctions in hESC 216
Cell Proliferation Assay
　gap junctions in hESC 216, 223–224
Cells staining 88
Cell suspension 126
Cell vitrification method 47
CF-1 derived mouse embryonic fibroblasts 314
CG, see Cortical granules

K. Turksen (ed.), *Human Embryonic Stem Cell Protocols,* Methods in Molecular Biology 584,
DOI 10.1007/978-1-60761-369-5, © Humana Press, a part of Springer Science+Business Media, LLC 2006, 2010

Chemical closure
    gap junctions in hESC 215, 222
Chondrogenic cells, in vitro derivation of
    aggregate processing 329
    EB-derived cells, micromass cultures of 324
    embryoid body formation 322–324
    histological analysis 329–330
    immunostaining 326, 328
    isolation of 324–325
c-kit mutations 302
Collagenase
    R4 hESC engineering 236–237, 262
Collagenase IV 351, 430
Collagenase type I 430
Collagen I and II immunohistochemistry 329–330
Colloidal coomassie gel staining 157, 169
Colony growth assay
    gap junctions in hESC 225
Connexin mRNA expression
    RT-PCR for 216–218
Connexins 211–212
Connexion 43 phosphorylation states
    western blot analysis 219–220
Cortical granules 14
Cre-mediated HESC lines 390
Crossing point detection 146
Cryopreservative medium, for MEFs 303
Cx43 expression
    immunocytometry 221
Cx45 expression
    immunocytometry 221
Cytokines 347

D

Dextran 337
Dextran-based hydrogels 342–344, 351
Digital cameras 373
DiI-labeled acetylated lowdensity lipoprotein
    (DiI-Ac-LDL) 335
Dispase 225
Dispermic tripronuclear ova, TEM image 6
DMEM, see Dulbecco's Modified Eagle's Medium
Dopamine neuron 357
Dopaminergic neurons 363–364
Dulbecco's Modified Eagle's Medium 152
    feeder medium 24
    serum replacement 153
Dye Transfer Assay
    gap junctions in hESC 215, 221–222

E

EBs, see Embryoid bodies
ECM, see Extracellular matrix
ECM proteins, see Extracellular matrix proteins

ECs, see Endothelial cells
Ectopic expression of HOXB4, 388
EDNRB, see Endothelin-B receptor
Electroporation 392, 404–405, 420
    See also Genetic modification
Embryo-handling media 25
Embryoid bodies 111, 117, 309–310
    dissociation and migration 381
    formation 322–324
    medium 304
    micromass cultures 324
    See also Mouse embryonic fibroblasts
Embryonic outgrowths 45
    cryopreservation 22
    passaging 44–47
    putative embryonic stem cell line 46
Embryonic stem cells 1
    blastocyst 2
    cell cultures 15
    colony 18
    culture 3
    generation 16
    neuroglial differentiation 72
Embryo plating 41–44
Embryos
    abnormal multinucleated dispermic 10
    assessment
        blastocysts 16–18
        cleavage stage 15–16
        fertilization 14–15
    bisection blade positions, graphic presentation 43
    cell dispermic 9
    cleavage 2, 7
    degenerating 16
    embryo transfer, grading 12
    fragmented dispermic 9
    fragmented, two-cell 7
    growth 3
    human embryos culturing 3–4
    illustration of 2
    morula stage 14
    normal/abnormal, diagrams 10, 11
    normal bipronuclear 8
    phase contrast microscopy 14
    in vitro fertilization 4
Embryo transfer pipettes 26
Embryo vitrification 47
Endogenous MIXL1, expression 389
Endothelial cells
    characterization 335
    protein/RNAm markers for 346–347
    $VEGF_{165}$, role for 338
Endothelial nitric oxide synthase 335
Endothelial progenitor cells 347

Endothelin 3, 302
Endothelin-B receptor 302
eNOS, see Endothelial nitric oxide synthase
Epiblast, outgrowth 45
Epithelium-related cell–cell adhesion complexes 414
ESC, see Embryonic stem cells
E-selectin proteins 335
ExGen 500, 409
Extracellular matrix 336
Extracellular matrix compounds
    cell-adhesion 60
    conditioned media
        derivation of 57
        preparation of 61
    cytogenetic analysis 67
    cytokines, degradation protection 56
    fibroblast cells, preparation of 60–61
    fibronectin 57
    foreskin fibroblast cells 61
    glycoproteins 56
    growth factors 56
    hESC colonies, RNA extraction 64
    human-derived 56
    human fibroblasts 60
    human foreskin fibroblast, conditioned
        media derivation 56–57, 157
    human serum 98
    karyotype stability, analysis of 59
    plastic dishes coated 63
    preparation of 61
    samples, preparation of 60
    secretion of 56
Extracellular matrix proteins 55

F

FACS, see Fluorescent-activated cell sorter
Familial dysautonomia 284
Feeder cells 25, 47
    MEF, high density of 73
    mitomycin C treatment of 39
Feeder layer media 152
Feeder plates 25
Fertilization
    illustration of 5
    syngamy 6
Fetal calf serum 98
α-Fetoprotein (AFP) 356
Feulgen-staining 379
FFIEF, see Free-flow isoelectric focusing
Flamingo fluorescent gel staining 157
Flamingo staining 169–170
Fluorescent-activated cell sorter 388
    analysis, for stem cell markers 416, 421–422
    of CD31+ cells 335

for generating NC-like cells from hESCs 284
for vascular differentiation, of encapsulated
    hESCs 345–347
Fluorescent proteins 414
Foreskin fibroblast cells, SEM 61
FP medium 430
Free-flow isoelectric focusing 159

G

Gap junctional intercellular communication (GJIC) 211
Gap junctions
    in hESC 211–212
    methods to study hESC in 212–225
        cell apoptosis assay 215, 222–223
        cell pluripotency assay 216, 224
        cell proliferation assay 216, 223–224
        chemical closure of gap junctions 215, 222
        colony growth assay 225
        image and data analysis 216, 225
        immunocytometry 214–215, 221
        materials 212–216
        RT-PCR 213, 216–218
        Scrape Loading/Dye Transfer Assay 215,
            221–222
        Western Blot Analysis 213–214, 219–220
Gelatin solution 430–431
Gene
    expression monitoring 148
    housekeeping 136
    markers 103
    primers, reaction conditions 59
    transfer 395
Genejammer plasmid transfection 419–420
Genetic abnormalities 15
Genetic modification
    approaches and applications
        knock-down 390
        knock-in 389–390
        knockout 388–389
        overexpression of genes 387–388
    of hESCs in serum and feeder-free media 413
        cryopreserving cells 418–419
        culture hESCs on Matrigel-coated
            substrates 416
        electroporation 420
        by electroporation 414
        FACS analysis, for stem cell markers 421–422
        genejammer plasmid transfection 419–420
        hESCs to feeder-free culture,
            adaptation of 417–418
        immunofluorescence 422
        lentiviral infection and limitation 413–414
        si-RNA transfection 419
        stably transfected cells, selection of 420–421

Genetic modification (*continued*)
    use of drug-metabolizing enzymes 414
    viral transduction 420
  methods, for genetic manipulation
    electroporation 392
    infection 392–394
    short- vs long-term expression 395
    transfection 391–392
GFP-encoding plasmid pQBIpgk 273
Giemsa/trypsin/leishman (GTL) banding
  techniques 68
Glia cells 302
Glutaraldehyde 60
Glycolic acid 350, 351
Glycoproteins 346, 394
Green fluorescent protein (GFP) 388
Growth factors 347
  BMP2 and/or TGFβ1 effect on chondrogenic
    development of EBs 320
  controlled release of 350–351
  freeze-thaw cycles of 330
  modulation of chondrogenesis of
    EB-derived cells 328
  role in 3D scaffolding for hESCs 336

H

HA hydrogels, *see* Bioactive hydrogel-based scaffolds
Hank's buffered saline solution (HBSS) 304
hESC, *see* Human embryonic stem cells (hESC)
hESC colonies
  bulging, for 3D analysis 185
  manual isolation of 184
  shapes and pattern of 180
  three-dimensional reconstruction using
    3D-Doctor 188–191
    Mimics 182, 186–188
hESC culture medium 181
hESC markers
  agarose gel electrophoresis of 204
  in LRB010 hESC line 204
  microheterogeneity of expression of 180
Hexadimethrine bromide 429
hFFs, *see* Human foreskin fibroblasts
Hirschsprung disease 72
Hirsch-sprung's disease type 2, 302
HIV-1-based lentiviral vectors 394
HIV-1-based vectors 394
hMSC, *see* Human mesenchymal stem cells
Housekeeping gene 136, 137
hPDLFs, *see* Human periodontal ligament fibroblasts
HPRT1, *see* Hypoxanthine phosphoribosyl transferase 1
hrbFGF, *see* Human recombinant basic fibroblast growth
  factor
HS, *see* Human serum

Human dermal fibroblasts, preparation of 431–432
Human embryonic stem cell medium 303–304
Human embryonic stem cells culture 272–273,
  276–277, 290
  conditions, optimization 98
  and maintenance 57, 135, 401
  on Matrigel-coated substrates 416
Human embryonic stem cells (hESC) 21, 121, 269
  adaptation to feeder-free culture 417–418
  alignment 182, 185–186
  bulk scraping method 101
  cell medium 430
  cell surface pluripotency markers,
    detection of 62–63
  clinical applications 21
  coating dishes 110
    CELLstart^{TM} volumes 113
    Geltrex^{TM} qualified 112–113
  colocalization studies of 201
  colonies, *see* hESC colonies
  conditioned media, preparation of 60–61
  cryopreservation 22, 79
  cultivation and differentiation in vitro 323
  culture of, 276–277, *see* Human embryonic stem cells
    culture
  cytogenetic analysis 67–68
  derived neurons 369
  differentiation of 277–278
  differentiation removal 111, 115–116
  downregulation by RNAi in 390
  3D scaffolding for 336
  embryoid bodies 79–81, 111, 117–118
  embryonic stem cell derivation 40–47
    embryo plating 41–44
    passaging 44–47
  encapsulation and culture of
    characterization 345–348
    in photocured dextran-based hydrogels 342–345
    in photocured HA-based hydrogels 341–342
  expansion and maintenance
    medium preparation 110, 112
  extracellular matrix, preparation of 61
  feeder cell preparation 23–25
    freezing 36–38
    mitomycin C treatment 39–40
    passaging 35–36
    primary establishing 33–35
    thawing 38–39
  fibroblast cells, preparation of 60–61
  fibrogenic differentiation hPDLFs 270
  fluorescence imaging 85
  freezing 111, 116
  gap junctions in 211–212
  gene markers, calibration curve 145

generation
    NC-like cells from 284
    by transfection.. 391
genes expression.. 97
genetic changes in 133
genetic engineering of.................................... 387
graft of hESC into 375
growth 205
    under feeder-free conditions....................... 196–197
    techniques for.. 199
growth of.. 62
H7, low-density feeders..................................... 87
    neural markers, semiquantitative
        analysis of...................................... 88
H7, low-density MEF 88
on hPDLFs
    fibrogenic differentiation of.................. 270
    osteogenic differentiation of 271
H7, phase contrast images of............................. 81
HSF6, low-density MEF 88
HSF-6, neurectodermal lineages 87
HSF-6, neuroglial markers
    semiquantitative analysis of 90
IF analysis with feeder-free cultures................ 205–207
    localization of stem cell markers.......... 202
    materials.. 197
    primary antibodies for............................ 198
    signaling pathways in stem cells 202–203
image analysis.. 91–92
imagination-stimulating application of 367
immunocytochemistry.. 83–85
immunohistochemistry of
    antibodies for 181
induced melanocyte differentiation 311
karyotype stability, analysis of...................... 78
LDMEF, HSF-6, 86
localization of signaling pathways in 196
low- density mouse embryonic fibroblasts,
    differentiation of
    H7EB grown 82
    H7 EBs 82
    HSF-6, 83
markers, analysis of...................................... 57, 67
MEF feeder substrate 73
MEFs 109
methods to study gap junctions in................ 212–225
    cell apoptosis assay 215, 222–223
    cell pluripotency assay........................ 216, 224
    cell proliferation assay........................ 216, 223–224
    chemical closure of gap junctions 215, 222
    colony growth assay 225
    image and data analysis............................ 216, 225
    immunocytometry................................ 214–215, 221
    materials.. 212–216

RT-PCR.. 213, 216–218
Scrape Loading/Dye Transfer Assay................ 215
Western Blot Analysis 213–214, 219–220
morphology and cell surface markers........................ 63
morphology of.. 201
mouse embryonic fibroblasts
    preparation and inactivation 77–78
mouse embryonic fibroblasts feeder cells
    HSF-6 and H7 grown 80
murine feeder cells 109
neuroglial differentiation 72
objective of... 135
paraffin embedding of.. 181
    for 2D survey 182–183
    markers in 185
    strategies for.. 184–185
passaging 110, 152–153
    collagenase type IV 113–114
    EZPassage^{TM} disposable cell passaging
        tool... 114–115
    STEMPRO$^®$ SFM............................... 114–115
phase contrast images 85
plating and outgrowth passaging.......................... 25–27
pluripotent 72, 179
    growth of... 79
    H7 colony, immunocytochemical staining 81
potential use of... 179
preparation of.. 49
primary cilia in 196, 201
protein expression 160
proteomics-based analysis
    batch ion-exchange 165–166
    BIEX.. 154–155
    cell lysis/basic fractionation 154, 162–165
    definitive endoderm, generation........................ 153
    2D electrophoresis 155–156, 168–169
    differentiation protocol.. 161
    feeder layers... 159–161
    free-flow isoelectric focusing 170–171
    immunohistochemistry 153, 161–162
    in-gel protein digestion 158, 173
    passage... 152–153
    peptide desalting/concentration........................ 158, 173–174
    protein desalting/concentration................... 155, 166
    quantitative data... 171–172
    rotofor 157–158
    SDS-PAGE.. 155, 166–167
    spot cutter 172–173
    staining....................................... 156–157, 169–170
    western blotting 155, 167–168
and pseudogenes 203–204

Human embryonic stem cells(hESC) (*continued*)
  putative
    vitrification of 47–49
    warming of 49–50
  real-time PCR (RT-PCR) technique 135
  real-time RT-PCR of
    cDNA synthesis 198–199, 207–208
    total RNA extraction 197–198, 207
  R4 engineering, *see* R4 hESC engineering
  retroviral, and lentiviral gene transfer into 406
  RNA analysis
    pitfalls 203
    primer pairs 205
    transcripts 204
  RT-PCR analysis, samples preparation
    cDNA Synthesis 65
    RNA, extraction and quantification 64–65
  SEM, samples preparation 60, 64
  single cell enzymatic dissociation 122
    cell colony 128
    cells transfer 125–126
    characterization of 130
    culture dishes, gelatin coating of 124
    feeder cells, culture and handling of 122–123
    hFF culture 124
    hFFs, mitomycin C treatment of 124
    hFFs, thawing of 123–124
    medium change 126
    passage procedure 126–129
    seeding of mitomycin C-treated
      hFFs 124–125
  source of 22
  STEMPRO® hESC SFM, grown 115
  telomerase activity assay 65–67
    amplification protocol 65
    electrophoresis, separation 66–67
  thawing 111, 116–117
  transcription factors 196, 202
  transfection of 277
  undifferentiated 180
  undifferentiated and differentiated 272
  and vascular differentiation 334–336
  viral transduction 420
  vital staining of 374
  vitrification and warming 22–23, 27–28
    embryo vitrification 29–31
    embryo warming 31–32
  in vitro neural differentiation
    cell culture and cryopreservation 74–75
    imaging and image analysis 76–77
    immunocytochemistry 75–76
Human embryonic stem cells markers
  alkaline phosphatase detection kit 58
  immunocytochemistry, pluripotency
    cell surface 57–58
  RT-PCR analysis, intracellular 58–59
  telomerase activity 59
Human embryonic stem cells pluripotency
  checking
    alkaline phosphatase staining 103
    morphology 102–103
    nuclear markers 103
    taqman low density arrays 103
  chromosomal abnormalities 100
  insulin receptor, detection of 171
  iPS lines 100
  matrix preparation 100–101
  mouse embryonic fibroblasts 151
  TGF-β superfamily 98
Human embryos, culturing
  in vitro involvement
    cleavage stage 4–11
    extended to blastocysts 11–14
    fertilization assessment 4
Human fibroblasts
  ECM derived 63
  ECM samples 60
Human foreskin fibroblast feeders 122
  Iscove's modified Dulbecco's medium 123
Human foreskin fibroblasts 56, 60,
    61, 121
  feeder-free culture conditions 122
Human-induced pluripotent stem cells (hiPS cells),
    see Human iPS (hiPS) cells
Human iPS (hiPS) cells 426
  characterization of 440–441
  establishment of 433–438
  feeder- and serum-free production of 434
  freezing and thawing 439–440
  propagation of 438–439
Human mesenchymal stem cells 99
  matrix preparation 100
Human periodontal ligament fibroblasts 270
Human postblastocyst
  in culture 17
Human recombinant basic fibroblast growth
    factor 123
Human Rosa26 locus 414
Human serum 56
Hyaluronic acid (HA) 337, 338
Hydrogels, *see* Bioactive hydrogel-based scaffolds
Hypoxanthine phosphoribosyl transferase 1 .... 389, 414
Hypoxia 338

I

ICM, *see* Inner cell mass
ICSI, *see* Intracytoplasmic sperm injection

Image and data analysis
    gap junctions in hESC 216, 225
IMDM, *see* Iscove's modified Dulbecco's medium
Immobilized pH Gradient 156
Immunocytochemistry
    integrase-mediated R4 hESC
        engineering 246–248
Immunocytometry
    gap junctions in hESC 214–215
    expression of Cx43, Cx45, and pluripotency
        markers 221
Immunohistochemistry
    antibodies for 181
Induced pluripotent stem (iPS) cells 97, 426
Infection
    generation, genetically modified HESCs 393
    protocol timetable 405
    retrovirus/lentivirus production 405
    studies with, RNA and DNA viruses 392–394
    See also Genetic modification
Inner cell mass
    blastocoele 12
    blastocysts 2, 7, 13
    isolation of 16
    stem cells 17
Insemination 4
International stem cell Initiative 122
Intracytoplasmic sperm injection 4
In vitro derivation of chondrogenic cells
    aggregate processing 329
    EB-derived cells, micromass cultures of 324
    embryoid body formation 322–324
    histological analysis 329–330
    immunostaining 326, 328
    isolation of 324–325
In vitro fertilization 4
IPG, *see* Immobilized pH Gradient
iPS, *see* Induced pluripotent stem (iPS) cells
Iscove's modified Dulbecco's medium 123
IVF, *see* In vitro fertilization

K

Karyotype analysis
    integrase-mediated R4 hESC
        engineering 245–246
Keratinocyte growth factor 152
Keratinocytes 302
KGF, *see* Keratinocyte growth factor
Kit tyrosine kinase receptor (c-Kit) 302
KNOCKOUT™ Serum Replacement 111
KO™-DMEM 280
Kruppel-like factor 4 *(Klf4)* 425
KSR, *see* KNOCKOUT™ Serum Replacement

L

Lactic acids 350
L-cell medium 304
L-conditioned medium 310
LDMEF, H7s grown on 84
LD-/ND-MEF
    H7s 85
Lentiviral vectors (LVVs) 393, 394
Lesch-Nyhan syndrome 389
LIF/gp130 pathway 152
LightCycler® program
    data analysis
        absolute quantification 145
        relative quantification 146–147
        specificity assessment 145–146
    relative quantification 137
LightCycler® system 136
LightCycler program for real-time quantification
    of hESC transcripts 209
Lipofectamine 2000 reagent 392
L-Wnt3A conditioning medium 304

M

MALDI-TOF fingerprint data 173
Matrigel-coated plates 418, 431
Maturing neural cells, markers of 74
MEF, *see* Mouse embryonic fibroblasts
MEF-conditioned medium (MEF-CM), feeder-free
    culture
    R4 hESC engineering 244–245
MEF derivation medium 303
MEF feeder
    R4 hESC engineering in
        culturing 239–244
        materials 232–233
        medium composition 240
MEF medium 303
Melanocytes 284, 291
    defects in 302
    development 301
    differentiation 312–313
        hES-induced, morphological changes 311
        novel feeder-free approach for 303
    expression of markers 315
    in mutant offspring 302
    primary function of 302
Melanosomes 303
Mel-1 medium 304, 313
Melting curves
    amplicon specificity 145
    Oct4 primers 146
β-Mercaptoethanol 430
mESCs, *see* Murine embryonic stem cells

Mesenchymal cells 284
Methylcellulose 335
β2-Microglobulin 390
Microsurgery, for transplants 374
    exposure of embryo, for surgery 374–375
    follow-up, of grafts 377
    incubation, of embryos 374
    preparation of embryos
        for grafts of hESC into somites 375–376
        of hESC in place of somites 376
    transfer and implantation of hESC to 376–377
    visualizing embryo 374
        See also Transplantation, of hESCs and derivatives
MilliQ Plus water purification system 152
MiniRNA Isolation Kit 139
Mitomycin C 271, 277, 314, 399, 400, 416
Mitomycin C-treated MEFs
    R4 hESC engineering in 240–242
Monospermic fertilization, pronuclear stage 14
Mouse embryonic fibroblasts 72, 74, 121, 151, 276
    derivation of 305–306
    and derivatives transplanted 368
    freezing 306–307
    generation, EBs from 309–310
    H7, low-density feeders
        neural markers, semiquantitative analysis of 89
    irradiation, for hES cell seeding 308
    L-conditioned medium 310, 312
    passaging hES cells onto irradiated 308–309
    splitting 307
    thawing 307
    Wnt3a-conditioned medium
        confirmation of 312
        for Mel-1, 310
    See also Melanocytes
Mouse embryonic stem cells 47
Mouse Rosa26 locus 390
MultiSite Gateway® Technology
    for integrase-mediated R4 hESC
        engineering 229–230
        constructing the retargeting expression
            vector 250–254
Murine embryonic fibroblasts 109
Murine embryonic stem cells 109

N

N-cadherin 357
NCAM, *see* Neural cell adhesion molecule
NEAA, *see* Non-essential amino acids
Neural cell adhesion molecule 94
Neural crest cells generation from hESCs
    generation and propagation, of NSPs 291
    human ES cell culture 290
    immunostaining 295–296
    induction of NC and PSNs from 291–293
    microscopic analysis 296
    nucleic acid analyses 296–297
    stromal cell culture 290–291
    using FACS sorting 284
Neural crest (NC) 284
Neural crest stem cells (NCSC) 369
Neural precursors, differentiation
    dopaminergic neurons 363–364
    ESC aggregates 359, 361
    morphologically distinct steps 360
    neuroepithelial cells 361–362
        enrichment and expansion of 362
    neurons differentiation, from 362–363
Neuroectodermal transcription factors 357
Neuroepithelia 357
Neuroepithelial cell
    differentiation 361–362
    enrichment and expansion of 362
Neurospheres (NSPs) 291
    from hESCs using PA6 co-culture 292
NIC, *see* Nicotinamide
Nicotinamide 152
Nitrocellulose staining 170
Non-essential amino acids 111

O

Oct4 experiment, absolute quantification 146
Oct4 promoter 389
Oculocutaneous 302
Oocytes 4

P

PA6-induced neurospheres 284, 285
PA6 mouse stromal cells 290
Paraffin embedding of hESCs 181
    for 2D survey 182–183
    markers in 185
    strategies for 184–185
PBS, *see* Phosphate buffered saline
PCR
    BG01v/R4 hESC platform line 248–249
    to screen retargeted hESC/ R4 Cells 259–261
PCR amplification, crossing point 137
PDQuest software 172
PECAM1, *see* Platelet endothelial cell adhesion molecule-1
Periodontal apparatus 269
Periodontal ligament fibroblasts, hESCs differentiation on
    hESCs, culture of 276–277
    hESCs, transfection of 277
    hPDLFs, culture of 275–276
    immunohistochemistry 278–279
    RT-PCR 279
    SEM 279
Periodontal regeneration processes 270

Peripheral somatic sensory neurons
  differentiation from hESCs 284
  generation, from hESCs, *see* Neural crest cells
    generation from hESCs
  for modeling, PNS disease and injury 284
  progenitors, in amniote embryos 284
Peripheral sympathetic neurons 284
PGD, *see* Pre-implantation genetic diagnosis
Phenylmethylsulfonyl fluoride 154
PhiC31 integrase-mediated chromosomal targeting 230
  R4 hESC engineering 229–261
    materials 231–234
    methods 234–261
    quality control assays 233, 245–248
    troubleshooting 264–267
Phosphate buffered saline 74
Photography, of fluorescence 380
Photopolymerizable hydrogels 337
Physiological xenofree
  molecularly defined medium
    cells 100
    chemicals and plastic 99
    TLDA cards 98
    transcriptomics 100
Piebaldism 302
pJTI™R4 DEST vector 251–252, 256
Platelet endothelial cell adhesion molecule-1 ... 335, 347
PLGA polymers 350
Pluripotency markers, expression
  immunocytometry 221
PMSF, *see* Phenylmethylsulfonyl fluoride
pol genes 394
Polyethylenimine (PEI) 392
Porcine gelatin 152
POU class 5 homeobox 1 *(Pou5f1/Oct4)* 425
Preimplantation embryos 3
Pre implantation genetic diagnosis 21
Primary antibodies for IF protocol 198
Primary cilia in hESCs 196, 201
Primer sequence, for RT-PCR 288–289, 327, 328
Primers for hESC genes 200
Priodontal ligament 269
ProCell™ transfection system
  R4 hESC engineering 233–234, 256–257
Pronase 26
Protein dilution buffer 157
PSNs, *see* Peripheral somatic sensory neurons
Puromycin 429
Putative embryonic stem cell line 46

Q

Quality Control Assays
  integrase-mediated R4 hESC
    engineering 233, 245–248

R

Real-time PCR (RT-PCR) 326
  analysis of PA6-NSPs in suspension 297
  Col 1 and *Col 2* gene expression, analysis ... 326
  differentiation marker expression,
    detection of 271, 279
  efficiency correction 136
  expression levels, of hyaluronidase
    isomer 349, 350
  gap junctions in hESC 213
    connexin mRNA expression 216–218
    reaction mix 218
  gene expression 136
  gene expression, quantification of 148
  LightCycler® system 136
  monitoring approach
    agarose electrophoresis 138
    calibrator, selection 141
    cDNA, preparation of 139–140
    cells, collection of 137
    genes, primers for 142
    hESC Cultures 138
    primers 137–138
    primers design 140–141
    real-time kit 138
    reference genes, selection of 141
    retrotranscription 137
    RNA extraction 137
    RNA, preparation of 139
    target genes, selection of 142
  nucleic acid quantification 136
  protocol establishment
    data analysis 145–147
    gene expression, quantification of 148
    gene markers, calibration curve 145
    LightCycler Instrument 142–143
    LightCycler protocol 142
    PCR mix, preparation 142
    quantification, accuracy of 144
  sequence of primers and conditions 327
  sequence of primers used for 328
  use of 136
Relative Quantification Monocolor, steps 147
Replication competent retroviruses (RCR) 429
Retargeting hESC/ R4 Cells
  integrase-mediated chromosomal
    targeting 254–258
  PCR for screening 259–261
  selection and expansion 258–259
Retroviruses 429
Reverse transcription 296–297
Reverse transcription buffer
  for cDNA synthesis 217

rev genes 394
RGD peptide 338
R4 hESC engineering
  by PhiC31 integrase-mediated chromosomal
    targeting 229–261
    materials 231–234
    methods 234–261
    quality control assays 233, 245–248
    troubleshooting 264–267
Ribosomal genes 141
RNA interference (RNAi) 390
RNAm markers 346
Roche system, see LightCycler® system
ROCK inhibitor Y-27632, 430, 442

S

Scanning electron microscopy 274, 279, 343
SCED, see Single cell enzymatic dissociation
Schwann cells 284
Scrape Loading
  gap junctions in hESC 215, 221–222
SDIA-PA6 cell-induction method 284
SDS-polyacrylamide gel electrophoresis (SDS-PAGE)
  gap junctions in hESC 219–220
SEM, see Scanning electron microscopy
Serum albumin-containing media 98
Severe combined immunodeficient (SCID) 335
Silver staining 156–157, 169
Single cell enzymatic dissociation 121, 122
  human ES cells cultured 129
Skin-dwelling melanocytes 301
small interfering RNA molecules (siRNAs) 390
  transfection 415, 419
Smooth muscle cells (SMCs) 335
Sox1, neural marker 357
SRY (sex determining region Y)-box 2 (Sox2) 425
Statistical analysis
  gap junctions in hESC 225
Stem cell factor (SCF) 302
Stem cell research 195
Stem cells, trypsinization of 209
StemPro® hESC SFM 111, 112
  R4 engineering
    collagenase preparation 236–237
    freezing 237–239
    media components 236
    propagating 231–232, 235–239
    thawing 231, 235
SYBR Green 140
Syngamy 6

T

Taqman low density arrays 103
TEM, see Transmission electron microscopy

Texas-Red fluorescence 384
TFA, see Trifluoroacetic acid
Thermocycler program 209
Three-dimensional reconstruction of hESC colonies
  using 3D-Doctor 188–191
  using Mimics 182, 186–188
Tissue culture 395–397, 399
  clonal assay to test competence and 402
  freezing and thawing HESCs 401–402
  isolation of MEFS 399–400
  maintenance of HESCs and clones 401
  mitomycin-C Inactivation of MEFS 400–401
  subculture of 401
  See also Genetic modification
TLDA, see Taqman low density arrays
Toxic cryoprotectants 47
Transcriptional remnants 203
Transcription factors OCT3, 416
Transfection 391–392
  by calcium phosphate 403
  DNA preparation for 402
  efficiency 280
  by Exgen 500, 404
  HESC cell culture, on day of transfection 407
  protocol timetable 403
  rate 408
  R4 hESC engineering 233–234, 254–258
  See also Genetic modification
Transmission electron microscopy 2, 14
Transplantation, of hESCs and derivatives
  to chick embryo 368
    culture of ES cells 373–374
    fixation and embedding 377–381
    grafting undifferentiated hESC to 369
    NIH guidelines 368
    transplanted CD34+ HSPCs into 369
    transplants, microsurgery for 374–377
Trifluoroacetic acid 158
TrypLE Express 226
Trypsin-ethylenediaminetetraacetic acid 99
Tuberculin syringe 383
Tyrode's acidic solution 42

U

Undifferentiated hESC markers 212
Undifferentiated hESCs 349
  characterization of 182
UV visualization and documentation system 279

V

Vascular endothelial-cadherin (VE-CAD) 335, 347
Vascular endothelial growth factor receptor
  2 (VEGFR 2) 335
$VEGF_{165}$, to induce differentiation of ESCs 338

Vesicular stomatitis virus (VSV-G) 394
Vitiligo 302
von Willebrand factor (vWF) 335

W

Waardenberg-Shah syndrome 302
Western Blot Analysis
    gap junctions in hESC 213–214
        of connexion 43 phosphorylation states 219–220
Western blotting 312, 315

Wnt3a-conditioned medium 312
Wnt, genes of 105

X

Xenofree, definition 98

Z

Zygotes 6, 11
Zymo extraction columns 137

Printed in the United States of America